T0296170

THERMIONIC VALVES

THEIR THEORY AND DESIGN

**

THERMIONIC VALVES

THEIR THEORY AND DESIGN

BY

A. H. W. BECK

B.Sc. (Eng.), A.M.I.E.E.

Standard Telephones & Cables Ltd.

CAMBRIDGE
AT THE UNIVERSITY PRESS
1953

CAMBRIDGE
UNIVERSITY PRESS

University Printing House, Cambridge CB2 8BS, United Kingdom

Cambridge University Press is part of the University of Cambridge.

It furthers the University's mission by disseminating knowledge in the pursuit of education, learning and research at the highest international levels of excellence.

www.cambridge.org
Information on this title: www.cambridge.org/9781107502246

First published 1953
First paperback edition 2015

A catalogue record for this publication is available from the British Library

ISBN 978-1-107-50224-6 Paperback

PREFACE

In writing this book I have attempted to give a theoretical account of the behaviour of thermionic high vacuum devices. Such an account should primarily be of use to graduates with a first degree in physics or electrical engineering and those who are starting independent work in this field, either in industry or in the various post-graduate courses on electronics. A reader who has mastered the contents should be able to make a critical appreciation of current research papers. I hope that the book will also prove useful to practising valve engineers and others engaged in the industry. I also hope, though faintly, that it may help the users of valves to understand the physical factors which limit valve performance and thereby encourage them to restrict their demands to what is possible.

I have assumed that the reader is familiar with the ordinary uses of valves as amplifiers and oscillators which are the subject of many recent textbooks. However, what might be termed the second order effects in valve theory, matters such as noise and the effects of initial emission velocities, have been treated rather fully. These questions are as important practically as they are interesting theoretically. Gas discharge devices are not treated, and, in view of the copious modern literature on the subject, cathode ray tubes are ignored.

The book is divided into three parts, the first dealing with the fundamental physics of thermionic emission, etc. In this section C.G.S. units are used. The second part considers the general theory of the fields set up by charged conductors and the electron motions resulting therefrom. The last and longest section applies all this material to the study of various classes of valve. M.K.S. units are used in parts two and three. This need create no confusion and the student should, in any case, gain confidence with several systems of units. The second and third parts may be read before the first if desired.

The book is strongly biassed towards microwave valves. There are several reasons for this; first, the existing textbooks deal very adequately with ordinary gridded valves. Secondly, an extremely wide range of interesting and useful phenomena is encountered in this field, and lastly, because of my personal interests.

I have written almost nothing about the practical aspects of valve manufacture. This is simply because an adequate treatment of valve technique would fill a much longer book. It does not mean that I consider technique unimportant. In fact, quite the most important equipment a valve engineer can possess is a really sound knowledge of the materials and processes involved in making all types of valves. It is harder to acquire this knowledge than it is to learn the necessary theory.

In the sections dealing with well-established material I have only included references to the most important papers and those which are classics in their domain. In the newer sections a much wider range of references has been given. The body of this book was written in 1949–51. The appendices were prepared in 1952, and the references have been brought up to that date. These details are given to ease the task of searching the various abstracts of current literature.

I now turn to the pleasant task of thanking the numerous persons and organizations who have helped me. My particular thanks are due to my friends and colleagues in Standard Telephones and Cables, Ltd., especially to Mr. A. B. Cutting who read the manuscript and made many corrections and improvements. Any remaining errors are my sole responsibility. I am indebted to the late Mr. W. T. Gibson for his continuous advice and interest. The Bell Telephone Laboratories, the Institute of Radio Engineers, the American Physical Society and the Cambridge University Press have allowed the reproduction of data and illustrations, for which I thank them.

To my wife my thanks are due, not only for typing the manuscript, but also for her help and support in the whole project. Finally, I wish to thank the management of Standard Telephones and Cables, Ltd. for permission to publish this book.

HIGHGATE A. H. W. B.
February 1953

CONTENTS

Chapter 2. SECONDARY, FIELD AND PHOTO-ELECTRIC EMISSION

Chapter 10. MULTIGRID RECEIVING VALVES

Chapter 11. TRANSMITTER VALVES

Chapter 12. VELOCITY-MODULATED VALVES

LIST OF TABLES

LIST OF SYMBOLS

A_0 = Richardson's constant.

B = Susceptance.

$\mathbf{B}$ = Magnetic flux density.

C = Capacitance. In Chapter 14 a beam parameter defined by Pierce.

E = Energy, in Pt I of the book.

$\mathbf{E}$ = Electric field strength.

e = Electronic charge.

G = Conductance. Power gain.

$\mathbf{H}$ = Magnetic field strength.

h = Planck's constant. Debunching parameter in Chapter 12.

$\hbar = h/2\pi$.

I, i = Current.

I_n, K_n = Bessel functions with imaginary argument.

J, j = Current density.

J_n, Y_n = Bessel functions of the first kind.

k = Boltzmann's constant.

L = Normalized length, equal to transit angle of unmodulated electron through the field.

l = Gap length.

ln = Napierian logarithm.

M = Mutual inductance.

m = Electronic mass.

P = Efficiency parameter defined in Chapter 12.

Q = Quality factor of resonator.

R = Resistance or normalized radius.

r = Radius.

S = Screening fraction of grid. Normalized drift length in Chapter 12.

S_0 = Distance travelled per cycle by an unmodulated electron.

s = Drift length in metres.

T = Temperature.

t = Time. (*N.B.*—Dots represent differentiation with respect to time, i.e., $\dot{x} = dx/dt$, etc.)

u = Velocity.

V_0 = D.C. voltage.

W = Power.

X = Reactance.

Y = Admittance.

Z = Impedance.

Z_0 = Resonant impedance.

z = Axial coordinate in cylindrical system.

α = Depth of modulation at V.M. gap. Also used for attenuation constant.

β = V.M. coefficient, relating velocity gained by electrons to velocity gained in an infinitely narrow gap. Also used for phase constant.

LIST OF SYMBOLS

β_e = Phase velocity of electron beam.

Γ = Propagation constant, a complex number $\alpha + j\beta$.

γ = V.M. coefficient used to calculate diode loss. Also used for propagation constant.

δ = Secondary emission coefficient or phase angle.

ϵ_0 = Permittivity of vacuum.

ζ = Fermi level in a solid.

η = Efficiency.

θ = Bunching parameter = $\frac{1}{2}S\alpha\beta$ in Chapter 12. Transit angle in earlier chapters.

λ = Wave-length, with appropriate suffices.

μ, μ_0 = Permeability of material and of vacuum.

ν = Frequency, used in Pt. I.

ρ = Charge density.

Σ = Cross-sectional area.

σ = Conductivity.

τ = Transit time.

φ = Work function of surface. Transit angle.

ω = Angular frequency.

ω_L = Larmor frequency.

ω_p = Plasma frequency.

VALUES OF PHYSICAL CONSTANTS

$e = 1{\cdot}60 \times 10^{-19}$ coulomb.

$m = 9{\cdot}11 \times 10^{-31}$ kg.

$\mu_0 = 4\pi \times 10^{-7}$ henry/metre.

$\epsilon_0 = 36\pi \times 10^{-9}$ farad/metre.

$h = 6{\cdot}62 \times 10^{-34}$ joule sec.

$k = 8{\cdot}61 \times 10^{-5}\ e.V/{}^\circ K$.

PART I

THE PHYSICAL THEORY OF ELECTRONICS

Chapter 1

THERMIONIC EMISSION

1.1. Thermionic emission from pure metals

1.1.1. *The modern theory of metals and insulators.* The quantum theory of metals, largely due to Sommerfeld, looks on them as crystalline solids formed by a regular network of positive ions. The ions are formed from the atoms by the loss of the valence electrons which exist as a cloud of charges drifting in all directions through the lattice. The cohesive force in metals is due mainly to the attraction between the ions and the electrons. When no external field is applied, the mean number of electrons drifting to the right through a plane is equal to the mean number drifting to the left. The numbers in very short time intervals are, however, subject to a statistical fluctuation, and it is this fluctuation current which gives rise to the phenomenon of thermal noise in metallic resistors. When a field is applied, a slow drift in the direction of the field is superposed on the rapid thermal motions, and this unidirectional flow constitutes the electric current.

In the theory of thermionic emission we are mainly concerned with two aspects of the quantum treatment: first, the way in which the regular spacing of the atomic nuclei in a perfect lattice breaks up the energy spectrum into permitted and forbidden zones; and secondly, the statistics obeyed by the free electrons which are those of Fermi and Dirac. In dealing with emission from metals we shall only have to deal with the second aspect, but in the consideration of emission from oxide-coated cathodes the first is of primary importance. It would be out of place to do more than indicate the physical basis for the ideas used here, so the reader is referred to the works mentioned below for complete accounts.† The first problem

† Mott and Jones, *The Theory of the Properties of Metals and Alloys*, Oxford University Press, 1936. Brillouin, L., *Wave Propagation in Periodic Structures*, McGraw Hill, 1946.

arises as follows. In quantum theory the free electrons are each assigned a wave function $\psi(x, y, z)$ proportional to the probability that the electron shall be in any specified volume element. The metal is represented by a three-dimensional array of positive ions, i.e. the potential in a plane through the centres of a row of atoms varies as shown in fig. 1.1. If the electron is to be free to move it cannot have an arbitrary value of energy because the regularly spaced potential wells act in a manner analogous to regular discontinuities in an electric transmission

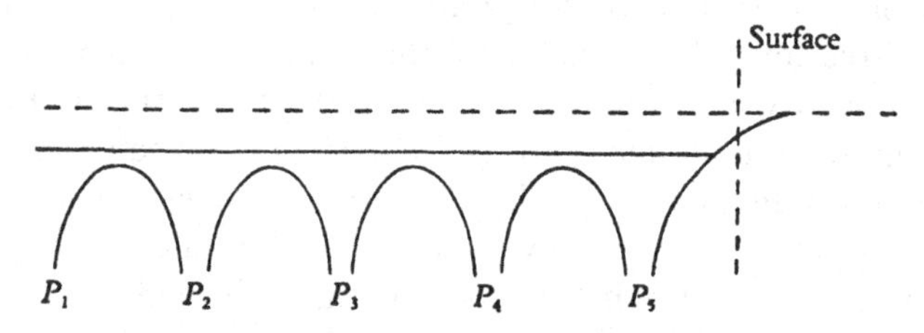

Fig. 1.1. Potential diagram for a metallic crystal.

line; in the latter case only waves in certain well-defined frequency bands can be transmitted, in the former case only electron waves in certain definite energy bands can propagate. In addition to the electrons which move freely there are the inner shell electrons bound to the nucleus. These occupy the energy levels specified by their quantum numbers, very nearly as if the ionized atom existed alone, for the effects due to the average potentials of the other ions and the free electrons are fairly small. They are, however, sufficiently large for the perturbation introduced by their presence to split up the sharply defined energy levels of the unperturbed system into energy bands, so that the correct number of electrons can be accommodated without violation of the Pauli exclusion principle. Returning to the free electrons, these will all have energies very considerably above the ground state, since the bound electrons fill all the lowest levels. Using these ideas, Mott† gives a simple derivation of the number of states in a given energy range which is reproduced below.

† Mott and Jones, op. cit. pp. 51 *et seq.*

We consider a metal cube of side L and assume that the metal is monovalent. The true potential variation of fig. 1.1 is approximated by the potential well of fig. 1.2. The

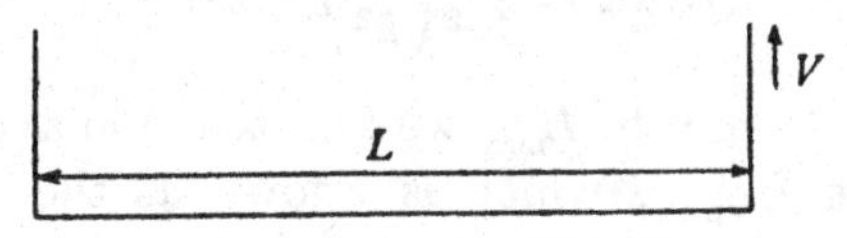

Fig. 1.2. A potential well.

Schrödinger equation is

$$\nabla^2\psi + \frac{2m}{\hbar^2}E\psi = 0, \tag{1}$$

together with the boundary conditions $\psi = 0$ at $x = 0, L$; $y = 0, L$; $z = 0, L$; i.e.

$$\psi = \sin\frac{\pi l_1 x}{L}\sin\frac{\pi l_2 y}{L}\sin\frac{\pi l_3 z}{L}, \quad l_1, l_2, l_3 \text{ integers.}$$

For E this gives $\qquad E = \dfrac{\hbar^2}{2m}\dfrac{\pi^2}{L^2}(l_1^2+l_2^2+l_3^2).$

The maximum energy which an electron can have at the absolute zero of temperature, where all the lowest levels are filled, is $E_{\text{max.}}$. If there are N electrons there will be $\frac{1}{2}N$ occupied levels. $\frac{1}{2}N$ must therefore equal the number of integral sets of l_1, l_2, l_3, such that

$$\frac{\hbar^2}{2m}\frac{\pi^2}{L^2}(l_1^2+l_2^2+l_3^2) < E_{\text{max.}}. \tag{2}$$

Since l_1, l_2, l_3 are all positive this number of sets is an eighth of the volume of a sphere whose radius is $(l_1^2+l_2^2+l_3^2)^{\frac{1}{2}}$. That is

$$\frac{N}{2} = \frac{\pi}{6}\left[\frac{2m}{\hbar^2}\frac{L^2}{\pi^2}E_{\text{max.}}\right]^{3/2},$$

or

$$E_{\text{max.}} = \left(\frac{3}{\pi}\right)^{2/3}\frac{\pi^2\hbar^2}{2m}\left(\frac{N}{\Omega}\right)^{2/3}, \tag{3}$$

where Ω = volume of the metal. By the same argument the number of states with energy $< E$ is

$$\frac{\Omega}{6\pi^2}\left[\frac{2m}{\hbar^2}E\right]^{3/2}.$$

Differentiating with respect to E, we find for the number of states with energy between E and $E+dE$ in the volume Ω

$$N(E)dE = \frac{\Omega}{4\pi^2}\left(\frac{2m}{\hbar^2}\right)^{3/2} E^{\frac{1}{2}} dE. \tag{4}$$

By integrating from 0 to $E_{max.}$ we find that the average energy per electron is $\frac{3}{5}E_{max.}$, which is known as the mean Fermi energy.

If now we replace the potential well of fig. 1.2 by some periodic potential representing fig. 1.1 we find that the resulting electronic wave functions† are of the form

$$\psi = \exp\left[j\left(\frac{\pi \mathbf{nr}}{L}\right)\right] u_k(x, y, z),$$

where u is a function which has the same periodicity as the lattice and L has the same meaning as before. If the energies are plotted as a function of $\pi n/L$ we find that the curve is discontinuous, having energy jumps $L = N_1 A$, where A is the lattice constant of the crystal. These energy jumps exist whatever the precise form chosen to represent the periodic potential. The forbidden energies occur at the wave-lengths for which Bragg reflexion from the lattice planes prevents the electron wave from penetrating the lattice, just as regularly spaced discontinuities prevent electromagnetic waves of special frequencies from flowing along a transmission system. Thus, in any crystal, the permissible electron energies are confined to well-defined bands, separated by what are termed 'forbidden' zones. The difference between a metal and an insulator is simply that in metals the permitted levels are not all filled, whereas in an insulator all the levels right up to the lower level of a forbidden zone are filled. In the latter case electrons can only move if they are given sufficient energy by heating, by application of an intense field, by collision or other means to jump over the forbidden zone into a higher conduction band. In diamonds the forbidden zone is about 5 eV. broad (5 eV. corresponds to *c.* 60,000° K.), so that thermal activation is very improbable, but many substances have forbidden zones of

† Bloch, *Zeits. f. Phys.* **52**, (1928), 555.

c. $\frac{1}{2}$ eV., in which case thermal activation is relatively probable. Pure substances in which the energy gap is very low are classified as intrinsic semi-conductors. However, it is doubtful if such pure substances exist, as it seems likely that such low values of the energy gap are due to traces of impurity. We shall discuss the properties of semi-conductors in more detail in later sections.

In this introduction we have tried to indicate the main ideas of quantum mechanics on the solid state. These ideas are of fundamental importance in improving our knowledge of the mechanism of emission and are thus important to the whole subject of thermionics; for, as will be made clear later in the book, the main obstacle to further progress in valve design lies in the limited current density available from even the best emitters known to-day.

1.1.2. *The emission equation for pure metals.* We next proceed to deduce the emission equation from the quantum statistical viewpoint. This equation was originally deduced classically by Richardson and Dushman.

If we consider the statistics of a system consisting of an extremely large number of indistinguishable particles in a phase space divided into volumes of magnitude h^3 and introduce the restriction that each elementary volume can either contain only one particle or be empty, we obtain the following expression for the probability that the state of energy E is occupied:

$$f(E) = \frac{1}{e^{(E-\zeta)/kT}+1}. \tag{5}$$

In this expression k is Boltzmann's constant and ζ is a constant which will later be determined by requiring that there shall be just enough states to accommodate all the electrons. The corresponding energy-distribution function is obtained by multiplying (5) by (4),

$$F(E) = \frac{2\pi\Omega(2m)^{3/2}E^{1/2}dE}{h^3(e^{(E-\zeta)/kT}+1)}. \tag{6}$$

These are known as the Fermi-Dirac distribution functions, and they are applicable to the electrons which constitute the free electron 'gas' in a metal, since we already know that the energies of these electrons are quantized and that they obey the exclusion principle. Figs. 1.3 and 1.4 show these functions plotted against the relative energy.

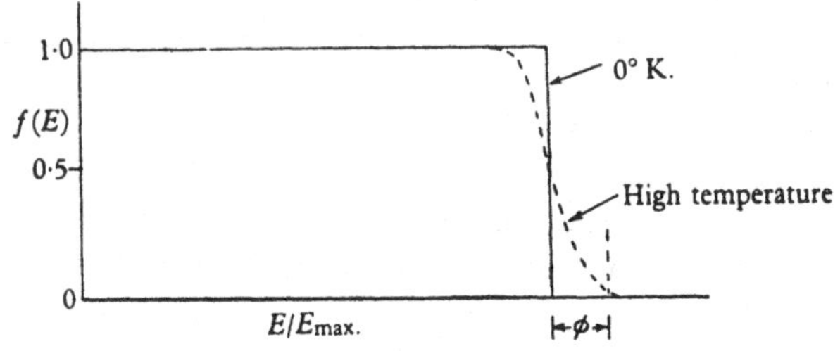

Fig. 1.3. Fermi-Dirac distribution for the probability that the cell with energy between E and $E+dE$ is occupied.

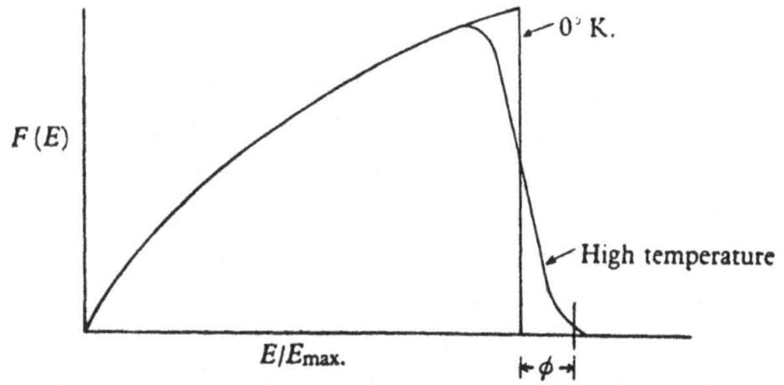

Fig. 1.4. Fermi-Dirac distribution for the number of electrons in the energy interval E to $E+dE$.

We now consider a heated sheet of metal, enclosed in a vacuum. The surface of the sheet is normal to the z axis and the dimensions of the sheet are l_1, l_2, l_3, the volume being therefore $\Omega = l_1 . l_2 . l_3$. Inside the metal, free electrons are moving in all directions with a distribution of energies given by eqn. (6) evaluated at the temperature of the metal, $T°$ K. Some of the electrons which impinge on the surface of the metal from the interior have enough z-directed energy to overcome the surface forces and escape. These electrons form the thermionic current, and it is their number which we wish to find. For this purpose it is convenient to convert the distribution function from an energy basis to a velocity basis. To do this we write down expressions relating the velocity increments to increments in the quantum numbers. Each z-directed quantum number together with a pair of x, y quantum numbers defines a new state, so that the total number of new quantum states is the product of the three increments in the quantum numbers. However, the energy being quadratic in the velocity, the calculation so far takes no account of the sense of the velocity. It is equally probable that an electron will be moving to the right or to the left, so that half the z quantum numbers will correspond to positive z velocities, half the y quantum numbers to positive y velocities, etc. Thus we find that one-eighth of the total new states correspond to positive velocities. Multiplying this number by eqn. (5) gives the number of electrons in the range du_z at u_z, du_y at u_y, du_x at u_x, which is the required distribution function. Carrying this out, we have

$$E = \tfrac{1}{2}m(u_x^2 + u_y^2 + u_z^2)$$

for the total kinetic energy. This is quantized, so that for translatory motion

$$E = \frac{h^2}{8m}\left(\frac{n_1^2}{l_1^2} + \frac{n_2^2}{l_2^2} + \frac{n_3^2}{l_3^2}\right),$$

where n_1, n_2, n_3 are the quantum numbers, l_1, l_2, l_3 the dimensions of the sheet which, as always, are very large in comparison with h. The increase in velocity du_x is thus equal to a change dn_1 in n_1, where $dn_1 = 2m/hl_1 du_x$, with similar expressions for dn_2 and dn_3. The total number of new states is $dn_1 dn_2 dn_3$, but only one-eighth correspond to positive du. The number of new positive states is

$$\frac{dn_1 dn_2 dn_3}{8} = \frac{m^3}{h^3}\Omega\, du_x du_y du_z,$$

and the velocity distribution function is

$$dN = 2\Omega\frac{m^3}{h^3}\frac{du_x du_y du_z}{1 + \exp[-\zeta/kT]\exp[m(u_x^2 + u_y^2 + u_z^2)/2kT]}.$$

An additional factor of 2 appears because we are discussing electrons which are allowed two values of spin in addition to the translatory quantum numbers.

We next suppose that the surface forces are equivalent to an energy $E' = \frac{1}{2}mu_{z0}^2$, so that a z-directed electron must have $u_z > u_{z0}$ if it is to be emitted. The u_x and u_y components may have any value whatever. Also, if an electron with velocity u is to be emitted in a specified unit time, say the single second between 0 and 1, it must be within u cm. of the surface, so that the current emitted from the x, y surface of a cm. cube in the sheet is

$$J = \frac{2m^3 e}{h^3} \iint_{-\infty}^{+\infty} du_x du_y$$

$$\times \int_{u_{z0}}^{\infty} \frac{u_z du_z}{1 + \exp[-\zeta/kT] \exp[m(u_x^2 + u_y^2 + u_z^2)/2kT]}, \qquad (7)$$

the symbol J denoting that this is a current density.

The double integral can be evaluated by introducing polar coordinates $u_x = u_r \cos\theta$, $u_y = u_r \sin\theta$, and becomes

$$4\int_0^{\frac{1}{2}\pi} d\theta \int_0^{\infty} \frac{u_r du_r}{1 + \exp[mu_r^2/2kT] \exp[-(2\zeta - mu_z^2)/2kT]}.$$

By elementary means we obtain the integral

$$\frac{2\pi kT}{m} \ln\{1 + \exp[(2\zeta - mu_z^2)/2kT]\}.$$

Therefore

$$J = \frac{4\pi e m^2 kT}{h^3} \int_{u_{z0}}^{\infty} u_z \ln\{1 + \exp[(2\zeta - mu_z^2)/2kT]\} du_z.$$

Figs. 1.3 and 1.4 show that the emitted electrons are an extremely small proportion of the whole; or, more concretely, for the emitted electrons $\frac{1}{2}mu_z^2 \gg \zeta$, so that the exponential term is very small and the logarithm can be expanded in series. Retaining only the first term

$$J = \frac{4\pi e m^2 kT}{h^3} e^{\zeta/kT} \int_{u_{z0}}^{\infty} u_z \exp[-mu_z^2/kT] du_z, \qquad (8)$$

$$= \frac{4\pi e m k^2 T^2}{h^3} \exp\left[\frac{\zeta - E_c}{kT}\right],$$

where we have written E_c for the critical energy corresponding with u_{z0}. We have now to determine ζ, which is done by requiring that the integral of the distribution function over all values of energy must equal the number of electrons per c.c. That is $2\int N(E)f(E)dE = N$. From the diagrams it is obvious that $\zeta = E_{max.}$ at $T = 0°$ K. At higher temperatures this is not quite true, but it is good enough for our purposes. It is easily verified that $\zeta = \frac{h^2}{2m}\left(\frac{3N}{8\pi}\right)^{\frac{2}{3}}$ in this approximation. We have already assumed that $E_c > \zeta$, which is experimentally verified by the fact that there is no emission at low temperatures. $E_c - \zeta$ corresponds to the work which has to be done in overcoming surface forces, i.e. moving an electron from just inside the metal to just outside it. It is called the work function of the surface and is a specific quality of the metal under consideration which can be calculated from the atomic constants of the material† The work function is usually denoted by the symbol ϕ and is measured in electron volts.‡ Then, finally,

$$J = A_0 T^2 e^{-\phi/kT}, \tag{9 a}$$

where
$$A_0 = \frac{4\pi me\,k^2}{h^3}$$

a universal constant.

The numerical value of A_0 is 120·4 amp./cm.^{2} degree^2. Eqn. (9 *a*) is often referred to as Dushman's equation, while the following equation is called the Richardson equation:

$$J = AT^{\frac{1}{2}}e^{-\phi/kT}. \tag{9 b}$$

Both equations can be deduced classically, and because the exponential term is dominant it is not possible to decide experimentally between them. The quantum derivation leaves no doubt that the T^2 form is correct and also assigns a value to A_0, which cannot be done from classical considerations only.

† Wigner and Bardeen, *Phys. Rev.* **48** (1935), 84. Wigner and Seitz, *Phys. Rev.* **43** (1933), 804; **46** (1934), 1002; **47** (1935), 400.

‡ The same units must be used for ϕ and k, i.e. when ϕ is measured in eV., $k = 8{\cdot}61 \times 10^{-5}$ eV./°K. The exponent then becomes $-11{,}606T^{-1}$.

Let us return for a moment to the question of the validity of assuming $\exp[(2\zeta - mu_z^2)/2kT]$ or $e^{-\phi/kT} \ll 1$. Caesium has the lowest ϕ of any pure metal, i.e. 1·81 eV. At 500° K., $\phi/kT \doteqdot 40$, so $e^{-\phi/kT}$ is extremely small. For tungsten $\phi = 4{\cdot}54$ eV. and the melting-point is 3655° K. At $T = 3600°$ K., $\phi/kT \doteqdot 15$ and $e^{-\phi/kT} < 10^{-6}$. The assumption is thus well founded.

Some comments on the temperature dependence of eqn. (9 *a*) are now in order. We have stated above that $\zeta \neq E_{max}$ for temperatures above the absolute zero. This introduces an extra temperature-dependent term into the work function, since a second approximation to ζ is

$$\zeta = E_{\text{max.}} - \frac{\pi^2}{6}(kT)^2\left(\frac{d\ln N}{dE}\right)_{E=E_{\text{max.}}}.\dagger$$

Moreover eqn. (3) shows that $E_{\text{max.}}$ depends on the number of electrons per c.c. As the metal expands N gets smaller and so the work function increases. This effect gives a temperature coefficient of increase in work function of about 10^{-4} eV./degree, while the T^2 term is only about 10^{-6} eV./degree. In addition, E_c may be temperature-dependent, so we must regard the usual assumption that ϕ does not depend on T with some caution.

1.1.3. *The energy distribution of the emitted electrons.* It is clear from eqn. (8) that the electrons which have energies greater than that corresponding to the work function possess the whole range of energies up to infinity. In several problems it is useful to know the distribution functions for the electrons outside the metal surface. An electron loses the energy $\frac{1}{2}mu_{z0}^2$ in crossing the surface barrier, so that the number of electrons having energy $\frac{1}{2}mu_z^2$ outside the metal equals the number with energy $\frac{1}{2}m(u_z^2 + u_{z0}^2)$ within. Thus the volume distribution of electrons with energy $\frac{1}{2}mu_z^2$ or velocity u_z is eqn. (8) $\div\, eu_z$ or

$$dN_{u_z} = \frac{4\pi m^2 kT}{h^3} e^{\zeta/kT} \exp\left[-\frac{1}{2}\frac{m}{kT}(u_z^2 + u_{z0}^2)\right] du_z,$$

† Mott and Jones, op. cit. p. 178.

but $$e^{-\phi/kT} = \exp\left[\frac{\zeta}{kT} - \frac{1}{2}\frac{m}{kT}u_{z0}^2\right],$$

therefore

$$dN_{u_z} = \frac{4\pi m^2 kT}{h^3} e^{-\phi/kt} \exp\left[-\frac{1}{2}\frac{m}{kT}u_z^2\right] du_z,$$

$$= \frac{J}{e}\frac{m}{kT}\exp\left[-\frac{1}{2}\frac{m}{kT}u_z^2\right] du_z. \tag{10a}$$

Then $$N = \frac{J}{e}\sqrt{\left(\frac{2m}{kT}\right)}\int_0^\infty e^{-r^2}\,dr = \frac{J}{e}\sqrt{\left(\frac{2m}{kT}\right)}\frac{\sqrt{\pi}}{2},$$

and $$dN_{u_z} = N\sqrt{\left(\frac{2m}{\pi kT}\right)}\exp\left[-\frac{1}{2}\frac{m}{kT}u_z^2\right] du_z. \tag{10b}$$

Eqn. (10*b*) will be recognized as an ordinary z-directed Maxwellian distribution, so the emitted electrons have a normal z distribution. It is more important to know how the current to an electrode will vary with negative voltage. For the plane parallel diode this can be obtained directly from eqn. (8). If the anode is at $-V$ volts to the cathode, the bottom limit of the integration is increased from $u_z(0)$ to a higher value such that $u_z^2 = 2e(V_c + V)/m$. If J_0 = saturation density,

$$J = A_0 T^2 \exp\left[\frac{\zeta - V_c - V}{kT}\right],$$

$$= J_0 e^{-V/kT}, \tag{11}$$

or $$\ln\left(\frac{J}{J_0}\right) = -\frac{V}{kT}. \tag{12}$$

In eqn. (12) V is the total potential opposing the electron motion, not merely the battery voltage applied to the anode, i.e. $V = V_a + \phi_a - \phi_k$. We see that in the retarding field region a parallel plane diode characteristic plotted on log paper becomes a straight line with slope of $1/kT$.

In the case of the cylindrical diode, matters are a little more complicated; for, unless the cathode is nearly the same diameter as the anode, the anode current depends not only on the radial velocities but also on the tangential velocities. Schottky† has treated this case and gives the following expression:

$$J = \frac{2}{\sqrt{\pi}} J_0 \left[\sqrt{\left(\frac{V}{kT}\right)} e^{-V/kT} + \int_{V/kT}^{\infty} e^{-x^2} dx \right].$$

This expression is correct within $\frac{1}{2}$ per cent for a Maxwellian distribution, if the cathode radius is less than one-thirtieth of the anode radius and if the retarding field is sufficiently great. In most cases the second term is negligible in comparison with the first. When this is so

$$\ln J = \text{const.} + \tfrac{1}{2} \ln \frac{V}{kT} - \frac{V}{kT}. \tag{13}$$

For values of V around 0·1 V. the log term is negligible and the plot is linear. In most structures space-charge limitation would be still in force at such low retarding voltages. In conclusion, we should note that these retarding field equations depend on the energy distribution of the electrons just outside the emitter, and there is no *a priori* reason to expect that they should apply to emitters other than pure metals.

1.1.4. *The measurement of thermionic emission from metals.* The thermionic emission is measured in a tube of the type shown in fig. 1.5. A filament F of the metal to be studied is tightly stretched along the axis of three co-axial cylinders G, A, G. The two outer cylinders act as guard rings and the centre one as an anode. The filament is heated to a known temperature by a current, and a sufficiently high potential is applied to A to collect the saturated current, the guard rings being held at the same potential. F is made long enough so that

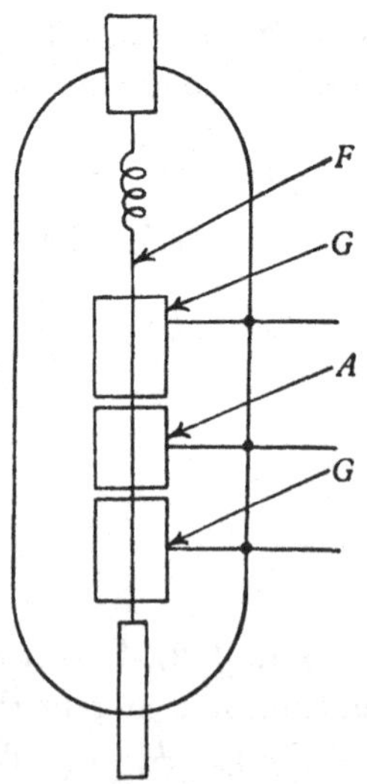

Fig. 1.5 Guard ring diode for emission measurement.

† Schottky, *Ann. Phys., Lpz.* **44** (1914), 1011.

conduction cooling at the supports does not cause the temperature of the part used for measurements to vary along its length. The temperature is determined either by optical pyrometry, knowing the spectral emissivity, or by resistance measurements, in which case potential leads may be added to the tube. The variation in potential along F is either taken into account or is eliminated by commutating the heating current and only measuring the emission when no heater current is flowing.

To obtain repeatable measurements great care is necessary. The tube must be very carefully outgassed and a prolonged heat treatment applied to F to ensure that the figures really relate to a clean surface. Tungsten is rather easy to outgas because of its high melting-point, but platinum is very difficult, and it was not until 1933 that fairly reliable values of ϕ and A_0 were obtained for this material. In the case of nickel it is probable that completely reliable values are not known even to-day.

1.1.5. *Emission in strong fields.* In the analysis of § 1.1.2 it was assumed that the field at the metal surface was zero. In the experiment described above this is not the case, and we have to allow for this. Also as the electron leaves the surface it induces an image in the surface, which modifies the field very near to the surface. The emission formula can be easily corrected for these effects. The lowering of the work function due to electric field is called 'Schottky effect'. The image force on an electron at a distance z from the surface is $e^2/4z^2$ attracting the electron into the metal, i.e. reducing the work function. This force is the gradient of a potential $-e^2/4z$. The potential due to V volts on the anode (fig. 1.6.) is eEz, where $E = V/d$. The total reduction in work function is then $eEz+e^2/4z$. This has a maximum when $z = a = \frac{1}{2}(e/E)^{\frac{1}{2}}$, and the value is $e^{\frac{3}{2}}/E^{\frac{1}{2}}$. Thus

$$\phi' = \phi - e^{\frac{3}{2}}E^{\frac{1}{2}}, \tag{14}$$

and eqn. (8) becomes

$$J = A_0 T^2 \exp\left[-\frac{\phi - e^{\frac{3}{2}}E^{\frac{1}{2}}}{kT}\right]$$

$$= J_0 \exp\left[\frac{e^{\frac{3}{2}}E^{\frac{1}{2}}}{kT}\right]. \tag{15}$$

To eliminate this effect the current values in the measurement of § 1.1.3 are taken for several anode voltages at each temperature. A plot of $\log J$ against $E^{\frac{1}{2}}$ is made and should be a straight line. It is extrapolated back to $E = 0$ to give the value of J_0. By this means fields strong enough to overcome

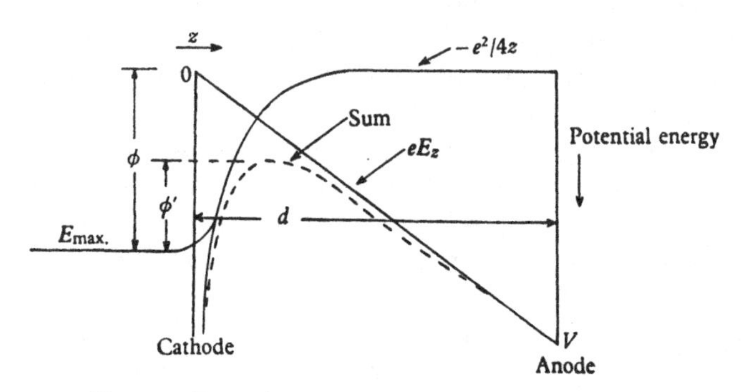

Fig. 1.6. Potential distribution at an emitting surface.

space-charge effects can be used without introducing any error in the measurement of the zero-field emission.

1.1.6. *The experimental determination of the emission constants.* The apparatus of fig. 1.5 is used to obtain a series of values of J_0 as a function of T, using the method of § 1.1.5 to correct for the surface field. Taking the log of equation (9 *a*) we have

$$\ln J_0 = \ln A_0 + 2 \ln T - \phi/kT,$$

or

$$\ln (J_0/T^2) = \text{const.} - \phi/kT. \quad (16)$$

A plot of $\log (J_0/T^2)$ against $1/T$ is known as a Richardson line. Clearly the slope is equal to ϕ/k and A_0 can then be determined. The experimental errors are fairly large, but ϕ can usually be determined to the second decimal place, the error in A_0 being much larger. The values of A_0 thus determined are not in very good agreement with the theoretical value of 120·4, as the figures in Table 1 show.

Three explanations of the discrepancy are given. One postulates the existence of a surface potential barrier which reflects some of the electrons, introducing a term $(1-r)$ into the emission law. The trouble about this is that the values of r which such barriers give are much too small if any reasonable barrier thickness and shape are assumed. Moreover, it is not clear how reflecting barriers can arise unless the surface is contaminated, as it has been shown that smooth barriers of image-force type reflect only about 5–10 per cent of the incident electrons. It is as certain as anything can be that truly uncontaminated tungsten surfaces have been obtained experimentally, and yet the values of A_0 are found to lie between 60 and 100. A much more probable explanation lies in the temperature-dependence of ϕ which we have mentioned. For most metals the range of temperature over which emission measurements can be made is small, and it is difficult to detect any temperature variation experimentally. However, for tungsten the figure of $6{\cdot}5\times10^{-5}$ eV. per degree has been given.† This would account for an A value of about 60, as can be seen by putting $\phi=\phi_0+\alpha T$ in the emission equation.

The third explanation is one that applies to all experimental work carried out on polycrystalline wires. It is now known that the work function is different for the various crystal faces of the metal, because of the different distances between the centres of force. Again taking tungsten as example, the (111) face has the lowest ϕ, 4·36 eV., while the (112) face has the highest, 4·66 eV. A consideration of a surface with two ϕ values $\phi_1<\phi_2$ covering areas of a and $1-a$ shows that if the area of low work function is dominant

$$\phi_{\text{obs}}=\phi_1,\quad A_{\text{obs.}}=aA_0.$$

If the high value of ϕ is dominant

$$\phi_{\text{obs.}}=\phi_2,\quad A_{\text{obs}}=(1-a)A_0.$$

† Potter, *Phys. Rev.* **58** (1940), 623.

This shows that the usual experimental results give smoothed values of ϕ and A. Experiments on single crystals are much more suitable for theoretical interpretation; but, on the other hand, the measurements on polycrystalline materials are more

TABLE 1. EMISSION DATA FOR PURE METALS

$$I_s = AT^2 e^{-\phi/kt}$$

Metal	Thermionic work function (eV.)	A (amp./cm.^{2} deg.^{2})	Source
W	4·54	60–100	1
Mo	4·15	55	1
Ta	4·10	60	1
Ni	5·03	1380	2
Ni	4·61		3
Pt	5·40	170	4
C	4·35	48	5
C	4·39	15	6
C	4·60	4·6	7
Ba	2·11	60	1
Ca	3·20†		8
Sr	2·74†		9
Cs	1·81	162	1

† Denotes a photo-electric value.

Sources : (1) Preferred value from Reimann, *Thermionic Emission.*
(2) Fox and Bowie, *Phys. Rev.* **44** (1933), 345.
(3) Whalin, *Phys. Rev.* **61** (1942), 509.
(4) Van Velzer, *Phys. Rev.* **44** (1933), 831.
(5) Reimann, *Proc. Phys. Soc.* **50** (1938), 496.
(6) Braun and Busch, *Helv. Phys. Acta,* **20** (1947), 33.
(7) Ivey, *Phys. Rev.* **76** (1949), 567.
(8) Rentschler *et al.*, *Rev. Sci. Instrum.* **3** (1932), 794.
(9) Anderson, *Phys. Rev.* **54** (1938), 753.

An extensive list of work functions will be found in Michaelson, *J. Appl. Phys.* **21** (1950), 536. Unfortunately, there are some mistakes in this list, the values quoted not being those given in the original references.

closely related to the conditions obtaining in practical uses of thermionic emission.

The work function can be measured by other means. A method which has been used with some accuracy is to make the filament one arm of a bridge which is balanced when no electron current is drawn to the anode and guards. When current is taken, the filament is cooled and the bridge unbalanced. The experimental errors are considerable, but some experimenters have obtained good agreement with the first method.

The thermionic work function of a pure metal or, more strictly, the room-temperature value of the work function, should equal the photo-electric work function, and when the latter is carefully determined by Fowler's method, good agreement is actually obtained, particularly for the high melting-point materials. Values of the thermionic constants for some common metals are given in Table 1. The thermionic constants for nickel cannot be regarded with confidence, although the figures quoted were obtained in very careful work and the agreement between $\phi_{Th.}$ and $\phi_{p.E.}$ is good. This agreement provides a good check on thermionic measurements and some assessment of the importance of the temperature coefficient.

1.1.7. *Contact potential difference.* The existence of the work function and the Fermi energy raises the question of the zero from which our physical measuring instruments measure potential. The potential of the most tightly bound electrons only appears in phenomena such as X-ray emission, and all ordinary measurements relate to the surface potential. What then happens when two metals of differing work functions are in thermal equilibrium in a vacuum? Clearly the metal of lower work function will emit more electrons than it gains from the material of higher work function, so that the surface potential will rise relative to that with the large ϕ until the opposing field is just sufficient to bring the two currents into equilibrium. Let the work functions be ϕ_1 and ϕ_2 (fig. 1.7). Then for equilibrium

$$A_0 T^2 \mathrm{e}^{-\phi_1/kT} = A_0 T^2 \mathrm{e}^{-(\phi_2+V)/kT},$$

where V = potential difference between the surfaces. Solving, we find

$$V = \phi_1 - \phi_2.$$

Thus the difference in potential between the surfaces is equal to the difference between the work functions, and the sense is such that the lower work function surface is more positive. A more fundamental analysis shows that the general condition for equilibrium is that the Fermi levels, ζ_1 and ζ_2, come to the same potential. In this form the condition can be extended to insulators and semi-conductors.

Contact potentials always exist in valves, and we shall discuss their bearing on the experimental characteristics elsewhere. From the practical point of view they would be unimportant if they remained constant. This is not the case, however, as the work function of a surface depends so much on the degree

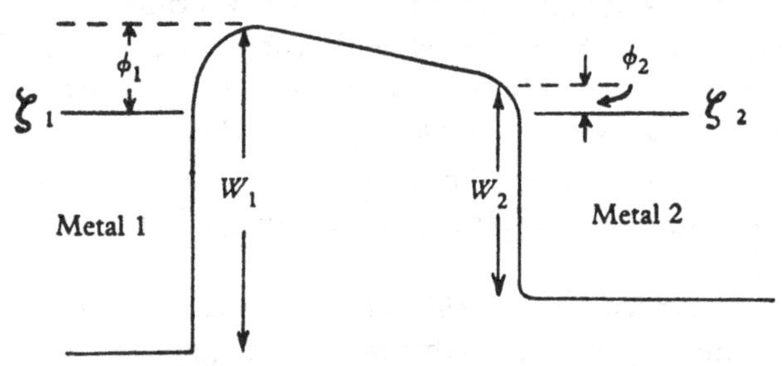

Fig. 1.7. Potential between two metal surfaces closely approaching one another.

of contamination of the surface by gas, evaporated cathode material and so on. The observed drift of the static characteristics, particularly in high slope valves where the grid is very close to the cathode, is very frequently due to a change in contact potential grid-cathode as the grid surface becomes more or less contaminated. It should be noted that in a valve, the contact potential difference is not equal to the work function difference of the corresponding clean metal surfaces but has to be ascertained experimentally. The grid-cathode contact potential difference may be either positive or negative, and one may expect it to be usually less than 0·5 V.

1.1.8. *Practical uses of pure metal emitters.* The only pure metal widely used as an emitter is tungsten. For experimental purposes molybdenum and tantalum are sometimes used, and it is likely that the use of tantalum will increase because it can be readily obtained in sheet form, allowing one to fabricate unipotential cathodes to be heated by electron bombardment. Such cathodes are convenient in high-current tubes using electron guns. However, tungsten, due to its wide use in high-power transmitting valves, is now vastly more important. Tungsten has the further great advantage that all its oxides are readily volatile, while those of tantalum and molybdenum are not.

Several properties are of importance in assessing a material as a useful emitter. For instance, the emission per watt of heating power is an important economic factor. Then the rate of evaporation determines the highest temperature that can be used for a desired life. The resistivity and temperature coefficient are important, because, other things being equal, one prefers to use the minimum heating current so as to ease the design of the filament leads. Finally, cheapness and availability are important. Tungsten is, broadly speaking, the best available metal, and all the necessary information on its properties, over a very wide range of temperatures, is tabuated by Espe and Knoll.†

In transmitter valves the main considerations to be borne in mind are the following. It has been found experimentally that a tungsten filament breaks when its diameter has been reduced 10 per cent by evaporation. This means that the life is directly proportional to the diameter. The temperature is fixed by the required total emission, allowing a factor of safety of about five. In medium-wave tubes it is usually easy enough to use a filament with sufficient area to give the required emission at a low enough temperature to ensure long life; but in short-wave tubes this is not the case, so that one must either use temperatures which are too high for long life, or go over to materials of higher specific emission, such as thoriated tungsten or oxide cathodes. Given the temperature and the total area, one must next decide on the diameter and length. For long life, the diameter should be great; but enlargement in diameter is limited by the current-carrying capacity of the seals and the fact that it is usually easier and more economical to supply

† Espe and Knoll, *Werkstoffkunde der Hochvakuumtechnik*, J. Springer, Berlin, 1936.

a fairly high voltage at a lower current. A good discussion of the economic factors involved is given by Vormer.†

To conclude this section we show in fig. 1.8 plots of the saturated emission in amp./cm.² and the rate of evaporation against $1/T$ for tungsten, molybdenum and tantalum.

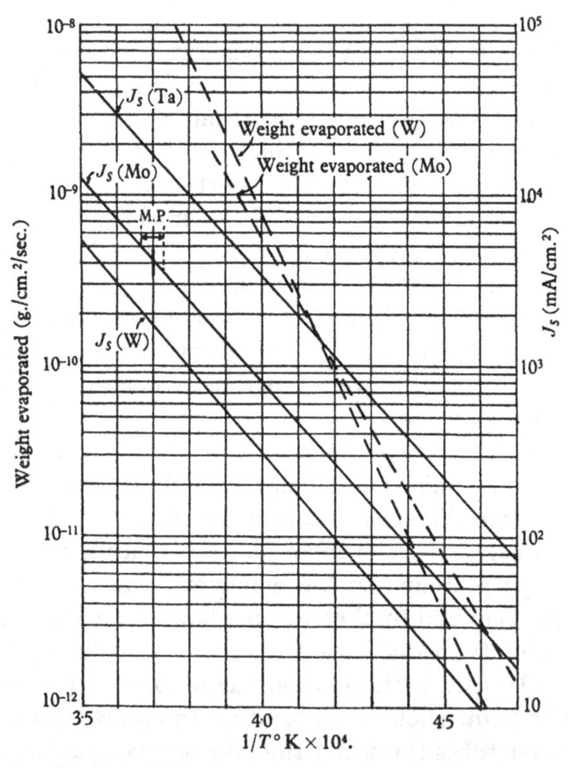

Fig. 1.8. Richardson plots for refractory metals.

1.2. Thermionic emission from thin films

Very early in the history of the valve, Langmuir discovered that vastly increased emissions could be obtained, by proper heat treatment, from tungsten filaments containing about 1 per cent of thoria. This addition was originally made to inhibit crystal growth. Since the original discovery, a great deal of

† Vormer, *Proc. Inst. Radio Engrs., N.Y.* **26** (1938), 1399.

work has been done on emitters of this type, known as thin film or contaminated metal emitters, but thoriated tungsten is the only one in use to-day and we shall confine ourselves here to its properties.

The experimental phenomena observed were as follows. The filament was flashed to a high temperature, 2500–2600° K., for a short time, then the temperature was lowered to a value in the range 2000–2300° K., and emission drawn from the filament. The emission rose, during the period of running, up to a certain value and then remained nearly constant. The temperature was then dropped to about 1500° K., at which value the emission was completely steady. The thermionic constants of such a surface are $A \doteqdot 3{\cdot}0$ amp./cm.^{2}, $\phi = 2{\cdot}63$ eV., i.e. nearly 2 eV. less than pure tungsten or 2×10^5 as much emission at 1000° K. For pure thorium $\phi = 3{\cdot}38$ eV., so the emission cannot be due to a thick layer of pure thorium.

The explanation of the increase in emission is simple in principle but in detail rather elaborate, so we shall only indicate the main features here. Reference to the standard texts on thermionic emission mentioned at the end of the chapter will provide the details. The process is due to the adsorption of positive thorium ions on the face of the tungsten. These ions induce a negative image in the metal, so that an electric double layer results with the positive side outwards. Crossing such a double layer causes a sharp fall in potential, i.e. the work function is lowered by the layer potential. An energy-level diagram is shown in fig. 1.9. The change in potential in crossing the double layer is $4\pi\sigma\mu$, where σ is the number of dipoles per unit surface and μ is the dipole moment, or more strictly the z component of the dipole moment. The experiments can then be explained as follows. The preliminary flash reduces some of the thoria to thorium, but the temperature is so high that any thorium reaching the tungsten surface immediately evaporates. Running at the intermediate temperature keeps the rate of diffusion of thorium through the complex of tungsten crystals at a value high enough to cover the tungsten surface with a monomolecular layer of dipoles in a reasonable time, because the rate of evaporation has been much reduced.

The emission builds up, as more and more of the surface is covered; but when one layer is completed only second-order changes in ϕ result, because the second layer is shielded from the metal. Also the forces holding a second layer to the metal are small in comparison with those holding the original layer, and it is therefore much less stable. When the monomolecular layer has been formed, the diffusion to the surface can be stopped, and the filament is stable at temperatures below the activating temperature.

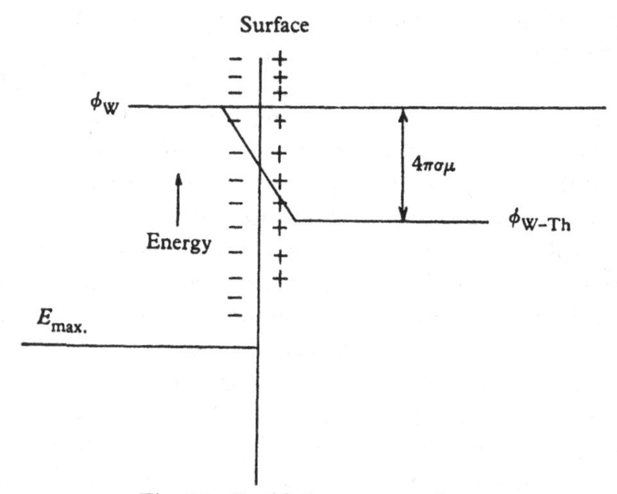

Fig. 1.9. Double layer on a surface.

In the above it has been implied that the adsorption takes place in the form of ions. In the view of de Boer,† thorium atoms also play an important role. This is because the ions which are adsorbed at the beginning of the process lower the work function of the surface to such an extent that, if the polarization of the atoms by the ions is taken into account, an atom can be adsorbed. Eventually each ion is surrounded by four atoms. The induced dipole moment in the atom has a component away from the surface, so they make a contribution to the lowering of the work function. The last stage of the process occurs when atoms are adsorbed into interstitial

† De Boer, *Electron Emission and Adsorption Phenomena*, Cambridge University Press, 1935.

positions between groups of atoms. This stage produces a slight increase in the work function. These ideas have been developed to explain the experimental fact that the maximum lowering of the work function corresponds with only about 70 per cent coverage of the tungsten surface by the monolayer, if ions alone are assumed present. Owing to the greater diameter of the atoms, fewer atoms will cover the surface, and it is possible to reconcile theory and experiment.

An important question evidently relates to the diffusion. This has been shown to take place along the faces of the tungsten crystals, and there is no evidence that any diffusion takes place through the tungsten lattice. This naturally leads one to inquire whether the thorium does not concentrate in certain active areas near the crystal boundaries. However, it has been shown that, if a tungsten ribbon is activated on one face by glowing a thorium filament near it and is then maintained in the operating temperature range for some tens of hours, the rear face shows the same activity as the front face. This means that the rate of transport of thorium over the surface is sufficient to maintain uniform coverage.

Similar effects are found when other contaminating layers than thorium are used. Barium and caesium on tungsten have both been the subject of extensive experimental investigation. The latter surface has a work function of about 1·36 eV., but it is only stable up to about 850° K., so that it does not compare favourably with the modern oxide cathode with a work function of 1·0 eV. and temperature limit of 1100–1200° K.

The emission constant A is much lower than A_0 in the case of thin film emitters, probably because the work function depends much more strongly on temperature than in the case of pure metals, and the measurements do not cover a large enough range of T to disclose the fact.

1.2.1. *Thoriated tungsten.* The modern technique for using thoriated tungsten differs in detail from the early work described above. The main difference is that the filament is carburized by flashing in a mixture of hydrogen and hydrocarbon vapour. The filament is mounted in an iron or steel bell-jar, into which a stream of dry hydrogen, which has been bubbled through a suitable hydrocarbon such as benzene or xylene, is allowed to pass. The filament is heated to about 2300° K. by passing current, and a chemical reaction takes place in which tungsten carbide, W_2C, is formed. The progress of the reaction is checked by observing the resistance of the filament which is a nearly linear function of the thickness of the tungsten carbide layer. When the layer is about 10 per cent of the radius the process is complete. Further carburization makes the filament very brittle and has no compensating advantages.

Filaments which have been carburized have several advantages over uncarburized ones. They are much less sensitive to traces of gas, and the forming or activating temperature is lower, presumably because the carbide reacts with the thoria, reducing it to thorium. A reaction of this type will supply much more thorium than will thermal dissociation. A minor disadvantage is that the thermal emissivity is increased by about 20 per cent, and therefore the power for a given temperature is somewhat greater than for ordinary thoriated wire.

Reasonable emission figures for modern thoriated filaments are: $T = 2000°$ K., $J = 3$ amp./cm.^{2}, dissipation = 25–30 W./cm.^{2}, efficiency = 125–100 mA./W.

1.2.2. *Uses of thoriated tungsten.* Thoriated tungsten was used extensively as an emitter for receiving tubes in the twenties, but was soon superseded by oxide-coated emitters. It then found a use in small transmitting tubes, but, owing to the sensitivity to traces of gas and to erratic life figures, it was not used in high-power valves. The technique has been gradually improved, however, particularly in Germany. Nowadays it is being used for higher power valves, since the carburized filaments will stand operation under severe conditions, where formerly bright tungsten was used. The much lower heating power and current is, of course, a great advantage. Another advantage which goes hand in hand with the lowered heating current is the possibility of heating the filament with a.c. The magnetic field due to the heater current produces a modulation of the anode current and a resultant hum. For this reason big transmitter valves are usually heated with d.c. The use of thoriated tungsten makes it possible to eliminate the costly low-voltage d.c. machines in favour of transformers.

1.3. Oxide-coated emitters

In this section we shall first describe the modern technique of manufacture of oxide-coated cathodes, then describe some of the crucial experiments which have been made on them, and finally describe the various attempts which have been made to give a theory of their operation.

1.3.1. *Preparation of oxide-coated emitters.* Nowadays, the cathode is always prepared from alkaline earth carbonates, though in the early days hydroxides and oxides were used. These salts are not stable in the atmosphere and the inert carbonates are preferable. The materials used are barium and strontium, sometimes with an admixture of calcium. The first step is to prepare a pure carbonate powder. This is often done by co-precipitation of barium and strontium carbonate from the nitrates by treatment with carbon dioxide. This yields a mixed carbonate, usually near the proportions 50 : 50 per cent by weight, although more barium is sometimes used. The carbonate powder is carefully washed and dried before being made into the spraying 'sludge'. The preparation of the sludge consists in mixing the required amount of powder with a mixture of organic solvents and an organic binder, placing the mixture in a ball mill and milling it until a fine suspension is obtained. The sludge is then stored in bottles which are agitated until the contents are required for use.

The choice of organic solvents and binder is largely governed by the requirements of the ordinary paint spray guns used for the spraying. These are very easily clogged unless care is taken to ensure that the carbonate particles do not acquire electric charge. Valve manufacturers nearly always use their own recipes, but in this country the typical basis is amyl acetate, with about 1 per cent of collodion shreds to act as binder. Some ethyl oxalate is often added to reduce the rate of evaporation. Distilled water with collodion has been used and so has alcohol. The solvents and binder are not important in respect of the emission from the finished product, but obviously they are very important in mass production.

When the sludge has been well milled and agitated, it is loaded into the spray gun. The cathode sleeves, usually 'O' nickel, are sprayed in a booth with an extractor to remove the unpleasant and dangerous solvent fumes. The technique depends on the purpose for which the cathode is intended; e.g. in valves with close spaced grids, a thin coating with an even surface is aimed at. The main point is to avoid leaving the coating too wet; for although this gives a smooth surface, the whole coating often lifts during activation.

After spraying, the nickel cathode sleeves are placed in an air oven at 100–200° C. to dry and harden. They are then ready for mounting into the valves.

The technique of activation is as follows. After the valve has been baked and the metal parts heated by eddy currents to outgas them, the heater volts are increased from zero to about twice the operating voltage. It is not necessary to do this unduly slowly, about 30 sec. is a good figure. The carbonates start to decompose to oxide with the evolution of carbon dioxide, which is observed either with a Pirani gauge or with the aid of a leak tester. The gas clears away in about a minute, and the heating is continued for a little longer to ensure that no more gas is evolved. The heater voltage is then dropped to a value somewhat above the operating figure and the processing of the metal completed. The emission appears immediately the gas clears away; and if the valve parts have been properly cleaned and the processing is correct, no further treatment is necessary to obtain a high and constant emission.

The above description has been written with experimental valves in mind. In machine exhausting of receiver valves the stages are telescoped together, but the process is essentially the same.

Directly heated, coated, filaments are not prepared by spraying but by dragging the wire through baths of a somewhat denser coating than is used for spraying. Each bath is followed by an oven, and a typical coating installation uses three or four stages of this type.

Another process has been used for coating, for which great claims were made on its first introduction. This is cataphoretic coating which is analogous to electroplating. If a colloidal suspension of carbonates is placed between two electrodes, carbonate particles are deposited on the cathode, and in this way a very fine, even coating can be obtained. It was claimed that the small grain size ($< 1\mu$) gave an improvement in emission, but this is now known not to be the case. Cataphoresis has a great advantage in that it is much less wasteful of the expensive carbonate sludge than is spraying, but it has not proved reliable under factory conditions and is not often used.

1.3.2. *The measurement of the emission and thermionic constants.* Emission measurements can be made with a tube of the type already shown in fig. 1.5, the filament usually being replaced by an indirectly heated unipotential cathode. The early experiments showed that the emission temperature results could be fitted to either (9 *a*) or (9 *b*) equally well. As (9 *a*) came to be accepted as the correct equation for metals, experimenters gradually gave up the use of (9 *b*), so that most values of A and ϕ quoted to-day refer to (9 *a*). One important difference between oxide cathodes and metals is that the saturation of the emission in an oxide cathode is very much less marked than in a metal. We have seen that in a metal Schottky's law is obeyed, and in the case of an oxide the log emission plotted against $\sqrt{E}$ is usually linear, but the slope of the line is much greater than is given by the theory. This fact, which is of theoretical importance, is also of very real practical importance, because it makes the measurement of the saturated emission very uncertain. The difficulty can be overcome by various means, but these only amount to agreeing on a definition of saturation and do not necessarily give identical results. Sometimes the plot of log I against V is drawn and the linear portion extrapolated back to $V = 0$. $\sqrt{V}$ is also used. Another technique is to measure the heater power required to give specified emissions, e.g. of 0·2, 0·5, 1·0, 2·0, 5·0 mA., at a specified fixed anode voltage well above the 'saturation' value. These points are plotted on special paper called a power emission chart in which the coordinates are warped according to the Dushman equation. A straight line, which can be extrapolated to full heater power, is obtained. This test is more common in the United States of America than here. It is much more useful for making a comparison between different valves of the same species than it is for quantitative work. This is because (*a*) the form of emission equation is assumed, and (*b*) no attempt to find the zero field saturation emission is made.

It seems best to adopt a purely empirical attitude, i.e. to use the simplest type of plot, the $\log I \sim V$ curve with extrapolation back to $V = 0$. There is no good reason why $\sqrt{V}$ should be used and no better straight lines are obtained by its use, so there appears to be no justification for attempting to thrust the experimental results into the straitjacket of an inapplicable theory.

We have not yet discussed the major experimental difficulty in measurements on oxide cathodes, namely, the construction of an accurate temperature scale. In the case of a metal, two main approaches are possible and can be used to check one another. The spectral emissivity is either known or can be measured, so that optical pyrometry provides a sound method of temperature measurement. Secondly, resistance thermometry can be used, with the usual precautions. In the case of the oxide, the spectral emissivity cannot be measured once and for all, because it varies according to the preparation, coating thickness, etc. Resistance thermometry can be used to determine the core temperature, but a further determination, the thermal conductivity, is necessary to approximate to the surface temperature. This again is not a unique determination, although it probably varies less than the spectral emissivity.

The best that can be done is to measure the spectral emissivity of a fairly large number of samples prepared under standard conditions, taking care that the coating is thick enough to eliminate completely any effects due to the core. The mean spectral emissivity used with optical pyrometry then provides a reasonable temperature scale. A sufficient number of samples must then be tested for emission to provide for the overall accuracy required. Insufficient attention to statistical questions of this sort is possibly one of the reasons for the great divergences in the experimental values of A and ϕ quoted in the literature.

For technical purposes it is usual to discuss the emission per cm.^{2} at a given temperature. Modern receiver valves with indirectly heated cathodes usually operate in the range 900–1050° K. The saturated emission would range from 0·2 to 10 amp./cm.^{2}. In manufacture the temperature is a variable which is adjusted until nearly all the valves have a saturated emission greater than, say, ten times the mean working current. For this reason output valves and rectifiers have hotter cathodes than input valves. Transmitter valves and very high slope valves are run even hotter, but at a cost in life. We shall see later in the book that the saturated emission available limits the performance of high-frequency valves, and that further progress is not possible in some fields until better cathodes are available.

1.3.3. *Experiments on the emission mechanism in oxide-coated cathodes.* It was soon found that experimental data on the emission as a function of temperature could be fitted to the Richardson equation, giving a low work function and also a very small A value; e.g. Koller† gives $\phi = 1{\cdot}04$ eV., $A = 10^{-3}$ for a mixed BaO–SrO cathode; other experiments at about the same time gave substantially higher work functions, although most results were lower than the figure for an ordinary barium surface. In spite of this, and in spite of the fact that the vapour pressure of barium is as high as 3×10^{-2} mm. Hg at 1000° K., it was at first thought that the emission was due to a bulk layer of barium on the cathode surface, the coverage being an equilibrium between the amount appearing from the oxide and the amount evaporated or oxidized. The way in which the thermionic constants varied throughout the activation process was then studied, with contradictory results. It

† Koller, *Phys. Rev.* **25** (1925), 671.

should be pointed out here that low-temperature activation has to be used to slow the process down enough to study it by stages, and the results do not necessarily correspond with the high-temperature methods actually used. Apart from this, some investigators† found that ϕ remained constant and A increased during activation, while others‡ found that both quantities altered. It had also been found that the emission was sensitive to minute traces of some gases such as oxygen and chlorine, which lent colour to the picture of a barium film, as it would clearly react very quickly with these gases.

Experiments were therefore directed to establishing the presence or otherwise of free barium in the coating. (We speak of barium alone, although strontium may be produced.) It was soon established that the cathodes evolved oxygen during the activation process and also (by chemical means), that there was free barium in the coating. Furthermore, inactive cathodes could be activated by distilling barium on to their surfaces. To explain these facts Koller§ advanced the view that the barium covered not only the surface, but all the faces of all the crystals forming the coating, and that the layer was monatomic, so that the adsorption forces held the barium in place in the same way that they hold a thorium film on tungsten.

This theory would have been acceptable but for the fact that in many experiments it was found that the core material had a marked effect on the saturated emission. In particular, alloys of nickel with reducing agents such as magnesium, silicon and titanium gave better results than pure nickel. Lowry's|| experiments are classical in this field. Filaments of 'konel' (nickel, cobalt, iron, and titanium) gave as much emission at 1050° K. as platinum-iridium filaments at 1250° K., a very large effect. He therefore postulated that the barium film was formed on the surface of the core and not on the oxide crystals. It would reduce the work function of the core metal by an amount independent of the material, and therefore

† Espe, *Wiss. Veröff. Siemens-Konz.* **5** (1926), 29.

‡ Huxford, *Phys. Rev.* **38** (1931), 379.

§ Koller, loc. cit.

|| Lowry, *Phys. Rev.* **35** (1930), 1367.

the work function of the core would be involved in the emission.

It was soon shown that the effect of the core was not nearly as direct as Lowry supposed, by making valves in which the coating could be removed under vacuum, after the valve had been fully activated. This removal should not have altered the emission had Lowry's theory been correct, but in fact the emission dropped by a factor of about 5000. Some degree of return to the theory of Koller was therefore indicated.

At about this time Wilson and Fowler had deepened our knowledge of the solid state, and, in particular, Wilson had introduced the idea of semi-conductors in which conduction was due to the presence of impurities in the crystal lattice. Many substances which were earlier thought to be solid electrolytes were shown to be semi-conductors, i.e. the charges were carried by electrons and not by ions. Reimann and his co-workers had already studied the electric conductivity of the oxide coating, and they found that it could be expressed in the form $\sigma = a\,e^{-b/T}$. Also, in the activation process, the conductivity increased as the emission increased. The original explanation was based on electrolytic conduction, and Reimann postulated a barium circulation in which the positive barium ions were drawn towards the core, gained an electron near the core-oxide interface, and diffused back to the surface as atoms, and there were ionized to go through the cycle again. The observations, however, agree equally well with semi-conductor theories, which also account for the fact, established by Becker, that only about $\frac{1}{2}$ per cent of the current is carried by ions.

Another phenomenon which must be mentioned is the decrease of emission with time when fairly heavy currents are being drawn from the cathode. Several types of effect can be isolated. One which is very well established is the slow drift in emission after a change of cathode temperature. On lowering the temperature it is often observed that the emission decreases to a minimum at the instant the temperature changes and then increases to a new equilibrium value. Such phenomena are explicable if we regard the cathode as being in a state of equilibrium between activating and poisoning processes.

The most interesting phenomenon is the decay of current during pulsed application of voltage. Two qualitatively different stages appear. The first, in which the pulses are of very short duration, less than say 5μ sec., we shall deal with in detail somewhat later. The other, which has been known much longer, is characterized by very slow changes as current is drawn. It can be observed if saturated currents are drawn continuously at not too high a temperature, but pulses of the order of 100μ sec. present the best experimental method. The effect has been investigated fairly recently by Blewett† and Sproull.‡ Sproull has worked out a theoretical explanation based on the migration of barium away from the cathode surface when current is drawn. His theory of cathode emission was based on the monolayer picture and is therefore not in accordance with more modern views on the subject, which make it seem unlikely that a barium monolayer exists, or needs to exist, to obtain a low-surface work function. Wagener has very recently pointed out that the form of Sproull's equations, which agree with the experimental results, are unaltered if the picture is revised to that of oxygen diffusing from the inside of the cathode to poison the emission. On any theory an oxygen layer will increase the surface work function, so this view is not in conflict with present ideas.

In the past there has been a great deal of discussion as to whether ϕ or A changed during the activation process. We have already seen that a temperature-dependent term in ϕ will appear as an apparent change in A. In a later section we shall find that such temperature-dependent terms appear in emission equations for oxide cathodes and that they are more important than in metals. Such discussions are therefore rather unrealistic. A very clear discussion of the correlation between theoretical constants and those derived from Richardson plots is given by Becker and Brattain.§ In addition to the effects considered by them, we must include the effect of variations in work function from point to point on the surface. It is

† Blewett, *J. Appl. Phys.* **10** (1939), 668 and 831.
‡ Sproull, *Phys. Rev.* **67** (1945), 166.
§ Becker and Brattain, *Phys. Rev.* **45** (1934), 694.

hardly surprising, therefore, that the values of ϕ and A quoted in the literature on oxide cathodes have a very wide spread.

It should be emphasized that it is difficult to get consistent experimental results on oxide-coated cathodes, particularly when vacuum conditions are not all they should be. In view of the rapid improvement in vacuum technique which has taken place, it is not surprising that there is a considerable disagreement between experimental results. It is almost true to say that contradictory evidence can be cited regarding every critical experiment. In the above account we have only described work which is in agreement with what one observes in cathodes prepared by modern methods. Before carrying the discussion of oxide cathodes further, we must describe the most fundamental ideas on semi-conductors.

1.3.4. *Semi-conductors.* We have already said that in some substances the gap between filled and empty bands is very narrow, giving rise to a certain conductivity at room temperature. As the temperature increases, more electrons have sufficient energy to jump the narrow gap and the conductivity increases (whereas, of course, in a metal it decreases as T increases). Such substances are called intrinsic semi-conductors. More important are the 'impurity' semi-conductors. These are pictured as being substances in which a small number of impurity atoms exist, either in interstitial positions or else as an excess or deficiency of one component of the substance. The classic example is cuprous oxide. If this compound is prepared in stoichiometric proportions, the log conductivity $\sim 1/T$ plot is a straight line, whereas, if there is a small excess of oxygen, the conductivity jumps by several orders of magnitude at low temperatures, and then is linear to about 550° K., where there is an abrupt change in curvature bringing the conductivity back to the original curve. Fig. 1.10 shows the form of the curves. Wilson postulated that the impurity centres contribute energy levels which lie in the forbidden zone, e.g. of cuprous oxide, but near to either the upper, empty conduction band or to the lower, filled band. In the former case, the impurity atoms can contribute electrons which have sufficient thermal energy to make the transition into the conduction band. Once in the conduction band they are free to move, and give rise to a conductivity. In the second case the electrons in the filled band can jump into the impurity levels, leaving empty spaces in the filled band. These 'holes' act as though they were positive charges, free to move, and therefore give rise to a conductivity. In both cases the

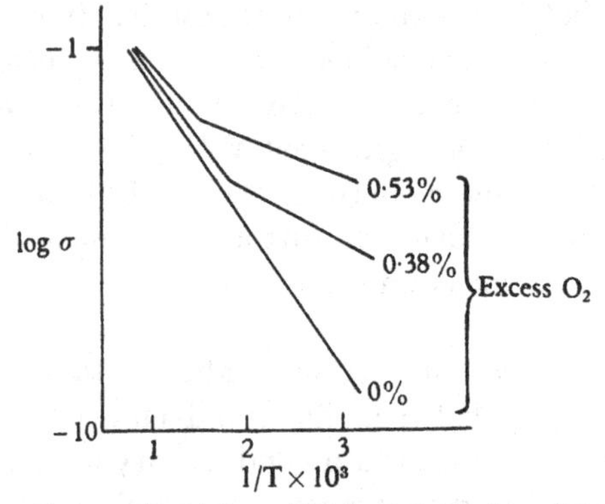

Fig. 1.10. Log conductivity $\sim I/T$ for a typical semi-conductor.

conductivity increases with temperature, since a rise in temperature increases the number of electrons able to bridge the gap. Clearly the conductivity will also depend on the number of impurity centres. Semi-conductors in which the impurities lie just below the empty band are called 'excess' or 'N-type' semi-conductors. When the impurities lie near the full band, the material is termed a 'defect' or 'P-type' conductor. The N and P nomenclature refers to the sign of the charge on the carrier, N = negative = electrons, P = positive = holes. Fig. 1.11 shows the relative energy levels for intrinsic, N-type and P-type semi-conductors. In addition to these three basic types of conduction, materials are known in which two or more processes exist simultaneously. In fact, an intrinsic semi-conductor can be looked on as a combination of N-type and P-type conduction, because the relatively small number of excited electrons leave vacant holes in the full band. The type of conduction is determined experimentally from the sign of the Hall coefficient or the sense of the thermo-electric potential, the Hall coefficient being negative for N-type semi-conductors and positive for P-type. The type of conduction seems to be a specific property of the material, for instance, cuprous oxide is a P-type semi-conductor, while zinc oxide is N-type, and it is not possible to prepare N-type cuprous oxide or P-type zinc oxide. There seems to be no theoretical reason for this, and in the case of silicon and germanium, which have small energy intervals in the pure state but are usually treated with impurities to increase the conductivity, it is possible to prepare either N- or P-type material.

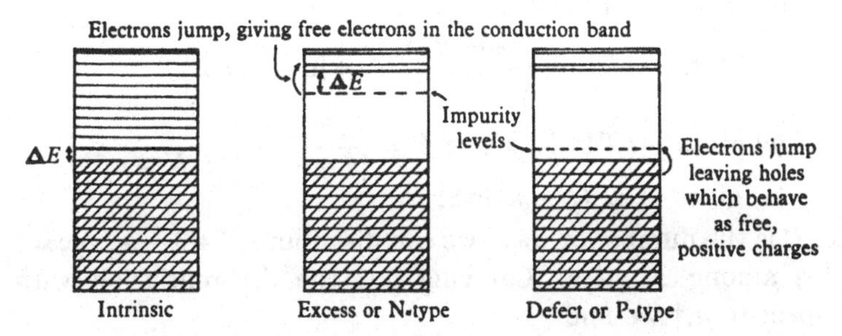

Fig. 1.11. Energy levels in metals and semi-conductors.

The theory of the properties of semi-conductors is by no means fully developed, but it is possible to deduce expressions for the electric conductivity and the thermionic emission, and this we shall now do.

The electric conductivity $\sigma = nev$, where n = number of free electrons, e = electronic charge, v = electronic mobility. We therefore have to calculate n and v. Following Mott and Gurney,† we assume that in a volume of V there are N impurity centres and N_e electrons of which n are excited to the conduction band at temperature T. The remaining $N_e - n$ electrons are frozen at their position in the impurity level. Although the calculation is made explicitly for electrons, the results are equally applicable to holes. The difference in energy between the impurity levels and the conduction band is ΔE. The calculation is made by applying the condition that the crystal is in thermal equilibrium at T. This is done by

† Mott and Gurney, *Electronic Processes in Ionic Crystals*, Oxford, 1940; 2nd ed. 1949.

calculating the free energy (F) as a function of n and T and putting

$$\left(\frac{\partial F}{\partial n}\right)_T = 0.$$

F consists of terms due to (a) the energy gained in the transition to the conduction band, (b) the energy of the free electron gas, which is given by the classical statistics because of the small number of electrons involved, (c) the energy of the vacant sites in the impurity level, and lastly (d) the spin contribution of the vacant sites:

(a) gives a contribution $n\Delta E$,

(b) gives $-nkT\left[\ln\left(\frac{2\pi mkT}{h^2}\right)^{\frac{3}{2}} + 1 + \ln\frac{V}{n} + \ln 2\right]$,

(c) gives $-kT\ln P$, where $P = \frac{N!}{(n+N-N_e)!\,(N_e-n)!}$,

(d) gives $-kT(N-n)\ln 2$;

i.e. P is the number of ways we can distribute $N+n-N_e$ vacant sites among N sites. Carrying out the differentiation with respect to n,† we find

$$\frac{n(N-N_e+n)}{N_e-n} = V\left[\frac{2\pi mkT}{h^2}\right]^{\frac{3}{2}} \exp[-\Delta E/kT]. \tag{17}$$

To obtain useful results this has to be simplified, so we assume $n \ll N_e$. There are then two limiting cases

(1) $n \ll N-N_e$

$$n = \frac{N_e V}{N-N_e}\left[\frac{2\pi mkT}{h^2}\right]^{\frac{3}{2}} \exp[-\Delta E/kT], \tag{18}$$

(2) $n \gg N-N_e$

$$n = N_e^{\frac{1}{2}} V^{\frac{1}{2}}\left[\frac{2\pi mkT}{h^2}\right]^{\frac{3}{4}} \exp[-\Delta E/2kT]. \tag{19}$$

It can be shown that the mobility v is nearly constant for $T \ll \Delta E/2k$. (In practical cases $\Delta E > 0{\cdot}2$ eV., corresponding to $T \doteqdot 1200^\circ$ K.) so that we can put

$$\left.\begin{aligned} \sigma &= \sigma_0 e^{-\Delta E/kT}, \\ \text{or} \qquad &= \sigma_0 e^{-\Delta E/2kT}. \end{aligned}\right\} \tag{20}$$

† Sterling's formula ($\ln N! \to N\ln N$ for large N) has been used.

ΔE is often referred to as the activation energy. Eqns. (18) and (19) provide a method of measuring this important quantity, but care has to be taken to make sure which equation applies. Fig. 1.12 shows a type of curve obtained when a wide temperature range is explored, T_c being the temperature at which $n = N - N_e$. The situation is further complicated by the fact that there may be other mechanisms contributing to

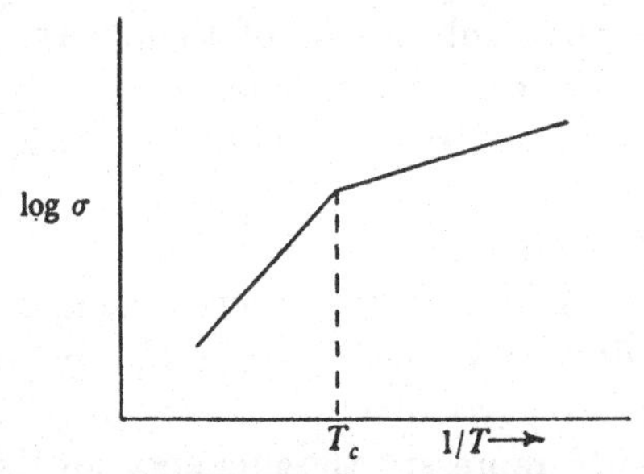

Fig. 1.12. Transition point in the conductivity diagram.

the conductivity, in which case

$$\sigma = \sum_n A_n \exp[-\Delta E_n/kT].$$

Furthermore, the impurity level is probably not well defined, especially in the case of high concentrations of impurity, and it should probably be replaced by an impurity band. However, the experimental fact is that σ varies exponentially with inverse temperature, although great variations occur between different samples of the same material. In the last section it was pointed out that Reimann determined a law of the type (20) for oxide cathodes before the theory of semi-conductors had been developed.

From the conductivity we turn to the question of the thermionic emission. This has been dealt with by Fowler,† Herrmann and Wagener‡ and other authors. The derivation given here is a synthetic one, which should not be regarded as rigorous, but has the virtue of simplicity. The first step is to calculate the relative potentials of the semi-conductor and the metal

† Fowler, *Statistical Mechanics*, Cambridge University Press, 2nd ed. 1936.
‡ Herrmann and Wagener, *Die Oxydkathode*, J. A. Barth, Leipzig, vol. II, 1944.

core. When the cathode is heated to the temperature T and no current is drawn from it, the electrons are distributed partly in the conduction band, where the density of conduction levels per unit volume is

$$D(E) = \frac{4\pi\,(2m)^{\frac{3}{2}}}{h^3} E^{\frac{1}{2}} dE,$$

and partly in the impurity level. There are N impurity centres, and we assume that only a few of them have lost electrons by excitation to the conduction band, i.e. $n \ll N$. The number of electrons, always per unit volume, with energy E is $\dfrac{D(E)}{\exp[(E-\zeta_1)/kT]+1}$ according to the Fermi-Dirac statistics. The integral of this over all the allowed energies must equal n and so we can determine the constant E_1. Below the conduction band only the particular value of energy ΔE is allowed, so that we have to represent these energy levels by a Dirac δ function, a function zero everywhere except at $-\Delta E$ (taking the bottom of the conduction band as zero for the moment), where it goes to infinity in such a way that the integral equals N. The number of electrons in the impurity band is then given by

$$\begin{aligned} N &= \int_{-\infty}^{0} \frac{\delta\, dE}{\exp[(E-\zeta_1)/kT]+1} \\ &= N \frac{1}{\exp[-(\Delta E-\zeta_1)/kT+1]}, \end{aligned} \tag{21}$$

from a well-known result of operating with δ functions. The number of electrons in the conduction band is

$$\frac{4\pi(2m)^{\frac{3}{2}}}{h^3} \int_0^{\infty} \frac{E^{\frac{1}{2}} dE}{e^{(E-\zeta_1)/kT}+1},$$

and we know that this is much smaller than N, also the exponential term is much larger than 1, as it was in the derivation of the Richardson equation. We get therefore

$$\begin{aligned} n &\doteq 4\pi \frac{(2m)^{\frac{3}{2}}}{h^3} e^{E_1/kT} \int_0^{\infty} e^{-E/kT} E^{\frac{1}{2}} dE \\ &= \frac{2\,(2\pi mkT)^{\frac{3}{2}}}{h^3} e^{\zeta_1/kT}. \end{aligned} \tag{22}$$

The sum of (21) and (22) must equal N, assuming that each impurity can only lose one electron.

Therefore $$N = \frac{\mathrm{N}}{\exp[-(\Delta E - E)/kT]+1} + \frac{2(2\pi m kT)^{\frac{3}{2}}}{h^3} e^{\zeta_1/kT}.$$

or $$N \frac{\exp[-(\Delta E - \zeta_1)/kT]}{\exp[-(\Delta E - \zeta_1)/kT]+1} = \frac{2}{h^3}\left(\frac{2\pi m kT}{h^3}\right)^{\frac{3}{2}} e^{\zeta_1/kT}.$$

The left-hand side can be rewritten as

$$N \frac{1}{1+\exp[(\Delta E + \zeta_1)/kT]},$$

and we have already limited our analysis to the case in which this term is much greater than 1, so we can put

$$N \exp[-\Delta E/kT] = \frac{2(2\pi m kT)^{\frac{3}{2}}}{h^3} \exp[2\zeta_1/kT];$$

therefore $$\zeta_1 = -\frac{\Delta E}{2} + \frac{kT}{2} \ln \frac{Nh^3}{2(2\pi mkT)^{\frac{3}{2}}}. \tag{23}$$

We can now test the goodness of the approximation obtained by neglecting 1 in comparison with

$$\exp[(\Delta E + \zeta_1)]/kT \doteqdot \exp[\Delta E/2kT].$$

If $\frac{1}{2}\Delta E = 0{\cdot}5$ eV., $T = 1000°$ K., the exponential term is over 300. These figures are representative of conditions in an oxide cathode. In germanium crystals $\frac{1}{2}\Delta E \doteqdot 0{\cdot}1$ eV., $T = 300°$ K. and the exponential is about 50. The approximation is thus justified. Equation (23) gives the value of the energy ζ_1, relative to the bottom of the conduction band. When the semiconductor is in contact with a metal the relative energy levels adjust themselves so that the electrochemical potential in both materials is equal, i.e. ζ_1 is equal to the Fermi level of the metal. Fig. 1.13 shows the arrangement at zero temperature, and we have left a space between the metal and the semiconductor to represent the actual contact because, as we shall show in the sequel, matters are more complicated there.

The second term of eqn. (23) is usually small compared with the first, so we have

$$\zeta_1 \doteqdot -\tfrac{1}{2}\Delta E. \tag{24}$$

Inserting values for the constants and inverting the logarithm we have

$$\zeta_1 = -\frac{\Delta E}{2} - \frac{kT}{2}\ln\left(4{\cdot}89\times 10^{15}\frac{T^{3/2}}{N}\right),$$

The second term is zero when $T_1^{3/2} = \frac{N}{4{\cdot}89}\times 10^{-15}$. For most semi-conductors studied N is of the order 10^{20} so $T \doteqdot 500°$ K. If T is doubled or halved the second term is $\mp 0{\cdot}045$ eV. For an oxide-coated cathode $\frac{1}{2}\Delta E \doteqdot 0{\cdot}6$ eV., so the correction term is about 7 per cent.

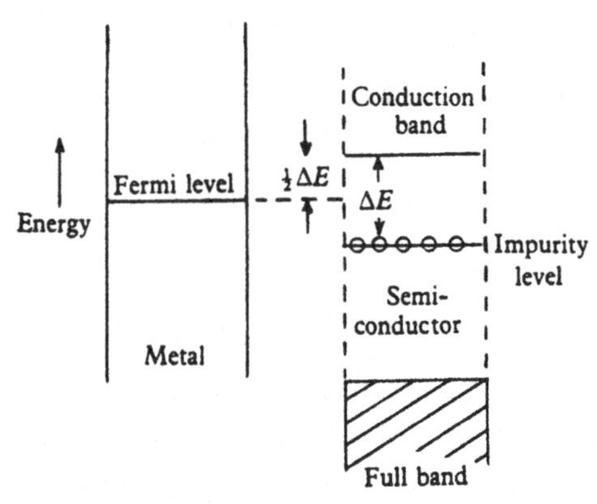

Fig. 1.13. A contact between metal and semi-conductor.

Having determined ζ_1 we can now obtain an explicit expression for the electron distribution. If eqn. (23) is used for ζ_1 in the Fermi distribution and unity neglected in comparison with the exponential, we get

$$n(E) = N^{1/2}\left[\frac{8\pi^{1/2}}{h^3}\left(\frac{2m}{kT}\right)^{3/2}\right]^{1/2} e^{-\Delta E/2kT}\, e^{-E/kT}\, E^{1/2}\, dE. \tag{25}$$

In classical statistics the distribution law is

$$\frac{2}{\sqrt{\pi}}\frac{N(v)}{kT^{3/2}}\, e^{-E/kT} E^{1/2}\, dE, \tag{26}$$

where $N(v)$ = number of electrons in the whole volume. It will be seen that eqns. (25) and (26) are identical if we identify $N(v)$ as

$$N^{1/2}\left[2^{1/2}\left(\frac{2\pi m\, kT}{h^2}\right)^{3/4}\right] e^{-\Delta E/2kT},$$

a quantity which has an exponential temperature dependence. This is equivalent to saying that, because of the relatively large energy required to excite an electron from the impurity levels to the conduction band, few electrons will make the transition, so that the electron gas is non-degenerate and the Maxwell-Boltzmann statistics apply.

We can now derive an emission equation for the semi-conductor. If the surface work function is ϕ only electrons with an energy greater than ϕ will be able to escape. The expression analogous to eqn. (22) giving the number of electrons with a z-directed energy in dE at E is

$$n_{(z)} = \frac{N(v)}{2\sqrt{\pi}} \frac{E^{-\frac{1}{2}}}{(kT)^{\frac{1}{2}}} e^{-E/kT} dE.$$

The number which will reach the surface per second is $n_{(z)}\sqrt{2E/m}$, and so the total number reaching the surface with $E > \phi$ is

$$\frac{N(v)}{(2\pi mkT)^{\frac{1}{2}}} \int_{\phi}^{\infty} e^{-E/kT} dE = \frac{N(v)k^{\frac{1}{2}}}{(2\pi m)^{\frac{1}{2}}} T^{\frac{1}{2}} e^{-\phi/kT}.$$

The current is e times this last expression or, inserting the value of $N(v)$,

$$J = N^{\frac{1}{2}} e \left[\frac{2k^{\frac{5}{2}}(2\pi m)^{\frac{1}{2}}}{h^3}\right]^{\frac{1}{2}} T^{\frac{5}{4}} \exp\left[-\frac{2\phi + \Delta E}{2kT}\right] \tag{27}$$

$$= 1{\cdot}8 \times 10^{-6} N^{\frac{1}{2}} T^{\frac{5}{4}} \exp\left[\frac{-2\phi + \Delta E}{2kT}\right] \text{ amp./cm.}^2. \tag{28}$$

Eqn. (28) shows that the emission from a semi-conductor is proportional to the square root of the number of impurity centres, to $T^{\frac{5}{4}}$ and to $\exp -(\phi + \frac{1}{2}\Delta E)$. ϕ is the surface work function and

$$\frac{\Delta E}{2} - \frac{kT}{2} \ln \frac{Nh^3}{2(2\pi mkT)^{\frac{3}{2}}}$$

is the inner work function, i.e. the energy required to excite an electron to the conduction band. It is interesting that the temperature-dependent part of ζ_1 does not appear in the exponent but only modifies the emission constant. As we have already said, thermionic emission measurements are not capable of deciding the correct power of T, and in view of the form of the approximation (24) it is not surprising that $J = AT^2 e^{-B/kT}$

gives a very good fit to the experimental points. The T^2 law has usually been used to discuss the emission, as it was natural to attempt to fit an expression which was well founded for pure metals.

Eqn. (28) could have been deduced, to within a numerical factor of $\sqrt{2}$, from eqn. (19). It has been deduced above by a method which is more intuitive and which also gives us some relations which we shall need later. However, it is natural to inquire what the significance of an emission equation derived from eqn. (18) is. This equation is

$$J = \frac{N_e}{N-N_e}\frac{2\pi mk^2}{h^3}T^2\exp\left[-\frac{\phi+\Delta E}{kT}\right],$$

$$= \frac{N_e}{N-N_e}A_0T^2\exp\left[-\frac{\phi+\Delta E}{kT}\right]. \quad (29)$$

This equation has a much greater work function than eqn. (28). It corresponds to the case in which most of the impurity centres are ions, which cannot contribute an electron to the conduction band. In view of the higher work function it is tempting to assume that eqn. (29) gives the emission at the beginning of activation, and that activation consists in forming the impurity centres, de-ionizing them and increasing their number. This would mean that the work function would fall sharply as the expression (29) changes to (28), and thereafter A would increase with more or less constant work function. This does not appear too probable, but it may be that the free barium is first formed by electrolysis and is ionized and that the role of current carrier is soon taken over by the electrons.

Other emission equations can be derived from slightly different assumptions and approximations. Fowler† gives several and others are given by Busch.‡ As a matter of fact, the details of the emission formulæ are of very little importance at the present time because our knowledge is not sufficiently advanced to allow us to calculate, for instance, ΔE or to find the number of impurity centres at all accurately. They are

† Fowler, op. cit.

‡ Braun and Busch, *Helv. Phys. Acta*, **20** (1947), 33.

only of interest in so far as they show the general behaviour of the system and indicate in what directions research should proceed. The important features are that the emission depends (*a*) on the surface forces, through the surface work function, (*b*) on the energy interval between impurity levels and the conduction band, i.e. ΔE, and (*c*) on the concentration of the impurity.

We now proceed to discuss the application of the semi-conductor picture to an oxide cathode.

1.3.5. *Oxide-coated cathodes as semi-conductors.* In the last section we have briefly discussed semi-conductors and have derived expressions for their conductivity and the electron emission under certain assumptions. We have not shown that these ideas can be applied to oxide cathodes and, in fact, it is not absolutely certain that they are entirely applicable. However, a physical model which accounts for most of the observed facts can be built up.

When the carbonates are broken down to oxides by heating a certain amount of free barium is formed. It is not known precisely what reactions are involved, but the following are all possible and occur to a greater or less extent depending on the experimental circumstances:

(1) Electrolytic dissociation.

(2) Reaction of barium oxide with de-oxydizers in the core metal.

(3) Reaction of barium oxide with carbon left in the coating by the organic binder.

(4) Some of the carbon dioxide evolved dissociates to carbon monoxide which takes up more oxygen from the oxide, leaving free barium.

The resulting free barium diffuses throughout the coating, and some of it is taken into the lattice either in interstitial positions or else removes a certain amount of oxygen from the lattice giving rise to vacant sites. The energy levels of the impurity centres fall somewhat below the bottom level of the conduction band and the system functions as an excess or

N-type semi-conductor. By analogy with sodium chloride which has the same crystal structure as barium oxide, Schottky† proposes the arrangement shown in fig. 1.14, where it is seen that the impurities consist of barium atoms in some of the sites usually occupied by barium ions and a neighbouring vacant oxygen site so as to maintain the condition of zero electric charge. Clearly if the concentration of impurities becomes sufficiently great, they will form a regular pattern of their own with the appropriate system of allowed and forbidden bands in the impurity level. The amount of barium

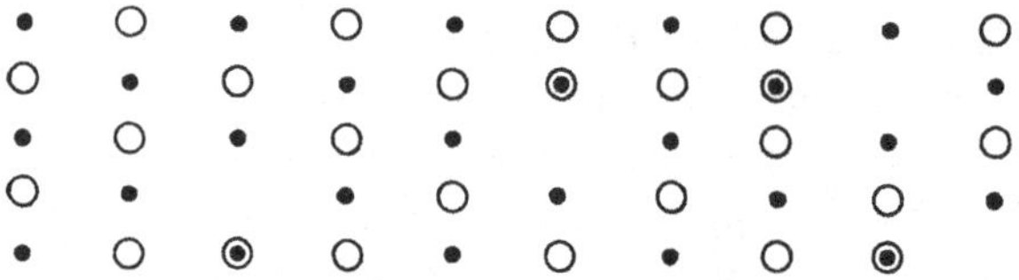

Fig. 1.14. Lattice for semi-conducting barium oxide. • $=Ba^{++}$; ○ $=O^{--}$; ⊙ = Ba atom.

which can be taken into the crystal lattice is not known, nor do we know whether diffusion takes place through the lattice or whether it only occurs along the crystal faces. Presumably, when the lattice has taken up as much excess barium as it can, the remainder evaporates fairly easily from the surface. The emission is that from a heated semi-conductor and the good properties of barium oxide–strontium oxide are due to a favourable combination of surface work function and ΔE.

On this model two types of poisoning are possible. First, a surface layer can form which acts so as to increase the surface work function. Oxygen will form such a layer. Secondly, the removal of the excess barium by chemical reaction will render the cathode inactive. The first type should be overcome by heating in a good vacuum, but the second may not be recoverable, since it may not be possible to prepare a second supply of excess barium. Similarly, life may end because of the formation of a surface oxygen or chlorine layer, the oxygen being evolved from the electrodes and the chlorine from the glass, or because of chemical reaction with the free barium.

† Schottky, *Z. phys. Chem.* B, **29** (1935), 335.

It will be noticed in the above that we make no mention of a surface monolayer of barium. This is because careful experiments do not disclose its presence and also because calculations by Wright† seem to indicate that the oxide has in any case a very low surface work function. A much more interesting question relates to the interface between the core and the oxide. It is known that high-resistance layers are formed there, and as we shall see in the next section, this can give rise to important effects.

1.3.6. *Pulse emission from oxide-coated cathodes.* During the 1939–45 war it was found that the oxide-coated cathodes used in pulsed cavity magnetrons were capable of relatively enormous saturated emissions on short pulses. At first this was thought to be due to secondary emission, but it was soon found that pulsed diodes showed the same effect. It is now known that emissions of up to 150 amp./cm.^{2} have been observed from pulses of about 1 μ sec. duration. If the pulse duration is longer than 10–20 μ sec. the emission falls to its ordinary value. The main phenomena observed are the following.

(*a*) In a good pulsed cathode, i.e. a cathode that obeys the power law up to very high emissions, the limit of emission is set by the field strength at the cathode surface, which causes sparking.

(*b*) Good continuous cathodes are not necessarily good pulsed cathodes, but good pulsed cathodes are invariably good continuous emitters.

(*c*) If continuous current is drawn in addition to pulses, the maximum allowable field strength is increased.

(*d*) Electron-diffraction studies indicate that the surface layers, to a depth of about 200 atoms, are pure strontium oxide. It is not possible to say from these studies whether there is a barium surface layer or not.

(*e*) Probes inserted in the cathode show that the metal-oxide interface in general has a high resistance.

† Wright, *Proc. Phys. Soc.* **60** (1948), 13 and 22.

(*f*). There is considerable I.R. drop in the coating itself; however, it is usually less than in the interface layer.

(*g*) At some point in the life of the cathode, it often shows an anomalous Schottky effect in which the emission against voltage curve has a plateau and then continues to rise beyond the plateau.

Fig. 1.15 illustrates the normal and anomalous Schottky effects, the current I_L, which is the limit of operation in the space-charge regime, and the sparking limit at (I_S V_S).

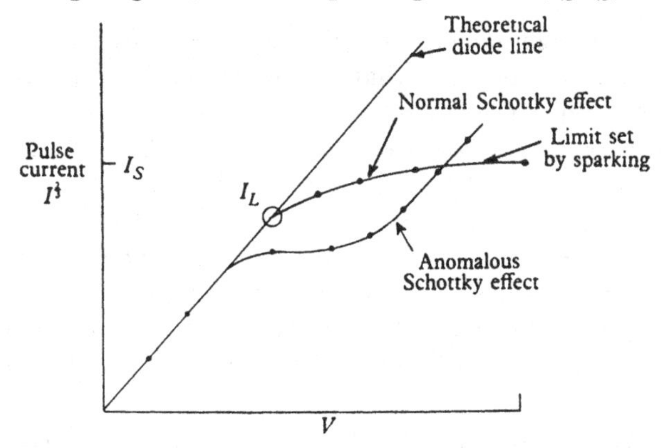

Fig. 1.15. Phenomena in pulse diode characteristics.

Perhaps the most important observation from the point of view of interpreting the operation under pulsed conditions is (*e*). It is further found that the pulsed emission depends on the core material in a different manner from the continuous case. Pulsed cathodes made on pure nickel sleeves are better than those made on the nickel with activating additions, such as silicon, manganese, etc., which gives the best results in receiving tubes. However, some additions seem to be beneficial, since cathodes with a large percentage (5 per cent) of chromium perform as well as pure nickel. It has also been reported that chromium-plating on the sleeve is beneficial. These findings are subject to a considerable uncertainty because opinions differ very much from laboratory to laboratory, and there is some evidence that the representative oxide cathodes of commercial firms are often better than those taken as reference standards in some research institutes. However, the point of vital importance is that the interface layer plays an important role in pulsed emission. Semi-conductor† theory predicts that, if we attempt to draw electrons from a metal into an N-type semi-conductor, the contact will have a high resistance. In the reverse direction the resistance is low. This is the rectifier effect made use of in crystal rectifiers and metal rectifiers.

Wright‡ and others have proposed a theory of pulsed cathodes on the following lines. In the equilibrium condition the heated oxide contains a number of free electrons determined by its volume and the number of impurities per c.c. When the pulse comes on, the free electrons are pulled out from the oxide, but, owing to the high resistance of the interface layer and to its capacitance, the supply of free electrons cannot be replenished from the metal. Calculations show that, assuming reasonable figures for the impurity density, the maximum current per μ sec. would be about 100–200 amp./cm.^{2}, in agreement with observation. If the pulse is made longer, the current cannot be maintained and the current decays while the voltage across the interface builds up, so that the current can be maintained from the metal. As the pulse is made still longer, oxygen diffusion to the surface begins to take effect, and the emission falls to the continuous level.

It should be emphasized that this picture is certainly very deficient in detail, and may even be completely false, but it is the best one can do to explain the experiments at present. It should be remembered that, as Fowler pointed out, our theories of semi-conduction and rectification are based on states in statistical equilibrium. A condition in which a current of 100 amp./cm.^{2} is drawn for 1 μ sec. can hardly be an equilibrium state. It is to be expected that a very great deal more theoretical and experimental work must be done before we understand pulsed emission at all fully.

1.3.7. *Thoria cathodes.* Another type of semi-conducting cathode has recently been used in magnetrons. This is the

† Mott and Gurney, op. cit.
‡ Wright, *Proc. Roy. Soc.* A, **190** (1947), 394.

thoria cathode in which the emitter is thorium oxide, ThO_2. The thoria is sometimes used as a coating on tantalum or molybdenum wire or strip, but more commonly a thoria tube is made by extrusion and sintering, or by building up successive layers of thoria, in a suspension with suitable binder, on to a mandrel. The whole is sintered between applications, and when the thickness is sufficiently great the tube is removed from the mandrel and mounted. These thoria tube cathodes can be heated by an internal tungsten heater, but great care has to be taken in the mechanical design to prevent the cathode from cracking under thermal stresses. Another method of heating is to include a fairly large proportion (25 per cent or more) of finely powdered metal, such as molybdenum, tungsten or tantalum, in the suspension so that the electric conductivity of the cathode is high enough at room temperatures to allow direct heating to be used. In a magnetron there is usually sufficient back bombardment to allow the heating current to be switched off when the tube has been in oscillation for a short time.

The thoria cathode does not give such high current densities as first-class oxide cathodes, and it has to be operated at about double the temperature, but it is robust, easy to handle and is less prone to field sparking than are oxide cathodes. It has therefore found application in magnetrons, where the working current density is partially due to secondary emission, and in a few other very heavy duty tubes. Contrary to one's initial expectation, the life is not very good, as the figures given below indicate.

Wright† gives the following account of the thermionic properties of thoria. Temperature figures are based on an assumed spectral emissivity of 0·35.

(1) A 0·1 mm. layer of thoria on tantalum.

Continuous current tests gave $\phi = 2{\cdot}5(4)$ eV., $A = 2{\cdot}5$ in the emission law $I = AT^2 e^{-\phi/kT}$.

This gives an emission of 2·5 amp./cm.² at 1900° K.

Pulse figures were $\phi = 2{\cdot}62$ eV., $A = 7{\cdot}5$.

† Wright, *Nature, Lond.* **160** (1947), 129.

Life tests for 500 hr. showed a fall of emission of about 25 per cent depending on the temperature.

(2) Extruded and sintered tube, 0·5 mm. wall.

$$\text{Pulse figures } \phi = 2{\cdot}5(5),$$
$$A = 3{\cdot}5.$$

200 hr. at 1930° K. reduced the emission by 40 per cent.

The electrical conductivity was 30 Ω-cm. at 1900° K., and the activation energy of the conductivity 1·1 eV. At 1900° K. the thermal dissipation was 40 W./cm.^{2}.

As the ordinary oxide cathodes used in receiver tubes yield, at a conservative estimate, 1 amp./cm.^{2} at 2 W./cm.^{2}, the emission efficiency of thoria is only about one-tenth as great as that of the oxide cathode.

The emission from thoria does not fall off with increase in pulse length, as does the emission from oxide cathodes, and probe measurements indicate that there is no barrier layer. These observations lend strength to the conclusion of the last section, which was that the pulse behaviour of cathodes depends mainly on the core oxide interface.

The thoria cathode requires no activation process. The cathode is outgassed on the pump and then held at the operating temperature while current is drawn. The current rises by a factor of 3–4 in this process, but this is a minute change by comparison with the increase in emission from oxides during activation. The fact that the cathode never has to be taken above the normal operating temperature may be of use in some applications, e.g. where the heating current is limited by seal design to a value only slightly above the operating value.

In conclusion, one cannot give a reliable estimate of the future of thoria cathodes until more is known of their performance, but much development will have to be done before they are widely used in commercial valves.

1.3.8. *De Boer's emission theory.* To conclude this chapter we must give an account of the theory of emission from oxides due to de Boer† which is the only other theory which to-day‡

† De Boer, op. cit.
‡ But see Appendix 1.

has any claims to compete with the theory outlined in previous sections.

De Boer bases his theory not on defects in the crystal lattice, but on the adsorption of barium on the faces of the crystallites throughout the whole volume of the oxide. The oxide coating is very rough and has deep fissures, and he pictures barium adsorbed on any and all free faces, purposely leaving the precise details of the location rather vague. These barium atoms are then supposed to lose electrons by ionization. The emission will not only depend on the energy of ionization, but also on the rate at which electrons from the core can flow to the barium ions, neutralize them and be emitted to take part in the thermionic emission. This picture leads to an emission equation

$$I = A\,e^{-\phi_i/kT},$$

which clearly can be fitted to the experimental results. In this equation A depends on the number of atoms adsorbed per $cm.^2$ of true surface and on the conductivity of the bulk of the oxide layer, while ϕ_i is the ionization energy. Since A depends on the conductivity, A must be of the form $A = A'\,e^{-\Delta E/kT}$ or $I = A'\,e^{-(\phi_i+\Delta E)/kT}$, and the 'work function' will have two parts, one of which will change during activation, while the other (ϕ_i) will not. (De Boer has said that ϕ_i depends on the impurity concentration, but this is an effect that one would only expect at very high concentrations.)

Friedenstein, Martin and Munday‡ prefer this picture to that of Wilson, and it should be emphasized now that if we are to accept this theory, not merely the emission laws, but the whole of Wilson's semi-conductor theory must be abandoned. Their most fundamental reason is that excess barium built into the lattice by reason of absence of oxygen should be observable as 'F' centres, whereas on present knowledge this is not the case. Both theories can be made to fit the experimental facts of thermionic emission, and since our theoretical knowledge of the energy levels in oxides is not advanced enough to allow us to make unequivocal calculations of the activation energy,

‡ Friedenstein, Martin and Munday, *Rep. Phys. Soc. Progr. Phys.* **11** (1946–7), 298.

it is not possible to decide between the theories. Apart from the fact that the Wilson semi-conductor theory explains many other phenomena, it seems to me that de Boer's theory cannot be entirely correct. Experimental work does not seem to show any such direct correlation with roughness of coating as this theory would lead us to expect. Many competent experimenters have held that the smoothest coatings were the best, and in any case the crystallites grow during life, and there seems to be no direct correlation with the emission. Furthermore, the existence of surface barium is not proven. Then again, one would expect very little difference in emission between barium oxide–strontium oxide and barium oxide–calcium oxide on de Boer's theory, which is not the case.

It seems probable that the Wilson theory is substantially correct for individual crystallites. It is possible that the mechanism of de Boer also contributes to the emission in real cathodes, with all their departures from the perfect lattice treated in the theory, but it is certainly not the determining factor. It would seem to be possible to decide between the mechanisms by measurements on single crystals and, until more experimental evidence has been obtained, it is unprofitable to attempt a final decision between the two theories.

ADDITIONAL REFERENCES

Richardson (1921). *Emission of Electricity from Hot Bodies.* Longman.
Reimann (1934). *Thermionic Emission.* Chapman and Hall.
Seitz (1940). *Modern Theory of Solids.* McGraw Hill.
Eisenstein, A. S. (1949). *Advances in Electronics.* Vol. i. ed. Marton. *Articlean Oxide Cathodes.* Academic Press, Inc.
Wright, D. A. (1950). *Semi-conductors.* Methuen.
Lindsay, R. B. (1941). *Introduction to Physical Statistics.* John Wiley.

Chapter 2

SECONDARY, FIELD AND PHOTO-ELECTRIC EMISSION

2.1. Secondary emission, basic phenomena

2.1.1. *Metals.* It has been known since the early years of the century that, if a metal is bombarded by a stream of electrons, a secondary radiation is produced. This is known as secondary emission, and it is of considerable importance in thermionic valves, usually as an undesirable effect.

Secondary emission is best measured in an apparatus of the type shown in fig. 2.1. A tungsten filament F serves as a

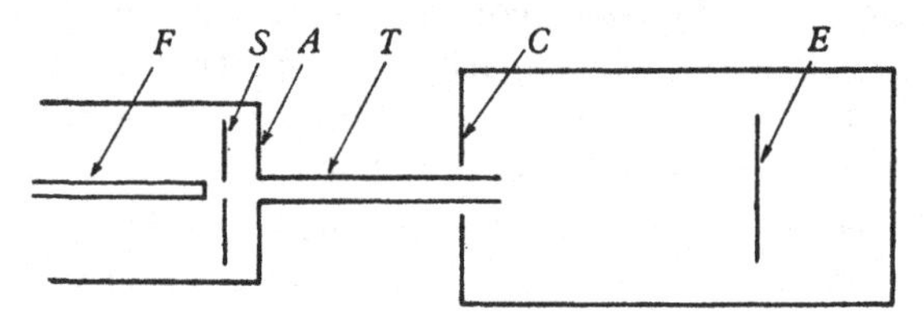

Fig. 2.1. Tube for measurement of secondary emission.

source of electrons, the current being adjusted by the potential on the screen S. An anode A has a long narrow tunnel T attached, through which the electrons pass, so that they are well collimated into a narrow beam. On emerging from T they impinge on a plate E, of the material whose secondary emission is to be measured. The secondary electrons are collected on C. T, E and C are so proportioned that few secondaries can return to T. C is held a few volts, e.g. 6, positive to E, so that all the secondaries are collected. If the primary current is i_p and the secondary emission coefficient, i.e. the number of secondaries emitted per primary, is δ, then the current to $C = \delta i_p$, and the current to $E = i_p(1-\delta)$, therefore

$$\frac{i_C}{i_E} = \frac{\delta}{1-\delta}, \quad \text{or} \quad \delta = \frac{i_C}{i_E + i_C}. \tag{1}$$

If δ is measured as a function of the energy of the incident electrons, it is found to rise to a maximum value and then to decrease again. In the case of pure metals the maximum occurs at 200–600 V. and the value of δ is, very roughly, 1·5. The obvious interpretation of this is that low-energy electrons do not penetrate far enough into the metal to give off many secondaries, and that as the energy increases the number of secondaries increases also. The maximum is explained by the fact that some of the secondaries which are formed deep inside the metal are absorbed again on their way to the surface, and that the rate of recombination eventually becomes greater than the rate of increase of formation. Fig. 2.2 shows δ as a function of V_p for some metals commonly used in tube construction. It should be remembered that this figure relates to very pure surfaces and that the δ may be completely different for

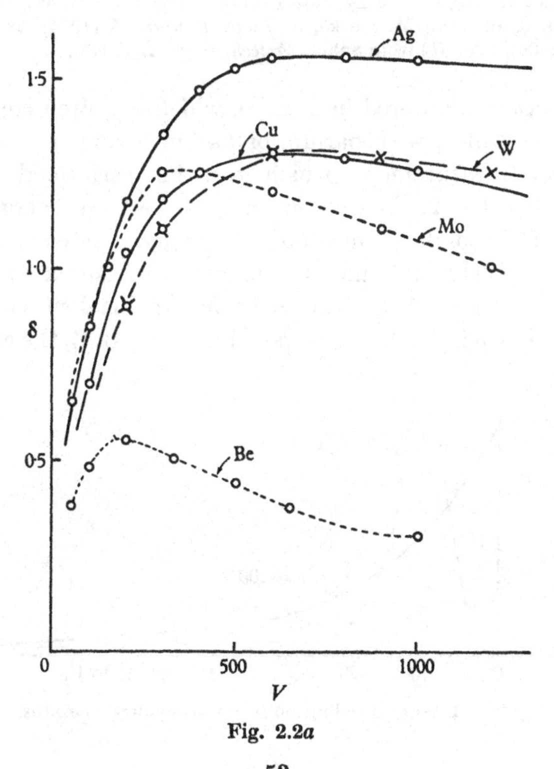

Fig. 2.2*a*

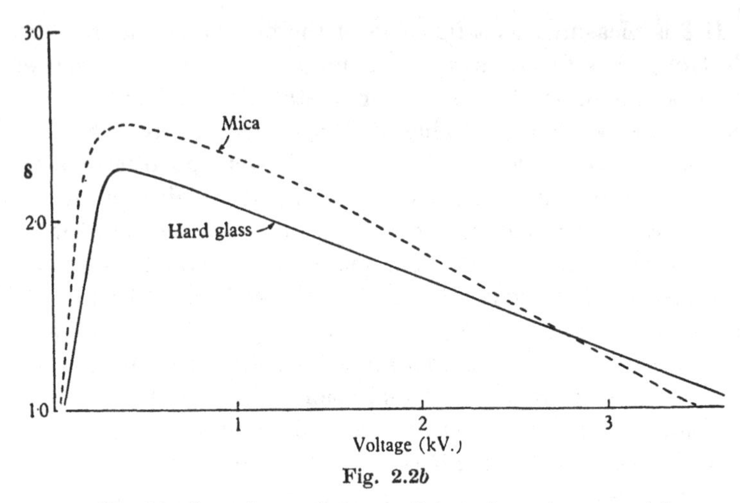

Fig. 2.2*b*

Fig. 2.2. Secondary emission coefficients for various materials. Data: (*a*) Ag, Cu and Be from Bruining, *Physica*, 5 (1938), 17; W and Mo from Warnecke, *J. Phys. Radium*, 5 (1934), 267; 7 (1936), 270. (*b*) from Salow, *Z. tech. Phys.* 21 (1940), 8.

the surfaces encountered in a valve, which are often contaminated, for instance, with barium or barium oxide.

Another measurement which can be performed in the apparatus of fig. 2.1 is the determination of the energy spectrum of the secondary emission. To do this, the collector C is biased negatively to E and the current is measured as a function of this bias. A typical curve for an incident energy of 100 V. is shown in fig. 2.3. For positive values of V_C the emission

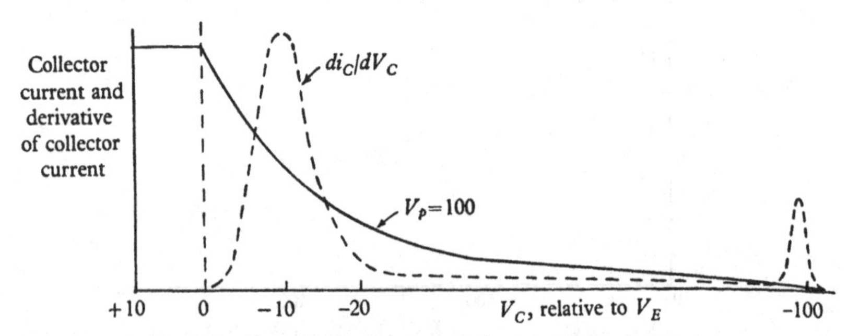

Fig. 2.3. Energy distribution of the secondary electrons.

is saturated, while for negative values it decreases rapidly at first and then slowly, until finally it drops abruptly to zero at a voltage equal to minus the accelerating voltage. The relative number of electrons in each energy group can be determined from this curve by graphical differentiation. This curve, shown dotted, indicates that more than 90 per cent of the secondaries are emitted with low energies, < 20 eV., and that about 5 per cent of the electrons have the full energy of the bombarding beam. Both δ and the distribution are relatively independent

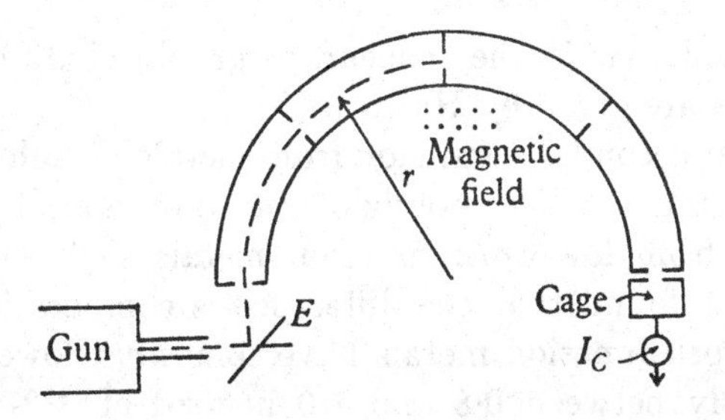

Fig. 2.4. A magnetic velocity analyser.

of temperature. It is usual to term the low-energy secondaries the true secondary emission, and to refer to the high-energy component as reflected primaries. The distribution in energy of the true secondaries is independent of the bombarding energy. This is important in making measurements on current distribution in valves because it means that if two electrodes are maintained at a p.d. of 20 V., the current to the electrode at higher potential will be the incident current plus the secondary emission from the lower potential electrode. In many types of beam systems the lower potential electrode is only taking a small current, so the true current to the high-potential electrode can be measured.

Another method of measuring the energy distribution is capable of much better accuracy. This is to pass the secondary emission from the target into a magnetic or electric velocity sorting field. The current to a Faraday cage is measured as a function of the sorting field, which gives the energy distribution directly. A magnetic apparatus is sketched in fig. 2.4.

In a uniform magnetic field H, an electron of velocity v performs a circular orbit radius r, where

$$\frac{e}{m}Hr = v. \tag{2}$$

If the slit width is Δr it is easy to show that

$$I_C = N(E)\frac{e}{m}H\,\Delta r,$$

or

$$\frac{I_C}{H} = \text{const.}\,N(E). \tag{3}$$

This method avoids the inaccurate graphical differentiation. Full details are given by Haworth.†

When the secondary emission from metals of high-work function, tungsten, nickel, molybdenum, tantalum, is compared with that from low-work function metals such as the alkali metals, it is found that the difference is comparatively small. The low-work function metals have somewhat lower values of $\delta_{max.}$, usually between 0·8 and 1·0 instead of 1·2–1·5 for the first group, but the curves are very similar in shape. This is to be expected, since secondary emission is a volume effect rather than a surface effect and the influence of the work function is indirect. Further confirmation of this view is derived from experiments in which a base plate of, say, nickel is covered with succeeding layers of another metal, e.g. lithium or beryllium. The secondary emission coefficient changes very slowly from the value for nickel to that of the surface material as the thickness of the latter is increased. If there were a large surface effect, a sudden change would result. Naturally, care must be taken to ensure that the metals used are clean and free from oxide, since an oxygen dipole layer can produce very different results.

Surface roughness is another factor governing the secondary emission. A rough surface gives a lower value of δ than a smooth one because the secondaries can be trapped in the surface irregularities. For this reason rough carbon surfaces are sometimes used to suppress partially the secondary emission. Lamp-black suspended in a suitable binder is more useful than

† Haworth, *Phys. Rev.* **48** (1935), 88; **50** (1936), 216.

colloidal graphite, because the particles are large and irregular in shape rather than flat plates. Fig. 2.5 shows some results of Bruining† on this question.

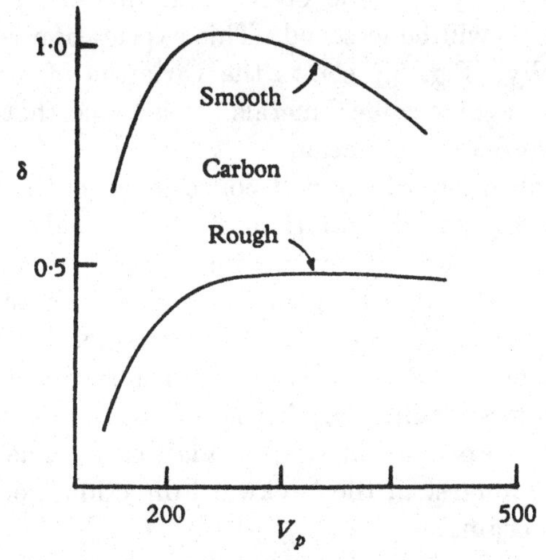

Fig. 2.5. Effect of surface roughness.

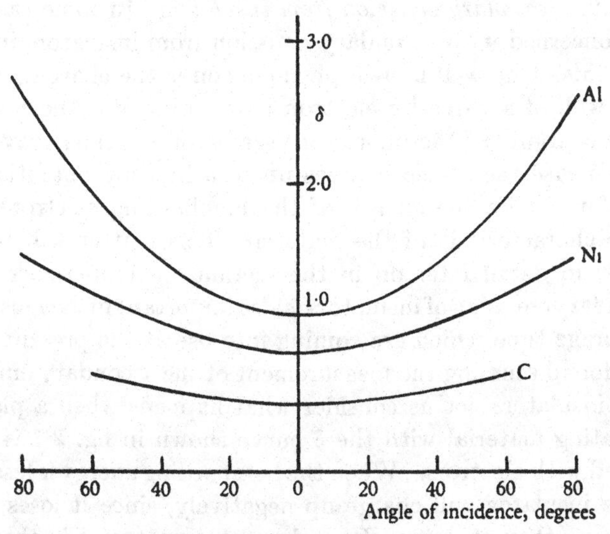

Fig. 2.6. Effect of varying angles of incidence. Data from Muller, *Z. Phys.* **104** (1937), 475.

† Bruining, *Physica*, **9** (1936), 1046.

From the foregoing it will be clear that the angle of incidence of the primary beam will influence the secondary emission, since a large angle of incidence will prevent the beam from penetrating deeply into the body of the material, and therefore absorption will be lessened. This expectation is borne out experimentally. Fig. 2.6 shows the variation of δ with angle of incidence for some typical metals. It is found that $\delta \propto 1/\cos\theta$ to a fairly close approximation.

The last question which will concern us in this section is that of the angular distribution of the secondary emission. This is measured directly by moving a Faraday cage round a semicircular path with the point of incidence of the beam as centre. The current to the cage at any position is proportional to the number of electrons emitted in a range of angle $\Delta\Theta$ about the corresponding Θ. $\Delta\Theta$ is, of course, determined by the size of the aperture in the Faraday cage. The secondary emission is strongest in the backward direction, i.e. at 180° to the incident beam.

2.1.2. *Secondary emission from insulators.* In some cases we are concerned with secondary emission from insulators instead of metals. One well-known phenomenon is the charging of the glass wall of a valve by electron bombardment. Another case is the charging of the fluorescent screen of a cathode-ray tube. In this case the screen charges up to a limiting potential and therefore limits the energy of the bombarding electrons to a value characteristic of the material. This matter will be discussed in detail later on in the section on fluorescence. The secondary emission of insulators is also important in various types of storage tube which are coming into use at the present time.

Before discussing the measurement of the secondary emission from insulators, let us consider what happens when a plate of insulating material with the δ curve shown in fig. 2·7 is bombarded with electrons. When the bombarding energy is less than V_1 the insulator will charge up negatively, since it loses fewer electrons than it gains. Its only stable potential is therefore cathode potential because it then receives no additional electrons. $\delta = 1{\cdot}0$ at $V_p = V_1$, and the potential then jumps to a potential corresponding with that of the adjacent metallic electrodes. V_1 is not a stable point because if V_p increases slightly δ increases and the plate tends to go more positive. At V_2, however, there is a stable equilibrium because if the plate potential rises above V_2, δ decreases, and the plate potential is depressed again. Thus, however great V_p is made, the potential of the insulator surface will never increase above V_2, which is often called the 'sticking' potential. For many glasses and fluorescent materials V_2 lies between 2 and 6 kV. The maximum bombarding energy is therefore limited to quite low values,

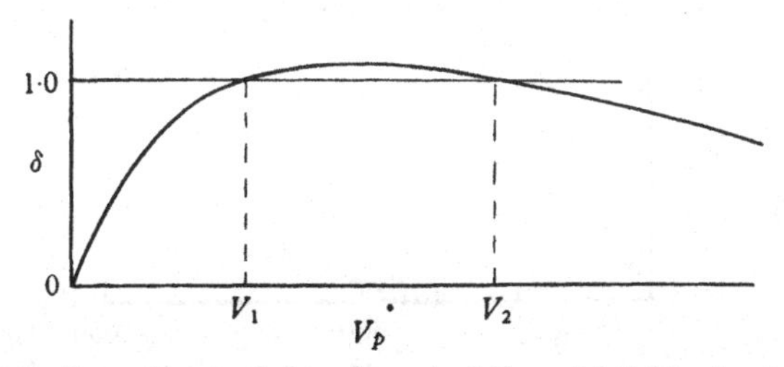

Fig. 2.7. Secondary emission characteristic and 'sticking' potential.

and the light output from fluorescent screens does not increase in proportion to the beam energy above this limit.†

It is comparatively simple to determine the potentials V_1 and V_2 by backing the insulator with a metal plate and measuring the potential of the resulting condenser with a pulsed beam, varying V_p. The measurement of δ is more difficult. Salow‡ has accomplished it by using two beams, one pulsed with an energy such that $\delta > 1$ and the other continuous and with an energy for which $\delta < 1$.

If $V(t)$ is the potential of the insulating plate of capacity C

$$V(t) = -\frac{(\delta-1)}{C}\int_0^t i_1\,dt + ki_2 t,$$

where t = pulse duration. Differentiating we find the pulsed current in the condenser lead i_z, which can be measured with an oscilloscope,

$$i_z = -(\delta-1)i_1. \tag{4}$$

† In modern cathode ray tubes a thin layer of aluminium is evaporated over the screen to eliminate this effect.

‡ Salow, *Z. tech. Phys.* **21** (1940), 8.

The curve of $\delta \sim V_p$ obtained in this way for glasses and phosphors is very much the same as that for high-work function metals. Values are more appropriately discussed when we come to specific applications. In the case of some insulating crystals such as sodium chloride and potassium chloride, δ is large, i.e. 7 or 8, indicating that there is an important difference in behaviour.

2.1.3. *Secondary emission from complex surfaces.* Complex surfaces differ from metals and insulators in having much

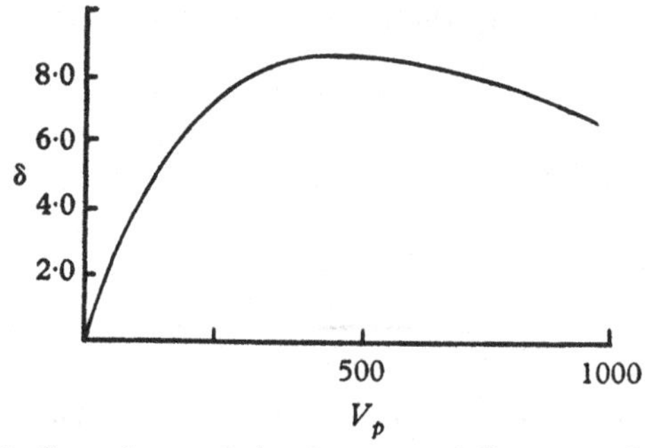

Fig. 2.8. Secondary emission from a specially prepared surface.

higher values of δ. They are therefore of technical importance in secondary emission photocells, or multipliers, and in television pick-up tubes. In addition to their practical value they have considerable theoretical interest, as their behaviour throws light on the structure of complex films. For these reasons complex surfaces have been the subject of very extensive research, and it will not be possible to present more than the barest outline of the subject here.

Some of the earliest surfaces to be studied were those used in photocells, particularly the Ag–CsO–Ag–Cs type. Here we are dealing with a silver plate which has been oxidized, covered with caesium and then heat-treated until the photo-electric activity is a maximum. In this process the silver oxide reacts with the caesium, so that it is thought that the result is a layer of caesium oxide which contains silver particles with absorbed caesium atoms on the surface. Such surfaces have values of δ between 5 and 10. The shape of the $\delta \sim V_p$ curve is much the same as for a metal (see fig. 2.8). It is found that the conditions for maximum photo-emission are not the same as those for maximum secondary emission, which requires longer heat treatment, more caesium being removed thereby. For this reason Timofaev and Piatnitsky† consider that the secondary emission is due to caesium atoms embedded in the caesium oxide layer, the photo-electric emission being a surface effect. Other metal bases can be used but silver gives the highest δ, magnesium being only slightly inferior. Rubidium and potassium can be used for the coating but again give smaller values of δ.

Other surfaces which have high δ values can be prepared much more simply. These are alloys on which a complex layer can be prepared by heating in air or oxygen. Instances are beryllium copper and duralumin. If beryllium copper is heated in air to a bright red heat for a few minutes, an adherent oxide layer is obtained which will survive the ordinary processes of outgassing in a vacuum tube. Such layers have fairly high coefficients, above 5. Duralumin‡ exhibits the same behaviour, giving a somewhat greater value of δ. Zworykin§ and his co-workers have developed alloys of silver magnesium giving high δ values.

Yet another type of complex surface whose secondary emitting properties are important is the oxide-coated cathode. In this case there is a considerable divergence of opinion on the experimental facts, particularly the variation of δ with temperature, which is obviously difficult to measure because of the thermionic emission. At room temperature δ is about 10 and the energy distribution is the same shape as in metals. According to Pomerantz the temperature variation is given by $\delta_T = \delta_{\text{room}} + \text{const.}\, e^{-\Delta E/2kT}$, where ΔE has the same significance as in the section on oxide cathodes. Other authors have found that δ is nearly constant with temperature in agreement with the findings on metals. An interesting technique for making such observations has been worked out by Johnson.|| This consists in pulsing the current to the secondary emitter so that

† Timofaev and Piatnitsky, *Phys. Z. Sowjet.* **10** (1936), 518.
‡ Le Boiteux, *Rev. Techn. C.G.C.T.-L.M.T.* **1** (1941), 21.
§ Zworykin *et al.*, *J. Appl. Phys.* **12** (1941), 697.
|| Johnson, *Phys. Rev.* **73** (1948), 1058.

the secondary emission pulses appear on a 'pedestal' due to the thermionic emission, and therefore the effects are separable. Johnson, moreover, finds that δ *decreases* from 14 at room temperature to 10 at 400° C. Above 400° C. a phenomenon, which he terms enhanced thermionic emission, occurs. This is an increase of emission throughout the pulse, persisting after the end of the pulse so that it cannot be secondary emission. This enhanced thermionic emission obscures the behaviour of δ above 400° C. Pomerantz† could not observe this phenomenon and suggests that it was some space-charge effect; but this seems less likely than the supposition that there was a difference of behaviour between the cathodes studied by the two authors. We shall see later that on purely theoretical grounds one would not be surprised if δ decreased with temperature in oxide cathodes.

2.1.4. *The Malter effect.*‡ This effect was observed by Malter when testing oxidized aluminium surfaces which were subsequently activated with caesium. The necessary condition is that a thin insulating layer should be between the emitter and the base-plate. Not only did such surfaces possess enormous values of δ, e.g. 1000, but the emission persisted after the removal of the primary beam. The emission rises fairly slowly after the instant of switching on. If the collector voltage is switched off, the emission ceases immediately.

It is supposed that the emission is field emission caused by the formation of a large positive surface charge on the insulator, which is only about 1000 Å. thick, by the initial secondary emission from the caesium or caesium oxide surface.

2.2. The theory of secondary emission

The quantum-mechanical theory of secondary emission was first given by Fröhlich,§ whose results were criticized and extended by Wooldridge.|| A quite different theory was given

† Pomerantz, *Phys. Rev.* **70** (1946), 33; *Proc. Inst. Radio Engrs., N.Y.* **34** (1946), 903.

‡ Malter, *Phys. Rev.* **50** (1936), 48.

§ Fröhlich, *Ann. Phys., Lpz.* **13** (1932), 229.

|| Wooldridge, *Phys. Rev.* **56** (1939), 562.

by Kadyshevich at about the same time as the latter author. Here we shall not give a full discussion of either theory, but shall content ourselves by indicating the physical arguments used and showing in a general way how secondary emission is bound up with the energetic structure of the emitter. This will show that, while there is relatively little difference in the secondary emission from all metals and semi-conductors, the emission from insulators will in general be much higher.

2.2.1. *The Fröhlich-Wooldridge theory.* After penetrating the surface the primary beam moves through the electron gas inside the metal, and some of the beam electrons pass near enough to the valence electrons to give them a considerable amount of energy. A simple calculation shows that this energy is in the direction of the normal to the primary electron path, so this process alone cannot be directly responsible for secondary emission. The Fröhlich-Wooldridge theory is based on the fact that the valence electrons are not completely free, but are acted on by binding forces which make it possible for them to gain energy in the direction anti-parallel to the primary beam. The binding terms are considered to be negligible as far as the relatively high energy of the primaries are concerned. The quantum-mechanical problem is then set up as follows. The initial state is that in which we have an incident beam travelling through the lattice where there is located a single electron. There is a coulomb interaction between the electrons and one can calculate the rate at which the initial state varies into other states, some of which lead to the emission of the valence electron as a secondary. The exclusion principle is taken into account (*a*) by using Fermi statistics, (*b*) by forbidding all transitions which leave the electron in an occupied state. The Bloch eigenfunctions $U_k(\bar{r})\exp[i(\bar{k}.\bar{r})]$ are used in the metal. The unperturbed probability amplitude for the two-particle system is

$$\psi_{\bar{k},\bar{K}} = \frac{1}{\Omega} U_k(\bar{r}) \exp[i(\bar{k}.\bar{r}) + i(\bar{K}.\bar{R})],$$

where $\bar{K}$ = momentum of primary, $\bar{k}$ = momentum of valence electron, $\bar{R}$ = coordinate of centroid of volume element about primary, $\bar{r}$ = coordinate of centroid of volume element about valence electron. The perturbing force is the coulomb repulsion

$$U = \frac{e^2}{|\bar{R}-\bar{r}|}.$$

A lengthy calculation leads to a value for the time rate of transitions to the new (emitted) state which depends very little on the bombarding energy. This is contrary to the facts, so the rate of loss of energy by the beam is studied, and it is found that the most serious source of loss is in secondary formation, 25–30 eV. being lost at each 'collision'. This leads to the idea that the mean free path of the primaries is considerably smaller than that of the secondaries. If this is the case, the number of secondaries actually observed will increase more or less linearly with bombarding energy up to the depth of penetration which equals the mean free path of low-energy electrons. When this is reached many of the secondaries will suffer collisions before being emitted, and the rate of increase in δ will decrease markedly. From these considerations it is possible to obtain a theoretical value of δ/δ_∞, where δ_∞ is the limiting δ for high energies. This cannot be taken directly from a curve because effects not in the calculation come into play, but a value somewhat above the observed maximum can be assumed and the curves fitted by using a scale factor. The expression is

$$\frac{\delta}{\delta_\infty} = \left\{1-\exp\left[-\frac{\frac{1}{2}\left\{1-\left(\frac{E_c}{E_0-\zeta}\right)^{\frac{1}{2}}\right\}}{\delta_\infty}\left(\frac{E_p}{E_0}-1\right)\right]\right\}$$
$$+\frac{1}{2}\left\{1+\left(\frac{E_c}{E_0+\zeta}\right)^{\frac{1}{2}}\right\}\int_0^A \frac{e^{-v}\,dv}{\left[\frac{E_p}{E_0}-\frac{\delta_\infty v}{\frac{1}{2}\left\{1-\frac{E_c}{E_0+\zeta}\right\}}\right]^{\frac{1}{2}}} \qquad (5)$$

where $E_0 = \left(\frac{\hbar^2}{2m}\right)\left(\frac{2\pi}{a_0}\right)^2$ = mean energy lost per collision, a_0 = lattice spacing, E_p = primary energy, A = the argument of the exponential term, and E_c and ζ have the same meaning as in Chapter 1. The agreement with experiment is reasonable for silver and copper, fair for aluminium and magnesium, but poor for the light metals lithium, beryllium, barium and caesium. In one point the theory disagrees with experiment for all materials, i.e. theoretically there is zero emission below about 30 V. bombarding energy; it may be that the observed emission at such low energies is a surface effect or, more specifically, is due to reflexion from the surface barrier. It can be seen that the work function does not influence the emission strongly, as was to be expected, since secondary emission is a volume effect. The conditions for large δ are (*a*) large lattice constant (small E_0) and (*b*) small value of ζ.

Wooldridge applies the theory to barium oxide and gets reasonable agreement with the results of experiments by Bruining. It is hard to believe that the agreement is more than fortuitous, the main factor being the experimental value of δ_∞.

2.2.2. *Kadyshevich's*† *theory.* This is a semi-classical theory being once more based on the fact that the primaries have much greater energies than the secondaries. The primaries enter the metal and there excite some of the electrons in the Fermi electron gas. The primaries may make elastic collision with the lattice ions, which alter the primary direction. Moreover, the secondaries may likewise make elastic collisions with the lattice, also suffering deflexions thereby. This corresponds to the binding of Wooldridge's theory. The calculation starts from the classical Rutherford scattering formula which yields the following expression for the number of secondaries per primary:

$$n = \frac{A \sin^2 \alpha}{E_a}, \tag{6}$$

where

$$A = -242 m^{\frac{3}{2}} e^{\frac{7}{2}} \left[\frac{\zeta}{\sqrt{(\zeta+\phi)}} + 2\sqrt{(\zeta+\phi)} \ln\left(1 - \sqrt{\frac{\zeta}{\zeta+\phi}}\right) + 2\sqrt{\zeta} \right],$$

α = angle of incidence,

E_a = incident energy,

and the other symbols have their earlier meanings.

The number of electrons emitted is *n*, less the number absorbed by inelastic collisions with other conduction electrons.

† Kadyshevich, *J. Phys. U.S.S.R.* **2** (1940), 115; **4** (1941), 341; **9** (1945), 431, 436.

If the emission is W_0, Kadyshevich determines it in series form

$$W_0 = W_{00} + W_{01} + \dots,$$

where
$$W_{00} = \frac{2n\Lambda_1\Lambda_2 a^3 \sin\alpha}{\Lambda_2 a \sin\alpha + \Lambda_1 \cos\alpha}. \tag{7}$$

The other terms are much smaller than W_{00}. W_{00} is clearly zero for normal incidence. It is then shown that the effect of elastic collisions made by the primaries with the lattice ions has a negligible effect on the calculation. However, the elastic collisions between the secondary electrons and the conduction electrons give rise to a non-vanishing value for the emission at normal incidence. This value is

$$\delta = \pi\frac{A}{Ea}\frac{\lambda_2}{l_2}\frac{\Lambda_1\Lambda_2 a^3 a_1^4}{\Lambda_1 + a a_1 \Lambda_2}. \tag{8}$$

Here λ_1, l_1 = mean free paths of primaries for inelastic and elastic collisions respectively,

λ_2, l_2 = mean free paths of secondaries,

$$\frac{1}{\Lambda_1} = \frac{1}{\lambda_1} + \frac{1}{l_1}, \quad \frac{1}{\Lambda_2} = \frac{1}{\lambda_2} + \frac{1}{l_2},$$

$$a = 0{\cdot}73, \quad a_1 = 0{\cdot}77.$$

Eqn. (8) is subject to a small correction term which is investigated in the original paper. Λ_1 is given by the Born approximate formula as

$$\Lambda_1 = \frac{xE_a^2}{\ln mE_a}, \quad x = \frac{1}{2\pi Ne^4}, \quad m = \frac{\sqrt{1{\cdot}35}}{1{\cdot}03\phi}.$$

The magnitude λ_2/l_2 is unknown and is adjusted to fit the experimental data at one energy. When this is done, the curve of δ against incident energy is in good agreement with the experimental results for metals. δ is a maximum when $\Lambda_1/\Lambda_2 = 0{\cdot}56$, which agrees with the supposition of Wooldridge.

It will be seen that the theories, although carried through by very different means, agree on the fundamental fact about the emission, namely, that the electron gas is not completely free, but is lightly bound. The Kadyshevich theory also gives the emission as a function of angle of incidence and the energy distribution of the emitted electrons, and these results are in general agreement with experiment. Both theories are inaccurate for very small primary energies, and the observed emission is almost certainly reflexion by the surface barrier.

It is easy to see from the Kadyshevich theory why, in the case of dielectrics, the emission is in general greater than for metals. In this case, in order to be emitted, an electron must be raised from the full band to the empty band, but, provided the energy gap is not very large, this does not make much difference to the number of secondaries found inside the insulator. Once in the empty band, the electrons only make elastic collisions with the lattice, and the absorption is very small. Semi-conductors occupy an intermediate status, behaving as metals when the energy gap is small and the number of conduction electrons large, and as insulators when the reverse is the case.

It would be idle to pretend that the theory of secondary emission is nearly as complete as that of thermionic or photoelectric emission. While the main features described above are almost certainly correct, there are many gaps in our knowledge. Instances of these are our lack of theoretical understanding of complex surfaces and of the Malter effect.

2.3. Practical secondary emitting surfaces

In later chapters we shall describe several devices making use of surfaces especially prepared to yield high values of secondary emission. In this section we shall give some information on the preparation of such surfaces. The technique of preparation differs considerably from laboratory to laboratory, so we shall go into little detail.

The modern tendency is to use either alloys of high δ, or oxide layers, in preference to the complex surfaces which were used earlier, as the preparation is much simpler. Zworykin† and his collaborators describe alloys of silver–magnesium with high emissions. One of the simplest and best surfaces can be prepared by raising ordinary commercial beryllium copper sheet to a bright cherry red heat in air for a period of a few

† Zworykin, loc. cit.

minutes. The surface should be cleaned by de-greasing and washing in chromic acid solution before heating. A film of very adherent oxide is obtained by this means; it is sufficiently durable to withstand the heat treatments necessary in the evacuation of the valve. Values of δ between 5 and 6 are obtained. This surface, and in fact that of all good secondary emitters, is very easily poisoned by operation in a tube with an oxide-coated cathode. This fact is responsible for the lack of success of thermionic multiplier tubes.

The complex surfaces have somewhat higher values of δ, but their preparation is so much more difficult that it is often economic to use slightly worse emitters which are easy to process. The Ag–AgO.Cs type is perhaps the best known. The base plate is of silver or is silver-plated. The tube is evacuated and baked in the ordinary way so as to clean the glass and metal surface; the base is then oxidized by initiating a discharge in oxygen, e.g. by an r.f. coil held near the bulb. The thickness of the oxide layer is important and can be gauged by the colour changes of the surface. The process is often stopped when the silver surface appears blue for the second time. The caesium is then distilled into the bulb or, more commonly nowadays, it is produced by the eddy current heating of a getter flag carrying a reaction pellet which breaks down to produce an excess of metallic caesium. These reaction pellets can be prepared in several ways, a stable caesium salt being compacted with a very finely granulated metal powder; for instance, caesium dichromate and powdered aluminium or zirconium have been used. Caesium oxide is also used. When the caesium has been introduced it is necessary to activate the surface. This is done by baking the whole bulb gently to about 200° C. At this temperature the caesium reacts with the oxide and also distributes itself evenly over the emitting surface. The progress of the activation can be determined by applying test voltages to the tube and measuring the emission. Care must be taken not to initiate a gas discharge in the high concentration of caesium vapour which is present. If the activation is carried out under carefully controlled and specified conditions, it will not be necessary to check more than the first few samples during manufacture. The activation of subsequent tubes can then be reproduced on a time and temperature basis.

Care must be taken in the design of the tube that the caesium is prevented as much as possible from being deposited on the lead wires and insulators, or else a leaky tube will result. Considerable ingenuity is necessary in the design and processing of more elaborate tubes using secondary multiplication.

We have said above that the operation of secondary emitters is difficult in the vicinity of oxide cathodes. This is because the δ factor decreases very rapidly and the life of the surface may be only a few hours. Some improvement is obtained if the tube is so constructed that the secondary emitter is shielded from barium ions, but life is still below good commercial standards. The probable reason for the poisoning of the emission is clear from our theoretical knowledge. The complex surface is a semi-conductor with a low concentration of conduction electrons, a low work function, etc., required to give a high δ value. If barium ions are deposited on such a surface, they constitute additional impurity centres, increase the density of conduction electrons, increase the internal absorption of secondaries and reduce the emission. This picture is certainly not correct in detail, as it is hard to imagine that barium should constitute an impurity suitable for increasing the conduction density in so many complex surfaces, but it seems to be essentially correct.

2.4. Field emission

In the last chapter we discussed the effect of moderate fields on the emission from heated surfaces. If the field is increased to values of the order of 10^6 V./cm., a relatively large electron current can be drawn from the cold surface, and it is found that the magnitude of the current is independent of temperature but depends exponentially on the field. The experimental results fit a formula of the type

$$J = CE^2 e^{-D/E},$$

where E = field strength.

The explanation of this effect is very simple on a quantum-theoretical basis. In quantum mechanics if we treat the problem of an electron of given energy located in a rectangular potential well, we find that there is a small but finite probability that the electron will be found outside the well, which is, of course, impossible in classical mechanics. Similarly, if the electron impinges on a potential barrier whose height is greater than the value corresponding to the kinetic energy of the

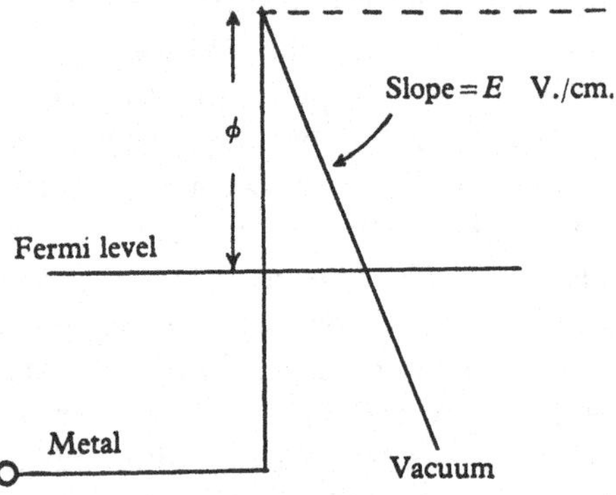

Fig. 2.9. Potential diagram for surface with large external field.

electron, there is a finite probability that the electron will penetrate the barrier. The probability of this 'tunnel effect', as it is called, is very small unless the barrier is exceedingly thin, but in this latter case quite large transmissions result. The phenomenon of field emission is due to this tunnel effect; for in metals the work function is a few volts and in fields of about 10^6 V./cm. the barrier thickness will be only 10^{-5}–10^{-6} cm. at the Fermi level and the transmission coefficient will be fairly large.

The image effect can be neglected and the shape of the barrier will be a saw tooth as shown in fig. 2.9. We have to calculate the transmission probability for an electron of energy E_z, where $0 < E_z < \phi$, and then calculate the total number of such electrons incident on the barrier from the left. The product of these two quantities gives the emission. The transmission through potential barriers is well dealt with by A. H. Wilson†

† Wilson, *Theory of Metals*, Cambridge University Press, 1936.

and by Rojansky,† so we shall not give the procedure here. Fowler and Nordheim have dealt with the present case, and give the transmission coefficient as

$$T_{(E_z)} = \frac{4\sqrt{[E_z(\phi - E_z)]}}{\phi}\exp\left[-\frac{\left\{4\sqrt{\left(\frac{8\pi^2 m}{h^2}\right)}(\phi - E_z)^{\frac{3}{2}}\right\}}{3E}\right].$$

The energy distribution of the electrons incident on the barrier is given by eqn. (7) of Chapter 1, so, by the same reasoning as was there employed, we can write for the field emission

$$J = 2e\left(\frac{m}{h}\right)^3 \int_0^{(2\zeta/m)^{\frac{1}{2}}} T_{(E_z)}\, u_z du_z$$

$$\times \int\int_{-\infty}^{+\infty} \frac{du_x du_y}{1+\exp[\{\frac{1}{2}m(u_x^2+u_y^2+u_z^2-\zeta\}/kT]}.$$

The double integral can be evaluated, as in Chapter 1, by inserting polar coordinates. We put $u_x = \dot{r}\cos\theta$, $u_y = \dot{r}\sin\theta$. The integral then becomes

$$-4\int_0^{\frac{1}{2}\pi} d\theta \int_0^{\infty} \frac{\dot{r}\,d\dot{r}}{1+\exp[(m\dot{r}^2/2kT)-(2\zeta - mu_z^2)/2kT]},$$

which by elementary means integrates to

$$\frac{2\pi kT}{m}\ln\left[1+\exp(\zeta - \tfrac{1}{2}mu_z^2)/kT\right].$$

Thus

$$J = \frac{4\pi em^2 kT}{h^3}\int_0^{(2\zeta/m)^{\frac{1}{2}}} T_{(E_z)}\, u_z \ln(1+\exp[(\zeta - \tfrac{1}{2}mu_z^2)/kT]).$$

The exponential term is much greater than 1 for all the electrons for which $\zeta - \frac{1}{2}mu_z^2 > kT$. (This is very different from thermionic emission where $\frac{1}{2}mu^2 > \zeta$.) As ζ is several eV. this includes very nearly all the emitted electrons, therefore

$$J = \frac{4\pi em^2 kT}{h^3}\int_0^{(2\zeta/m)^{\frac{1}{2}}} T_{(E_z)}\, u_z(\zeta - \tfrac{1}{2}mu_z^2)/kT\, du_z.$$

† Rojansky, *Introduction to Quantum Mechanics*, Blackie, pp. 150 *et seq.*, 214 *et seq.*

To evaluate this integral we let $(\zeta - \frac{1}{2}mu_z^2) = \xi$ and expand in terms of this variable. The result is

$$J = \frac{e}{2\pi h} \frac{\zeta^{\frac{1}{2}}}{(\zeta+\phi)\phi^{\frac{1}{2}}} E^2 \exp\left[-\left\{\frac{4}{3}\sqrt{\left(\frac{8\pi^2 m}{h^2}\right)}\phi^{\frac{3}{2}}/E\right\}\right].$$

It should be noted that this result depends on the assumed shape of barrier only through the constants. Other barriers give $J = \text{const.}\, E^2 \exp[-\text{const.}\, \phi^{\frac{3}{2}}/E]$. The experimental results agree well with the field variation, but their magnitude is generally considerably greater than the theoretical value. This is quite understandable because it is not possible to prepare surfaces which are ideally smooth. In the immediate vicinity of roughnesses, cracks and similar defects, the field is enormously greater than the calculated field, and therefore very high currents are concentrated on such spots.

Field emission has been used to study the surface properties of metals, and it is easy to demonstrate the differences in work function between crystal faces by its use. Benjamin *et al.*† have obtained very beautiful pictures in this way. It has also been used as a means of providing electrons in some types of electron-diffraction apparatus. In general, however, field emission shows itself as a phenomenon which the valve technician has to avoid. Instances of its appearance as a deleterious effect are common in high-voltage valves where all metallic surfaces have to be provided with smooth, rounded corners, and care must be taken to avoid field concentration to prevent arcs and leakage due to field emission. In manufacture the high voltages are usually applied slowly through limiting resistors so that small roughnesses, pieces of swarf, etc., are burned out by the passage of the intense field currents without permanently impairing the vacuum in the device. This is one of the main reasons for high-voltage ageing or 'spot knocking', as it is often called.

2.5. Photo-electric emission

The emission of electrons from metals under the influence of visible light has been known since about the beginning of

† Benjamin and Jenkins, *Rep. Progr. Phys.* **9** (1943), 177.

the century. The effect has historical importance in that the application of the quantum theory by Einstein in 1905 was one of the first successes for the new theory. For present-day physics the photo-electric emission is mainly important for the information on the electronic structure of the emitter which can be obtained by its study.

The basic photo-electric phenomena are the following. If two metal plates are supported apart in a vacuum, or low-pressure gaseous atmosphere, and a voltage is maintained between them, a current is observed to flow when the cathode is illuminated. The current depends on the intensity and on the frequency of the illumination. In fact, there is a threshold frequency below which, no matter how intense the illumination, no current will flow. Also, if the velocity of emergence of the photo-electron is studied by retarding fields, it is found that the maximum velocity of emission is related to the frequency. The critical frequency for emission is given by the Einstein law as

$$h\nu_0 = \phi.$$

Numerically, this yields $\lambda_0 = \frac{12{\cdot}36}{\phi} \times 10^3$ Å., if ϕ is in electron volts. The maximum velocity is given by

$$\tfrac{1}{2}mu_m = h(\nu - \nu_0).$$

In the modern theory of metals we shall see that these relations are correct at 0° K. but not at higher temperatures. In addition, we shall determine the energy distribution and spectral distribution as functions of temperature. This leads us to the Fowler method of measuring ν_0 and thus determining ϕ. We note here that in pure metals the work function photo-electrically determined should equal the zero temperature value of the thermionic work function.

The Einstein relation for the minimum frequency of emission is very easily justified at 0° K. Recall the Fermi-Dirac distribution given in fig. 1.3 and 1.4. We there assumed a critical energy for emission E_c, and the Fermi-Dirac distribution drops abruptly to zero at a lower value of energy ζ. The photons bombarding the surface have energy $h\nu$, so that the minimum energy required to release an electron is $h\nu_0 = E_c - \zeta = \phi$. If the incident energy $h\nu > h\nu_0$, some of the electrons will receive more than sufficient energy to release them and thus will possess kinetic energy after passing through the surface. The resulting velocity is clearly given by

$$\tfrac{1}{2}mu_m^2 = h\nu - (E_c - \zeta),$$
$$= h(\nu - \nu_0).$$

These relations will no longer be true at temperatures above the absolute zero because the Fermi distribution has then a

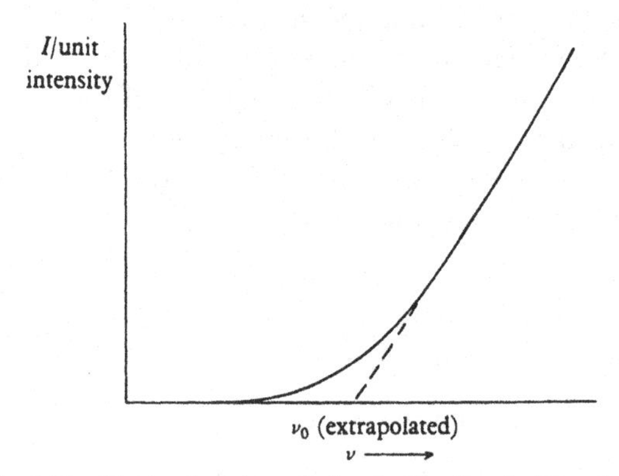

Fig. 2.10. Photo-electric emission in the threshold region.

long tail and some electrons have energies considerably above ζ. We now proceed to calculate the correct expressions.

2.5.1. *The spectral distribution.* Here we require to calculate the photocurrent per unit light intensity as a function of frequency and temperature. A typical experimental curve is shown in fig. 2.10, with an extrapolated value for ν_0 which is clearly subject to considerable error, in fact the curve is asymptotic to the axis of frequency. The computation of the distribution function should commence by a calculation of the probability that a quantum of energy $h\nu$ will be absorbed by an electron with velocity u. Then, since the distribution of electron velocities is known, the distribution of excited electrons can be computed. This, multiplied by the transmission coefficient, can be integrated to give the emission. This calculation has been carried out, but the mathematical difficulties are formidable, and Fowler made a great step forward by recognizing that the important question was the distribution in the immediate vicinity of the threshold. This means that the emitted electrons initially occupy a rather narrow band of energy, and it is therefore reasonable to assume that the probability of excitation is constant for the emitted electrons.

In the last section we obtained an expression for the number of electrons of velocity u_z incident on the x, y plane. Rewriting this in terms of the normal energy, we have

$$N_i = \frac{4\pi \mathrm{m}\, kT}{h^3}\int_{E_c - h\nu}^{\infty} \ln\{1+\exp[(\zeta - E_z)/kT]\}\, dE_z.$$

The probability of excitation is assumed to be a constant α. Then the photocurrent density is $J = \alpha e N_i$. If we change the variable to $u = \mathrm{e}^{(\zeta - E_z)/kT}$ and put $\mathrm{e}^{[\zeta - (E_c - h\nu)]/kT} = u_0$, we get

$$J = \frac{\alpha\, 4\pi \mathrm{m} e\, k^2 T^2}{h^3}\int_0^{u_0} \frac{\ln(1+u)}{u}\, du.$$

The evaluation of the integral depends on whether the upper limit u_0 is greater or less than unity. For $u_0 \leqslant 1$, the integral is $\mathrm{e}^{-x} - \frac{\mathrm{e}^{-2x}}{2^2} + \frac{\mathrm{e}^{-3x}}{3^2}$, etc., where $x = h(\nu_0 - \nu)/kT$; while for $u_0 > 1$, the integral is $\frac{x^2}{2} + \frac{\pi^2}{6} - \left(\mathrm{e}^{x} - \frac{\mathrm{e}^{2x}}{2^2} + \frac{\mathrm{e}^{3x}}{3^2}, \text{ etc.}\right)$.

The final result is $I = \alpha A_0 T^2 g(x)$, $g(x)$ being defined by the two series above. A_0 is the thermionic emission constant 120·4 amp./cm.^{2} degree^2. Fig. 2.11 shows $g(x)$, a function often called the Fowler function. Fowler's method of interpreting the experimental data consists in plotting $\log J/T^2$ against $h\nu/kT$. The resultant curve is the same shape as the theoretical one, but is displaced along the $h\nu/kT$ axis by an amount $h\nu_0/kT$. This procedure gives excellent agreement between the experimental points and the theoretical curve. The values of ν_0 determined from the shift are independent of temperature, and the values of the photo-electric work function calculated

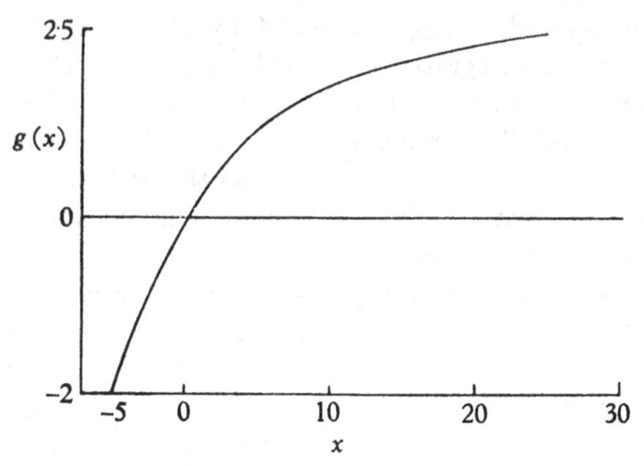

Fig. 2.11. Fowler's method for threshold determination.

from them are in good agreement with the thermionic work function. The thermionic work function is usually a few hundredths of an electron volt higher than the photo-electric work function. This is in the direction to be expected from the considerations we discussed earlier on the temperature-dependence of ϕ, but smaller in magnitude. The indication is that E_c decreases with increasing temperature.

2.5.2. *The energy distribution.* Two special cases of the energy distribution are particularly important, the distribution of normal energy which is observed by measuring the retarding field characteristic in a plane parallel system (fig. 2.12) and the total energy distribution which applies when the electrodes

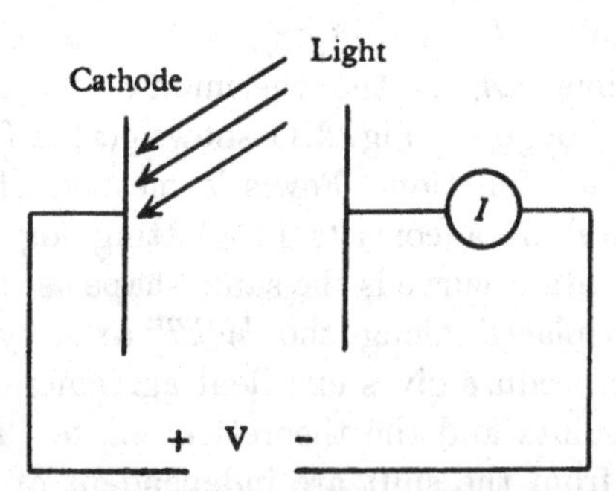

Fig. 2.12. Method for measuring velocity distribution.

are concentric spheres. The latter is a convenient arrangement when the cathode is small, as it is when studies are made on single crystals.

(*a*) *Normal energies.* If the anode potential is V volts negative to that of the cathode, an electron will have to possess the energy (E_c+E_n), $E_n = eV$ if it is to be emitted and to reach the anode. It is easily seen by altering the lower limit in the integration of the last section that

$$J(v) = \alpha A_0 T^2 g(x^1),$$

where $$x^1 = x - E/kT.$$

At $T = 0°$ K. it is easily shown that

$$J = \tfrac{1}{2}\alpha A_0 (E-E_m)^2,$$

where E_m = cut-off energy. The current-voltage curve is thus a parabola with its vertex at $E = E_m$, in agreement with the original Einstein equations. At higher temperatures the curve is asymptotic to the voltage axis, showing that the maximum energy is ill defined. The actual distribution, which is obtained from the current-voltage curve by differentiation, is theoretically deduced as follows:

$$F(v) = \frac{\alpha\, 4\pi \mathrm{m}\, kT}{h^3} \int_{E_c+E-h\nu}^{\infty} \ln\left(1+\mathrm{e}^{(\zeta-E_z)/kT}\right) dE_z.$$

$F(v)$ is the number of electrons which escape against the field. Differentiating, we get

$$\frac{\partial F(v)}{\partial V} = f(E_n) = \frac{\alpha\, 4\pi \mathrm{m}\, kT}{h^3} \ln\left(1+\mathrm{e}^{(E_m-E_n)/kT}\right).$$

At $T = 0°$ K., $$f(E_n)_0 = \frac{\alpha\, 4\pi \mathrm{m}}{h^3}(E_m - E_n).$$

The theoretical distribution is sketched in fig. 2.13.

(*b*) *Total energies.* The derivation of this result is more difficult because the components of velocity parallel to the cathode have to be included. The derivation is given by Du Bridge.†

† Du Bridge, 'New theories of the photo-electric effect', *Actualités Scientifiques et Industrielles*, no. 268 (1935).

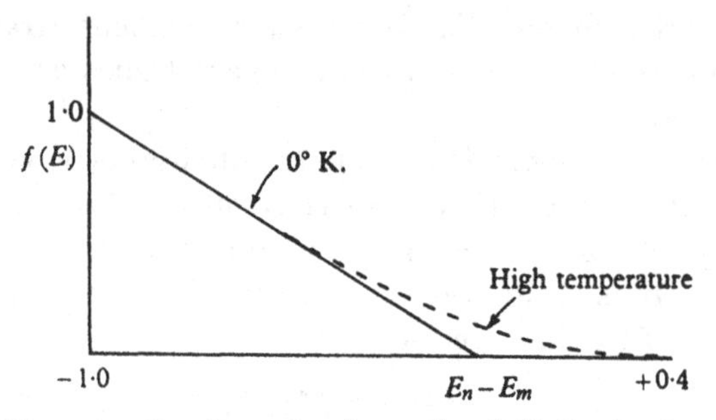

Fig. 2.13. One-dimensional retarding field distribution.

The result is

$$f(E) = \frac{\alpha\, 4\pi \mathrm{m}}{h^3} \frac{E dE}{\mathrm{e}^{(E-E_m)}+1}.$$

We have now presented the theoretical basis for the more important photo-electric phenomena in pure metals. However, the emission from pure metals is of little practical interest for the cut-off frequencies are too low, and the spectral distribution varies too rapidly. For instance, at 296° K. silver has a

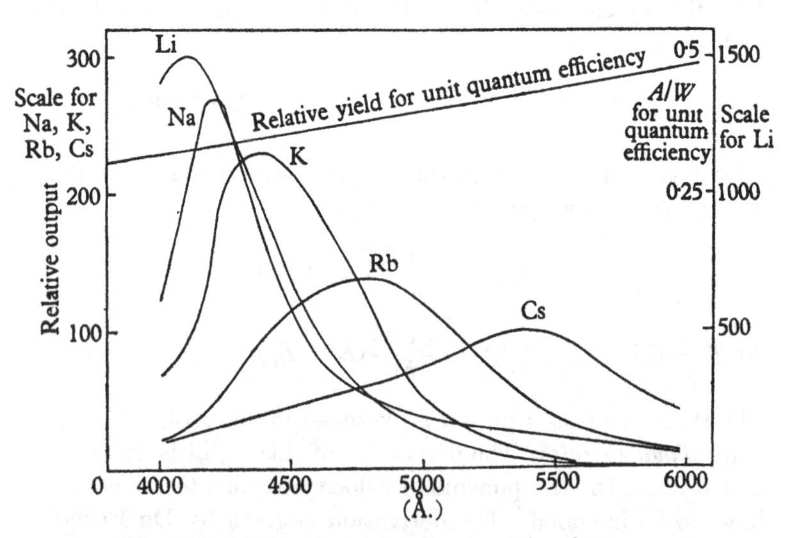

Fig. 2.14. Photo emission from pure metal surfaces.
Note. **The yield for the alkali metals is 10^{-4}–10^{-5} electron/photon, so the absolute yield for Li is about 33μA./W. at λ = 4100Å.**

limiting wave-length of 2620 Å. Caesium, with its very low ϕ, has $\lambda_0 \doteqdot 6800$ Å. and is the most useful metal for this purpose, but complex surfaces have very much better performances. We shall next consider some typical practical photo-surfaces.

2.6. Photo-electric surfaces

In fig. 2.14 we show a plot of the relative photo-electric current as a function of wave-length for a number of surfaces. It can be seen that the emission rises sharply from the threshold wave-length to a maximum and then decreases again. The maximum is partly due to the method of plotting because the energy per photon increases as the frequency of the incident light increases,† so that there are fewer photons per unit light flux. The yield would therefore show a maximum even if the excitation probability were constant over the whole spectrum. The probability is not constant with frequency and also depends on the direction of polarization of the incident beam, phenomena which are usually studied under the title of selective photo-effect. We shall not concern ourselves any more than to note the existence of these effects, whose theory is not yet completely understood.

It will be seen from the figure that, although the alkali metals, particularly potassium, rubidium and caesium, have photo-electric thresholds well up in the visible spectrum, their yield is very poor. Numerically, between 10^4 and 10^5 photons are required to excite one photo-electron from these metals. The complex surfaces such as Ag–AgO–Cs are very much better from this point of view, in addition to having a much longer threshold wave-length. Only about 3×10^2 photons are required to excite one photo-electron in this case.

The preparation of the Ag–AgO–Cs photo-surface is very similar to that of the Ag–AgO–Cs secondary emitter, the principal difference being that the optimum thickness of caesium is not the same, the optimum photo-electric layer being somewhat thicker. The optimum thickness corresponds to about 0·25 mg./cm.², so it is extremely difficult to devise

† Numerically A/W is equivalent to $2{\cdot}26 \times 10^{18}$ photons at $\lambda = 4500$ Å, and the number of photons varies directly as λ.

repeatable methods for introducing the caesium. Methods used include carefully made reaction pellets containing just sufficient caesium, pellets containing excess caesium which is removed by chemical sorption when the appearance of the cathode is correct, and distillation from side tubes. The tube is then re-baked at a temperature of between 420 and 520° K. for a time depending on the temperature. The emission is studied during this bake, and the tube is cooled as soon as the maximum is reached. The best results are obtained if a final thin silver flash is deposited on the cathode after the caesium has been baked on.

Physically speaking, a cathode of this type most probably consists of the following layers: base silver–AgO–Ag + CsO + Cs. It is probable that we are dealing with a semi-conducting oxide, i.e. caesium oxide with excess caesium in which minute silver particles increase the conductivity still further. If this view is correct, then the emission can be accounted for in the same way as for metals, with the exception that the energy distribution will be somewhat different owing to the different form of the law giving the number of filled states in the semi-conductor. De Boer gives a different picture based on his 'adsorption on inner surfaces' theory. It is even more difficult to decide between the theories than it was in the case of thermionic emission, as we are here dealing with surface effects, and several experimental methods which give useful information on thermionic cathodes become useless. On general grounds, we prefer the semi-conductor theory.

The complex caesium surface gives useful results over the visible spectrum, but many other complex surfaces are used for special purposes such as matching the spectral curve of the human eye, giving maximum sensitivity to tungsten light, to blue light, in the infra-red and so on. We shall not deal with these further at this point.

An interesting cathode is the antimony caesium alloy cathode† which has a threshold wave-length of about 7000 Å. (fig. 2.15) but a relatively enormous quantum yield, i.e. 1 photo-electron per 20 photons at 5000 Å. In this case antimony is

† Sommer, *Photo-electric Cells*, Methuen, 1947.

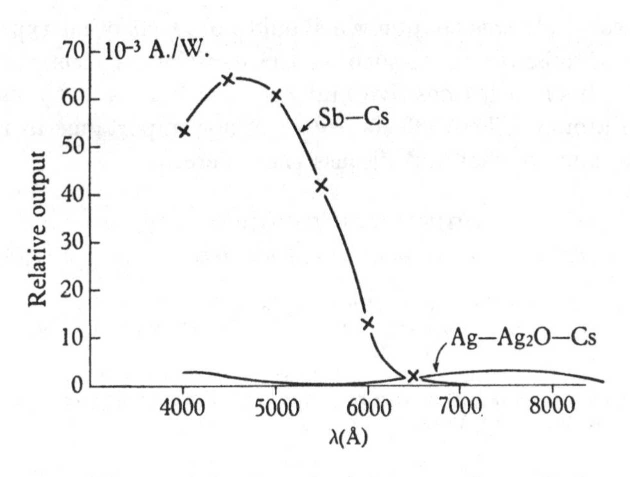

Fig. 2.15. Photo emission from the antimony-caesium cathode. Data from Sommer, *Proc. Phys. Soc.* 55 (1943), 145.

evaporated on to a base metal and a caesium layer put down on top of it. The tube is baked at a temperature of about 450° K. to alloy the two metals, and the bake is stopped at the time of maximum yield. Light surface oxidization improves the yield still more. Alloys similar to this, such as arsenic selenide, tellurium selenide, etc., are known to be semi-conductors, and the antimony caesium alloy is probably the same.

In the infra-red region practical interest has shifted from photo-electric surfaces to photoconductive substances, i.e. the radiant energy is absorbed as heat in a very thin film of semi-conductor. The heating increases the current flow in the film. Since the films are very thin and can be cooled to the temperature of liquid air, good sensitivities and signal-noise ratios are obtained. Typical examples are semi-conducting lead sulphide and lead telluride.

These necessarily brief remarks can only indicate the general methods of preparing photo-electric surfaces. When we subsequently deal with tubes using photo-surfaces, many examples of their application will be given.

2.7. Other types of emission

To conclude this chapter we should note that other types of electron emission exist, such as the emission of electrons by bombardment with positive and negative ions, and by metastable atoms. These effects are of minor importance in hard valves, and we shall not discuss them here.

ADDITIONAL REFERENCES

BRUINING (1941). *Sekundärelektronenemission fester Körper.* J. Springer, Berlin.

ZWORYKIN *et al.* (1949). *Photoelectricity.* John Wiley.

MCKAY, K. G. (1949). *Advances in Electronics,* ed. Morton, vol. I, Secondary Emission. Academic Press Inc.

HUGHES and DU BRIDGE (1934). *Photoelectricity.* McGraw Hill.

MASSEY and BURHOP (1952). *Electronic and Ionic Impact Phenomena,* chap. v. Oxford University Press.

Chapter 3

FLUORESCENCE AND PHOSPHORESCENCE

3.1. The properties of phosphors

Many solids and liquids exhibit the property of absorbing energy from an exciting source, e.g. light, X-rays, α-particles, or electrons, and re-emitting some of the absorbed energy at a frequency characteristic of the material. The re-emission which

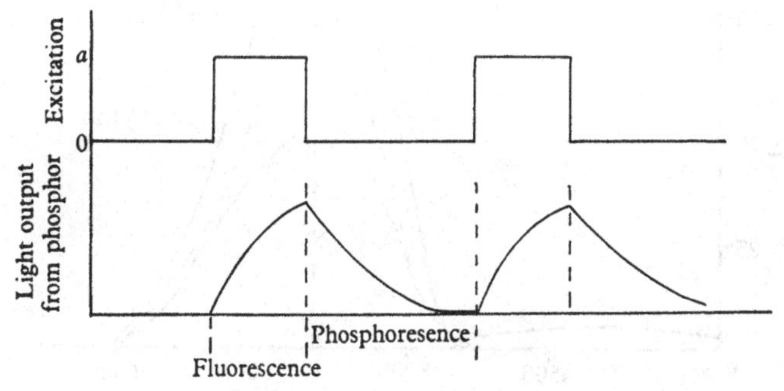

Fig. 3.1. The phenomenon of cathodoluminescence.

occurs during the period of excitation is called fluorescence, while that part of the emission which persists after the initial source has been extinguished is called phosphorescence, the phenomenon as a whole being called luminescence. These terms are illustrated in fig. 3.1. In this chapter we shall be mainly concerned with the phenomena of cathodoluminescence, i.e. excitation by electrons, although the response of phosphors to other types of stimulation is clearly of great importance in gaining a knowledge of the basic laws involved.

It is found that a few solids exhibit a fairly strong luminescence in their natural state, e.g. willemite or zinc orthosilicate (Zn_2SiO_4). It is usual to find that purification of such substances decreases the intensity of the luminescence. The most active phosphors are, however, substances which have been synthesized with the greatest possible degree of purity and then mixed with a small percentage of controlled impurity known as the 'activator'. Activators are usually multivalent metals such as copper, silver and manganese. Other metals, such as iron, cobalt, nickel and lead, act as inhibitors or poisons on the luminescence of the phosphor. Typical phosphors are artificial willemite, activated with about 1 per cent manganese, and zinc sulphide, activated with a much smaller percentage,

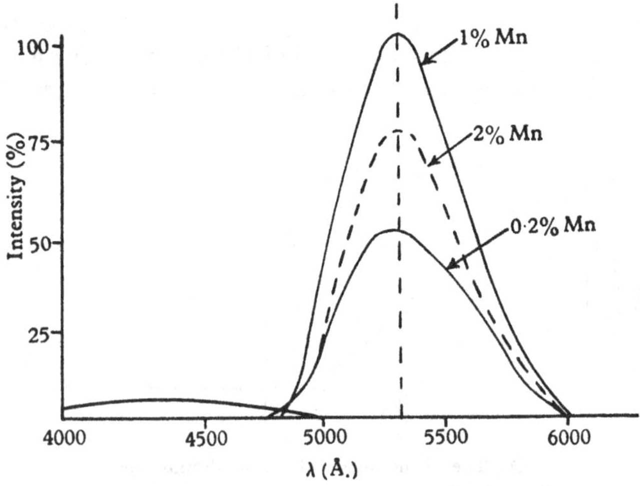

Fig. 3.2. Effect of increasing impurity concentration.

0·1–0·01 per cent, of silver or copper. The addition of more than one activator usually decreases the intensity but may shift the emission spectrum to a more desirable position. Fig. 3.2 illustrates the manner in which the addition of manganese alters the behaviour of pure willemite, increasing the intensity by about 10 times and shifting the spectral maximum from about 4100 to about 5200 Å. The fact that varying the percentage of manganese does not alter the position of the spectral maximum is important since it suggests that the maximum is characteristic of the activator, and the fact that it has shifted to a longer wave-length suggests that additional energy levels have been introduced into a forbidden zone. Other phosphors behave in a similar manner.

The behaviour of the phosphor under electron bombardment is clearly of the first importance in our study. The surface brightness is usually proportional to the current density for small densities, but saturation soon sets in. The brightness is proportional to a power of the voltage between 1 and 2 over the range of anode voltages between which the screen potential is the same as the final anode potential (see the discussion under secondary emission). If the power input to the screen is great enough to increase the temperature above a level dependent on the phosphor in use, the screen undergoes chemical change or burning and the properties alter drastically for the worse.

The influence of temperature of the phosphor on the luminescence is considerable. If the temperature is increased above about 500° K. the intensity of fluorescence decreases, but the effect on the phosphorescence is much more pronounced, the intensity and the duration both diminish. If a phosphor is excited, cooled to liquid-air temperature, and then warmed again, the phosphorescence becomes fainter and disappears only to reappear again as the temperature once more rises. Phosphorescence may be frozen into a material for a very considerable time by this means.

One of the most important means of studying the physics of phosphors is by the observation of the rate of decay of phosphorescence, which yields information on the lifetime of excited states and on the types of interaction taking place. Two types of decay curve are observed, sometimes in superposition. The first type exhibits exponential decay. Let I = intensity,

$$I(t) = I_0 e^{-\alpha t}. \tag{1}$$

However, the second takes longer to decay and obeys the law

$$I(t) = \frac{I_0}{(\beta t + 1)^n}, \tag{2}$$

where α, β and n are constants, $n \doteqdot 2$. In phosphors of this class β is often an exponential function of temperature, while in phosphors of the first class α is similarly temperature-dependent. Willemite obeys eqn. (1), while alkaline earth sulphides obey eqn. (2). Laws (1) and (2) suggest analogies with monomolecular and bimolecular reactions respectively.

These brief remarks can do no more than indicate the more important phenomena of luminescence. General reviews are given in *Luminescence* (Faraday Society, 1938); Riehl, *Physik und Anwendung der Lumineszenz* (J. Springer, 1941); Garlick, *Luminescent Materials* (Oxford, 1949); Pringsheim, *Luminescence of Liquids and Solids* (Interscience, 1946); while work done during the war is dealt with in Soller *et al.*, *Cathode-ray Tube Displays* (McGraw Hill, 1948), the symposium, *Solid Luminescent Materials*, edited by Fonda and Seitz (John Wiley, 1948), and Kroger, *Luminescence of Solids* (Elsevier, 1948).

3.2. The theory of luminescence

The theory of luminescence is not yet completely understood, and its development is still far from the stage of quantitative application to real substances; but, by the aid of the band structure which we have already discussed in earlier chapters, we can obtain a certain insight into the more important processes.

The substances used as phosphors are transparent crystals which are either insulators or extremely low conductivity semiconductors. The explanation of fluorescence in such a crystal is fairly simple: the primary electrons enter the crystals and excite electrons from the filled band into the empty conduction band. Some of these electrons escape as secondaries, but the others refill the holes in the filled band and give out a quantum of energy $h\nu$, where ν is the emitted frequency, in so doing. By studying the light absorption of phosphors it is found that an absorption band exists and is centred at a higher frequency than the maximum of the luminescence/frequency curve. This means that the excited electrons lose some energy to the lattice while in the conduction band, as the most probable gain in energy on excitation is greater than the most probable loss in energy on re-emission. This process has a low quantum efficiency, and the actual process in good artificial phosphors must be a great deal more efficient and must show the possibility of much longer delay times.

Consider the system of fig. 3.3; near the filled band there are impurity centres with filled electron levels, while near the empty band there are so-called traps with vacant electron levels. Electrons from the filled band or, more rarely, from the impurity centres are excited to the conduction band. Once in the conduction band they can either be emitted as secondaries, return to the filled band with emission of light or fill one of the electron traps. The electron trap is similar to a metastable state in the case of a gas ion, in that a transition from the trap

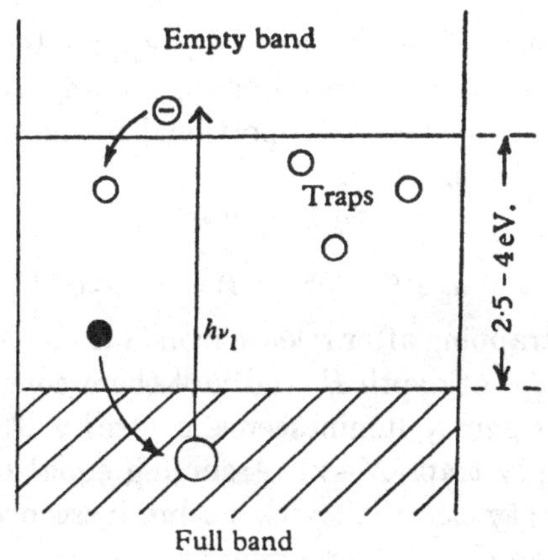

Fig. 3.3. A possible energy scheme for luminescence.

level to the ground state is relatively improbable, while a transition to an activation centre is much more probable. The vacant position in the filled band may be filled by an electron from the lower impurity centres in which case the hole becomes trapped at an impurity centre. We then have the situation that an electron is trapped at a higher level impurity and a hole at a lower one. The transition between the levels involved is energetically more probable than that between the higher centre and the filled band, but the two traps must come near to one another for the transition to take place. The mobility of the traps is due essentially to rearrangements of the molecules forming the lattice and is therefore a slow, diffusion type process with marked temperature-dependence, and so it is possible to account for long delays between the absorption of excitation and phosphorescent emission. The presence of two types of impurity centre may be explained by the added activator on the one hand and by a stoichiometric excess of one constituent of the phosphor on the other. Other surface faults, etc., may also be capable of trapping electrons. The general picture of the phosphorescence process given above is due to Johnson,† but much work on these lines was performed by Randall‡ and his colleagues at Birmingham.

We can now apply these ideas to the calculation of decay curves.

Case 1. Simple luminescence, no re-trapping by empty traps. The number of electrons recombining at any time from the conduction band is directly proportional to the number in the conduction band, i.e.

$$\frac{dn}{dt} = -\alpha n,$$

therefore $$n = n_0 e^{-\alpha t}, \quad \text{or} \quad I(t) = K\alpha n_0 e^{-\alpha t}. \tag{3}$$

Case 2. Re-trapping after release from a trap. At any instant there are N traps of depth E, and n of them contain electrons. The number of empty luminescence centres will be n and the number of empty traps $N-n$. Assuming equal capture cross-sections for empty electron traps and luminescence centres, the probability of an electron emerging from a trap and going to a luminescence centre is $N/n = \dfrac{n+N-n}{n}$. The rate of escape from the traps, due to thermal agitation, is $ns\, e^{-E/kT}$. Therefore

$$I(t) = -\frac{dn}{dt} = \frac{n^2}{N} s\, e^{-E/kT}.$$

Solving the second equation for n, we obtain, putting $n = n_0$ at $t = 0$,

$$I(t) = \frac{n_0^2 s\, e^{-E/kT}}{N[1+(n_0/N)\, ts\, e^{-E/kT}]^2}. \tag{4}$$

If all the traps are filled at $t = 0$, i.e. $N = n_0$,

$$I(t) = \frac{Ns\, e^{-E/kT}}{(1+st\, e^{-E/kT})^2}, \tag{5}$$

† Johnson, *J. Opt. Soc. Amer.* **29** (1939), 387.

‡ Randall and Wilkins, *Proc. Roy. Soc.* A, **184** (1945), 366; **184** (1945), 406; **188** (1947), 485.

and for very large delays

$$I(t) \rightarrow \frac{N\,e^{E/kT}}{st^2}. \tag{6}$$

Case 3. Bimolecular reactions. In this case the intensity is proportional to the product of the number of electrons in the conduction band and the number of vacant impurity centres to which they can return. These numbers are equal, so

$$\frac{dn}{dt} = -bn^2,$$

therefore

$$n = \frac{n_0}{1+btn_0},$$

or

$$I(t) = \frac{\text{const.}}{(1+btn_0)^2}. \tag{7}$$

Comparison of eqn. (7) with eqns. (5) and (6) shows that the shape of delay curves alone does not give unambiguous answers to the question: What type of reaction is involved? It is interesting to note that expressions (5) and (7) show that the time for the phosphorescence to decay to half its initial intensity is inversely proportional to the root of the initial intensity, i.e. high intensities mean fast decay. De Groot† has shown that copper-activated zinc sulphide exhibits this behaviour. In general, though, real phosphors depart rather markedly from the theoretical laws.

To conclude this section we shall discuss the possible reasons why some materials luminesce and others do not; for, from what has been said up to now, there is no reason why all insulators and semi-conductors should not exhibit the phenomenon. Seitz‡ gives the following explanation. If we plot the energy of an impurity centre (*a*) in the normal state, (*b*) in an excited state, against any configurational coordinate such as the displacement of the centre from its mean position, we obtain a diagram such as fig. 3.4*a* or 3.4*b*. Consider the first case. The centre will normally be at the equilibrium position *A*, and according to the Franck-Condon principle the nucleus

† De Groot, *Physica*, **6** (1939), 275.
‡ Seitz, *Trans. Faraday Soc.* **35** (1939), 74.

remains at rest in an optical transition so that the energy $h\nu_1$ is required to excite an electron. The equilibrium position for the electron in the excited state is at C, so the electron moves

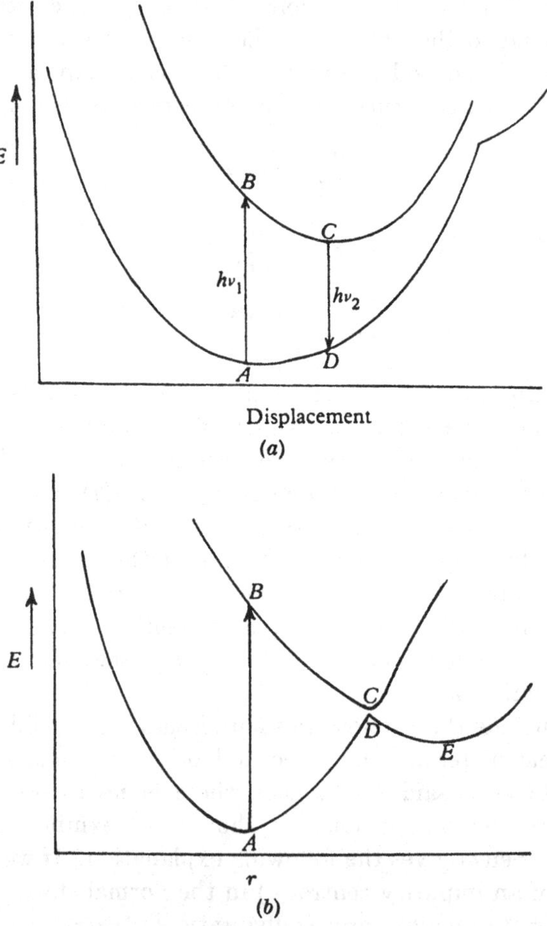

Fig. 3.4. Illustrating the Frank-Condon principle.

to C losing ΔE_1 as heat. From C the electron returns to the unexcited state with radiation of energy $h\nu_2, \nu_2 < \nu_1$, and finally the equilibrium position A is regained with further heat production. Thus far we have described the simplest case of fluorescence. If, however, the situation is not that sketched in fig. 3.4*a* but that of fig. 3.4*b*, where the energy levels 'cross', actual crossing is forbidden unless the states are noncombining, so the energy curves are those of fig. 3.4*b*. The absorption is as before, but when the thermal motion has brought the system to *C* the transition to *D* can take place with the radiation of a very small energy, i.e. fluorescence in the far infra-red. Having arrived at *D* the system can either return to equilibrium at *A* or to quasi-equilibrium at *E*. In the latter case some heat energy can be absorbed to bring the system from *E* to *D* and thence to *A*. Seitz claims that the case of fig. 3.4*b* is typical for non-fluorescent materials. Peierls† and Frenkel,‡ however, have pointed out that the transition *C–D* in fig. 3.4*a* is possible without radiation if the energy can be taken up by setting neighbouring atoms in vibration. The probability of this process has not been calculated, but Peierls shows that it increases with increasing temperature, in agreement with the observation that fluorescence decreases with rising temperature and with the fact that many substances normally inactive fluoresce at very low temperatures.

It will be seen from this brief survey that the theory of luminescence is not very far advanced. It is, however, possible to hope for more rapid progress in the future, for, in spite of the complexities of the subject, there are so many experimental methods by which information can be obtained, in contrast to the position in research on oxide cathodes where there are very few. For more detailed accounts of the theory, the reader is referred to Mott and Gurney, *Ionic Crystals*, and Kröger, *Luminescence of Solids*.

3.3. Some typical phosphors

Here we shall give a brief qualitative account of a few typical phosphors. Quantitative results would be of little value as phosphors differ very much according to their origin and precise physical state. We have already given some figures on willemite. When activated with about 1 per cent manganese the spectral maximum is at 5230 Å. and the light output about 3 c.p./W.

† Peierls, *Ann. des Phys.* **13** (1932), 905.
‡ Frenkel, *Phys. Rev.* **37** (1931), 17 and 1276; *Phys. Z. Sowjet*, **9** (1936), 158.

Willemite is fairly stable, will stand baking up to the softening point of hard glass, and, because the percentage of activator is rather high, is easy to prepare. The usual green screen measurement cathode-ray tubes are provided with willemite screens. The sticking potential is about 6·5 kV.

Zinc sulphide and cadmium sulphide are also extremely widely used. They are similar in their characteristics. The spectral maximum of zinc sulphide–silver is about 4400 Å., cadmium sulphide about 6300 Å. These materials form solid

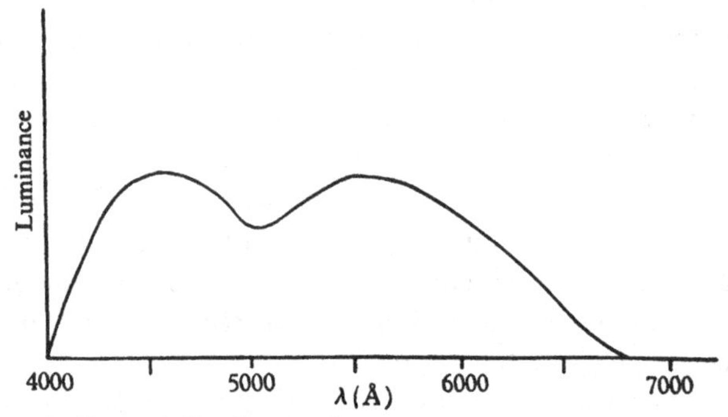

Fig. 3.5. Spectral distribution for a mixed phosphor, giving a 'white' light.

solutions in almost any proportions, so that a phosphor for any wave-length between these values can be obtained. The output is high, about 5 c.p./W. The sulphides are much more difficult to prepare and handle than willemite, but the manufacturing difficulties have been overcome. The sticking potentials are higher than that of willemite, about 10 kV. The 'white' television screens are obtained by a mixture of zinc sulphide and cadmium sulphide, the spectral response being roughly as shown in fig. 3.5.

Many other phosphors are described in the works of Fonda and Seitz and Kröger already cited.

The preparation of phosphors is largely a matter of extremely painstaking chemical preparation, so as to reach the necessary high degree of purity, and will not be described here. Some particulars are given in Fonda and Seitz and by Zworykin and Morton in *Television* (John Wiley, 1940).

PART II

THE MATHEMATICAL THEORY OF ELECTRONICS

Chapter 4

ELECTROSTATIC FIELDS

In the study of electron flow inside vacuum tubes we are very frequently presented with the problem of determining the path of an electron inside a system of charged conductors, or with the inverse problem of determining the shape of the conductors to produce a desired electron path. The first problem has received a great deal of attention both in the development of the general mathematical theory of potentials and in the theoretical side of electrostatics, so beloved of English theoreticians of a few decades ago. Unfortunately, this work is of very little value to us in our specialized application, because the cases which permit analytical solution are limited to a few special coordinate systems. We shall, however, note that it is sometimes useful to synthesize results, i.e. if electron paths calculated in some soluble case have valuable properties, it is usually easy enough to find a set of electrodes of manufacturable form which will yield substantially the desired field. In this chapter we shall study the methods which are used to evaluate the potential inside practical electrode systems.

4.1. Some theorems on potential

We quote without proof some useful theorems on the electric potential.†

THEOREM I. The potential, at any point in free space, satisfies the Laplace equation $\nabla^2 V = 0$. The potential and its derivatives exist and are continuous at all points.

THEOREM II. The integral of the normal derivative of a harmonic, continuously differentiable function vanishes when taken over the boundary of a closed region in space.

† Extended treatments can be found in any text-book of potential theory, e.g. Kellogg, *Foundations of Potential Theory* (J. Springer), or Ramsay, *Newtonian Potential* (Cambridge University Press).

THEOREM III. If a harmonic and continuous function vanishes at all points of the boundary of a closed region, it also vanishes at all points inside the region.

THEOREM IV. A harmonic, continuous function is uniquely determined at all points of a region by its value on the boundary of the region.

THEOREM V. If a function is single-valued, continuous and harmonic in a closed region, and if its normal derivative vanishes at every point of the boundary of the region, it has a constant value in the region. Furthermore, any such function is determined to an additive constant, by the values of the normal derivative on the boundary.

THEOREM VI. If the well-behaved function V satisfies $\partial V/\partial n + V = \text{const.}$ on the boundary of a closed region and is never negative, V is the only solution satisfying such conditions.

These theorems may appear somewhat dull and obvious. They are quoted here mainly as a salutary reminder that, when designing electrode configurations to give desired electron paths, one must not attempt to impose more than a sufficient number of conditions.

4.2. Analytical solutions of Laplace's equation

We now briefly discuss some analytical solutions of the Laplace equation. The reader can easily verify for himself that the potential between two parallel planes at potentials V_1, V_2, distant d apart, is

$$V = V_1 + \frac{V_2 - V_1}{d} z, \tag{1}$$

$$E_2 = -\text{grad}\, V = -\left(\frac{V_2 - V_1}{d}\right). \tag{2}$$

For concentric cylinders the expressions are

$$V = V_1 + V_2 - V_1 \frac{\ln r/r_1}{\ln r_2/r_1}, \tag{3}$$

$$E_2 = -\frac{(V_2 - V_1)}{r \ln r_2/r_1}, \tag{4}$$

In symmetrical cylindrical coordinates the Laplace equation becomes

$$\frac{\partial^2 V}{\partial z^2}+\frac{1}{r}\frac{\partial}{\partial r}\left(r\frac{\partial V}{\partial r}\right) = 0. \tag{5}$$

If we solve this by separation, putting

$$V(r,z) = R(r)Z(z),$$

we get the following two expressions:

$$\frac{1}{Z}\frac{\partial^2 Z}{\partial z^2} = k^2,$$

$$\frac{1}{rR}\frac{\partial}{\partial r}\left(r\frac{\partial R}{\partial r}\right) = -k^2, \quad k^2 = \text{separation constant.}$$

These equations have the well-known general solutions

$$Z(z) = a' \exp(kz) + b' \exp(-kz) = a \cosh kz + b \sinh kz,$$

$$R(r) = cJ_0(kr) + dY_0(kr).$$

The solution to (5) is then

$$V(r,z) = \sum_k c_k R(r) Z(z);$$

or, writing the sum as an integral,

$$V(r,z) = \int c_k R(r) Z(z)\, dk. \tag{6}$$

If $V(r,z)$ is known along a line of constant r, or z, for example, if $V(0,z)$ is known, the coefficients may be determined by the usual methods of Fourier expansions, or by expansions in other orthogonal functions. This method is restricted by the fact that the required integrations are not always known, and by the fact that the resulting series often converge too slowly to be of much use. Moreover, there is often a difficulty in finding acceptable boundary conditions. An example of this type of difficulty arises in the case of coaxial cylinders of equal diameter shown in fig. 4.1. If the gap is infinitely small and the walls thin, it is a soluble problem, but for finite gaps and thick walls the problem becomes very complex.

As an example of the method, consider fig. 4.2. A cylindrical box radius $r = a$, depth l, is closed at one end by a plate insulated by a very narrow air gap and held at potential V_0, the rest of the box being earthed. Since the origin is included in the region of integration the Y_0 functions which $\to\infty$ as arg $Y_0 \to 0$ cannot form part of the solution. Further, $\cosh kz = 1$ at $z = 0$, and we require that $V = 0$ at this point, so only the sinh need

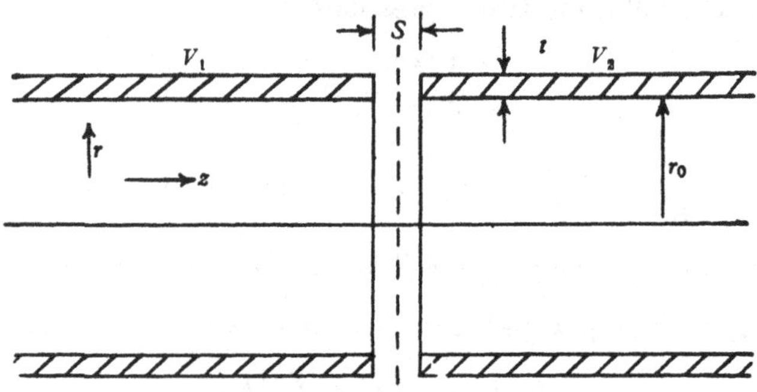

Fig. 4.1. Coaxial cylinders.

be considered. The solution is then

$$V(r, z) = \int_0^\infty A_k \sinh(kz) J_0(kr)\, dk. \tag{7}$$

The remaining conditions, that $V(a, z) = 0$ and $V(r, l) = V_0$, are satisfied by putting $k = p_n/a$ where p_n are the zeros of the Bessel function and by expanding the function $V = V_0$, $0 < r < a$; $V = 0$, $r > a$; in a series of Bessel functions.

The general coefficient of a Bessel expansion is

$$b_n = \frac{2}{a^2 J_1^2(p_n)} \int_{-\infty}^{+\infty} r f(r)\, J_0\!\left(\frac{p_n r}{a}\right) dr. \tag{8}$$

For our case $f(r) = V_0$, $0 < r < a$; therefore

$$b_n = \frac{2V_0}{a^2 J_1^2(p_n)} \int_0^a r\, J_0\!\left(\frac{p_n r}{a}\right) dr$$

$$= \frac{2V_0}{p_n J_1(p_n)}. \tag{9}$$

Thus $$V(r,l) = \sum_{1}^{\infty} \frac{2V_0}{p_n J_1(p_n)} J_0\left(\frac{p_n r}{a}\right),$$

therefore $$A_{kn} = \frac{2V_0}{p_n J_1(p_n) \sinh(p_n l/a)},$$

and $$V(r,z) = \sum_{1}^{\infty} \frac{2V_0}{p_n J_1(p_n) \sinh(p_n l/a)} \sinh\left(\frac{p_n z}{a}\right) J_0\left(\frac{p_n r}{a}\right). \quad (10)$$

This expression can be used to obtain the potential anywhere in the box. It exhibits one important feature of all potential distributions, which is that the potential only depends on the relative dimensions of the box and not on the absolute dimensions, i.e. on r/a and z/a not on r and z. This means that if V_0 is made, for example, 100 V. and the potential determined as a percentage of V_0, the same diagram can be made to serve for all values of V_0 and any choice of a. In other words, if we scale all the linear dimensions by the same factor, the potential remains unaltered. The field strength does not possess this quality, as is easily demonstrated by partial differentiation of (10).

Bertram† has given a method based on the use of Green's function which gives results on some useful cylindrical geometries. For the case of fig. 4.2 and with l/a large (in

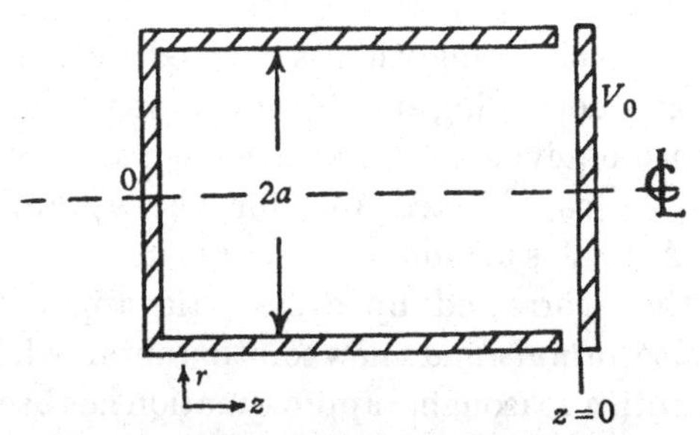

Fig. 4.2. A flat plate on a cylindrical cup.

practice l/a greater than 1·5 is large) the axial potential is given by

$$V(z,0) = V_0 \tanh\left(\frac{1{\cdot}32z}{R}\right).$$

† Bertram, *Proc. Inst. Radio Engrs., N.Y.* **28** (1940), 418; *J. Appl. Phys.* **13** (1942), 496.

The utility of the above method can be extended by the use of numerical integration. Zworykin *et al.* give many examples in their *Electron Optics and the Electron Microscope.*

4.3. Approximation methods for the solution of potential problems

There are many well-established procedures for the solution of potential problems based on the replacement of the differential equation by finite-difference equations. In all these

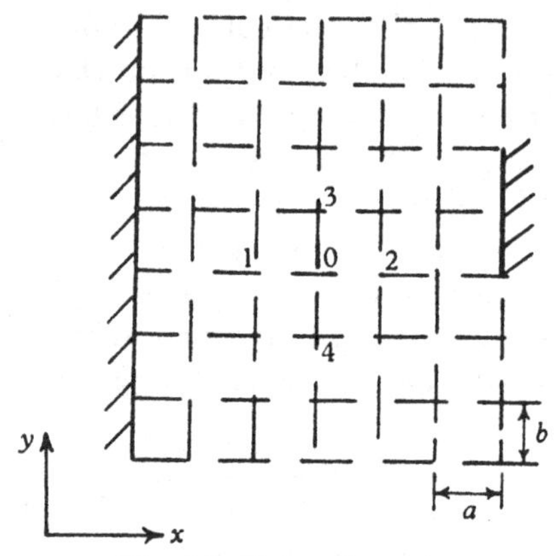

Fig. 4.3. Relaxation net.

methods the domain in question is covered by a net of points, the mesh shape depending on the shape of the domain. The finite-difference equivalent of Laplace's equation is written in the appropriate coordinates and for the appropriate mesh dimension. A trial solution is then assumed. The assumed values are then corrected by systematic application of the finite-difference formulæ and a new solution obtained. The process is carried on until a reasonable approximation has been obtained.

The version which has been most highly developed and seen the widest use in recent years is the relaxation method of Southwell†. In view of the full description given in the reference a very brief account will suffice us.

Consider a rectangular domain, fig. 4.3, covered by a mesh

† Southwell, *Relaxation Methods in Theoretical Physics*, Oxford University Press, 1946.

of sides a and b. For these coordinates Laplace's equation is

$$\frac{\partial^2 V}{\partial x^2}+\frac{\partial^2 V}{\partial y^2}=0. \tag{11}$$

Select a mesh point for which the potential is $V = V_0$, say. Then expand in a Taylor series

$$V = V_0+\left(\frac{\partial V}{\partial x}\right)_0\Delta x+\left(\frac{\partial V}{\partial y}\right)_0\Delta y+\frac{1}{2}\left(\frac{\partial^2 V}{\partial x^2}\right)_0\Delta x^2+\frac{1}{2}\left(\frac{\partial^2 V}{\partial y^2}\right)_0\Delta y^2$$
$$+2\left(\frac{\partial^2 V}{\partial x\,\partial y}\right)\Delta x\,\Delta y+O(\Delta)^3. \tag{12}$$

Then for the numbered points

$$V_1 = V_0-a\left(\frac{\partial V}{\partial x}\right)_0+\tfrac{1}{2}a^2\left(\frac{\partial^2 V}{\partial x^2}\right)_0,\quad V_2 = V_0+a\left(\frac{\partial V}{\partial x}\right)_0+\tfrac{1}{2}a^2\left(\frac{\partial^2 V}{\partial x^2}\right)_0,$$
$$V_4 = V_0-b\left(\frac{\partial V}{\partial y}\right)_0+\tfrac{1}{2}b^2\left(\frac{\partial^2 V}{\partial y^2}\right)_0,\quad V_3 = V_0+b\left(\frac{\partial V}{\partial y}\right)_0+\tfrac{1}{2}b^2\left(\frac{\partial^2 V}{\partial y^2}\right)_0.$$

Then if (11) is to be obeyed

$$V_1+V_2+\frac{a^2}{b^2}(V_3+V_4)-2V_0\left(1+\frac{a^2}{b^2}\right)=0. \tag{13}$$

For the trial solution this will not be true. Suppose that

$$V_1+V_2+\frac{a^2}{b^2}(V_3+V_4)-2V_0\left(1+\frac{a^2}{b^2}\right)=\delta. \tag{14}$$

The Southwell method is to choose the point with the greatest δ and there replace V_0 by $V_0+\dfrac{\delta}{2(1+a^2/b^2)}$ and repeat the calculation. The greatest δ is then eliminated, and the process can be repeated until all the δ's are as small as required. It will be seen that the process is simple but laborious, unless the initial solution is very luckily chosen.

The usefulness of this technique can be much extended by transforming the initial boundaries to a new coordinate system in which they have a simpler geometrical form, by a conformal transformation. We take up the discussion of conformal transformations in the next section.

Many examples of relaxation solutions have now been published, those of Motz† apply particularly to problems on the electrostatic potential. In my opinion the relaxation method is not nearly so useful in the solution of the Laplace equation as it is in the solution of other equations in mathematical physics, such as the wave equation and fourth-order equations in the theory of vibrations, because the electrolytic trough provides a quicker and more flexible method of obtaining the same results, particularly if the effect of small variations in the various dimensions is to be determined, in addition to a straightforward plot.

4.4. Conformal transformations

Conformal transformations provide a powerful weapon for the analytical solution of two-dimensional potential problems in their own right, as well as providing an aid to the application of relaxation methods. They are thoroughly dealt with in books on the complex variable, so we can be brief.

If we subject some function z of a complex variable $(u+iv)$ to a transformation to another complex plane with the requirements that two corresponding curves in each of the two planes intersect at the same angle and the rotation from one curve to the other is in the same sense, the transformation is said to be conformal. It is shown, in books on the complex variable,‡ that a solution of Laplace's equation in one plane is a solution in any plane derived from the first by a conformal transformation. The technical importance of these transformations derives, then, from the possibility of transforming a given domain, in which the solution of Laplace's equation is unknown, into some simple domain in which the desired solution is known. It is easily shown that if a function is transformed by the relation $w=f(z)$ from the w to the z plane, the linear magnification is $|f'(z)|$.

We now consider a few examples of conformal transformations.

† Motz and Klanfer, *Proc. Phys. Soc.* **58** (1946), 30. Motz, *J. Instn. Elect. Engrs.* **95** (1948), 295, pt. III, no. 36.

‡ Copson, *Functions of a Complex Variable*, Oxford University Press, 1935. Phillips, *Functions of a Complex Variable*, Oliver and Boyd, 1940.

I. $$w = z^2,$$
$$w = u+iv = (x+iy)^2 = x^2-y^2+2ixy,$$
$$u = (x^2-y^2), \tag{15 a}$$
$$v = 2xy. \tag{15 b}$$

If we put $u = a^2$ we see that eqn. (15 a) defines a rectangular hyperbola while $b^2 = 2xy$ is another hyperbola orthogonal to the first. If we put $u_1 = a_1^2$, $u_2 = a_2^2$ two hyperbolas are mapped as straight lines parallel to the v axis on the w plane. Together with the rules for linear magnification this yields all the information necessary for the solutions of potential problems involving hyperbolæ.

II. A more important case is $w = a\log z+b$.

If we put $z = r\,e^{i\theta}$,
$$w = u+iv = a\log r+b+ia\theta.$$
This transformation plots coaxial circles in the z plane as parallel lines in the w plane.

We shall meet some other examples in our consideration of fields in triodes.

An important special transformation is the Schwartz-Christoffel transformation which can be used to map a closed polygon in the z plane on the real axis and upper half-plane of the w plane. If the polygon has n vertices and $\alpha, \beta, \gamma, \ldots$ are the interior angles so that $\alpha+\beta+\gamma+\ldots = (n-2)\pi$, the vertices will be mapped as n points on the real axis of the w plane a, b, c, etc., $a<b<c$; by the transformation
$$\frac{dz}{dw} = K(w-a)^{\alpha/\pi-1}(w-b)^{\beta/\pi-1}(w-c)^{\gamma/\pi-1}, \text{etc.} \tag{16}$$
where K = a constant, real or complex.

The Schwartz-Christoffel transformation has been of considerable utility in the reduction of complicated boundaries to simple forms suitable for relaxation procedure. The works of Southwell and his collaborators show several examples of its use.

4.5. The rubber sheet

We next consider two important experimental methods for the determination of potential fields. The first, the rubber sheet, is, in principle, capable of being used to determine the equipotentials in any two-dimensional plane system. In fact, however, it is much more convenient to determine the electron paths direct. In the second method, that of the electrolytic trough, both plane and cylindrical systems can be handled and the equipotential plot is determined, the electron paths being computed from the plot by means which we shall discuss in the next chapter.

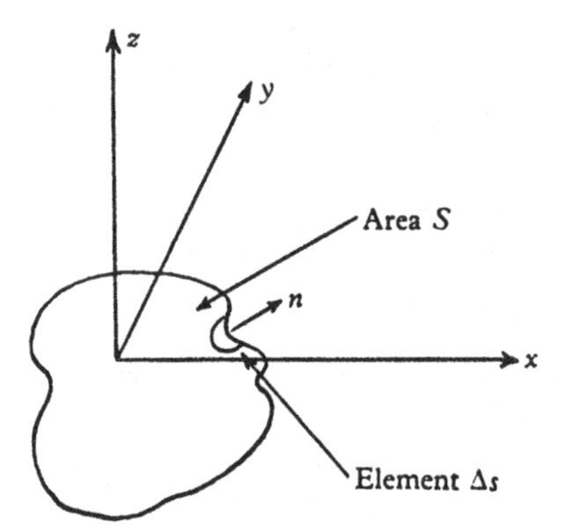

Fig. 4.4. Tension in a membrane.

Returning to the rubber sheet method, we assume that a scale model of the electrode system has been built in which the height of each electrode above an arbitrary base-line represents the potential of that electrode. A thin rubber sheet is carefully stretched between supports, taking care that the tension in x and y directions is as nearly equal as possible, and is then placed over the model so that it is brought into contact with all the electrodes along their upper edges. This means that the vertical scale (potential) must not be too great or it will not be possible to make the rubber touch all the boundaries without undue local stretching. If we consider the small area S in fig. 4.4, assuming a uniform tension T in the sheet, the element of force in the z direction is $T\frac{dz}{dn}\Delta S$ and the total force on S is

$$T\int\frac{dz}{dn}\,dS.$$

The membrane is in equilibrium under the action of the tension, so this force must equal zero. Applying Green's first identity we can reduce the line integral to a surface integral thus:

$$T\int \frac{dz}{dn}\,dS = T\iint \nabla^2 z\,dS = T\nabla^2 zS,$$

if S is an infinitesimal unit of area. The equilibrium condition thus becomes

$$\left.\begin{aligned} ST\nabla^2 z &= 0, \\ \text{or} \qquad \nabla^2 z &= 0, \end{aligned}\right\} \tag{17}$$

i.e. the vertical displacement of the membrane satisfies Laplace's equation, and therefore yields a scale model of the potential distribution between the assigned boundaries. It would be possible to determine the potential by determining contours on the model, but there is no very simple way of doing this. The usual procedure is to roll small steel ball-bearings from the cathode to the anode. Their paths can be determined either by inspection or by a succession of photographs taken under conditions of intermittent illumination so that the electron path appears as a row of dots or lines.

The rubber model is a very quick and convenient tool for the approximate determination of electron paths. It is very useful for the first experiments on an electrical optical system, but if quantitative results are required it is necessary to use photographic recording as described above, which is time-consuming and expensive. The great defects of the method are that it is quite impossible to allow for space charge, that it is not possible to handle cylindrical problems, and that the equipotentials are not recorded so that every problem has to be started from scratch.

4.6. The electrolytic trough

In this technique a scale model of the electrodes is immersed in a trough full of weak electrolyte. Alternating potentials are applied to the electrodes, in proportion to those required in the electron optical system under test, and the potential of any point inside the electrode system is determined by measuring the potential at the corresponding point in the electrolyte by means of an a.c. bridge.

The current flowing in the electrolyte all comes from the conductors, i.e. there are no sources or sinks of electricity in the interior of the electrode system and the equation of continuity $\nabla I = 0$ is therefore satisfied. If the conductivity is σ we further have

$$I = \sigma E,$$

therefore

$$\nabla \sigma E = 0;$$

but

$$E = -\text{grad } V,$$

therefore

$$-\sigma \nabla \text{ grad } V = 0,$$

or

$$\nabla^2 V = 0.$$

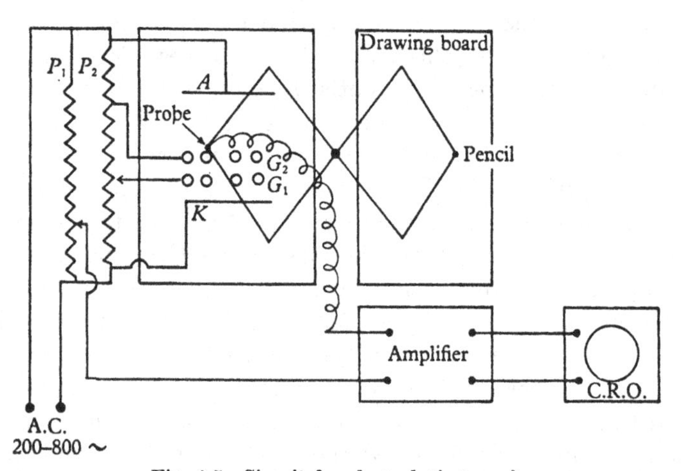

Fig. 4.5. Circuit for electrolytic trough.

We see that the potential distribution does in fact satisfy Laplace's equation.

The connexions to a simple trough are shown in fig. 4.5. The trough itself consists of a shallow watertight vessel as large as possible and at least 2 ft. by 3 ft. by 6 in. deep. A scale model of the electrodes is built on the bottom and the liquid level adjusted until it reaches to within about 1 in. of the top of the electrodes. The electrodes are set up on their required potentials by the tapped potentiometer P_2, while P_1 is another potentiometer divided into the desired steps, often one hundred equal divisions so that each step equals 1 per cent potential change. Alternating current is applied to the trough, and it should be of sufficiently high frequency to avoid any trace of polarization effects. We have generally used 400–500 ~ supplies from a valve oscillator. The movable probe, made from a platinum wire sealed into a glass tube so that 2–3 mm. projects from the glass, is carried on a pantograph system, which in our experience is much the best mounting for it. The other end of the pantograph carries a pencil or other marker which makes a dot on the paper when the bridge has been balanced. The detector system is connected between probe and P_1 so that by keeping the bridge balanced all the points at the potential corresponding to the setting of P_1 can be determined, and thus an equipotential line can be drawn by connecting the points.

There are several practical points which are important if good and consistent results are to be obtained. The electrodes should not contain soluble impurities. Copper, carefully cleaned by means of chemicals, satisfies the requirements well. The liquid surface should be clean and free from dust or oil. There is a wide choice of suitable liquids; in some districts tap water is quite suitable, but heavily chlorinated water has too high a conductivity unless a high-power a.c. source is available. Commercial distilled water is often quite suitable. Weak copper sulphate is sometimes used, or very weak sulphuric acid. The probe should be inserted sufficiently far below the surface to overcome surface effects. As regards the electrical system, we have found that the best presentation is on the screen of an oscilloscope, as it is easy to see the balance condition even in the presence of residual signals from other apparatus. In order not to overcomplicate the diagram we have not shown the Wagner earth which we always use. This is a third potentiometer, continuously variable, and with the tap earthed. The amplifier is arranged so that it can be switched from the probe to the tap on P_3. P_1 is set to the required potential, say n per cent, and P_3 adjusted for balance. This means that the n per cent equipotential is held at earth potential. The amplifier is then switched back to the probe and the n per cent equipotential plotted out in the trough. By this means a much more accurate balance is obtained, as well as greater freedom from pick-up troubles.

If the trough described above is used for work on cylindrical models, a great deal of time is expended on model-making, as one has to make a half-section model complete. This labour can be eliminated by a scheme due to Bowman-Manifold and Nicholls† shown in fig. 4.6, where an end-view of a system of concentric cylinders is shown. In the classical trough the semi-cylinders ABC and $A'B'C'$ would have to be made. By cutting

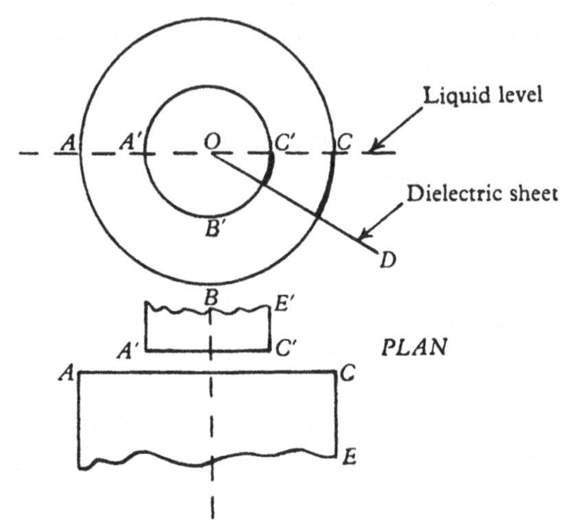

Fig. 4.6. Inclined plane type of trough.

the system along the axis with a dielectric sheet OD the model is reduced to the slightly curved sheets CE and $C'E'$. A glass plate tilted at an angle of 10–30° is placed in the trough. The intercept of the liquid on the plate forms the axis of the cylinder, and the electrodes become simple pieces of sheet bent to the required shape. It is convenient to rule the glass in squares to facilitate setting up and to grind it lightly so that the water-air-glass interface is clearly defined. The gain in simplicity is immense.

Several methods‡ for automatically determining electron paths directly from the trough have been described, most of

† Bowman-Manifold and Nicoll, *Nature, Lond.* **140** (1938), 39.

‡ Gabor, *Nature, Lond.* **139** (1937), 373. Langmuir, *Nature, Lond.* **139** (1937), 1066.

them before our knowledge of servo-mechanisms was as advanced as it is to-day. It may be expected that simpler and more convenient devices will soon be available.

The electrolytic trough is, in my opinion, the best and most flexible method of determining equipotential plots. Once a system has been studied, the plots are available and can be used in any subsequent work. The effect of small variations in dimensions is easily studied, since most of the time is taken in model making and setting up, the actual plot being a relatively quick process. It is not beyond the potentialities of even a small laboratory to build up a large enough store of plots to yield at least preliminary information for most new designs. A method of taking account of space charge by shaping the profile of the trough bottom has been described by Musson-Genon† but, judging by the description, it is very tedious in application. Step-by-step methods‡ for space-charge correction can be applied with somewhat less trouble, but are still laborious. In many electron-optical problems space charge is unimportant, and in these cases the electrolytic trough is the method *par excellence*.§

We have now described the methods of determining the potentials set up by a given electrode system. In the next chapter we turn to the consideration of electron motion in predetermined potential fields, and the use of fields as optical systems.

† Musson-Genon, *Onde élect.* **28** (1948), 236.

‡ Maloff and Epstein, *Electron Optics in Television*, McGraw Hill, 1938. Charles, *Ann. Radioélect.* **4** (1949), 33.

§ The use of the electrolytic trough for the solution of magnetic field problems is discussed by Peierls and Skyrme, *Phil. Mag.* **40** (1949), 249.

Chapter 5

ELECTROSTATIC AND MAGNETIC ELECTRON OPTICS OF FIELDS WITHOUT SPACE CHARGE

5.1. The basic laws of electron motion in electromagnetic fields

In the problems considered in this book we can always consider the electron as an extremely small particle of negative charge e equal to $1{\cdot}59\times10^{-19}$ coulomb and mass m equal to $9{\cdot}01\times10^{-31}$ kg., and in most cases it will suffice to neglect the relativistic variation of mass with velocity. If we consider for a moment that an electron is moving freely in a region where there is zero magnetic field and a constant electric field of E_z V/m, we obtain an expression for the velocity of the electron at any point by equating the potential and kinetic energies evaluated at the point. If the electron is d metres from the plane at zero potential we have

$$\tfrac{1}{2}mu_d^2 = -eE_zd = eV_d, \tag{1}$$

where V_d = voltage at plane d; therefore

$$u_d = \left(2\frac{e}{m}\right)^{\frac{1}{2}} V_d^{\frac{1}{2}}$$

$$= 5{\cdot}95\times10^5 V_d^{\frac{1}{2}}\ \text{m./sec.} \tag{2}$$

If the field has more than one component direction, the total velocity at the point is given by eqn. (2), and the equation of motion is

$$m\ddot{s} = e\,\text{grad}\,V$$

or

$$\left.\begin{aligned} m\ddot{x} &= e\frac{\partial V}{\partial x},\\ m\ddot{y} &= e\frac{\partial V}{\partial y},\\ m\ddot{z} &= e\frac{\partial V}{\partial z}.\end{aligned}\right\} \tag{3}$$

For potentials higher than about 10 kV. the relativistic variation of mass with velocity must be considered. The effective mass $m = \dfrac{m_0}{\sqrt{(1-\beta^2)}}$, where $\beta = u/c$, the ratio of the electron's velocity at the point is given by eqn. (2) and the equation of electron is then

$$(m-m_0)c^2 = m_0 c^2\left[\frac{1}{\sqrt{(1-\beta^2)}}-1\right]$$
$$= \tfrac{1}{2}m_0 u^2\left[1+\frac{3\beta^2}{4}+\ldots\right].$$

The velocity calculated from this expression agrees with eqn. (2) to within 1 per cent for voltages less than 7 kV.

The equation of motion, in the more general case in which there is a magnetic field acting as well as an electric one, is given by the Lorentz equation

$$\mathbf{F} = -e(\mathbf{E}+\mathbf{u}\times\mathbf{B}). \tag{4}$$

In the m.k.s. units we are using $B = \mu_0 H$, in air or vacuum. For electrons moving in rectangular coordinates the equations then become

$$\left.\begin{aligned}\ddot{x} &= \frac{e}{m}\left(\frac{\partial V}{\partial x}+\mu_0\mathscr{H}_z\dot{y}-\mu_0\mathscr{H}_y\dot{z}\right),\\ \ddot{y} &= \frac{e}{m}\left(\frac{\partial V}{\partial y}+\mu_0\mathscr{H}_x\dot{z}-\mu_0\mathscr{H}_z\dot{x}\right),\\ \ddot{z} &= \frac{e}{m}\left(\frac{\partial V}{\partial z}+\mu_0\mathscr{H}_y\dot{x}-\mu_0\mathscr{H}_x\dot{y}\right).\end{aligned}\right\} \tag{5}$$

As an example consider the system of fig. 5.1. The equations (5) become

$$\ddot{x} = 0,$$
$$\ddot{y} = +\frac{e}{m}B_x\dot{z},$$
$$\ddot{z} = \frac{e}{m}\frac{V}{d}-\frac{e}{m}B_x\dot{y}.$$

To save writing we put $a = eV/md$ and $\omega_0 = eB_x/m$. The justification for using the symbol for angular frequency for this last quantity will soon be obvious. It is usually called the Larmor frequency. The above equations integrate directly to give

$$\dot{x} = \text{const.},$$
$$\dot{y} = +\omega_0 z + \text{const.},$$
$$\dot{z} = at - \omega_0 y + \text{const.}$$

If the particle has zero initial velocity all the constants are

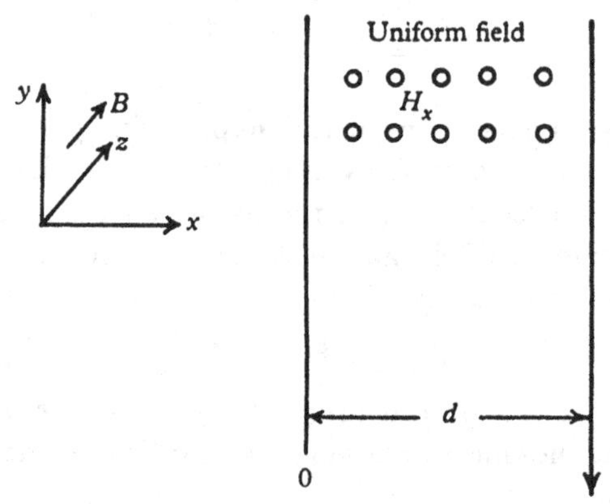

Fig. 5.1. Electron in a magnetic field. Notation.

zero, and inserting these expressions in the original equation we have

$$\ddot{y} + \omega_0^2 y = a\,\omega_0\, t,$$
$$\ddot{z} + \omega_0^2 z = a.$$

These are ordinary second-order differential equations with constant coefficients, and by standard methods the required solution is found to be

$$y = \frac{at}{\omega_0} - \frac{a}{\omega_0^2}\sin\omega_0 t,$$
$$z = \frac{a}{\omega_0^2}(1 - \cos\omega_0 t), \qquad (6)$$
$$x = 0.$$

These will be recognized as the equations of a cycloid in parametric form with the frequency $\omega_0/2\pi$. The case just dealt with is that of the so-called planar magnetron, and the electron paths are shown in fig. 5.2.

As a third example consider the passage of an electron stream between the deflexion plates of fig. 5.3 which are d metres apart. We let the velocity at entrance equal u_z and make the usual assumption that the transverse velocity gained in the deflexion is negligible by comparison with u_z.

The equation of motion is $m\ddot{y} = eV/d$.

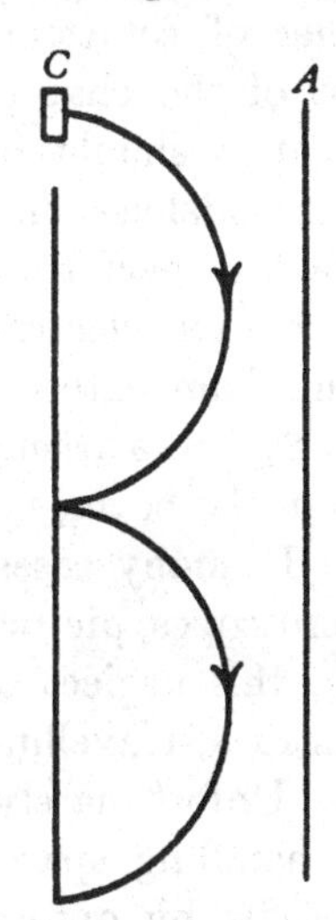

Fig. 5.2. The path of the electron.

Integrating, we obtain $\dot{y} = eVt/md$ and $y = eVt^2/2md$ if the initial velocity u_y is zero. The velocity at exit from the deflexion system is $u_y = eV\tau_1/md$, where $\tau_1 = l/u_z$

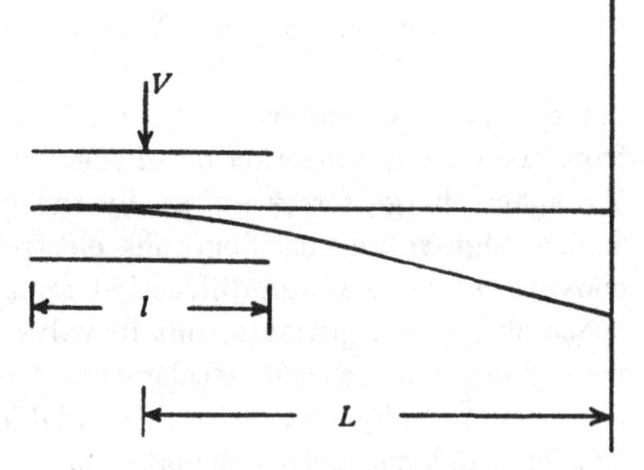

Fig. 5.3. Notation for electrostatic deflexion calculation.

= time of exit from deflexion plates. The electron arrives at the screen plane at $\tau_1 + \tau_2$ sec. after entering the system and $\tau_2 = L/u_z$. The total deflexion at the screen is then

$$\begin{aligned} y_s &= \frac{eV\tau_1^2}{2\,md} + \frac{eV\tau_1\tau_2}{md} \\ &= \frac{eV\tau_1}{md}\left(\frac{\tau_1}{2} + \tau_2\right) \\ &= \frac{eVl}{mdu_z^2}(L + \tfrac{1}{2}l) \\ &= \frac{Vl}{2V_0 d}(L + \tfrac{1}{2}l). \end{aligned} \tag{7}$$

The deflexion thus proves to be directly proportional to the deflecting voltage and inversely proportional to the accelerating voltage, while the tube geometry intervenes through the magnitudes l, d and L. A similar expression is readily deduced for magnetic deflexion.

In these paragraphs we have discussed simple cases of electron motion in planar fields. The subject of electron optics deals with electron motions in much more general fields, though in practical systems the fields are usually either of rotational symmetry or are two dimensional. The rest of the chapter deals with electron optics in some detail, but it should be emphasized before we start that the theoretical model assumed in electron optics differs in one very important respect from that encountered in practice. This difference is that real electron beams are subject not only to forces resulting from external fields, which are assumed to be given, but also to forces arising from the coulomb interaction of the electrons in the beam, i.e. the space-charge forces set up by the beam. In many cases, such as high-voltage oscillographs, electron microscopes, picture converters and electron-diffraction apparatus, this neglect of space charge is legitimate, but in valves, klystrons, travelling wave tubes and particle accelerators it is not. Unfortunately, the only technique at present available for handling space-charge problems is to calculate the beam profile by optical methods and then correct the paths with step-by-step computation, a laborious process. We shall speak of cases, in which space charge plays an important role, as heavy current optics, but before we come to them we shall consider the light current cases in which space charge can be neglected.

5.2. Introduction to electrostatic electron optics

Ordinary geometrical optics is developed from the following basic laws:

(1) Rectilinear propagation in regions of uniform refractive index.

(2) Snell's law of refraction.

(3) The reflexion law.

(4) The independence of the different rays of a beam of light.

We already know that electrons are propagated along straight lines in regions of uniform potential. Can analogies for the remaining three phenomena be discovered? Consider the system shown in fig. 5.4. An electron of velocity u_1 is

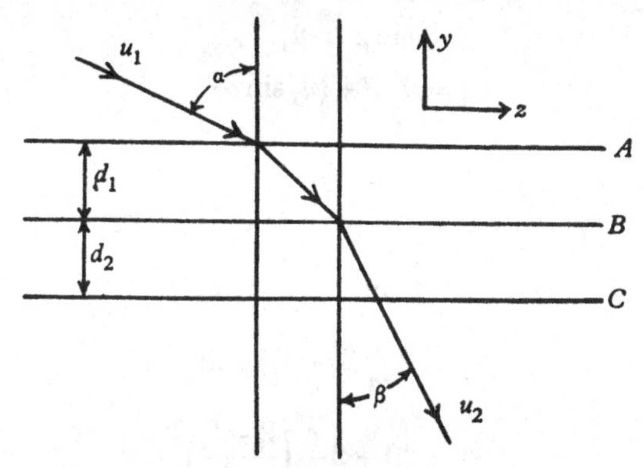

Fig. 5.4. Refraction of electrons.

incident at an angle α on the first of three small-meshed grids held at potentials V_1, V_2, $V_3 (V_2 > V_1)$. In the space AB the equation of motion is

$$m\ddot{y} = e\left(\frac{V_2 - V_1}{d_1}\right),$$

therefore
$$\dot{y} = \frac{e}{m}\left(\frac{V_2 - V_1}{d_1}\right)t + c.$$

If we take the time at plane A as origin, and insert the initial velocity $u_y = u_1 \cos\alpha$ at $t = 0$, we obtain

$$\dot{y} = \frac{e}{m}\left(\frac{V_2 - V_1}{d_1}\right)t + u_1 \cos\alpha, \tag{8}$$

also
$$\dot{z} = \text{const.} = u_1 \sin\alpha.$$

Integrating once more we get (space origin at plane A)

$$y = \frac{e}{m}\left(\frac{V_2 - V_1}{2d_1}\right)t^2 + u_1 t \cos\alpha,$$

therefore $$d_1 = \frac{e}{m}\left(\frac{V_2-V_1}{2d_1}\right)\tau^2 + u_1\tau\cos\alpha, \tag{9}$$

where τ = transit time from A to B. Now since the velocity component in the z direction remains unchanged we have

$$u_1 \sin\alpha = u_2 \sin\beta,$$

or $$\frac{\sin\alpha}{\sin\beta} = \frac{u_2}{u_1}.$$

Now $$u_2^2 = \dot{y}_{(\tau)}{}^2 + (u_1 \sin\alpha)^2.$$

Using eqn. (8)

$$u_2^2 = \left(\frac{e}{m}\right)^2\left(\frac{V_2-V_1}{d_1}\right)^2\tau^2 + 2\frac{e}{m}\left(\frac{V_2-V_1}{d_1}\right)\tau\, u_1\cos\alpha + u_1^2(\cos^2\alpha + \sin^2\alpha).$$

Substituting for $u_1\tau\cos\alpha$ from eqn. (9) we have

$$u_2^2 = 2\frac{e}{m}(V_2-V_1) + u_1^2,$$

$$\sqrt{\frac{u_2}{u_1}} = \left[1 + 2\frac{e}{m}\left(\frac{V_2-V_1}{u_1^2}\right)\right].$$

But $$u_1^2 = 2\frac{e}{m}V_1,$$

therefore $$\frac{u_2}{u_1} = \sqrt{\left(\frac{V_2}{V_1}\right)} = \frac{\sin\alpha}{\sin\beta}. \tag{10}$$

In geometrical optics the ratio of the sine of the angle of incidence to the sine of the angle of refraction is a constant, known as the refractive index. By analogy the electron optical refractive index is the square root of the voltage ratio. Since eqn. (10) is independent of d, it is true for a change of potential taking place very suddenly, e.g. across the metal of a grid. If we have the situation of fig. 5.5 where the lines represent perfect grids or equipotentials, the refractive index going from A to B is taken as $\sqrt{\left(\frac{V_3+V_2}{V_2+V_1}\right)}$, i.e. the mean potentials of the regions are used. The derivation has been given in this rather elaborate form to demonstrate that the law of electron refraction is exact and not an approximation and does not depend on any special assumptions about the way the potential changes, except that it has been assumed that the voltage ratio has been made sufficiently small to ensure that the equipotentials are parallel in the region traversed by the beam. In addition, we have shown that an optical law can be deduced by purely dynamical reasoning, an indication that there may be a very general analogy between dynamics and optics.

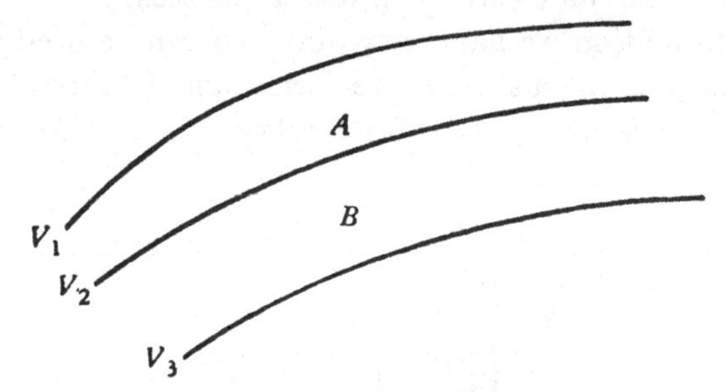

Fig. 5.5. Equipotentials.

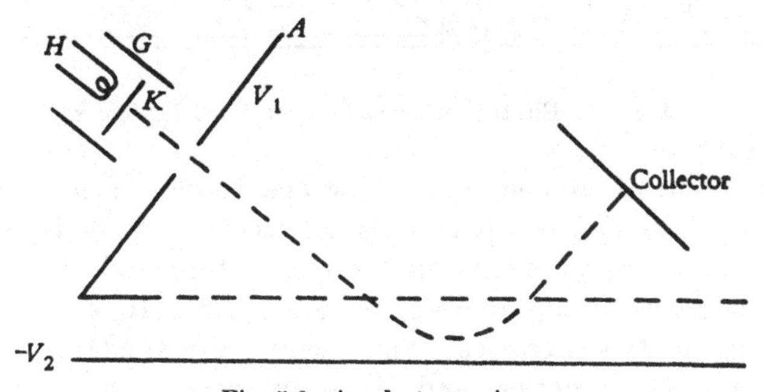

Fig. 5.6. An electron mirror.

In the case discussed above V_2 was larger than V_1 and the ray was refracted towards the normal. If $V_2 < V_1$, the refraction is away from the normal. If an electron is incident on a region wherein the potential decreases steadily until it reaches a value less than that of the cathode from which the electron emanated, the electron is steadily refracted away from the normal, through the position in which it is travelling parallel to the z-axis and is then turned back in the direction it came from. Fig. 5.6 shows this situation, which is clearly analogous to the use of a mirror. By application of the dynamical equations used above it is easily shown that the angle of incidence equals the angle of reflexion.

Finally, we note that if we deduce the paths of two electrons leaving the same point on the 'object' (cathode or crossover or whatever it may be), in different directions and determine the plane in which the paths cross once more, this will be the image plane of the electron optical apparatus.

The information we have now obtained can be used to solve the ordinary problems of image formation in electron optics, although much more powerful methods exist. To solve an

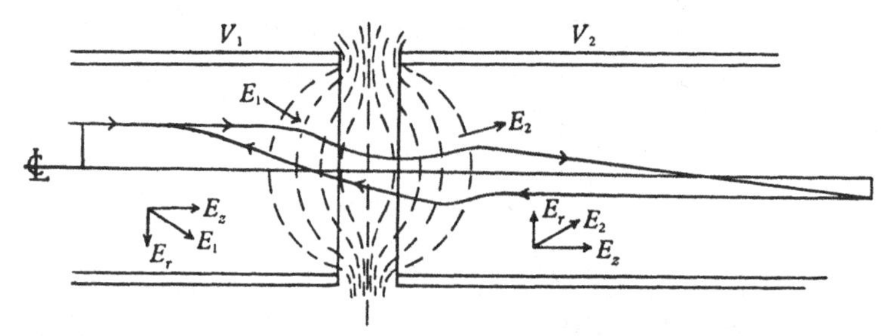

Fig. 5.7. Electrostatic lens between coaxial cylinders.

electron optical problem we can first make a plot of the electric field, drawing the equipotentials sufficiently closely together for them to be of nearly uniform curvature and then plot electron paths by repeated application of eqn. (10) every time an equipotential is crossed. This technique is fundamental in applied electron optics, and there are very many ways of refining the actual methods of plotting described in the literature. However, as in geometrical optics, there are certain problems which can either be solved more simply by analytical methods or be handled only by analysis. In the latter class come important questions such as the limit of resolving power of microscope objectives, and the generalized theory of aberrations.

To obtain a qualitative idea of the operation of a simple type of electron lens, the field between two coaxial cylinders at V_1 and V_2 is indicated in fig. 5.7. An electron enters the field from the left. On the left-hand side of the mean equipotential the radial component of the electric field is directed towards the axis, the electron being therefore deflected in this direction. On the right-hand side of the mean equipotential, E_r is always directed away from the axis and the electron is deflected upwards. However, the electron speed is increasing as it moves through the lens, and the time spent in the right-hand half of the lens is smaller than that in the left-hand half. The net deflexion is therefore towards the axis. If the electron is incident from the right on the lens it is deflected away from the axis at first and then towards the axis. It is, however, being slowed down, and the transit time in the converging part of the lens is greater than that in the diverging. In both cases $V_1 > V_2$, or $V_2 > V_1$, the lens is therefore convergent. This is an example of a general difficulty in obtaining diverging electron lenses. In general they can be obtained only by the use of space charge or by the use of high-frequency lenses. If, for instance, $V_1 > V_2$ at entry to the lens, but by the time of arrival at the lens centre the phase has changed and $V_1 < V_2$ then the net action will be divergent. High-frequency lens effects are of practical importance in high-frequency valves. In the next section we proceed to the study of electron lenses in more detail.

5.3. Axially symmetric electrostatic lenses

We consider an axially symmetric system such as that of fig. 5.7. If $V(r, z)$ is the potential at any point, we know that V must satisfy Laplace's equation. In symmetrical cylindrical coordinates this reads

$$\frac{\partial^2 V}{\partial r^2} + \frac{1}{r}\frac{\partial V}{\partial r} + \frac{\partial^2 V}{\partial z^2} = 0. \tag{11}$$

Expand $V(r,z)$ as

$$V(r,z) = V_0(z) + r^2 V_2(z) + r^4 V_4(z) + \dots, \text{ etc.}$$

(The odd terms drop out because of the symmetry.) Carrying out the indicated differentiations and substituting in eqn. (11), we determine the V_n by equating coefficients of r^n.

$$V(r,z) = V_0(z) - \frac{r^2}{2} V_0''(z) + \frac{r^4}{2^2 . 4^2} V_0''''(z) + \dots, \text{ etc.} \tag{12}$$

where $V_0'(z)$ means $\frac{\delta}{\delta z}.V_0(z)$, etc. Clearly $V_0(z)$ represents the potential along the axis, so we see that if this is measured or calculated the potential at every other point in the system is known.

The equation of radial motion, for an electron, is

$$m\ddot{r} = e\frac{\partial V}{\partial r}$$

or
$$\ddot{r} = \frac{e}{m}\left(-\frac{r}{2}V_0''(z) + \frac{r^3}{2^2.4^2}V_0''''(z)\ldots, \text{etc.}\right).$$

For 'paraxial' electrons, i.e. electrons so near the axis that r^2 is negligible compared with 1, we can approximate by

$$\ddot{r} = -\frac{e}{2m}r\,V_0''(z). \tag{13}$$

Now
$$\frac{d^2r}{dt^2} = \frac{d}{dt}\left(\frac{dr}{dt}\right) = \frac{dz}{dt}\frac{d}{dz}\left(\frac{dr}{dz}\frac{dz}{dt}\right) = \dot{z}\left(\dot{z}\frac{d^2r}{dz^2} + \frac{dr}{dz}\frac{d\dot{z}}{dz}\right).$$

But
$$\dot{z} = \sqrt{\left[2\frac{e}{m}V_0(z)\right]},$$

therefore
$$\ddot{r} = 2\frac{e}{m}V_0(z)r'' + \frac{e}{m}r'V_0'(z)$$

$$= -\frac{e}{2m}rV_0''(z), \quad \text{from (13).}$$

Therefore
$$r'' + \frac{V_0'(z)r'}{2V_0(z)} + \frac{V_0''(z)r}{4V_0(z)} = 0. \tag{14}$$

Eqn. (14) is a second-order differential equation with constant coefficients. Once the axial potential and its first and second derivatives are known through the lens region, eqn. (14) gives a point-by-point solution for the path. It is usually called the paraxial ray equation.

An excellent discussion of several methods of using eqn. (14) will be found in Maloff and Epstein,† and we shall not deal with these matters here, except to note that one of the more

† Maloff and Epstein, *Electron Optics in Television.*

fruitful approaches is to find approximate analytical expressions for the axial potentials in various cases, and to use these expressions for analytic solutions of eqn. (14). What is more important for demonstrating clearly the optical nature of the system is an analysis based on the Sturm-Liouville† theory of differential equations.

An equation $fr''+gr'+hr = 0$ is said to be self-adjoint when $g=f'$. If it is not self-adjoint it can be made so by the multiplier $\exp\int\frac{g-f'}{f}dz$.

Let $L(r)$ be the self-adjoint form of the equation, and r_1, r_2 be two independent solutions of the original equation. Then an important property of such a system is that

$$\int_a^b r_2 L(r_1)\,dz - \int_a^b r_1 L(r_2)\,dz = 0. \tag{15}$$

In the case of eqn. (14) the multiplier is

$$\exp\int -\frac{V_0'}{2V_0}dz = \exp\int -\frac{dV_0}{2V_0} = \frac{1}{\sqrt{V_0}},$$

therefore $$L(r) = \sqrt{V_0}\,r'' + \frac{V_0' r'}{2\sqrt{V_0}} + \frac{V_0'' r}{4\sqrt{V_0}} = 0,$$

or $$\frac{d}{dz}(\sqrt{V_0}\,r') + \frac{V_0'' r}{4\sqrt{V_0}} = 0.$$

Inserting this expression into (15) we have

$$\left[\sqrt{V_0}\left(r_2\frac{dr_1}{dz} - r_1\frac{dr_2}{dz}\right)\right]_a^b = 0,$$

or $$\sqrt{V_b}\,[r_2(b)r_1'(b) - r_1(b)r_2'(b)] - \sqrt{V_a}\,[r_2(a)r_1'(a) - r_1(a)r_2'(a)] = 0. \tag{16}$$

We now consider the trajectories of two electrons incident on the lens from right and left respectively, travelling parallel to the axis and displaced small distances d_1 and d_2 from it. These trajectories cross the axis after refraction, and in optical terminology the points where this occurs are called the focal

† Margenau and Murphy, *Mathematics of Physics and Chemistry*, pp. 253 *et seq.*

points (fig. 5.8). The angles made with the axis at the focal points are α_2 and α_1. Expressed mathematically

$$r_1(a) = d_1, \quad r_1'(a) = 0, \quad r_1(b) = 0, \quad r_1'(b) = \tan\alpha_2;$$

and $\quad r_2(a) = 0, \quad r_2'(a) = \tan\alpha_1, \quad r_2(b) = -d_2, \quad r_2'(b) = 0.$
Inserting these in (16) we get

$$V_b^{\frac{1}{2}} d_2 \tan\alpha_2 = V_a^{\frac{1}{2}} d_1 \tan\alpha_1. \tag{17}$$

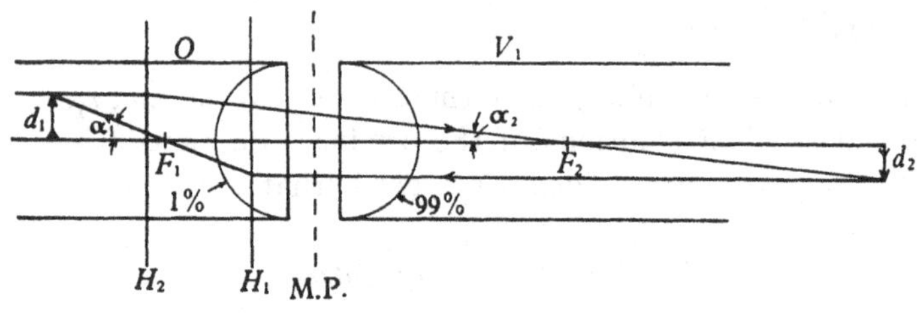

Fig. 5.8. Location of cardinal points in an electron lens.

Since the focal lengths are defined by the relation

$$f_1 = -\frac{d_2}{\tan\alpha_1}, \quad f_2 = \frac{d_1}{\tan\alpha_2},$$

$$\frac{f_2}{f_1} = \sqrt{\frac{V_b}{V_a}} = \frac{\mu_b}{\mu_c}. \tag{18}$$

Eqns. (17) and (18) are clearly completely analogous to the Lagrange law of optics. In fig. 5.8 we have inserted the focal points F_1, F_2, and the principal planes H_1, H_2. These are obtained by projecting the terminal trajectory of each electron back to cut the initial, axial part of the trajectory. The mid-plane or 50 per cent equipotential is also marked. In most tabulations and calculations of electron lenses the focal lengths are given as distances from the mid-plane. They are called mid-plane focal distances. If the mid-plane is known and the four focal distances given, the principal planes can be located in space. Once this has been done, the image corresponding to any object is very easily located, and no further consideration of the explicit form of the lens is necessary. The properties of coaxial lenses have been rather extensively tabulated; the references below give results for a considerable range of dimensions,† the results of Klemperer being the most reliable.

It remains only to deduce an expression for the magnification

$$m = \frac{h_2}{h_1} = \frac{\tan \alpha_1}{\tan \alpha_2}\sqrt{\frac{V_a}{V_b}}. \tag{19}$$

The cylinder lens used in cathode-ray tube construction has the highest potential on the image side, and therefore eqn. (19)

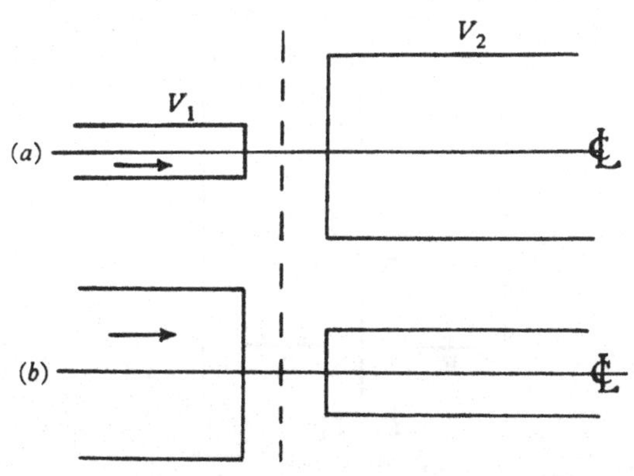

Fig. 5.9. Lenses with components of unequal diameter.

shows that the image will be smaller than the object, which is the desired result, since the spot is required to be as small as possible.

5.4. Types of electrostatic lens

In the last section we discussed explicitly the lens formed by equal concentric cylinders. The results are valid, however, for any system with axial symmetry. Another common type of lens is the cylinder lens with different diameters of tube on either side of the mid-plane. In this lens the equipotentials on the axis of the larger tube are less curved than those in the smaller. In fig. 5.9, the lens shown in (*a*) is therefore stronger than the equi-cylinder lens of the same voltage ratio: that in (*b*) is, *per contra*, weaker.

† Klemperer, *Electron Optics*, Cambridge University Press, 1939. Spangenberg, *Vacuum Tubes*, McGraw Hill, 1948. Maloff and Epstein, op. cit.

A very simple electron lens is the simple aperture between regions of different field strength as indicated in fig. 5.10. It has long been known that the focal length of such an aperture is given by

$$f = \frac{4V_1}{E_1 - E_2}, \tag{20}$$

where V_1 = voltage of apertured plate, E_1 = field on object side of lens, E_2 = field on image side of lens. It is interesting to note

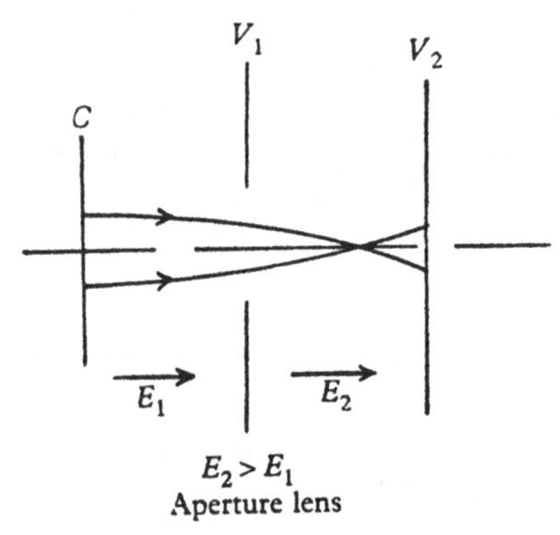

Fig. 5.10. The simplest electrostatic lens.

that the lens is diverging for the case $E_2 = 0$, i.e. when the electron enters field-free space.

When cylindrical lenses are considered instead of spherical lenses, the factor 4 in eqn. (20) has to be replaced by 2.

To conclude this discussion of types of lens, we must mention the symmetrical unipotential lens (German '*Einzellinse*'), which is widely used in electron microscopy. This lens is shown in fig. 5.11. It consists of three more or less elaborately profiled apertures, the two outer ones held at the same potential and the middle one at some higher or lower potential. In both cases the lens is converging, unless the lowest potential is less than the cathode potential, in which case the device acts as an electron mirror. In the lens, the electron trajectory is of type (*a*) when $V_2 < V_1$, the type (*b*) when $V_2 > V_1$. Many studies, both theoretical and experimental, have been made on this lens, and the reader is referred to the papers indicated for details.†

Many attempts have been made to treat the cathode-grid system of the cathode-ray tube as an analogue of the optical immersion lens. In this region the electrons are travelling slowly and space-charge forces are therefore more effective than at any other part of the beam. This fact renders the results

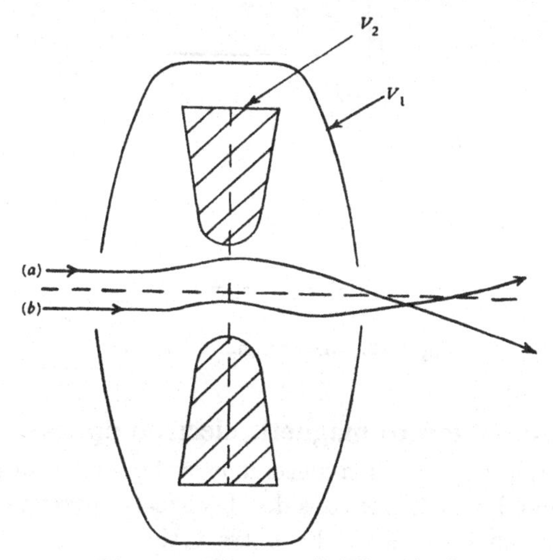

Fig. 5.11. Electrostatic '*Einzellinse*'.

obtained inaccurate, and while useful indications of behaviour can be obtained it seems more useful to treat the electron gun from the space-charge viewpoint developed in the next chapter. In the cathode-ray tube the function of the cathode-grid-screen system is to provide a short focus lens which produces a 'cross-over', i.e. a point at which all the electrons cross the axis, very near to the screen aperture. The main lens is then used to image this cross-over on the fluorescent screen. If space-charge forces are taken into account it is found that the electrons do not actually cross the axis but come in to a minimum radius at which point their motion is purely axial and then diverge again. A disk of least confusion is imaged on the screen.

† Johannson and Scherzer, *Z. Phys.* **80** (1933), 183. Plass, *J. Appl. Phys.* **13** (1942), 49. Regenstreif, *Ann. Radioélect.* **6** (1951), 51 and 114.

Fig. 5.12 illustrates (*a*) the electron optical, and (*b*) the real paths.

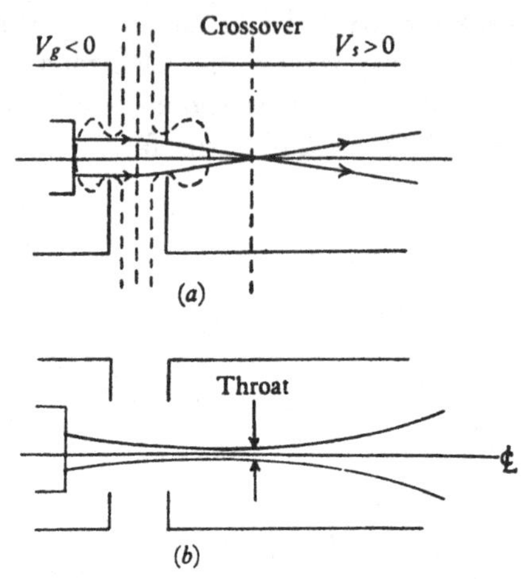

Fig. 5.12. Electron gun behaviour.

5.5. Introduction to magnetic electron optics

The simplest type of magnetic lens is the long solenoid. The short solenoid, which is of considerably greater practical importance, is much more difficult to treat theoretically, but it is interesting to note that the foundation of electron optics was the investigation of the properties of short solenoids by Busch† in 1926, although magnetic lenses had been used in experimental physics since the end of the nineteenth century. The long solenoid shown in fig. 5.13 will produce an erect image of unit magnification of an object from which the angle of divergence of electrons is small enough to make 1 a good approximation to $\cos\alpha$. These statements are verified as follows:

The initial velocity u is directed at an angle α to the field. If there were no component in the axial direction the electron would assume a circular orbit of radius

$$r = \frac{u \sin\alpha}{(e/m)B}, \tag{21}$$

† *Ann. Phys., Lpz.* **81** (1926), 974.

in time
$$\tau = \frac{2\pi r}{u \sin \alpha} = \frac{2\pi}{(e/m) B}, \tag{22}$$

where B is the magnetic induction. $(e/m)B$ is the Larmor frequency, so the time is proportional to the reciprocal of this frequency. If we now consider that the electrons have an axial component $u \cos \alpha \doteqdot u$, the circles are pulled out into a helix and the electron passes a second time through the hori-

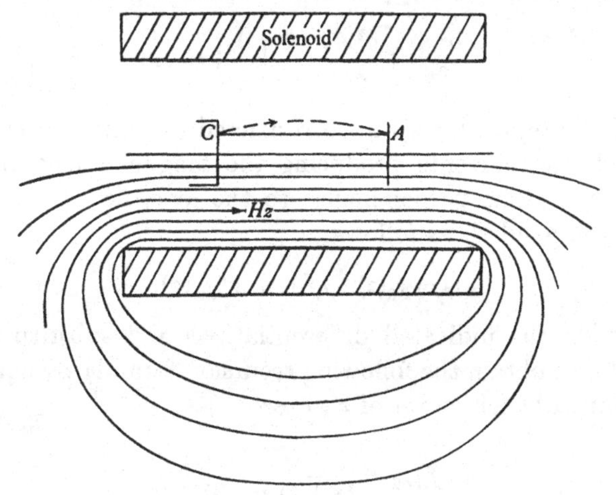

Fig. 5.13. A long magnetic lens.

zontal line through its initial position after time τ, i.e. after a distance τu:
$$\tau u = \frac{2\pi u}{(e/m) B} \tag{23}$$

Thus all electrons leaving the initial point inside the cone of semi-angle α are reunited after this distance, and an erect image of the object is formed there. It will be clear that the above development assumes a truly axial field of constant magnitude over the region occupied by the electrons, and therefore a solenoid considerably bigger than the electron beam device.

Turning now to the consideration of short magnetic solenoids, we have first to develop a little general theory of the magnetic field. An important point to remember is that a static magnetic field does not alter the kinetic energy of an electron but only the direction of motion. The field is usually handled by use of the vector potential A defined by

$$\left.\begin{aligned}\nabla \times \mathbf{A} &= \mathbf{B},\\ \nabla . \mathbf{A} &= 0.\end{aligned}\right\} \tag{24}$$

In charge-free space Maxwell's equations show that curl $\mathbf{B} = 0$, therefore $\nabla \times \nabla \times \mathbf{A} = \nabla \times \mathbf{B} = 0$. For the case of axial symmetry this last equation yields, on expansion,

$$\frac{\partial^2 A_\theta}{\partial z^2} + \frac{\partial^2 A_\theta}{\partial r^2} + \frac{\partial}{\partial r}\left(\frac{A_\theta}{r}\right) = 0. \tag{25}$$

In deriving eqn. (25) we have restricted ourselves to the case in which the currents producing the magnetic field flow in circles in a plane perpendicular to the axis, i.e. $A_r = A_z = 0$. We now expand A_θ as follows:

$$A_\theta = f_0(z) + r f_1(z) \ldots r^n f_n(z).$$

Performing the indicated differentiations and substituting in eqn. (25) we obtain the following recursion formula by equating coefficients of each power of r to zero:

$$f_{n+2} = \frac{f_n''}{(n+1)(n+3)}.$$

Now the vector potential has the same direction as the current producing it, and as this is in opposite directions on either side of the axis and zero on the axis, f_0 and all the even functions vanish. Then

$$A_\theta = r f_1(z) - \frac{r^3 f_1''}{2.4} + \frac{r^5 f_1''''}{2^2.4^2.6} \ldots, \text{ etc.} \tag{26}$$

From the first eqn. of (24) we obtain, for axial symmetry, the relation

$$\begin{aligned} B_z(r,z) &= \frac{1}{r}\frac{\partial}{\partial r}(r A_\theta) \\ &= 2 f_1(z) - \frac{r^2 f_1''(z)}{2} + \frac{r^4 f_1''''(z)}{2^2.4^2} \ldots, \end{aligned}$$

and

$$B_z(0,z) = 2 f_1(z). \tag{27}$$

Using eqn. (27) in (26),

$$A_\theta = \frac{r}{2}B(z) - \frac{r^3 B''(z)}{2^2.4} + \frac{r^5 B''''(z)}{2^2.4^2.6}\dots \tag{28}$$

Now the force on the electron in a combined electric and magnetic field is given by eqn. (4) as $m\mathbf{a} = e(\nabla . V + \mathbf{B} \times \mathbf{u})$, and, for our case, yields

$$\left.\begin{aligned} a_r &= \ddot{r} - r\dot{\theta}^2 &&= \frac{e}{m}\left[\frac{\partial V}{\partial r} - \dot{\theta}\frac{\partial}{\partial r}(rA_\theta)\right], \\ a_z &= \ddot{z} &&= \frac{e}{m}\left[\frac{\partial V}{\partial z} - \dot{\theta}\frac{\partial}{\partial z}(rA_\theta)\right], \\ a_\theta &= \frac{1}{r}\frac{d}{dt}(r^2\dot{\theta}) &&= \frac{e}{m}\frac{1}{r}\frac{d}{dt}(rA_\theta). \end{aligned}\right\} \tag{29}$$

The last of these equations integrates to ($A_\theta = 0$ on the axis)

$$r\dot{\theta} = \frac{e}{m}A_\theta.$$

Inserting this value into the first two equations of (29) we obtain

$$\left.\begin{aligned} \ddot{r} &= \frac{e}{m}\frac{\partial}{\partial r}\left[V - \frac{1}{2}\frac{e}{m}A_\theta^2\right], \\ \ddot{z} &= \frac{e}{m}\frac{\partial}{\partial z}\left[V - \frac{1}{2}\frac{e}{m}A_\theta^2\right], \end{aligned}\right\} \tag{30}$$

Now $dr/dz = \dot{r}/\dot{z}$, therefore

$$\frac{d^2r}{dz^2} = \frac{\dot{z}\ddot{r} - \dot{r}\ddot{z}}{\dot{z}^3} = \frac{\ddot{r} - r'\ddot{z}}{\dot{z}^2},$$

and

$$\dot{z}^2 = \frac{2eV_0}{m}.$$

Therefore the equation of motion becomes

$$\frac{d^2r}{dz^2} = \frac{1}{2V_0}\left[\frac{\partial}{\partial r}\left(V - \frac{1}{2}\frac{e}{m}A_\theta^2\right) - \frac{\partial r}{\partial z}\frac{\partial}{\partial z}\left(V - \frac{1}{2}\frac{e}{m}A_\theta^2\right)\right] \tag{31}$$

We can now insert (12) for V and (28) for A_θ^2 to obtain

$$\frac{d^2r}{dz^2} = \frac{1}{2V_0}\left[-\frac{r}{2}\left(V_0'' + \frac{1}{2}\frac{e}{m}B^2\right) + \frac{r^3}{2^2.4}\left(V_0'''' + 2\frac{e}{m}BB''\right) \text{ etc.}\right.$$

$$-r'\left\{V_0' - \frac{r^2}{2^2}\left(V_0''' + \frac{e}{m}BB'\right)\text{etc.}\right\}\Bigg],$$

or for the paraxial case (all terms in r^2 negligible),

$$\frac{d^2r}{dz^2} + \frac{V_0'}{2V_0}\frac{dr}{dz} + \frac{1}{4V_0}\left(V_0'' + \frac{1}{2}\frac{e}{m}B^2\right)r = 0. \tag{32}$$

This equation is very similar to eqn. (14) which gave the paraxial equation for the case of zero magnetic field. Eqn. (32) clearly reduces to (14) when $B(z) = 0$. When only a magnetic field is present we have

$$\frac{d^2r}{dz^2} + \frac{1}{8V_0}\frac{e}{m}B^2r = 0. \tag{33}$$

In m.k.s. units the constant has the value $2{\cdot}2 \times 10^{10}$.

We can see from the form of the equation that, as in the electrostatic case, if we plot the trajectories of two electrons the trajectory of any other electron may be determined. However, the treatment has obscured one rather important physical fact. The result was obtained by the elimination of the term $r\dot{\theta}$, so the final equation of motion represents the trajectory plotted on a plane which is rotating about the axis at the same rate as the electron does. In other words, the motion is helical, as it was in the case of the solenoid. The rotation of the image can be calculated as follows. We have

$$r\dot{\theta} = \frac{e}{m}A_\theta.$$

Then $$\dot{\theta} = \dot{z}\frac{d\theta}{dz} = \frac{e}{m}\frac{A_\theta}{r} = \frac{e}{m}\left(\frac{B_0}{2} - \frac{r^2}{2^2.4}B_0'\text{etc.}\right). \tag{34}$$

Restricting ourselves to a paraxial electron and inserting the value of $\dot{z}$ we get

$$\frac{d\theta}{dz} = \frac{1}{2}\sqrt{\frac{e}{2m}}\frac{B_0}{\sqrt{V_0}}$$

or $$\theta = \frac{1}{2}\sqrt{\frac{e}{2m}}\int_{z_1}^{z_2}\frac{B_0(z)}{\sqrt{\{V_0(z)\}}}dz, \tag{35}$$

where z_1, z_2 represent the bounding coordinates of the field. Since eqn. (35) is true for any paraxial electron the image is rotated as a whole, without distortion. This rotation is of no importance in most applications of magnetic focusing, but if it is objectionable in a particular instance, it can be removed by a simple device. Eqn. (32) shows that the focusing properties of the lens depend on B^2 while eqn. (35) shows that the rotation depends on B. Thus if we divide the focusing coil in half and connect it so that the current is in opposite directions in each half, the focal properties will be unaltered but the rotation will be eliminated. This arrangement is shown in fig. 5.14.

We have now indicated the main features of magnetic lens action and can proceed to discuss the various types of magnetic lens and their application. The main fields of application of short magnetic lenses are in electron microscopy and in cathode-ray tubes. Axial magnetic fields are used for focusing high-density electron beams and in one species of mass spectrometer.

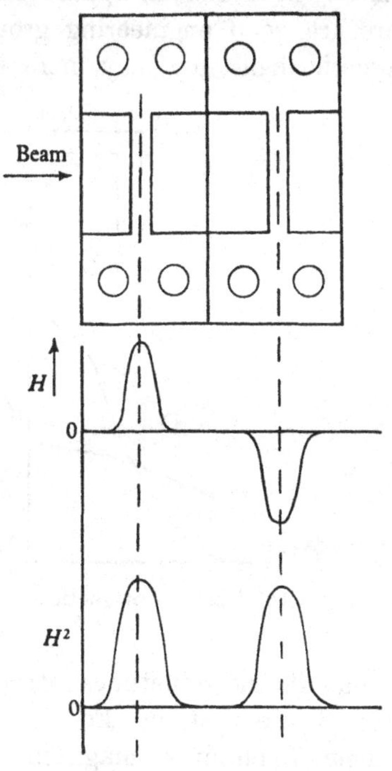

Fig. 5.14. Magnetic lens producing zero image rotation.

5.6. Types of magnetic lens

By far the most important magnetic lens is the short coil with magnetic screening used to restrict the axial spread of the field and to increase the maximum value of B. Fig. 5.15 illustrates the effect of building a soft iron shield round a coil. This process is taken to the limit in the magnetic immersion objectives of Marton and Hutter.† It should, perhaps, be mentioned here that the magnetic field is measured by

† Marton and Hutter, *Proc. Inst. Radio Engrs., N.Y.* **32** (1944), 3, 546.

oscillating or rotating a very small search coil in the field and measuring the voltage produced.

Many other types of magnetic field, or superposition of magnetic field and electric field, are used for special purposes such as mass spectroscopy, picture convertor tubes and so on. It is generally true that in apparatus designed for mass production there are good engineering grounds for avoiding the use of magnetic fields, for they increase the size and weight of the

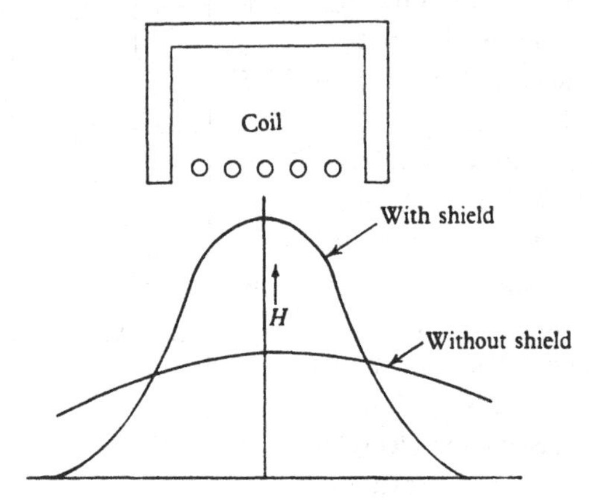

Fig. 5.15. Effect of magnetic shielding on a short magnetic lens.

equipment, and sometimes introduce unwanted effects on other parts of the system. For these reasons there is always a tendency to eliminate magnetic fields even when the equivalent electrostatic system is more complicated to design.

5.7. Ray tracing

In earlier sections we have derived the equations of motion for electrons in axially symmetric fields. We now have to consider the problem of how best to use these expressions to determine the trajectories, given the fields, and we have already indicated the simplest method of all, that of ray tracing using Snell's law. There is a large literature devoted to the problem of ray tracing in light optics, and most of this can be transferred to electron optics, but the central problem is rather different in the two cases. In light optics the shape of the refracting surfaces is given, i.e. one is dealing with glass lenses which can be figured to any desired degree of accuracy, to any shape the designer wishes. In electron optics the equipotentials have to be determined empirically given the electrode shapes. The electrode shapes can be modified to influence the curvature of the equipotentials in a given manner, but it is not, in general, possible to find electrodes which will produce exactly the desired shape. Another difficulty is the obvious one that in electron optics one is always dealing with media in which the refractive index varies continuously from point to point. For these reasons the electron optical problem is more complicated.

We consider first ray plotting in electric fields. The first possibility that comes to mind is the use of analytical methods to obtain a solution to the potential problem giving the values of $V_0(z), V_0'(z), V_0''(z)$ for insertion in eqn. (14). In some cases this can be done; for instance, Johannson and Scherzer† have found that the field in a symmetrical *Einzellinse* is given by $V_z = V_0 + A\,\mathrm{e}^{-bz^2}$, while Plass‡ has used $V_z = V_0(1-\frac{1}{2}\mathrm{e}^{-\frac{1}{2}z^2})$, and Bertram§ has shown that the axial potential in the equal diameter cylinder field is

$$V_z = V_0 + A \tanh 1{\cdot}32\,z.$$

Here z is measured in units of the cylinder radius. In making use of these formulæ eqn. (14) is usually converted to an expression for the focal length. The self-adjoint form of eqn. (14) was

$$\frac{d}{dz}(V^{\frac{1}{2}}r') = -\frac{V''r}{4\sqrt{V}}.$$

Integrating $$\Big[V^{\frac{1}{2}}r'\Big]_a^z = -\frac{1}{4}\int_a^z \frac{rV''}{\sqrt{V}}dz, \qquad (36)$$

where a = axial coordinate at entry. If the path is initially parallel to the axis this electron will pass through the focus.

† Johannson and Scherzer, *Z. Phys.* **80** (1933), 183.
‡ Plass, *J. Appl. Phys.* **13** (1942), 49.
§ Bertram, *Proc. Inst. Radio Engrs., N.Y.* **28** (1940), 418.

So, to determine the focal length, we use the further condition $r'(a) = 0$.

Then
$$r'(z) = -\frac{1}{4\sqrt{V_z}}\int_0^z \frac{rV''}{\sqrt{V}}\,dz. \tag{37}$$

We now solve eqn. (37) by successive approximation. Put

$$r' = r_0' + r_1' + r_2' + \ldots,$$

with
$$r_0' = r_a' = 0, \quad \text{and} \quad r_0 = r_a.$$

Then
$$r_1' = -\frac{1}{4\sqrt{V}}\int_a^z \frac{r_0 V''}{\sqrt{V}}\,dz$$

$$= -\frac{r_a}{4\sqrt{V}}\int_{-\infty}^z \frac{V''}{\sqrt{V}}\,dz;$$

therefore
$$r_1 = -\frac{r_a}{4}\int_{-\infty}^z \frac{dz}{\sqrt{V}}\int_{-\infty}^z \frac{V''}{\sqrt{V}}\,dz,$$

$$r_2' = \frac{1}{4\sqrt{V}}\int_a^z \frac{r_1 V''}{\sqrt{V}}\,dz = \frac{r_a}{16\sqrt{V}}\int_{-\infty}^z \frac{V''\,dz}{\sqrt{V}}\int_{-\infty}^z \frac{dz]}{\sqrt{V}}\int_{-\infty}^z \frac{V''\,dz}{\sqrt{V}},$$

and so on.

Now let $r'(b)$ = total deviation produced by the beam and V_b = potential of the image space. Then the focal length is

$$\frac{1}{f_2} = -\frac{r'(b)}{r_a},$$

and
$$r'(b) = r_a\left[-\frac{1}{4\sqrt{V_b}}\int_{-\infty}^{z_b} \frac{V''}{\sqrt{V}}\,dz\right.$$

$$\left. + \frac{1}{16\sqrt{V_b}}\int_{-\infty}^{z_b} \frac{V''\,dz}{\sqrt{V}}\int_{-\infty}^{z_b} \frac{dz}{\sqrt{V}}\int_{-\infty}^{z_b} \frac{V''\,dz}{\sqrt{V}}, \text{ etc.}\right];$$

therefore

$$\frac{1}{f_2} = \frac{1}{4\sqrt{V_b}}\int_{-\infty}^{+\infty} \frac{V''\,dz}{\sqrt{V}} - \frac{1}{16\sqrt{V_b}}\int_{-\infty}^{z_b} \frac{V''\,dz}{\sqrt{V}}\int_{-\infty}^{z_b} \frac{dz}{\sqrt{V}}\int_{-\infty}^{z_b} \frac{V''\,dz}{\sqrt{V}}, \text{ etc.} \tag{38}$$

f_1 can be determined from eqn. (18).

In early work the first term of this expression was often used as an approximation. The work of the investigators already mentioned has shown that it is not a good approximation but that two terms usually yield a value sufficiently good for practical purposes. Plass gives these results:

	Einzellinse	*Equal cylinders, 3:1 decelerating*
First term	$f = 21{\cdot}32$ cm.	15·85 cm.
First and second	$f = 12{\cdot}97$ cm.	10·75 cm.
First, second, third	$f = 12{\cdot}42$ cm.	10·85 cm.
Graphical	$f = 13{\cdot}35$ cm.	10·81 cm.

From this we see that it is fairly easy to obtain good values for the focal length if a sufficiently accurate expression for the axial field can be obtained. This method leads naturally to another in which the axial field is obtained graphically and is approximated by the dotted straight lines so that the potential is given by

$$V = V_n + \frac{V_{n+1} - V_n}{f}(z - z_n) \quad \text{and} \quad V'' = 0.$$

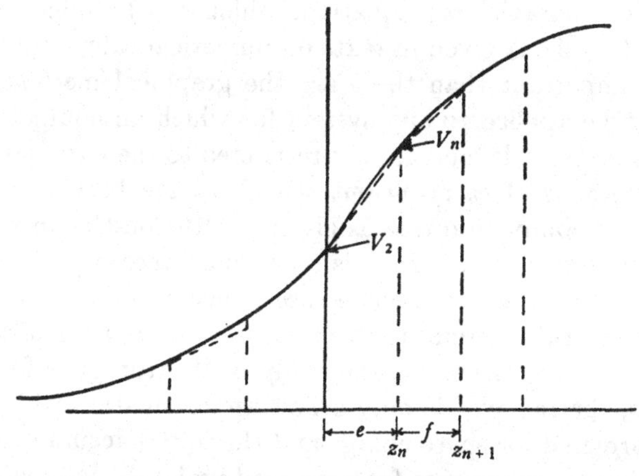

Fig. 5.16. Linear approximations to an electrostatic potential distribution.

Eqn. (36) then gives

$$[V^{\frac{1}{2}} r']_a^z = \text{const.}$$

or

$$r_f = r_e + \int_{z_n}^{z_{n+1}} \frac{c\,dz}{\left[V_n + \frac{V_{n+1} - V_n}{f}(z - z_n)\right]^{\frac{1}{2}}}$$

$$= r_e + \frac{2cf}{(V_{n+1} - V_n)}(\sqrt{V_{n+1}} - \sqrt{V_n}). \tag{39}$$

If the gradient is zero

$$r_f = r_e + r'(z_{n+1} - z_n). \tag{40}$$

At the boundary between two regions V'' is infinite, V and r are constant and V' and r' are finite and continuous. Then we have, integrating across the boundary,

$$\int_{z_{n-}}^{z_{n+}} r'' dz = -\int_{z_{n-}}^{z_{n+}} \left(\frac{V'}{2V} r' + \frac{V'' r}{4V}\right) dz,$$

or
$$r'(z_{n-}) - r'(z_{n+}) = \frac{V'(z_n) - V'(z_{n+1})}{4V_n} r_n. \tag{41}$$

Using the three eqns. (39), (40) and (41) we can plot the electron path through the field.

Another obvious method of solution is by numerical integration of the paraxial ray equation, which can be done by the standard methods given in texts on numerical calculations.

More important than these are the graphical methods, for they can be applied to any system for which an equipotential plot has been made, and are not restricted to the paraxial case. Many methods of carrying out the plot have been described and a very complete review is given by Musson-Genon.† It is not possible to say that there is a graphical technique which is best for all purposes. In some cases, such as microscopy and cathode-ray tube design, the paramount requirement is accuracy, whereas in others, e.g. the study of electron guns for high currents, the ray plot is only a first approximation which has to be corrected for space charge and the initial accuracy is less important, so one wishes for a method which is not too time-consuming. A simple method which has yielded good results in practice is that described by Jacob.‡ In fig. 5.17 we show an equipotential V_1 on which a ray AP_1 is incident to angle α. A scale is marked out so that $\dfrac{AB}{BC} = \dfrac{\sin\alpha}{\sin\beta} = \mu = \sqrt{(\text{voltage ratio})}$, and is so orientated that DP_1 is the perpendicular bisector of AC. If CP_1 is then produced it forms the refracted ray. This

† Musson-Genon, R., *Ann. Télécommun.* **2** (1947), 254.
‡ Jacob, L., *Phil. Mag.* **26** (1938), 570.

ray is then carried on to the next equipotential and the process repeated.

Other methods depending on the approximation of the equipotential by a spherical cap are also commonly used. The main difficulty in graphical methods lies in drawing the equipotentials so that they are sufficiently smooth and in locating the position of the normal.

In the case of magnetic lenses similar but not identical methods can be used. For instance, an equipotential plot can be obtained by plotting the function $V-(e/2m)A^2$ and the path plotted. This leaves out the effect of rotation. Another method is to replace the field by step functions. Further methods will be found in the works cited at the end of the chapter.

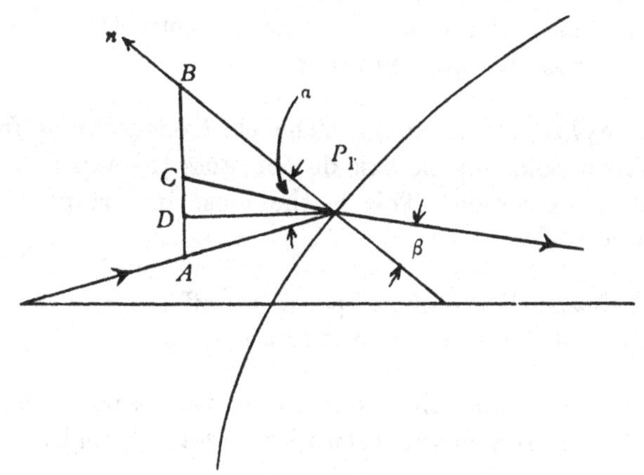

Fig. 5.17. A graphical construction for the refraction at an equipotential.

5.8. Aberrations

Up till now we have restricted ourselves to the consideration of paraxial rays or, what is the same thing, rays making only small angles with the axis, and we have also implicitly assumed that the electrons entering the lens all have the same initial velocity. In fact, actual electron optical systems depart from these assumptions more or less widely, and we must examine the effects produced by this departure. A detailed examination of aberration phenomena would take us too far away from our main subject, so we shall content ourselves by stating the general effects produced by the different types of aberration. The effects discussed here are of great importance in electron microscopy and of some importance in cathode-ray tube and picture convertor design, but in most instances they are not important in valve design. Here we are speaking only of distortions which are aberrations in the optical sense and not of space-charge effects which have no geometrical analogues.

In earlier sections we have assumed that $\cos\alpha = 1$, $\sin\alpha = 0$; these assumptions are good for angles of less than 10°. For bigger angles we must use $\sin\alpha = \alpha - \frac{\alpha^3}{3!}$, which is correct to 1 per cent for angles up to slightly over 50°. If we use the third-order correction it is found that in electrostatic lenses the following types of distortion occur.

5.8.1. *Spherical aberration.* The electrons coming from a single object point on the axis do not cross the axis in a single point after refraction. This is the most important type of aberration.

5.8.2. *Coma.* If the object point lies off the axis the image forms a comet-like shape, instead of a point.

5.8.3. *Astigmatism.* Electrons coming from a point displaced from the axis are focused to two lines at right angles.

5.8.4. *Field curvature.* Plane objects are imaged on to a curved surface rather than on to a plane.

5.8.5. *Distortion.* The image of points on the axis is shifted off the axis giving a square object the shape of a barrel or a pin-cushion.

In addition to these forms of aberration we have to consider chromatic aberration which is due to the fact that the electrons are distributed in velocity and not all of the same velocity. Chromatic aberration only produces large effects in lens systems where the electrons are travelling with low velocities.

In the case of magnetic focusing three extra distortions result from the rotation of the image. These are called anisotropic astigmatism, anisotropic coma and anisotropic distortion as they result from the anisotropic nature of the magnetic field. (This anisotropy is demonstrated at once by the fact that if an electron rotates to the right in going from u to v, say, it rotates to the left in going from v to u.)

The main importance of aberration lies, as has been said, in the design of microscope objectives. It is found that all the aberrations can be made small with the exception of spherical aberration. In light optics spherical aberration can be compensated by the use of negative lenses, but these do not exist in electron optics, at least in the static case. Spherical aberration limits the resolving power of microscopes using magnetic focusing to about 7 Å. in a 50 kV. instrument. Electrostatic lenses have resolving powers 1·3 times to twice as big. Practical figures are about four times as large as the theoretical values, mainly because it is extremely difficult to reach the desired degree of accuracy in alinement and centring. If there were lenses of zero spherical aberration, the limit imposed by the de Broglie wave-length (*c.* 0·05 Å. for 50 kV. electrons) could be approached much more closely. The figure of 7 Å. is, however, better than the best light microscopes by a factor of about 100.

Good general accounts of aberration and the explicit formulæ giving the value of each aberration in terms of the object diameter, aperture and lens constants are given in the books by Cosslett and by Zworykin *et al.*

5.9. Limitations on current density in electron optics

We now consider for a moment the limitations on current density which operate in electron optical systems. As we have explicitly placed space-charge effects out of the range of our present considerations, there is only the effect of thermal distribution in velocity (chromatic aberration) to be considered. From a physical point of view it is clear that the current density at the image plane will depend on the current density at the object plane, the magnification of the system and the distribution in angle of the electrons emitted from the object relative to the aperture of the lens. The calculation of the maximum current has been carried out from purely optical laws by D. B. Langmuir.† J. R. Pierce‡ has, however, deduced the same expression in a much more general form by statistical methods, and it is his treatment that we reproduce here. The reader should also refer to an important article by Gabor.§

The point of departure is Liouville's theorem. This theorem, which is fundamental in statistical mechanics, states that the total rate of change with respect to time of the density of representative points in phase space is zero. The expression 'phase space' refers to a $2n$-dimensional space in which each particle is represented by the point whose coordinates are the n generalized coordinates of the physical particle and the n momenta of the physical particle. Thus for mass points in ordinary space, the equivalent phase space is six-dimensional. As time passes the representative point traces out a path in phase space depending on its physical motion. Liouville's theorem then says that the density of points around a given point remains constant if the variation along the path is considered. Electrons obey the Liouville theorem even in magnetic and space-charge fields. Now we know from Chapter 1 that the electrons just outside the cathode have a Maxwellian velocity distribution. The Liouville theorem tells us that, apart from the action of the potential, the distribution elsewhere is the same. In general terms the Maxwell distribution is

$$dN = (2mkT)^{\frac{3}{2}} B \exp\{-[U_x^2 + U_y^2 + U_z^2 - \Phi]\} dU_x dU_y dU_z dx\, dy\, dz, \tag{42}$$

where U = generalized velocity,

$$= u\left(\frac{m}{2kT}\right)^{\frac{1}{2}};$$

$$\Phi = eV/kT = \text{generalized potential},$$

† Langmuir, *Proc. Inst. Radio Engrs., N.Y.* **25** (1937), 977.

‡ Pierce, J. R., *J. Appl. Phys.* **10** (1939), 715.

§ Gabor, 'Dynamics of electron beams', *Proc. Inst. Radio Engrs., N.Y.* **33** (1945), 792.

and B is a constant to be fixed by the electron density at the cathode.

If we now use eqn. (42) to write the expression for the x directed current density we obtain

$$dJ = \rho_u u_x = -4em(kT)^2 BU_x \exp\{-[U_x^2+U_y^2+U_z^2-\Phi]\} \times dU_x dU_y dU_z. \tag{43}$$

At the cathode surface we get, by integration,

$$J_0 = -2\pi emB(kT)^2;$$

therefore
$$dJ = \left(\frac{2}{\pi}\right) J_0 U_x \mathrm{e}^{-(U^2-\Phi)}\, dU_x dU_y dU_z. \tag{44}$$

Eqn. (44) allows us to calculate the current density at any other point in an electron beam.

5.9.1. *Point focus.* Here we use spherical coordinates U, Θ,

where
$$U^2 = U_x^2+U_y^2+U_z^2,$$
$$U_x = U\cos\Theta.$$

The volume element is $2\pi U^2 \sin\Theta\, d\Theta\, dU$.

If α is the maximum angle with the axis made by electrons reaching the focus the limits of Θ are $0, \alpha$. The minimum velocity is $U = \Phi^{\frac{1}{2}}$ and the maximum ∞. Then

$$J_{\mathrm{max.}} = 4J_0 \int_{\Phi^{\frac{1}{2}}}^{\infty} \int_0^{\alpha} U^3 \mathrm{e}^{-(U^2-\Phi)} \sin\Theta \cos\Theta\, d\Theta\, dU$$
$$= J_0(1+\Phi)\sin^2\alpha. \tag{45}$$

Eqn. (45) gives the limiting current and does not show how the current depends on the magnification. This can be introduced by considering that all the emission from a small area ΔA_c reaches an area ΔA_i, and that $\Delta A_i = M^2 \Delta A_c$. The result is

$$J = \frac{J_0}{M^2}\left[1-(1-M^2\sin^2\alpha)\exp\left(\frac{eV}{kT}\frac{M^2\sin^2\alpha}{1-M^2\sin^2\alpha}\right)\right] \tag{46}$$

For large M
$$J \to \frac{J_0}{M^2}; \tag{47}$$

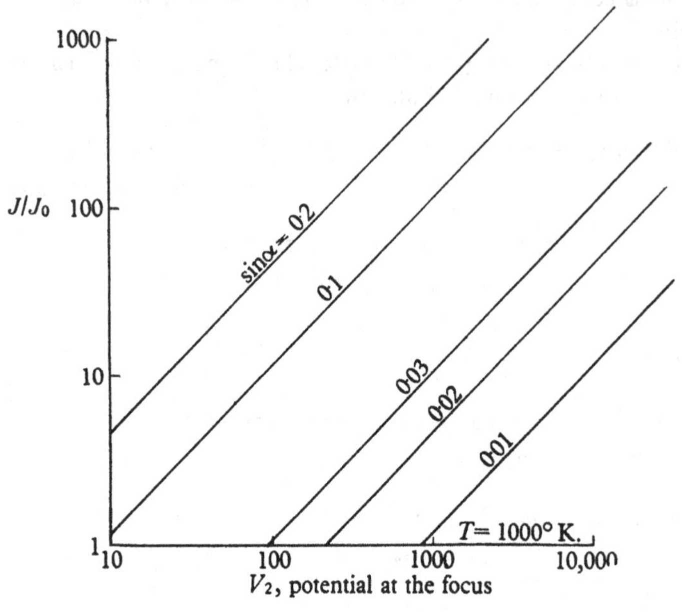

Fig. 5.18. The current focused to a point as a function of cathode current density and final voltage.

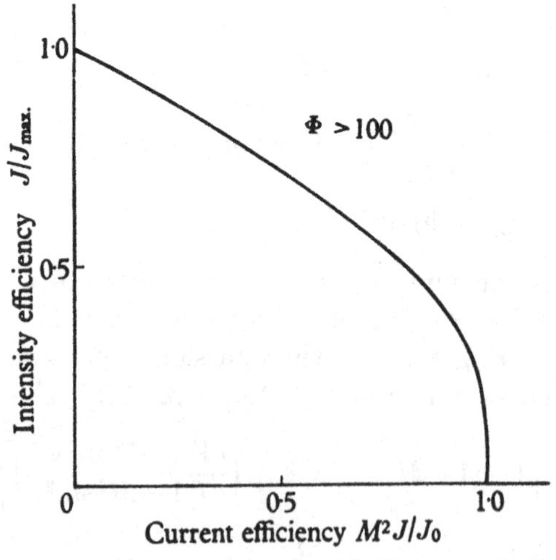

Fig. 5.19. The intensity efficiency versus the current efficiency.

for small M

$$J \to J_0(1+\Phi)\sin^2\alpha. \tag{48}$$

5.9.2. *Line focus.* Similar expressions can be obtained for the line-focus case. These are

$$J_{\text{max.}} = J_0\left\{2\left(\frac{\Phi}{\pi}\right)^{\frac{1}{2}} + e^{\Phi}(1-\operatorname{erf}\Phi^{\frac{1}{2}})\right\}\sin\alpha, \tag{49}$$

where

$$\operatorname{erf}\Phi^{\frac{1}{2}} = \frac{2}{\pi^{\frac{1}{2}}}\int_0^{\Phi^{\frac{1}{2}}} e^{-u^2}\,du,$$

$$J = \frac{J_0}{M}\left[\operatorname{erf}\left(\frac{\beta^2\Phi}{1-\beta^2}\right)^{\frac{1}{2}} + \beta e^{\Phi}\left\{1-\operatorname{erf}\left(\frac{\Phi}{1-\beta^2}\right)^{\frac{1}{2}}\right\}\right], \tag{50}$$

where $\beta = M\sin\alpha$.

For small M

$$J \to J_{\text{max.}},\ \text{i.e. (50)} \to \text{(49)},\quad \text{as before;}$$

for large M

$$J \to \frac{J_0}{M}.$$

We can plot curves of J/J_0, the current ratio, against V with the half-angle α as parameter, as has been done by Langmuir and also of the fraction of current used against the current efficiency as has been done by Pierce. This is done for the point focus in figs. 5.18 and 5.19, where Φ is taken very large. (This means that V is greater than about 10 V.)

These relations are useful in making calculations on devices where the space-charge forces are very low over most of the length of the electron paths, e.g. cathode-ray tubes and electron microscopes. Clearly, in the immediate region of the focus there will be appreciable space-charge forces in any electron beam, but these will not produce gross departures from the above formulæ unless high-current, nearly parallel beams are being considered. These circumstances are dealt with in Chapter 6. The application of the ideas discussed above in deriving an expression for the figure of merit for cathode-ray tubes is discussed by Pierce.†

† Pierce, *Proc. Inst. Radio Engrs., N.Y.* **33** (1945), 476.

5.10. Bibliography of books on electron optics

The works below contain extensive bibliographies of the original papers, especially those by Brüche and Scherzer and by Myers.

(1) *Introduction to Electron Optics*, V. E. Cosslett, Oxford University Press, 1946.

This is a useful introduction, with good theoretical treatment and material on aberration.

(2) *Electron Optics in Television*, Maloff and Epstein, McGraw Hill, 1938. It gives good treatment from first principles with many data on lenses, especially for cathode ray tubes.

(3) *Geometrische Elektronenoptik*, Brüche and Scherzer, J. Springer, 1934. It contains excellent bibliographies of original papers even though it is now a little out of date. The numerous beautiful illustrations make this a very useful book for obtaining a good physical idea of electron optical phenomena.

(4) *Electron Optics and the Electron Microscope*, Zworykin *et al.*, Wiley, 1945. This is the most full and comprehensive book on the subject. Part II (over 400 pages) is devoted to electron optics and deals very thoroughly with almost the whole field.

(5) *Electron Optics*, L. M. Myers, Chapman and Hall, 1939.

It has an excellent bibliography, but the text should be used with care as there are a large number of misprints and other errors.

(6) *Electron Optics*, O. Klemperer, Cambridge University Press, 1939.

A concise introduction for those with a good knowledge of the fundamental physics. Useful data on lenses are included.

(7) *Vacuum Tubes*, K. Spangenberg, McGraw Hill, 1948.

It contains chapters on electron optics, but its main importance lies in a collection of data on electrostatic lenses.

(8) *Theory and Design of Electron Beams*, J. R. Pierce, Van Nostrand, 1949. This recent book pays far more attention to problems met in heavy current devices than does any earlier work. It contains many examples of practical designs.

(9) *Grundlagen der Elektronenoptik*, W. Glaser, J. Springer, 1952.

A very full and rigorous treatment of theoretical electron optics especially aberration problems. The wave mechanical aspects of electron optics are fully treated. (Added in proof.)

Chapter 6

SPACE-CHARGE FLOW AND THE DIODE

In the last chapter the analysis was concerned with situations in which the presence of the electrons was assumed not to modify the fields produced by the electrodes. We now turn to the question of how the space-charge forces due to large currents alter the situation. It is clear that in practice we are nearly always unable to neglect space charge in valves because a valve is fundamentally a mechanism devised so that a weak signal can control a circuit with much more available power, i.e. a device of high-input impedance and low-output impedance. It is difficult to give criteria which allow an estimate of the probable importance of space-charge effects to be made, but a perusal of the cases to be considered should make it reasonably clear when space charge is likely to be of paramount importance.

In the presence of space charge the potential is no longer given by the solution of Laplace's equation, $\nabla^2 V = 0$, but by the Poisson equation:

$$\nabla^2 V = -\frac{\rho}{\epsilon}, \tag{1}$$

where ρ = space-charge density per unit volume, ϵ = permittivity of the medium. In a vacuum ϵ is naturally replaced by ϵ_0. Eqn. (1) can be derived from Maxwell's equations, one of which states that $\nabla.\mathbf{D} = \rho$. But

$$\mathbf{D} = \epsilon.\mathbf{E} \quad \text{and} \quad \mathbf{E} = -\nabla.V,$$

therefore

$$\epsilon.\nabla^2 V + \nabla\epsilon.\nabla V = -\rho. \tag{2}$$

Since $\nabla\epsilon = 0$ in a homogeneous medium eqn. (1) results. The charge density ρ is related to the current density as follows. Consider a beam of electrons of uniform cross-sectional area A and uniform velocity u, current I. In 1 sec. the charge passing through any given plane is $ne = It = I$. The whole of this charge is contained between two planes u metres apart, i.e.

in a volume uA cubic metres. Then, by definition,

$$\rho = -\frac{ne}{uA} = \frac{I}{uA}. \tag{3}$$

In some of our work it will be useful to have a special symbol for the current density. We shall use J for this quantity, and since $J = I/A$ we see that

$$\rho = -J/u. \tag{4}$$

It is sometimes useful to use Gauss's law instead of eqn. (1). To obtain this we substitute $\mathbf{E}$ for $-\nabla V$ and integrate to obtain

$$\int E_n \, dS = \frac{1}{\epsilon}\int \rho \, dv,$$

where E_n = normal outward component of field at ds and the integrals are over the surface and the volume respectively.

6.1. Space-charge flow between plane electrodes

As a first example of space-charge flow consider fig. 6.1, where C represents a plane from which electrons are copiously emitted with zero initial velocities, A being an anode at a fairly high positive potential. In the absence of emission the potential varies linearly from 0 to V_a across the distance s separating the two planes.

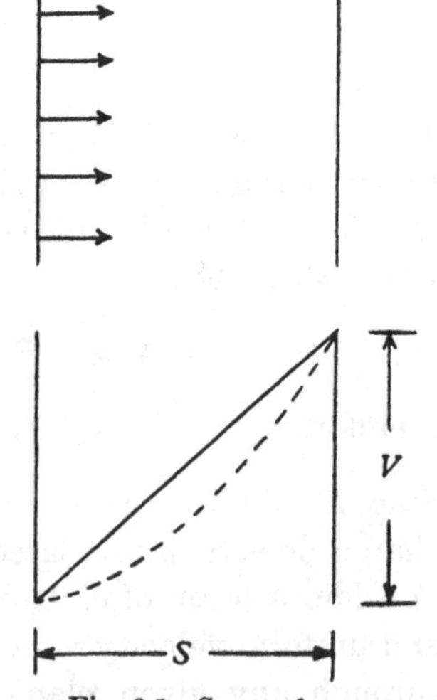

Fig. 6.1. Space-charge limited diode.

For the one-dimensional case we have, from eqn. (1)

$$\frac{d^2 V}{dz^2} = -\frac{\rho}{\epsilon_0}$$

$$= \frac{J}{u\epsilon_0}.$$

But $$u = \sqrt{\left(\frac{2e}{m}\right)} V^{-\frac{1}{2}},$$

therefore $$\frac{d^2V}{dz^2} = \frac{J}{\epsilon_0}\sqrt{\left(\frac{m}{2e}\right)}\, V^{-\frac{1}{2}}. \tag{5}$$

Multiplying both sides by $2(dV/dz)$ and integrating, we have

$$\left(\frac{dV}{dz}\right)^2 = \frac{4J}{\epsilon_0}\sqrt{\left(\frac{m}{2e}\right)}\, V^{\frac{1}{2}} + K_1.$$

Now $dV/dz = 0$ at $z = 0$ (where $V = 0$), when the equilibrium flow has been established, since any gradient at C would increase the number of electrons pulled away from it. Therefore $K_1 = 0$.

A second integration yields

$$\frac{4V^{\frac{3}{4}}}{3} = \sqrt{\left\{\frac{4J}{\epsilon_0}\sqrt{\left(\frac{m}{2e}\right)}\right\}}\, z + K_2.$$

But $V = 0$ at $z = 0$, therefore $K_2 = 0$; and $V = V_a$ at $z = s$, therefore

$$J = \frac{4\epsilon_0}{9\sqrt{\dfrac{m}{2e}}}\frac{V_a^{\frac{3}{2}}}{s^2} \text{ amp./sq.m.}$$

$$= 2{\cdot}336 \times 10^{-6}\frac{V_a^{\frac{3}{2}}}{s^2} \text{ amp./sq.m.} \tag{6}$$

It is obvious that if s is measured in cm., eqn. (6) gives the current density in amp./sq. cm. The potential, when space-charge flow has been established, is given by

$$V = 5{\cdot}68 \times 10^4 J^{\frac{2}{3}} z^{\frac{4}{3}}. \tag{7}$$

This potential is indicated by the dotted line in fig. 6.1. Eqn. (6) was originally derived by Child† in 1911 and is often referred to as Child's law. It is sometimes used in the discussion of real, plane-parallel diodes, but we shall see in later sections that considerable modifications have to be made to obtain good agreement with experiment. However, one important feature is brought out, which is that the current varies as the three halves power of the voltage. We shall find that this is true for all types of space-charge limited current flow.

† Child, *Phys. Rev.* **32** (1911), 492.

6.2. Cylindrical electrodes

In the case of cylindrical symmetry we have

$$\frac{\partial}{\partial r}\left(r\frac{\partial V}{\partial r}\right) = \frac{I}{2\pi\epsilon_0 l}\sqrt{\left(\frac{m}{2e}\right)}V^{-\frac{1}{2}}, \tag{8}$$

where l = cathode length. The solution of this equation was derived by Langmuir.†

Put
$$\frac{I}{l} = \frac{KV^{\frac{3}{2}}}{r\beta^2}, \tag{9}$$

where β^2 is a function of r only, which is tabulated at the end of the book.

Then, putting (9) into (8) we obtain

$$3\beta r^2\frac{d^2\beta}{dr^2}+r^2\left(\frac{d\beta}{dr}\right)^2+7\beta r\frac{d\beta}{dr}+\beta^2-1 = 0. \tag{10}$$

If we put $\omega = \ln(r/r_0)$, suggested by the charge-free potential variation, we get

$$3\beta\frac{d^2\beta}{d\omega^2}+\left(\frac{d\beta}{d\omega}\right)^2+4\beta\left(\frac{d\beta}{d\omega}\right)+\beta^2-1 = 0, \tag{11}$$

and a series solution can be found. This is

$$\beta = \omega-\frac{2\omega^2}{5}+\frac{11\omega^3}{120}-\frac{47\omega^4}{3,300}+\text{ etc.} \tag{12}$$

and
$$\frac{I}{l} = \frac{14{\cdot}66\times 10^{-6}V_a^{\frac{3}{2}}}{r_a\beta^2}\text{ amp./unit cathode length,} \tag{13}$$

or
$$J_c = \frac{2{\cdot}336\times 10^{-6}\,V_a^{\frac{3}{2}}}{r_c r_a\beta^2}\text{ amp./unit cathode area.} \tag{13 a}$$

Eqns. (12) and (13) are valid irrespective of whether the anode or the cathode is the inner cylinder.

In the next section we leave the simple types of space-charge flow for a while to consider the way in which space charge limits the maximum current that can be passed through a system.

† Langmuir, *Phys. Rev.* **2** (1913), 450; **22** (1923), 347.

6.3. The maximum current which can traverse a tube of rectangular cross-section

In fig. 6.2 we show an electron-producing system or gun which directs a beam of electrons towards a metal block A which is held at V volts above cathode potential and is pierced by a hole of rectangular cross-section. The dimensions of the

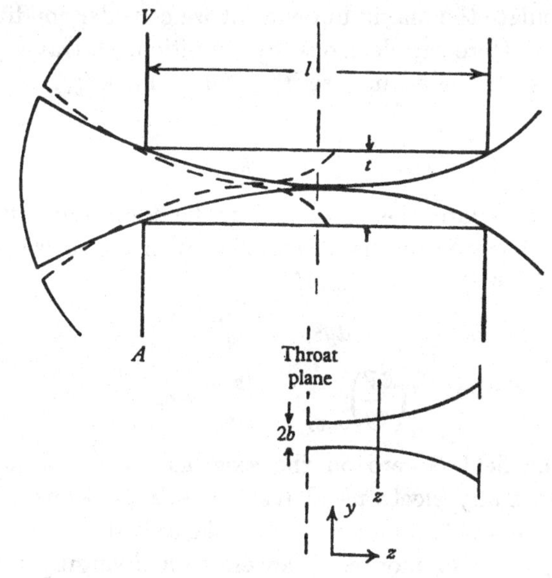

Fig. 6.2. Strip beam formation.

hole are l metres long, breadth t and height perpendicular to the paper h. Breadth t is assumed to be small compared with l and h, so that, except for a small region at entrance and exit of tunnel, the beam travels in space which is initially field-free. We further assume that the electrons which enter the tunnel are all travelling towards a common point, i.e. the vertical component of their motion is directly proportional to the distance from the axis. As the beam progresses inside the tunnel the vertical component of velocity is gradually destroyed by the increasing space-charge forces as the individual electrons come closer together. Eventually the electrons will travel parallel to the axis and then diverge from it as they move beyond the point at which the inward motion is reduced to zero. It is easily proved that motion in a space-charge field is symmetrical, and, this being so, the maximum current condition occurs when the minimum beam breadth is at the tunnel centre. If the current is increased beyond this value, the minimum will move towards the entrance and current will be intercepted near the exit as indicated by the dotted lines.

To calculate the maximum current we consider conditions at the throat. Here the electrons, by definition, all travel parallel to the axis. If the beam breadth is $2y$ we have ($y_{\text{min.}} = b$)

$$\rho = \frac{I}{u\,2y\,h}.$$

We first calculate the space-charge field on the outermost electron. There is no longitudinal component, so we have at plane Z (fig. 6.2)

$$\frac{d^2V}{dy^2} = -\frac{\rho_z}{\epsilon_0},$$

$$\left(\frac{dV}{dy}\right)_{y=y_z} = -\frac{\rho_z y_z}{\epsilon_0} + c. \tag{14}$$

Clearly the field is zero on the axis, at $y = 0$, as there are exactly as many electrons above the axis as below. No net force acts on an electron actually on the axis so $c = 0$.

The equation of motion of an electron diverging from the throat is

$$m\ddot{y} = eE_y.$$

But

$$t = z/u,$$

therefore

$$\frac{d^2y}{dz^2} = \frac{e}{m}\frac{E_y}{u^2}$$

$$= \frac{I}{4\epsilon_0\left(2\dfrac{e}{m}\right)^{\frac{1}{2}} V_0^{\frac{3}{2}} h} = \frac{KI}{V_0^{\frac{3}{2}} h}, \tag{15}$$

$$\frac{dy}{dz} = \frac{KIz}{V_0^{\frac{3}{2}} h} + c_1,$$

$$y = \frac{KIz^2}{2V_0^{\frac{3}{2}} h} + c_1 z + c_2.$$

Now $\qquad \dfrac{dy}{dz} = 0$ at $z = 0$ (plane of throat),

therefore $\qquad c_1 = 0.$

c_2 is determined by the requirements that $y = \frac{1}{2}t$ at $z = \frac{1}{2}l$ and that I be a maximum.

We have then that I is to be as large as possible for a constant y. Thus we must have $c_2 = 0$ (and $2b = 0$).

Then

$$\frac{t}{2} = \frac{K\hat{I}\,l^2}{8V_0^{3/2}h},$$

or

$$\hat{I} = \frac{4thV_0^{3/2}}{Kl^2}.$$

Numerically,
$$\frac{\hat{I}}{h} = 83{\cdot}3\times10^{-4}\frac{t}{l^2}V_0^{3/2}\,\text{amp./m.} \tag{16}$$

The fact that $2b = 0$ makes it appear that ρ becomes infinite at the throat. This is not correct, as an investigation of the density round the throat plane shows. A line focus is thus formed at the centre of the tunnel and the paths are parabolic.

Further remarks on the limitations imposed by space charge on attainable current density are given by Beck,† where references to the original papers may also be found. Later examples are given by Klemperer‡ and Wax.§ In view of the practical importance we quote the expression for the maximum current which can be passed through a circular tunnel, diameter d and length l:

$$\hat{I} = 38{\cdot}9\times10^{-6}\left(\frac{d}{l}\right)^2 V_0^{3/2}\,\text{amp.} \tag{17}$$

In this case the beam diameter is reduced to $d/2{\cdot}4$ at the throat. In practice currents of about 70 per cent‖ of the value given by eqn. (17) have been obtained. In the case of cylindrical lens systems, agreement is not nearly so good, as only about 35 per cent of the theoretical value has been obtained. For the

† Beck, *Velocity-Modulated Thermionic Tubes*, Cambridge University Press, 1948, ch. 4.

‡ Klemperer, *Proc. Phys. Soc.* **59** (1947), 302.

§ Wax, *J. Appl. Phys.* **20** (1949), 242.

‖ Beck, *J. Instn. Elect. Engrs.* **94**, pt. III (1947), 360.

same length and voltage eqn. (16) gives a current density about 70 per cent greater than eqn. (17), but the theoretical advantage is more than offset by this practical failure to reach the desired figures.

In the above work the space-charge forces cause the electrons, which are initially converging to a point, to travel parallel to the axis and then diverge so that the space-charge forces are reduced by increasing the volume occupied by the beam. Another type of space-charge limitation in beam density is

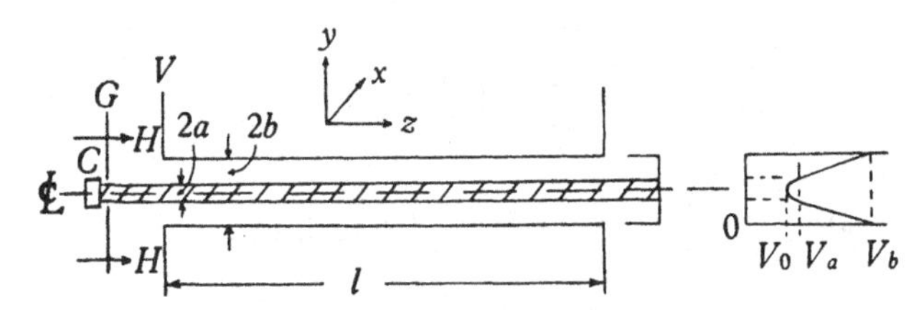

Fig. 6.3. Notation for magnetic collimation.

found when the electrons are prevented from diverging by a strong magnetic field parallel to their paths. In this case the space charge lowers the potential inside the beam. If the current is increased from zero the potential falls steadily until a point is reached at which it drops suddenly to zero with the formation of a virtual cathode and subsequent reflexion of electrons from the tunnel.

The arrangement to be studied† is shown in fig. 6.3. A cathode C in combination with a grid G produces a beam width $2a$, confined within a tunnel width $2b$ at potential V_b. The tunnel is l metres long, and the whole apparatus is in a strong field H parallel to the z axis. We again suppose that $l \gg b$, so that end-effects are negligible. The current density J is uniform across the beam.

Now, since we can assume that there is no longitudinal potential gradient, we can write for the region $0 < x < a$,

$$\left(\frac{d^2 V}{dx^2}\right) = -\frac{\rho}{\epsilon_0} = \frac{J}{\epsilon_0}\sqrt{\frac{m}{2e}}\,V^{-\frac{1}{2}}, \tag{18}$$

† Haeff, *Proc. Inst. Radio Engrs., N.Y.* **27** (1939), 586.

This integrates as eqn. (5) did, but since the potential is symmetrical about the centre line, the initial condition is $dV/dx = 0$, therefore

$$\left(\frac{dV}{dx}\right)^2 = kJ(V^{\frac{1}{2}} - V_0^{\frac{1}{2}}), \tag{19}$$

where

$$k = \frac{4}{\epsilon_0}\sqrt{\frac{m}{2e}},$$

and V_0 = potential on the centre line. Integrating a second time

$$\pm (kJ)^{\frac{1}{2}} x = \tfrac{4}{3}(V^{\frac{1}{2}} - V_0^{\frac{1}{2}})^{\frac{1}{2}} (V^{\frac{1}{2}} + 2V_0^{\frac{1}{2}}). \tag{20}$$

If we put $V = V_a$, at $x = a$ (edge of beam), and $(V_0/V_a)^{\frac{1}{2}} = p$, we obtain

$$\pm \tfrac{3}{4}(kJ)^{\frac{1}{2}} V_a^{-\frac{3}{4}} a = (1-p)^{\frac{1}{2}}(1+2p). \tag{21}$$

Outside the beam there must be a constant potential gradient, and the potential gradient must be continuous at the edge of the beam:

$$\left(\frac{dV}{dx}\right)_a^2 = kJ(V_a^{\frac{1}{2}} - V_0^{\frac{1}{2}}) = \left(\frac{V_b - V_a}{b-a}\right)^2. \tag{22}$$

Eqns. (21) and (22) can be rewritten

$$\frac{9}{16}\frac{kJab}{V_b^{\frac{3}{2}}} = \frac{b/a(1-p)(1+2p)^2}{[1+4/3(b/a-1)(1-p)(1+2p)]^{\frac{3}{2}}} \tag{23}$$

$$= F(p, b/a).$$

$$\frac{V_b}{V_a} = 1 + 4/3(b/a - 1)(1-p)(1+2p) \tag{24}$$

$$= G(p, b/a).$$

These functions can be plotted as functions of p for several values of the parameter b/a, but we are mainly interested in the value $\hat{F}$ corresponding with the biggest value of J. Putting $dF/dp = 0$ for a maximum, we have

$$\frac{b}{a} - 1 = \frac{3}{2}\frac{2p-1}{(1-p)(1+2p)},$$

$$\text{or } \hat{F} = \frac{(1-p)(1+2p)^2 + 3/2(2p-1)(1+2p)}{(4p-1)^{\frac{3}{2}}}, \tag{25}$$

$$(G)_{\hat{F}} = 4p - 1. \tag{26}$$

and from eqn. (23) we have

$$\hat{J} = \frac{16}{9}\frac{V_b^{3/2}}{k}\frac{\hat{F}}{ab} \qquad (27)$$

$$= 2{\cdot}336 \times 10^{-6}\hat{F}\frac{V_b^{3/2}}{ab}. \qquad (28)$$

For $b/a = 1$, $\hat{F} = 2$ and

$$\hat{J} = \frac{4{\cdot}672 \times 10^{-6}}{a^2} V_b^{3/2} \text{amp./sq. m.} \qquad (29)$$

For this value $p = \frac{1}{2}$ and therefore $V_0/V_b = 0{\cdot}25$. The potential on the axis is thus only 25 per cent of the potential of the walls when the virtual cathode forms. For many purposes this beam would have much too wide a spread in velocities, e.g. in velocity modulation tubes. In this case the potential depression should be kept to about 10 per cent and F equal to about 0·41; so the current must be divided by approximately five if this limit is not to be exceeded. Similar results are found for other values of b/a.

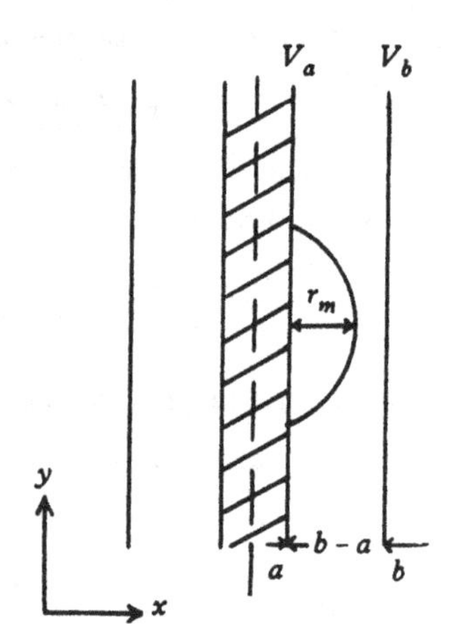

Fig. 6.4. Effect of radial velocities.

Having deduced the expressions for the current in terms of the potential depression, we must next consider the strength of magnetic field required to prevent electrons from reaching the side walls. Fig. 6.4 shows an end-view of the beam. If initial velocities in the x direction are ignored we can make the analysis as follows. The electrons move out from the edge of the beam in an electric field

$$E_x = \left(\frac{V_b - V_a}{b - a}\right).$$

The magnetic induction is B. Eqns. (6) of Chapter 5 can be slightly rearranged to give

$$\left.\begin{aligned} x &= -\frac{E_x}{\omega B}(\cos\omega\tau - 1), \\ y &= -\frac{E_y}{\omega B}(\sin\omega\tau - \omega\tau), \\ z &= u_z\tau. \end{aligned}\right\} \tag{30}$$

The maximum distance the electron can reach is then

$$r_m = \frac{2E_x}{\omega B} = \frac{2(V_b - V_a)}{(b-a)e/mB^2}. \tag{31}$$

If no current flows to the side walls

$$r_m < (b-a),$$

therefore
$$H_{\text{lim.}} = \sqrt{\left(\frac{2m}{e}\right)}\frac{(V_b - V_a)^{\frac{1}{2}}}{\mu_0(b-a)}$$

$$= 3{\cdot}36\times 10^{-2}\frac{V_b^{\frac{1}{2}}}{b}\frac{(1 - V_a/V_b)^{\frac{1}{2}}}{(1 - a/b)}. \tag{32}$$

Eqn. (32) shows that infinite field is required to prevent flow of electrons to the walls when $a = b$. If the beam is initially divergent similar expressions can be derived. If the angle of divergence is γ, i.e. $(u_x)_0 = u_0 \sin\gamma$,

$$H = H_{\text{lim.}}\left(1 + \frac{\sin^2\gamma}{(V_b/V_a - 1)}\right)^{\frac{1}{2}}, \tag{33}$$

where $H_{\text{lim.}}$ is given by eqn. (32).

As an example of the use of these formulæ let us calculate the maximum current that can be carried by a beam 1 mm. wide in a 2-mm. tunnel at 100 V., the tunnel height being 1 cm. and the allowed potential depression 10 per cent.

For this case $F = 0{\cdot}28$.

The total current in the beam is given by eqn. (28) multiplied by $2ah$, i.e.

$$I = 4{\cdot}672\times 10^{-6}\frac{V_b^{\frac{3}{2}}}{b}hF.$$

But $h = 10^{-2}$ m., $b = 10^{-3}$, $V_b^{\frac{3}{2}} = 10^3$, therefore

$$I = 13{\cdot}1 \text{ mA}.$$

$$H_{\text{lim.}} = 212 \text{ oersteds}.$$

It is interesting to note that the current maximum does not depend on the length of the tunnel l, very different from the inverse proportionality to l^2 observed in the electrostatic case. The current density for constant beam width is inversely proportional to tunnel width. This is easily understandable, since the farther away the walls are from the beam the smaller the electron density required to lower the potential by a given amount.

To conclude this section we quote the corresponding results for a circular tunnel. If the beam fills the tunnel

$$\hat{I} = 32{\cdot}5 \times 10^{-6} V_b^{\frac{3}{2}}.$$

If the potential depression is kept to 10 per cent

$$I = 6{\cdot}25 \times 10^{-6} V_b^{\frac{3}{2}}.$$

The derivation of this result will be found in Smith and Hartman† or Beck.‡

Space-charge effects usually limit the current available in valves, particularly in the long-beam devices used at high frequencies. The thermal limits discussed in the last chapter apply mainly to cathode-ray tubes and electron microscopes, although they may be important in very short beams used in extremely high frequency valves.

6.4. Space-charge corrections in electron optics

We next have to consider the effects produced by space charge in electron optical devices. In the last chapter we studied methods of determining electron trajectories in arbitrary fields. These trajectories depended only on the field and therefore only on the potentials applied to the boundaries and not on the beam current. In fact, as we have already said, this is not strictly true even for low-current high-voltage devices, and is grossly untrue for high-current low-voltage devices. In the event of space-charge limitation, however, the trajectories remain the same if all the potentials are multiplied by a constant. In this section we start off by assuming that the field has been determined and the trajectories deduced therefrom, and we

† Smith and Hartman, *J. Appl. Phys.* **11** (1940), 220.
‡ Beck, op. cit.

desire to calculate the trajectories in the presence of space charge. The problem divides itself into two parts:

(1) the determination of the current,

(2) the determination of a corrected potential.

In the interests of simplicity we consider the second problem first, assuming that the current has been determined experimentally and that the corrected trajectories are the interesting feature.

The general problem is to solve

$$\nabla^2 . V = -\rho/\epsilon_0, \tag{34}$$

with $V = V_B$ on the boundaries which are fixed. Let

$$V = V_1 + V_2,$$

where V_1 is the solution of $\nabla^2 . V = 0$, with the boundary condition $V = V_B$, i.e. V_1 is the potential in the absence of space charge. Then eqn. (34) reduces to

$$\nabla^2 . V_2 = -\rho/\epsilon_0, \tag{35}$$

with $V_2 = 0$ on the boundaries. This shows us that in general the correcting potential will be that due to the beam plus the effect of the charges induced in the electrodes. Physically this is equivalent to saying that the potential due to the beam would extend to infinity if the boundaries were absent. When they are present, positive charges are induced on them which reduce the potential outside the boundaries to zero. Eqn. (35) can be solved by approximation methods. Consider the electron gun sketched in fig. 6.5 *a*. The full lines show the boundaries of the beam determined by ray plotting, the dotted lines being the equipotentials. The beam is divided up into a large number of thin disks, in each of which the space charge and velocity may be considered constant and equal to the values evaluated at the centre of gravity of the disk. This is easily done, as I is given and the velocity at any point is taken from the plot. Fig. 6.5 *b* shows one of the disks on a large scale. The potential due to such a disk is well known.†

† Maloff and Epstein, *Electrical Optics and the Cathode-ray Tube*, McGraw Hill, 1938, p. 140.

The axial potential outside the disk is

$$V_2(z,0) = \frac{\rho}{4\epsilon_0}\left[(z+l)\sqrt{\{(z+l)^2+a^2\}} - (2-l)\sqrt{\{(z-l)^2+a^2\}} - 4lz + a^2 \ln\left(\frac{\sqrt{\{(z+l)^2+a^2\}}+(z+l)}{\sqrt{\{(z+l)^2+a^2\}}+(z-l)}\right)\right], \quad (36)$$

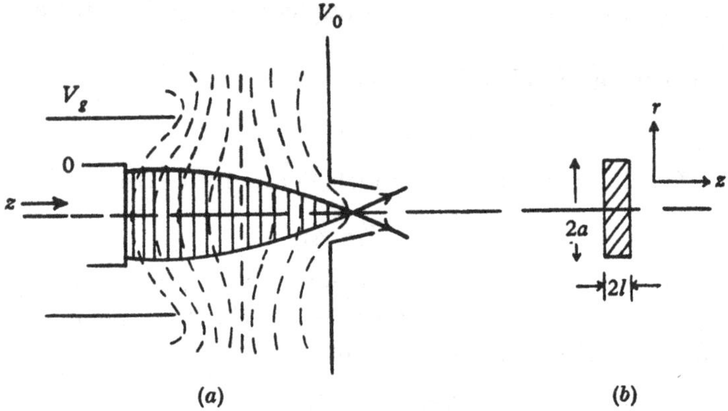

Fig. 6.5. The effects of space charge can be approximated by representing the beam by a series of disks of charge.

while inside the disk

$$V_2(z,0) = \frac{\rho}{4\epsilon_0}\left[(z+l)\sqrt{\{(z+l)^2+a^2\}} - (z-l)\sqrt{\{(z-l)^2+a^2\}} - 2(z^2+l^2) + a^2 \ln\left(\sqrt{\{(z+l)^2+a^2\}}+(z+l)\right) + a^2 \ln\left(\sqrt{\{(z+l)^2+a^2\}}-(z-l)\right) - 2a^2 \ln a.\right]. \quad (37)$$

When the potential on the axis is known, the potential at all other points outside the disk can be determined from the expansion given in eqn. (12) of Chapter 5. The off-axis potential inside the disk is given by

$$V(z,r) = V(z,0) + \tfrac{1}{4}[\rho/\epsilon_0 - V''(z,0)]\,r^2 + O(r^4), \text{ etc.}$$

Using these results the potential due to each disk at any point inside the electrodes can be determined. The total potential is then the sum of the potentials due to each elementary disk. Fig. 6.6 shows the sort of result which might be obtained for the axial potential ΣV. The potentials V_c and V_a are compensated by line charges which produce mirror-image potentials V_{ci} and V_{ai}. The potential due to space charge is then $\Sigma V - (V_{ci} + V_{ai})$, which is naturally zero at the boundaries. This potential can be evaluated for several current values and the

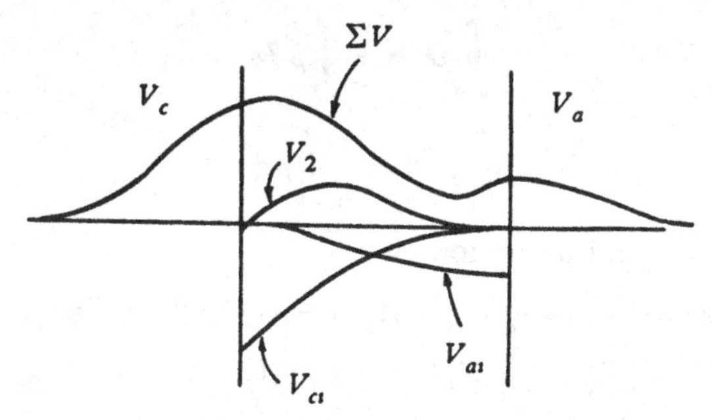

Fig. 6.6. Results of approximate space-charge correction.

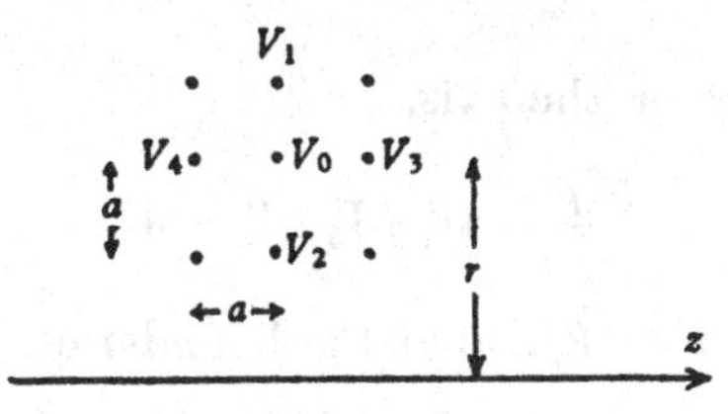

Fig. 6.7. Notation for a relaxation net used for space charge correction.

results used to correct the charge-free potential plot. The electron trajectories are then replotted, and if necessary the process can be repeated.

It will be seen that this is a very laborious business. Charles† discusses a technique which lightens the labour somewhat for cases where the initial beam is not very convergent. A better method is to apply the relaxation technique described briefly in Chapter 4. This has been described by Preddy in unpublished work. The analysis makes use of eqns. (34) and (35). A net of varying mesh size is used with fine mesh inside the beam and coarse mesh outside where the potential variation is nearly linear. In fig. 6.7 we indicate a square mesh of side a. V_0

† Charles, *Ann. Radioélect.* **4** (1949), 33.

corresponds to the potential of a ring element of charge width a lying between $(r+\frac{1}{2}a)$ and $(r-\frac{1}{2}a)$. (The mesh is small enough for ρ to be constant throughout the area.)

The element of charge is then $2\pi a^2 r\rho$ for an element off the axis and $\frac{1}{4}\pi a^3\rho$ for an element on the axis.

Gauss's theorem states that

$$\oint_s \mathbf{D}.\mathbf{n} = \int_v \rho\, dv,$$

or

$$\oint_s \mathbf{E}.\mathbf{n} = \frac{1}{\epsilon_0}\int_v \rho\, dv.$$

The total normal induction is

$$2\pi r[V_1+V_2+V_3+V_4-4V_0]+\pi a(V_1-V_2) \quad \text{off axis,}$$

$$(\pi a/4)[V_3+V_4-2V_0] \qquad +\pi a(V_1-V_0) \quad \text{on axis;}$$

therefore $$\frac{2a^2 r\rho}{\epsilon_0} = 2r[V_1+V_2+V_3+V_4-4V_0]+a(V_1-V_2), \tag{38}$$

and, for elements on the axis,

$$\frac{a^2\rho}{\epsilon_0} = 4V_1+V_3+V_4-6V_0, \tag{39}$$

$$\text{with } V_0 = 0 \text{ at the boundaries.} \tag{40}$$

These three equations suffice for a relaxation calculation of the space-charge potential. Once the correction has been determined the original plot can be adjusted as before, the trajectories redetermined and the process repeated if necessary. It is fairly rapid, since the space-charge potential is only a correction factor and need not be determined to a very high degree of accuracy.

The question of determining the current density given an equipotential plot can be done either by approximation or by the method due to Charles. In the approximation method the cathode surface is divided up into a number of regions over which simple current relations might be expected to apply; e.g. a disk with one or more surrounding rings over which the field is uniform so that plane parallel diode formulæ apply, or the formulæ for spherical caps.

The method of Charles consists in determining the space-charge potential V_2 for unit current and then calculating a multiplying factor so that

$$\left|\frac{dV_1}{dz}\right|_0 = \left|\frac{dV_2}{dz}\right|_0,$$

i.e. there is zero field at the cathode surface, which is the equilibrium condition. This is done for several values of r and the total beam current determined by integration.

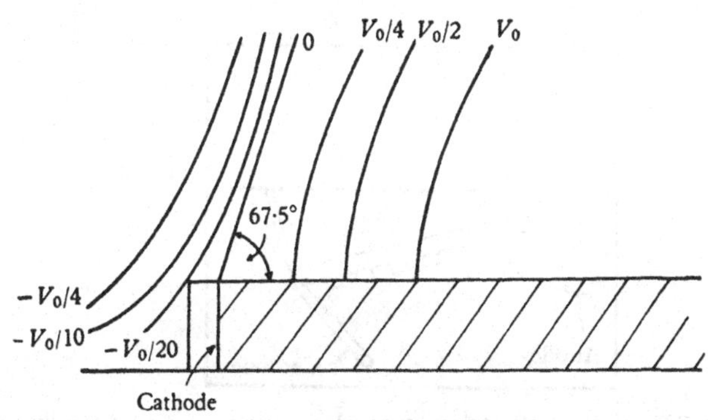

Fig. 6.8. Rectilinear flow.

6.5. Pierce guns and general trajectories in space-charge fields

By tackling the problem discussed in the last section from the opposite end Pierce† was able to prepare designs for a whole species of guns with extremely valuable properties. Instead of considering what trajectories result from a given field he asked the question what are the requirements to be satisfied by the field given that the trajectories are rectilinear?

Fig. 6.8 shows a half-section through a cathode which produces a rectangular beam, together with the equipotentials required to produce it. The equipotentials satisfy the following conditions:

(1) Inside the beam the potential obeys Child's law.

(2) The potential at the edge of the beam is continuous.

† J. R. Pierce, *J. Appl. Phys.* **11**, (1940), 548.

(3) There is no field in the direction normal to the axis at the edge of the beam, as if such a field were present the beam would not be rectilinear.

In the case of the parallel beam in a cylindrical field the potential problem can be solved by analytical means, but in general this is not possible. In particular, the solution for the case of spherical electrodes producing a point focus, which is of great practical importance, is not soluble, and another method must be found to deal with this problem. Pierce

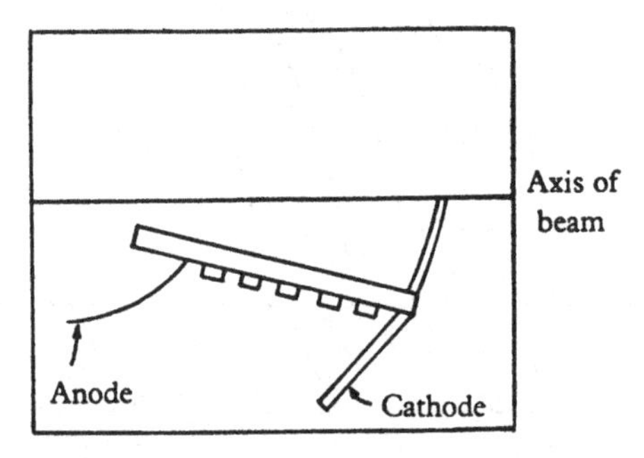

Fig. 6.9. Method of setting up an electrolytic trough for rectilinear flow gun.

realized that this could be done by using an electrolytic trough with insulating strips inserted to delineate the beam. Since no current enters or leaves the electrolyte at the insulator, there can be no normal potential gradient, and the electrodes can be adjusted until the potential variation along the strip follows the desired space-charge law. In practice the tilted trough, described in Chapter 4, is most suited to this work. Fig. 6.9 shows a plan view of the trough arranged for this work. The insulating strip representing the desired beam profile has several contacts of small area on it, each contact being taken to a valve voltmeter. The water-line represents the axis of the conical beam as usual, but in this case we are only concerned with fields outside the insulating strip. The cathode and anode are strips of metal curved to reasonable shapes to begin with. The adjustment consists in bending the anode and cathode until the valve voltmeters read the desired potentials. A plot of the field outside the strip can then be made and the remaining equipotentials determined.

In the spherical case the potential variation can be obtained from the formula for current flow between concentric spheres by Langmuir and Blodgett.† This is

$$I = 29{\cdot}36 \times \frac{10^{-6}\,V^{3/2}}{\alpha^2}\,(\sin^2 \tfrac{1}{2}\phi)\text{ amp.,} \tag{41}$$

where $\tfrac{1}{2}\phi$ = semi-angle of cone in which current flows,

α = a function of r_c/r_a and

$$V = \frac{1051\,I^{2/3}\,\alpha^{4/3}}{(\sin \tfrac{1}{2}\phi)^{2/3}}. \tag{42}$$

Values of α^2 are given in the tables at the end of the book. Electrode shapes for beams of several values of convergence have been determined in this way by Spangenberg‡ and his collaborators.

When it is desired to work with very high values of current, necessitating close spacings between anode and cathode, the hole in the anode through which the beam passes has a disturbing effect on the performance. Fig. 6.10 shows an example. Owing to the hole in A the field on the cathode at its centre is not as strong as it would be if there were no hole. This effect is important for close spacings as the theoretical current may be reduced by a large factor, 3 or more. The effect may be overcome by putting an open-mesh grid over the hole (if the power dissipation can be kept sufficiently low that the grid does not burn out), or by modifying the electrode shapes somewhat to allow for the hole. The hole in the anode also introduces another effect of practical importance as it acts as a diverging lens (see Chapter 5). This is indicated by the dotted lines in fig. 6.10. The beam initially converges along the solid lines, but inside the anode hole the diverging lens causes it to follow the dotted lines. This must be allowed for in designing the gun to

† Langmuir and Blodgett, *Phys. Rev.* **24** (1924), 49.

‡ Spangenberg *et al.*, *Electr. Commun.* **24** (1947), 101; Spangenberg, *Vacuum Tubes*, McGraw Hill, 1948.

give the maximum current through a tunnel, a topic which we now discuss.

Eqn. (17) gives the maximum current which can be passed through the tunnel. The tangent to the beam profile at the point of entry must pass through the centre of the tunnel so that, after the beam has been diverged by the anode aperture it must be travelling along the tangent,† i.e.

$$\tan\gamma = d/l. \tag{43}$$

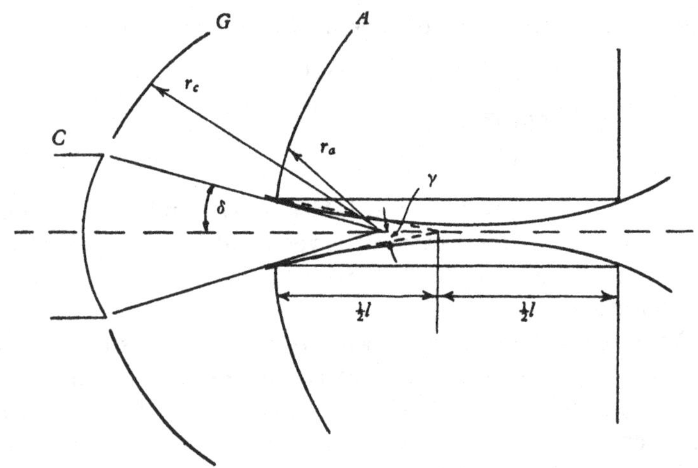

Fig. 6.10. Construction of an 'optimum' gun.

The lens at the aperture has the focal length

$$f = \frac{4V_0}{E_a},$$

where E_a = field strength at the anode, i.e. $\partial V/\partial r$ evaluated at the anode. Since α^2 is tabulated it is convenient to work in terms of this quantity instead of in terms of α:

$$\begin{aligned} E_a &= -\frac{dV}{d(\alpha^2)}\frac{d(\alpha^2)}{dr_a}\bigg|_{r_a} \\ &= \frac{dV}{d(\alpha^2)}\frac{r_c}{r_a^2}\frac{d(\alpha^2)}{d(r_c/r_a)}\bigg|_{r_a} \\ &= \frac{2}{3}\frac{V_0}{\alpha^2}\frac{r_c}{r_a^2}\bigg|\frac{d(\alpha^2)}{d(r_c/r_a)}\bigg|_{r_a}. \end{aligned} \tag{44}$$

† Huber, *Ann. Radioélect.* **4** (1949), 26.

The lens laws give

$$\frac{\sin\delta}{\sin\gamma} = \frac{1}{1-r_a/f'} \tag{45}$$

and

$$\sin\gamma \doteq \tan\gamma.$$

Then, using (17), (43) and (44),

$$\sin\delta = \frac{161\sqrt{(I/V)^{\frac{3}{2}}}}{1-\dfrac{1}{6\alpha^2}\dfrac{r_c}{r_a}\left|\dfrac{d(\alpha^2)}{d(r_c/r_a)}\right|_{r_a}}. \tag{46}$$

From (41)

$$\sin^2\tfrac{1}{2}\delta = \frac{I\alpha^2}{29{\cdot}36\times10^{-6}\,V^{\frac{3}{2}}}. \tag{47}$$

When we can put $\sin\delta \doteq \delta$, this gives

$$\alpha^2\left[1-\frac{6}{\alpha^2}\frac{r_c}{r_a}\left|\frac{d(\alpha^2)}{d(r_c/r_a)}\right|_{r_a}\right]^2 = 0{\cdot}19. \tag{48}$$

Eqn. (48) can be solved graphically by plotting the left-hand side as a function of r_c/r_a.

The solution is $r_c/r_a = 2{\cdot}15$.

Since we have already approximated by using $\tan\gamma \doteq \sin\gamma$, the replacement of $\sin\delta$ by δ does not restrict the analysis any further.

Solving for δ we get

$$\delta = 2{\cdot}1\times10^4\sqrt{\frac{I}{V^{\frac{3}{2}}}}\ \text{degrees}. \tag{49}$$

We have now determined all the values necessary for designing a gun to pass the maximum current through the given tunnel, and all that remains to be done is to determine the electrode shapes in the electrolytic trough which give the value of δ prescribed by eqn. (49).

It is interesting to note that r_c/r_a is practically constant. This means that the ratio of cathode area to anode entrance hole area is $(2{\cdot}15)^2 = 4{\cdot}62$. We stated in § 6.3 that the minimum beam diameter in the tunnel was $d/2{\cdot}4$. Thus the ratio of cathode area to throat area is $4{\cdot}62\times(2{\cdot}4)^2 \doteq 26$, and the current density at the throat is 26 times the current density at the cathode. We shall later see that current density is an important limiting factor in u.h.f. tubes, and it will be found that the possibility of obtaining high-beam densities is one of the important practical reasons for the use of electrostatic focusing.

It will be obvious that similar results can be obtained for other geometrical arrangements. The calculation has been performed for the cylindrical case by Klemperer.†

Of course, the rectilinear trajectory is not the only one which will produce a focused beam of electrons. Bunemann, Condon and Latter‡ have considered this question. After generalizing the results of Pierce they consider the properties of fields which produce hyperbolic trajectories. Here we shall only indicate

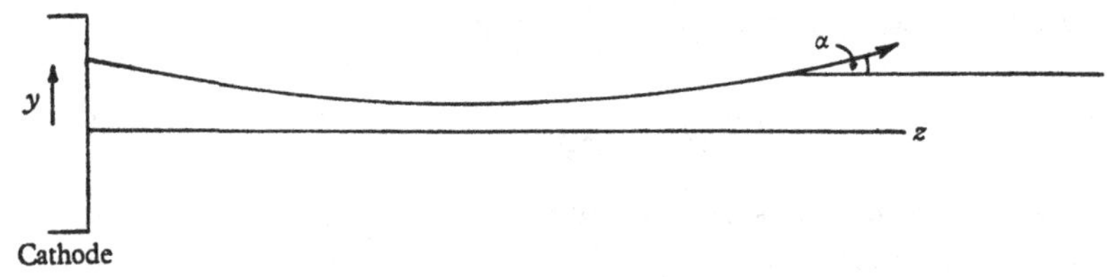

Fig. 6.11. Electron paths assumed by Condon *et al.*

the general nature of their analysis which is of very considerable interest.

Fig. 6.11 shows a cathode producing a beam of electrons whose motion is such that $(dy/dz)^2 \ll 1$ for all points. $Y(z)$ is the trajectory of the outer electron, the beam is symmetrical about the centre-line and the trajectories inside the beam are of the form $kY(z)$. $k = 0$ is a particular trajectory and the current density is uniform.

$V(y, z)$ is the potential when the space charge is present. $V(0, z) = U_z$. Then

$$\left.\begin{aligned} \frac{d^2z}{dt^2} &= \frac{e}{m}\frac{\partial V}{\partial z}, \\ \frac{d^2y}{dt^2} &= \frac{e}{m}\frac{\partial V}{\partial y}. \end{aligned}\right\} \tag{50}$$

† Klemperer, loc. cit.

‡ Bunemann *et al.*, U.S. Atomic Energy Commission Papers, MDDC 517 and 586.

The system is symmetrical, so $\partial V/\partial y = 0$ at $y = 0$. Since the spreading is assumed small, we can put

$$V(y,z) = U(z) + \tfrac{1}{2}y^2 W(z),$$

therefore $$\frac{d^2 z}{dt^2} = \frac{e}{m}\frac{\partial U_z}{\partial z} \quad \text{for the axial ray,}$$

or $$\left(\frac{dz}{dt}\right)^2 = 2\frac{e}{m}U_z. \tag{51}$$

It is assumed that eqn. (51) holds for the paraxial rays. Again,

$$\frac{d^2 y}{dt^2} = \frac{e}{m} y\, W(z); \tag{52}$$

eliminating t between eqns. (51) and (52), we get

$$\sqrt{U}\frac{d}{dz}\left(\sqrt{U}\frac{dy}{dz}\right) = \frac{y}{2}W(z). \tag{53}$$

Now let the beam current be I per unit length perpendicular to the paper, then

$$\rho = -\frac{I}{2Y(z)\sqrt{[2(e/m)U]}},$$

and Poisson's equation becomes

$$\frac{\partial^2 V}{\partial y^2} + \frac{\partial^2 V}{\partial z^2} = \frac{I}{2\epsilon_0 Y(z)\sqrt{[2(e/m)U]}} \tag{54}$$

But with our assumed potential

$$\frac{\partial^2 V}{\partial z^2} = U''(z), \quad \frac{\partial^2 V}{\partial y^2} = W(z).$$

Therefore eqn. (54) becomes

$$U''(z) + W(z) = \frac{I}{2\epsilon_0 Y(z)\sqrt{[2(e/m)U]}}. \tag{55}$$

Eqn. (55) can be used to eliminate $W(z)$ from (53) to obtain

$$\sqrt{U}\frac{d}{dz}\left(\sqrt{U}\frac{dy}{dz}\right) + \tfrac{1}{2}U''(z)\,y = \frac{I\,y}{4\epsilon_0\sqrt{[2(e/m)U\,Y(z)]}}. \tag{56}$$

Now, by hypothesis $y = Y(z)$ is a solution of eqn. (56) if this is regarded as a differential equation for $Y(z)$. Hence if we insert the desired form of trajectory $y = Y(z)$ we obtain a differential equation which must be satisfied by the potential along the axis of the beam. The result is

$$Y(z)\,U''(z) + Y'(z)\,U'(z) + 2Y''(z)\,U(z) = \frac{I}{2\epsilon_0\sqrt{[2(e/m)U]}}, \quad (57)$$

together with the boundary conditions $U(z) = 0$, $U'(z) = 0$ at $z = 0$. When $Y(z) = 1$, rectilinear flow, eqn. (57) reduces to

$$\frac{d^2V}{dz^2} = \frac{I}{2\epsilon_0\sqrt{[2(e/m)V]}},$$

which has already been discussed.

Eqn. (57) is an equation which has obvious analogies with the paraxial ray equations of Chapter 5.

The original papers referred to carry the discussion much further and give the necessary special functions for calculating the axial and boundary potentials for the hyperbolic family of trajectories. The question of setting up electrodes to give the required potential distribution is then discussed, but we cannot include the rather lengthy development here.

We have now carried the discussion of space-charge effects about as far as can usefully be done without further consideration of the effect of initial velocities. In the next section we reconsider the topics of §§ 6.1 and 6.2 including the effect of initial velocities.

6.6. The parallel plane electrode system with initial velocities

In § 6.1 we integrated the space-charge equation with the boundary condition $dV/dz = 0$ at the cathode surface. In Chapter 1, however, we found that the electrons from a thermionic emitter are emitted with initial velocities, the distribution in velocity being Maxwellian, i.e. the number of electrons emitted per unit time per unit area with normal velocities between u_1 and $u_1 + du_1$ is

$$dN_u = N\frac{mu_1}{kT}\exp[-mu_1^2/2kT]\,du_1. \quad (58)$$

where N = total number of electrons emitted $= J_0/e$.

Eqn. (58) is of the form shown in fig. 6.12, i.e. all the electrons have some initial velocity. Therefore if dV/dz were zero on the cathode surface all the electrons would just be able to leave the cathode and the saturation emission would flow. We know that the space-charge limited emission is very much less than

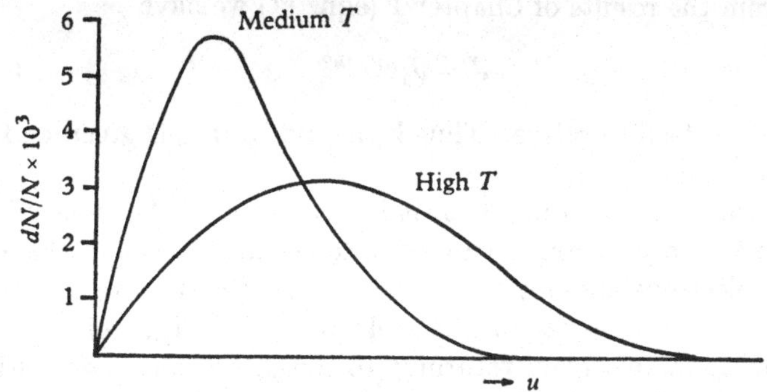

Fig. 6.12. Velocity distribution of emitted electrons.

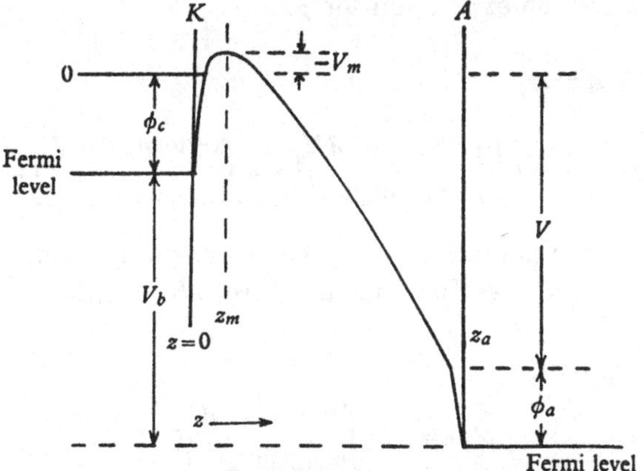

Fig. 6.13. Potential energy distribution in a diode.

the saturated emission and so we see that the true state of affairs is that there is a potential minimum just outside the cathode. All the electrons are emitted but only those with initial velocities sufficient to overcome the potential minimum can reach the anode. As the anode voltage increases, the depth of the potential minimum decreases and it moves nearer to the cathode until $dV/dz = 0$ at the cathode surface and the saturation emission is drawn. Fig. 6.13 shows the potential diagram.

ϕ_c is the cathode work function and ϕ_a the anode work function. V_m is the difference between the surface potential of the cathode and the space-charge minimum. The battery voltage V_b measures the difference in potential between the Fermi levels of the anode and cathode.

From the results of Chapter 1 (eqn. 11) we have

$$J = J_0 e^{V_m/kT}, \tag{59}$$

V_m being itself negative. Thus V_m is known if J_0 is given and J can be determined.

Turning again to fig. 6.13, it is clear that to the right of z_m there are only electrons travelling to the anode, but to the left of z_m electrons are moving in both directions. Also, to the right of z_m all electrons have initial velocities above V_m, while to the left of z_m the electrons returning to the cathode have velocities below V_m. These conclusions fix the limits of integration and we can write the expression for ρ.

6.6.1. $0 \leqslant z \leqslant z_m$.

$$\rho(z) = -e\left[\int_{[-2(e/m)V_m]^{\frac{1}{2}}}^{\infty} \frac{dN_u}{u} + 2\int_{[2(e/m)V]^{\frac{1}{2}}}^{[-2(e/m)V_m]^{\frac{1}{2}}} \frac{dN_u}{u}\right], \tag{60}$$

where V is the potential at z. The velocity u is given in terms of u_1 by $u^2 = u_1^2 + 2eV/m$ and therefore $udu = u_1 du_1$.

6.6.2. $z_m \leqslant z \leqslant z_a$.

$$\rho(z) = -e\int_{[-2(e/m)V_m]^{\frac{1}{2}}}^{\infty} \frac{dN_u}{u}. \tag{61}$$

Poisson's equation is

$$\frac{d^2 V}{dz^2} = -\frac{\rho(z)}{\epsilon_0} \quad \text{as before,}$$

and we can integrate once, between limits, to get

$$\left(\frac{dV}{dz}\right)^2 - \left(\frac{dV}{dz}\right)^2_{z=z_m} = -\frac{2}{\epsilon_0}\int_{V_m}^{V} \rho(z)\,dV.$$

Now $dV/dz = 0$ at $z = z_m$ and $V = V_m$ at $z = z_m$, therefore

$$\left(\frac{dV}{dz}\right)^2 = -\frac{2}{\epsilon_0}\int_{V_m}^{V} \rho(z)\,dV. \tag{62}$$

To carry out the integration Langmuir† introduced the notations

$$\eta = \frac{e(V-V_m)}{kT} = \frac{11{,}605(V-V_m)}{T},$$

$$2L = \frac{(2m\pi)^{\frac{1}{4}}}{(2kT)^{\frac{3}{4}}}\left(\frac{eJ}{\epsilon_0}\right)^{\frac{1}{2}} = 9{\cdot}186\times 10^5 T^{-\frac{3}{4}} J^{\frac{1}{2}}\,\text{metre}^{-1},$$

$$\xi = 2L(z-z_m).$$

Then for $0 \leqslant z \leqslant z_m$

$$\left(\frac{dV}{dz}\right)^2 = \left(\frac{2LkT}{e}\right)^2 \left[e^{\eta} - 1 + e^{\eta}\,\mathrm{erf}\,(\eta^{\frac{1}{2}}) - 2\left(\frac{\eta}{\pi}\right)^{\frac{1}{2}}\right], \tag{63}$$

and for $z_m \leqslant z \leqslant z_0$

$$\left(\frac{dV}{dz}\right)^2 = \left(\frac{2LkT}{e}\right)^2 \left[e^{\eta} - 1 - e^{\eta}\,\mathrm{erf}\,(\eta^{\frac{1}{2}}) + 2\left(\frac{\eta}{\pi}\right)^{\frac{1}{2}}\right], \tag{64}$$

where $\mathrm{erf}\,(\eta^{\frac{1}{2}}) = \dfrac{2}{\sqrt{\pi}}\displaystyle\int_0^{\eta^{\frac{1}{2}}} e^{-u^2}\,du.$

This is the ordinary error function of probability theory which we have used before. Inserting the expression for ξ and integrating we get

$$\xi = \int_0^{\eta} \frac{d\eta}{[e^{\eta} - 1 \pm e^{\eta}\,\mathrm{erf}\,(\eta^{\frac{1}{2}}) \mp 2(\eta/\pi)^{\frac{1}{2}}]}. \tag{65}$$

Here the top signs apply to the left of z_m and the bottom to the right. Eqn. (65) is not an analytic function, but tables of values have been prepared, the latest and most complete being those of Kleynen.‡

For negative ξ and when η is greater than 3

$$\xi = -2{\cdot}55389 + \sqrt{2}\,e^{-\frac{1}{2}\eta} - 0{\cdot}0123\,e^{-\eta} + \frac{1}{3\sqrt{2}}\left(\sqrt{\frac{\eta}{\pi}} + 1\right)e^{-\frac{3}{2}\eta}. \tag{66}$$

† Langmuir, *Phys. Rev.* **21** (1923), 419.

‡ Kleynen, *Philips Res. Rep.* **1** (1946), 81. Van der Ziel, *Philips Res. Rep.* **1** (1946), 91.

For positive ξ, and when η is greater than 8

$$\xi = 1{\cdot}25520\eta^{\frac{3}{4}} + 1{\cdot}6685\eta^{\frac{1}{4}} - 0{\cdot}50880 - 0{\cdot}1677\eta^{-\frac{1}{4}} + 0{\cdot}1441\eta^{-\frac{3}{4}}$$
$$- 0{\cdot}0145\eta^{-\frac{5}{4}} - 0{\cdot}069\eta^{-\frac{7}{4}} + 0{\cdot}036\eta^{-\frac{9}{4}} + 0{\cdot}083\eta^{-\frac{11}{4}}. \quad (67)$$

Using only the first term of (67) we obtain from the definitions of η and ξ

$$J = \frac{2{\cdot}336 \times 10^{-6}\,(V - V_m)^{\frac{3}{2}}}{(z_a - z_m)^2}\ \text{amp./sq. m.} \quad (68)$$

Eqn. (68) is very similar to eqn. (6). Using the first two terms of eqn. (67)

$$J = \frac{2{\cdot}336 \times 10^{-6}\,(V - V_m)^{\frac{3}{2}}}{(z_a - z_m)^2}\left[1 + \frac{2{\cdot}66}{\sqrt{\eta}}\right]. \quad (69)$$

In the last two expressions V is the effective voltage on the tube, i.e. $V = V_b + (\phi_c - \phi_a)$. Thus V is the algebraic sum of the battery voltage and the contact difference of potential. Since V_m is negative and z_m positive the accurate formulæ lead to higher currents than are given by eqn. (6). This is, of course, to be expected.

The method of using the results is the following. Suppose first that the saturation current, cathode temperature, anode voltage and current are given and it is required to calculate z_m and the separation. The steps are:

(1) Determine V_m from eqn. (59).

(2) Calculate η at the cathode, i.e. $\eta_c = 11{,}605 V_m/\mathrm{T}$.

(3) Look up ξ_c in the tables, opposite η_c.

(4) Calculate $2L$ from the known current.

(5) Calculate $z - z_m$ from $\xi_c = 2L(z_c - z_m) = 2Lz_m$.

(6) Repeat steps (2), (3) and (5) for the anode, or repeat steps (2) and (3). The spacing can then be calculated from $\xi_c + \xi_a = 2Ld$.

In most cases it will be accurate enough to apply eqn. (69) instead of using the tables.

When J_0, T, V and $z_c - z_a = d$ are given and J is required, the best procedure is to use eqn. (68) as a trial solution from which the correct current can be computed by successive approximation.†

To summarize this work we collect the formulæ:

$$V_m = -\frac{T}{11{,}605}\ln\frac{J_0}{J}, \quad \text{therefore} \quad \eta_c = \ln\frac{J_0}{J}. \tag{70}$$

$$V_a = \frac{T}{11{,}605}(\eta_a - \eta_c). \tag{71}$$

$$\xi_c = 2L.z_m, \quad \text{therefore} \quad z_m = \frac{\xi_c T^{\frac{3}{4}}}{9{\cdot}186 \times 10^5 J^{\frac{1}{2}}} \tag{72}$$

$$\xi_a = 2L(z_a - z_m) = 2Lz_a - \xi_c$$

$$= 9{\cdot}186 \times 10^5 T^{-\frac{3}{4}} J^{\frac{1}{2}} z_a - \xi_c. \tag{73}$$

As an example, consider a diode with a cathode area of 0·2 sq. cm., at temperature 1100° K., saturation emission 5 amp./sq. cm. and anode current 1·0 mA. The spacing equals 0·1 mm. What is the anode voltage and the distance to the minimum?

$$V_m = -\frac{1100}{11{,}605}\ln 10^3 = -0{\cdot}83\,\text{V}.$$

$$\eta_c = 6{\cdot}909, \quad \text{therefore } \xi_c = 2{\cdot}509 \quad \text{(from tables)}.$$

Then $z_m = \dfrac{2{\cdot}509\,(1100)^{\frac{3}{4}}}{9{\cdot}186 \times 10^5 \times (\cdot 005)^{\frac{1}{2}}} = 0{\cdot}068$ mm. 1st result.

Therefore $z_a = 0{\cdot}01 - 0{\cdot}0068 = 0{\cdot}0032$ cm.

$$\xi_a = \frac{3{\cdot}2 \times 9{\cdot}186 \times 10^3 \times (\cdot 005)^{\frac{1}{2}}}{(1100)^{\frac{3}{4}}} - 2{\cdot}509$$

$$= 8{\cdot}29.$$

Therefore $\eta_a = 8{\cdot}137$,

therefore $V_a = \dfrac{1100}{11{,}605} \times 1{\cdot}228 = 0{\cdot}116$ V. 2nd result.

† When using oxide-coated cathodes J is often $\ll J_0$. In this case we can assume that ξ_c takes its asymptotic value as $J_0/J \to \infty$. This value is $\xi_c = 2{\cdot}544$. Since η_a is given we can find ξ_a. Then L can be calculated from $L = \dfrac{\xi_c + \xi_a}{2d}$. If the resulting value of J is so large that the new ξ_c is different from 2·544 by an amount appreciable in comparison with ξ_c and ξ_a, the calculation can be repeated.

It is instructive to compare the result obtained by neglecting initial velocities.

Here
$$I = \frac{0{\cdot}2\times 2{\cdot}336\times 10^{-6}\times(\cdot 116)^{\frac{3}{2}}}{(0{\cdot}01)^2}$$
$$= 0{\cdot}187\ \text{mA}.$$

This is much less than the correct value of 1·0 mA. The figures in the example might apply to a u.h.f. diode with an oxide cathode. In the case of diodes with bigger spacings and smaller anode-current/saturation-current ratios the difference will be smaller, but considerable errors can result from the use of the simplified formulæ.

6.7. Cylindrical system with initial velocities

The exact solution to the problem for cylindrical systems seems never to have been found. Langmuir† gives the following expression:
$$\frac{I}{l} = \frac{14{\cdot}66\times 10^{-6}}{\beta^2 r}\left[V - V_m + \frac{V_0}{4}\left(\ln\frac{V}{\lambda V_0}\right)^2\right]^{\frac{3}{2}}. \tag{74}$$

Here V = effective voltage,

V_0 = average radial kinetic energy of electrons in volts
$$= \frac{3}{2}\frac{kT}{e} = \frac{T}{7733}\,\text{V.},$$

r = anode radius,

λ = a numeric between 1 and 2.

In discussing this equation Langmuir gives reasons for stating that the correction due to initial velocities is only 20–25 per cent as great for cylinders as it is for planes. Another treatment, not leading to closed formulæ, is given by Wheatcroft.‡

6.8. The diode

We have already discussed the separate features of diode characteristics in Chapter 1 and in the preceding sections. Here we collect the necessary results together, in view of the basic importance of the diode, and also derive some new results

† Langmuir, loc. cit.

‡ Wheatcroft, *J. Instn. Elect. Engrs.* **86** (1940), 473.

on the effect of emission variations, values of the conductance, etc.

Fig. 6.14 shows the I_a, V_a characteristic of a diode plotted on linear and semi-log scales. At the most negative end of the characteristic the shape is an exponential curve given by eqn.

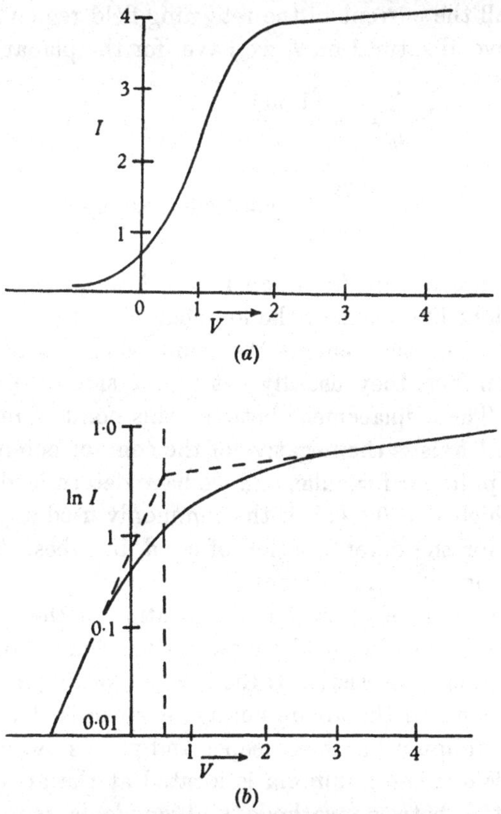

Fig. 6.14. I_a, V_a characteristics for a diode.

(12) of Chapter 1, or the similar expression for the cylindrical diode. This part of the characteristic gives a straight line on the $\log I \sim V$ plot. As the current increases space-charge limitation sets in and the characteristic becomes an approximate 3/2 power law, as determined from the formulæ of the last two sections. Finally, the current reaches the saturation value and, in the case of pure metals, increases beyond it, according to

$J = J_0 \exp\left(\frac{e^{\frac{3}{2}}E^{\frac{1}{2}}}{kT}\right)$, i.e. eqn. (15), Chapter 1. In the case of oxide cathodes it is found empirically that $\ln I \propto E$ or $E^{\frac{1}{2}}$, either expression being a reasonably good fit at low temperatures. It is usually simpler to take $\ln I \propto E$.

If we call the current in the retarding field region I_r and the current beyond saturation I_s we have, for the planar diode,

$$\ln \frac{I_r}{I_0} = -\frac{11{,}605 V}{T},$$

$$\ln \frac{I_s}{I_0} = \frac{BV}{s} \quad (s = \text{diode spacing}),$$

where B is a constant, i.e. when $V = 0$, $I_r = I_s = I_0$. Therefore, if the straight-line parts of the log characteristic are produced until they meet they cut at the point whose coordinates are $(0, \ln I_0)$. In fact, they usually cut to one side or other of the $\ln I$ axis. The displacement between this point of intersection and the $\ln I$ axis is the negative of the contact potential, since $V = V_b + \text{c.p.}$ in our formulæ, and we have determined the value of V_b for which $V = 0$. This is the commonly used experimental technique for the determination of c.p.d. in tubes. The result holds also for cylindrical diodes.

The next point of interest is the location of the coordinates at which the retarding field characteristic changes over to the space-charge characteristic. If the tube is working in the space-charge régime and the anode voltage is steadily decreased the potential minimum becomes deeper and moves away from the cathode. When the minimum is located at the anode surface the potential between cathode and anode is monotonically decreasing, and this anode potential and current represents the lower boundary of space-charge operation. This condition can be calculated, owing to the fortunate fact that ξ_c tends asymptotically to 2·554 as $\eta_c \to \infty$. For $\eta_c = 10$, ξ_c has already reached 2·544 or within 0·4 per cent of the final value. Now in any practical valve the ratio between I_0 and $\hat{I}_r$ is certain to be 10^3 at least, i.e. $\eta_c \doteqdot 7$; so the error in using $\xi_c = 2{\cdot}554$ to determine $\hat{I}_r$ will be small. Also for this case $z_m = s$.

As $\xi_c = 2L.s$, therefore

$$\hat{I}_r = 0{\cdot}245 \times 10^{-6}\left(\frac{T}{1000}\right)^{\frac{1}{2}}\frac{A}{s^2}\text{ amp.} \tag{75}$$

$\hat{I}_r$ is a little more than 1 per cent greater than the true boundary current for a saturation current/retarding-field current ratio of $10^4 : 1$. For a ratio of $10^3 : 1$ the error is about 3·5 per cent. It is simple to determine the correction, since

$$\frac{\hat{I}_r}{I_0} = \frac{I_r}{I_0}\left(\frac{2{\cdot}554}{\xi_c}\right)^2. \tag{76}$$

ξ_c is correctly the value obtained using I_r and I_0, but it will usually be sufficient to determine $\hat{I}_r$ from eqn. (75) and then use this to obtain ξ_c^1, from which eqn. (76) enables I_r to be determined. If $\hat{I}_r$ and I_r are sufficiently different, a repetition of the process using ξ_c may be necessary. The corresponding anode voltage is given by eqn. (70). Ferris,† in an interesting paper, has built up universal diode characteristics from these relations, together with Langmuir's analysis of the space-charge region.

Unfortunately, it is not possible to locate the point at which the potential minimum reaches the cathode surface by any similar means. The methods which can be used are familiar because they are used for the determination of saturation emission. In addition to producing the $\log I \sim V$ plot to cut the I axis, $I^{\frac{2}{3}}$ is sometimes plotted against V. In the space-charge region this is linear, but when saturation sets in the curve falls below the extrapolated line. The break point is often very poorly defined.

We must now discuss the conductance of the diode. Conductance is defined by the relation $\partial I/\partial V$ and is often spoken of as the slope of the tube.

In the retarding field region therefore

$$\frac{G}{I} = \frac{e}{kT}$$

$$= \frac{11{,}605}{T}\text{ amp./V./amp.} \tag{77}$$

† Ferris, *R.C.A. Rev.* **10** (1949), 134.

But $11,605/T$ is the slope of the retarding field characteristic on the $\log I \sim V$ plot, so the conductance is the slope of the characterisic multiplied by the current at the working point. Moreover, it is obvious by inspection of fig. 6.14 (*b*) that the slope of the retarding field characteristic is greater than the slope of any other part of the characteristic, so that eqn. (77) represents the maximum slope per unit current that can be obtained from a diode or, in fact, from any electronic device depending on space-charge control. The value is $11,605/T$ amp./V./amp. or mA./V./mA., and the only way in which the value can be modified is to use the coldest possible emitting surface.

In the space-charge limited region, differentiating eqn. (69) gives

$$\frac{G}{I} = \tfrac{3}{2}(V - V_m)^{-1}. \tag{78}$$

This shows that the highest relative conductance is obtained with small values of V, i.e. the conductance per unit current steadily decreases from the value given by eqn. (77) as the current and anode voltage increase.

In the saturated current region the conductance is zero to a first approximation. This region is of no practical importance. Ferris† gives a full treatment and universal curves for the conductance.

The last topic we shall discuss in this section is the influence of the anode and cathode work functions and temperature on the characteristics. First consider the retarding field condition shown in fig. 6.14. In Chapter 1 we deduced that

$$I = I_0 \exp\left(-\frac{eV}{kT}\right),$$

and clearly $V = V_b + \phi_a - \phi_c$.

This makes it appear that the characteristic depends on ϕ_c. This is incorrect as the following reasoning shows. Fig. 6.15 shows that to reach the anode an electron must have the total energy $V_b + \phi_a$ eV. at least. Now, if the work function of the

† Ferris, loc. cit.

cathode changes, the number of electrons with energies greater than $V_b + \phi_a$ remains the same, all that happens is that the critical energy for leaving the cathode surface alters. Thus, even if ϕ_c changes, I remains constant. What has happened is that although V has changed to $V \pm \Delta\phi_c$, I_0 has also changed to $I_0 \exp(\pm \Delta\phi_c)$, and therefore I remains constant. I depends only on the negative battery voltage, the anode work function and the temperature.

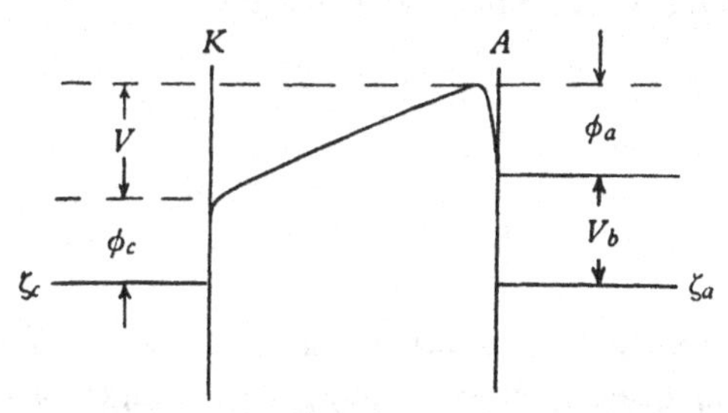

Fig. 6.15. Contact potential difference in a diode.

Now consider the space-charge limited case. Here the voltage which appears is $V - V_m$ and V_m is a negative quantity. From fig. 6.13 $V = V_b + \phi_c - \phi_a$ and we have $V_m = \dfrac{kT}{e} \ln \dfrac{J}{J_0}$.

If ϕ_c changes to $\phi_c - \Delta\phi_c$, V changes to $V - \Delta\phi_c$, J_0 changes to $J_0 \exp\left(\dfrac{e\Delta\phi_c}{kT}\right)$ and V_m changes to $V_m + \Delta\phi_c$, i.e. the potential minimum gets deeper. The algebraic sum $V - V_m$ therefore remains constant, and the current does not change in the approximations given by eqns. (68) and (69). What has been neglected is the effect of variations in J_0 on z_m. It can be proved, by preparing a curve of z_m against J_{anode} with J_0 as parameter, that the change in z_m is very small as long as $J_0 > 10J_a$, a condition that is obeyed in nearly all real valves. The I_a, V_a characteristic is thus independent of ϕ_c even in the space-charge regime.

Finally, we consider the effect of temperature change. It is clear, on physical grounds, that increasing the temperature must increase the anode current in the retarding field case, since more electrons will be able to reach the anode. In the less trivial case of space-charge limitation we obtain from the considerations of the last paragraph

$$V-V_m = V_b+\phi_c-\phi_a+\frac{kT}{e}\ln\frac{J_0}{J}$$

$$=V_b-\phi_a+\frac{kT}{e}\ln\frac{AT^2}{J}. \tag{79}$$

Differentiating with respect to T we get

$$\left.\frac{\partial(V-V_m)}{\partial T}\right|_{J\,\mathrm{const.}} = \frac{k}{e}\ln\frac{AT^2}{J}+\frac{2k}{e}$$

$$=8{\cdot}618\times 10^{-5}\ln\frac{AT^2}{J}+1{\cdot}73\times 10^{-4}. \tag{80}$$

For oxide cathodes, $A \doteqdot 0{\cdot}1$ amp./cm./sec., $T \doteqdot 1000°$ K., $J \doteqdot 0{\cdot}1$ amp./sq. cm., so eqn. (80) gives changes of a few parts in one thousand in the voltage for a given current. If the voltage remains constant the change in current is given by eqn. (80) multiplied by the diode conductance per unit cathode surface.

The expressions in this section relate explicitly to the planar diode. Similar ones may be obtained for the cylindrical diode, but these usually involve some approximation. The questions considered here are mainly important in u.h.f. diodes which are usually built in the planar form. Cylindrical diodes with anode/cathode diameter ratios less than 1·6:1 may be considered as planar diodes. All the necessary expressions for the design of the fairly widely spaced cylindrical diodes used at ordinary frequencies have been given. It is interesting to note that Hahn† has been able to give a fairly thorough discussion of the diode from the standpoint of kinetic theory.

6.9. Injection at high velocities

To complete our work on space charge we have to consider the case in which a high-velocity beam is injected into a space in which there is an electric field parallel to the direction of

† Hahn, *Proc. Inst. Radio Engrs., N.Y.* **36** (1948), 1115.

motion of the electron beam. A practical example of this situation is the screen-anode space of a beam tetrode. When the current is at its maximum the anode potential is low, well below the screen potential, and the relatively heavy current is injected into a region in which a strong decelerating field exists. This situation has been the subject of several analyses,† the earliest being that of Gill, usually assuming that the current injected into the system is constant. Let us first discuss this assumption. Fig. 6.16 shows a simple electron optical system consisting of a cathode and grid, screen and cup-shaped anode.

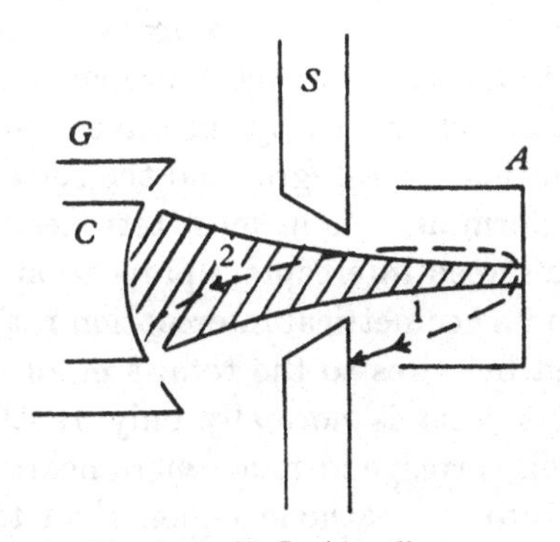

Fig. 6.16. Reflexion effects.

When the anode is held at a potential slightly above the screen potential, say $V_s + 20$V., all the electrons passing through the aperture will be collected and no secondaries will pass from A to S. It is, in principle, possible to make the gun efficiency $I_a/(I_s + I_a)$ equal to 100 per cent; we suppose that this is so. Then, if the voltage of A is made strongly negative to the cathode, I_a becomes zero and I_s increases from zero to some positive value. What is this value of I_s? In general I_s depends on the focusing properties of the reflecting field; if the hole in s is very small and the reflecting field is correctly arranged (path 1), I_s may be nearly equal to I_a, but, if the hole in s is

† Plato, Kleen and Rothe, *Z. Phys.* **101** (1936), 509. Gill, *Phil. Mag.* **49** (1925), 993; **104** (1937), 711. Salzberg and Haeff, *R.C.A. Rev.* **2** (1938), 336. Fay, Samuel and Shockley, *Bell Syst. Tech. J.* **17** (1938), 49.

fairly large and the reflector is arranged so that most of the current is focused back through the hole (path 2), $I_s \doteqdot \frac{1}{2} I_a$.

This last value may seem surprising, but it can be justified. The current drawn from the cathode depends on the space charge in the cathode-screen space. When A is positive the space charge in this region is due only to electrons going from right to left and the current is I_a. When electrons are reflected from A, now at a negative potential, there are two components of current in the cathode-screen space, I_e and I_r, say. The space charge in this region is therefore proportional to $I_e + I_r$, and if all the emitted current is returned through S we have $I_e = I_r$. Therefore the space charge is proportional to $2I_e$. Thus $2I_e = I_a$ and the current taken by the screen will be $\frac{1}{2}I_a$. If the screen S is an ordinary mesh grid and the screen-anode field is more or less uniform as it is in most tetrodes, it is natural to assume that the screen intercepts a proportion of the reflected current equal to its geometrical interception ratio, i.e. the ratio of the area filled by wires to the total area of the screen electrode. Since this ratio is normally only 10–20 per cent, the tetrode valve will correspond much more nearly to the case of total reflexion into the cathode region than to the first case mentioned. In general, therefore, we cannot consider that the injected current is constant in calculations on real valves. We therefore follow the analysis of Bull† which does not make this assumption.

Consider the plane parallel tetrode of fig. 6.17, and assume that the screen current is zero, the anode being held at a low positive potential. Then the charge density in the regions CS, SM depends on $I_e + I_r$, while that in MA depends on I_a only and $I_a = I_e - I_r$.

Then, clearly,
$$I_e + I_r = \frac{DV_s^{\frac{3}{2}}}{a^2}, \tag{81}$$

where
$$D = \text{diode constant}$$
$$= 2{\cdot}336 \times 10^{-6}.$$

† Bull, *J. Instn. Elect. Engrs.*, **95**, pt. III (1948), 17.

Also for the space SM,

$$I_e + I_r = \frac{DV_s^{3/2}}{b^2}, \tag{82}$$

and in MA

$$I_a = \frac{DV_a^{3/2}}{d^2}. \tag{83}$$

For this case $a = b$, but in the more general case where there is a control grid, V_s and a in eqn. (81) are replaced by the

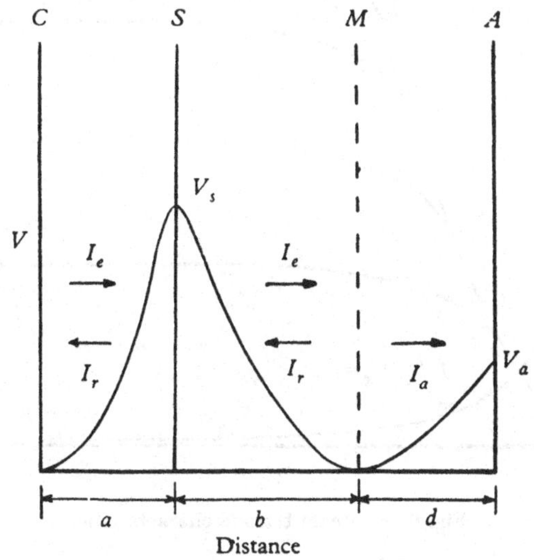

Fig. 6.17. Potential profile in tetrode with virtual cathode.

equivalent grid-plane voltage and the cathode-grid spacing respectively, while V_s in eqn. (82) is replaced by the equivalent screen-plane voltage. Now eqn. (82) shows that the distance from S to the minimum depends only on the sum of I_e and I_r. This is a constant value I_0, say. I_0 is evaluated by increasing V_a until no electrons are returned, i.e. until the I_a, V_a characteristic has flattened out. Thus b is a constant and the location of the minimum does not depend on V_a, but only on I_0 and V_s. When V_a is increased it eventually reaches the critical value V_a' at which $I_a = I_0$, and the virtual cathode disappears to be replaced by a potential minimum in the beam. The I_a, V_a

characteristic is therefore as sketched in fig. 6.18, where the injected current is varied by varying the grid voltage. The anode characteristic rises slightly beyond the 'knee' because the anode is not, in general, completely shielded from the cathode so that the current I_0 shows a slow increase with V_a.

When the anode voltage is above the highest value corresponding with a virtual cathode all the electrons reach the anode,

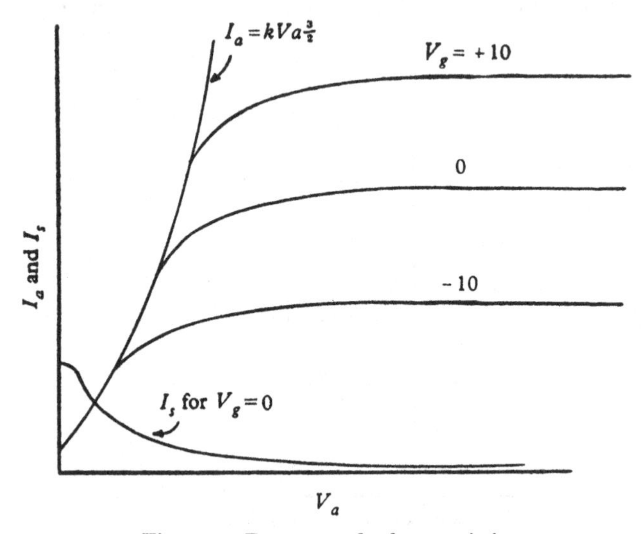

Fig. 6.18. Beam tetrode characteristic.

but there is a potential minimum in the anode-screen space. When the anode potential is only just above the virtual cathode value the minimum is actually at the anode. When $V_a = V_s$ the minimum is half-way between anode and screen, and when $V_a \gg V_s$ the minimum is on the screen. Fig. 6.19 demonstrates the variation of potential for six values of V_a.

In the beam tetrode the depression of potential by the space charge is used to prevent secondary electrons from the anode reaching the screen. If the anode potential lies between 0 and V_2 in fig. 6.19, secondary electrons emitted from the anode emerge into a strong retarding field and are turned back. In the immediate vicinity of V_3, secondaries can go from anode to screen but at a somewhat higher voltage, between V_3 and V_4

the minimum will have moved back from the anode and the secondaries will be prevented completely from reaching the screen when the potential depression V_m is greater than about 12 V. For all potentials above this critical value secondaries are again prevented from reaching the screen.

The potential distribution is easily calculated by assuming a variation which is symmetrical about the minimum. This has

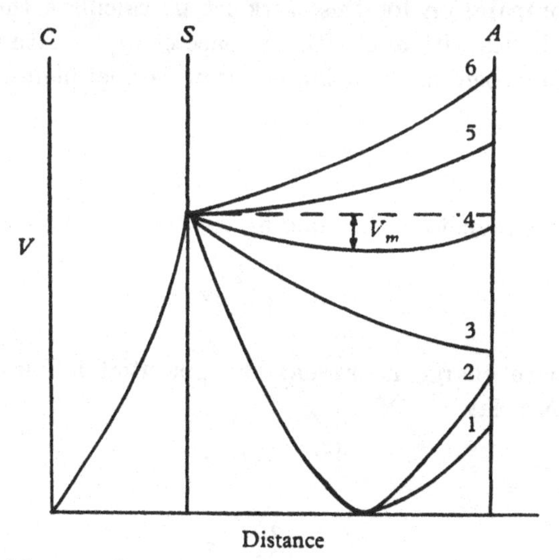

Fig. 6.19. Potential profiles for different values of V_a.

been carried through in the references cited and will not be repeated here.

The matters discussed in this section are practically important in the theory of the beam tetrode. We discuss tetrodes later on, but we should here emphasize that the theory of these devices is necessarily very much more complicated than has been indicated above, because they are not simple parallel beam devices but usually have a complex geometry. In addition, extra electrodes at cathode potential are often inserted into the screen-anode space which depress the charge-free potential below the values corresponding to the potentials of S and A alone. A particularly good agreement between simple theory and practice is not, therefore, to be expected.

6.10. Space charge and transit time

In the next chapter we shall study the effects of raising the operating frequency of valves so much that the electronic velocity is insufficient to allow the time interval between the instant of leaving the cathode and the instant of reaching the anode to be assumed short in comparison with the period of the impressed h.f. fields.

As a preparation for this work let us calculate the transit times in diodes with and without space charge. With no space charge the potential variation between parallel planes is linear and

$$\tau = \frac{2s}{u_a}, \tag{84}$$

where u_a = velocity corresponding to the anode voltage,

$$= \sqrt{\left(2\frac{e}{m}\right)} V_a^{-\frac{1}{2}}.$$

When space charge is present the potential follows a four-thirds power law,

$$V(z) = \left(\frac{z}{s}\right)^{\frac{4}{3}} V_a,$$

or

$$u(z) = \left(\frac{z}{s}\right)^{\frac{2}{3}} u_a.$$

Then

$$\tau = \int_0^s \frac{dz}{u(z)} = \frac{s^{\frac{2}{3}}}{u_a} \int_0^s z^{-\frac{2}{3}}\, dz = \frac{3s}{u_a}. \tag{85}$$

Eqn. (85) shows that space charge increases the transit time by 50 per cent. The derivation of eqn. (85) takes no account of the potential minimum. By suitable re-definition of s and u_a, eqn. (85) can be adjusted to give the time from the virtual cathode to the anode, and another calculation can be made to give the time from the cathode to the virtual cathode. Since the electrons are moving with their lowest velocity in this part of the tube the correction is by no means negligible. The magnitude can be assessed by determining the depth and position of the potential minimum by the methods of § 6.6. The retarding field between cathode and minimum can be assumed linear, and the transit time to the minimum of an electron which just passes the barrier is then

$$\tau_{\text{min.}} = \frac{2z_m}{u_m}, \tag{86}$$

where u_m = initial velocity of electron just able to surmount the barrier,

$$= 5{\cdot}93 \times 10^5 \sqrt{V_m}.$$

The total transit time is then

$$\tau = \frac{3(s - z_m)}{u_a} + \frac{2z_m}{u_m}. \tag{87}$$

The second term can be very important in closely spaced diodes, such as are used for microwave work.

The calculations in the case of the cylindrical diode are not quite so simple. With no space charge the potential variation is logarithmic, and it is easily shown that

$$\tau = \frac{r_c \sqrt{\{\ln r_a/r_c\}}}{u_a} \int_0^{\ln (r_a/r_c)} \frac{e^\omega \, d\omega}{\sqrt{\omega}}, \tag{88}$$

where $\omega = \ln(r/r_c)$. In the space-charge limited case

$$u = \left(\frac{r\beta_{rc}^2}{r_a \beta_{ca}^2}\right)^{\frac{1}{3}} u_a,$$

therefore

$$\tau = \left(\frac{r_a \beta_{ca}^2}{u_a}\right)^{\frac{1}{3}} \int_{r_a}^{r_c} (r\beta^2 r_c)^{-\frac{1}{3}} \, dr$$

$$= \left(\frac{r_a r_c^2 \beta_{ca}^2}{u_a}\right)^{\frac{1}{3}} \int_0^{\ln r_a/r_c} \left(\frac{e^\omega}{\beta_{rc}}\right)^{\frac{2}{3}} d\omega. \tag{89}$$

Ferris† has given curves which give these transit times in the form

$$\tau = \frac{s}{u_c} f\left(\frac{r_a}{r_c}\right). \tag{90}$$

† Ferris, *Proc. Inst. Radio Engrs., N.Y.* **24** (1936), 82.

The increase in transit time due to space charge is smaller in the case of the cylinder diode than it is in parallel plane diodes, but for closely spaced diodes which are actually used at u.h.f. the difference is small. This is to be expected, since the close-spaced cylinder diode must approximate to the parallel-plane diode.

In general, the effect of space charge is to increase the transit time in a given region, but, unless a virtual cathode is formed, the increase is not very great and in most cases it will be well below 50 per cent.

Chapter 7

TRANSIT TIME EFFECTS

Up to this point we have been concerned only with the d.c. and low-frequency behaviour of electrons moving in electromagnetic fields. We have now to consider some of the phenomena which take place when the time taken by the electrons to move from one electrode to another is comparable with the period of the alternating voltages applied to the electrodes. Before we can do this we must get a more precise idea of the nature of electric current, which has so far been defined as the rate of arrival of electrons at the electrode surface. Maxwell introduced the idea of the total current into electromagnetic theory. The total current is defined as the sum of conduction and displacement current, or in symbols

$$\mathbf{I} = \rho \mathbf{u} + \epsilon \frac{d\mathbf{E}}{dt}, \tag{1}$$

the first term being the conduction current and the second the displacement current. The displacement current is stored in forming the electric field and, averaged over a cycle, the value is zero. Clearly, the higher the frequency the greater the magnitude of the displacement current. Now, although the time average of the displacement current is zero, this does not mean that it produces no measurable effects. In fact we can show that

$$\nabla . \mathbf{I} = 0, \tag{2}$$

so that if we imagine a region in which u is suddenly changed eqn. (1) shows us that the field must change in such a way that the total divergence is zero.

7.1. Ramo's theorem

It is possible to deduce more useful expressions for the total current from eqn. (2), but a simpler physical picture results from following the work of Jen.† Before doing this it is worth while discussing the case of a single electron traversing the space between two plane-parallel electrodes (fig. 7.1 *a*). Imme-

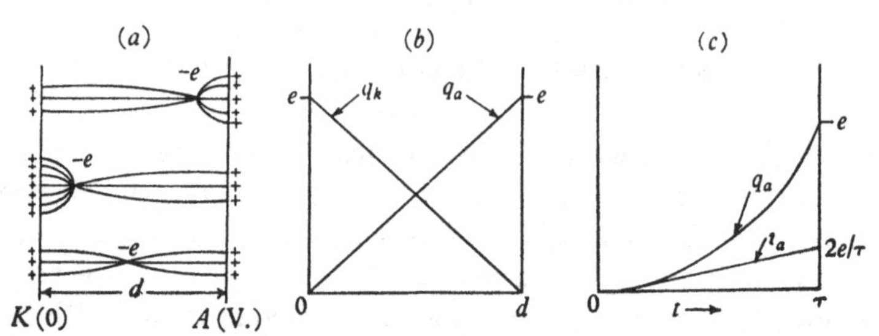

Fig. 7.1. Fields and currents induced by a moving electron.

diately the electron leaves the cathode plane K, it induces positive charges on the cathode and anode. The magnitudes of the charges are such that

$$q_k + q_a = e. \tag{3}$$

When the electron has moved a distance z from the cathode the anode charge has changed from 0 to $q_{a(z)}$, but to balance this the field has done work on the electron of amount Eez, therefore $Edq_{a(z)} = Eez$ or

$$q_{a(z)} = \frac{ez}{d},$$

and the charge on the anode is directly proportional to the z displacement. It is obvious that $q_k = e(1-z/d)$. The charges therefore vary as in fig. 7.1 *b*.

Now,

$$i_a = \frac{dq_a}{dt} = \frac{e\dot{z}}{d} = \frac{eu}{d}, \tag{4}$$

where u = velocity. Eqn. (4) can be written in a more informative way. Let u_a = velocity at anode and τ = transit time. Then $\tau = 2d/u_a$ and $i_a = \frac{2e}{\tau}\frac{u}{u_a}$, or, since the acceleration is uniform,

$$i_a = \frac{2et}{\tau^2}, \tag{4 a}$$

† Jen, *Proc. Inst. Radio Engrs., N.Y.* **29** (1941), 345.

and
$$q_a = \frac{et^2}{\tau^2}. \tag{4 b}$$

These are shown in fig. 7.1 *c*. A stream of electrons in a planar diode without space charge is therefore equivalent to a series of triangular current pulses.

Eqn. (4), or rather a generalization of it, is often called Ramo's† theorem. The general form is

$$i = E_u eu, \tag{5}$$

where E_u = field component in the u direction of the field which would exist in the space considered if the electron were removed, the given electrode raised to unit potential and all other electrodes grounded. The above expression assumes that all h.f. magnetic fields are negligible, and that the frequency is not so high that field propagation times have to be considered, i.e. the retarded potentials are approximated by the electrostatic potential.

The development of eqn. (4) is, of course, merely illustrative of the fact that immediately electrons enter a space surrounded by conductors, currents start to flow in the said conductors quite independently of whether the electron finally reaches the conductor or not. (The electron of fig. 7.1 *a* could escape through a minute hole in A without altering the behaviour during the transit K–A, though a current would be induced in A during any subsequent movement.) A particular case of practical importance is the fact that currents are induced in the negative grids of triodes when these are operated at high frequencies, and in some circumstances heavy damping of the input circuit is the outcome of such induction. This effect can only be overcome by making the transit time K–G very small, which usually has to be done by reducing the spacing; as the alternative, increasing the effective grid voltage, is usually not permissible.

As an example of the application of Ramo's theorem, consider the valve of fig. 7.2, where we assume that the electron beam from the gun has a density variation so that the current

† Ramo, *Proc. Inst. Radio Engrs., N.Y.* **27** (1939), 584.

entering the space between the grids is of the form $i = i_0 (1+\alpha \sin \omega t)$. Since $E_u = 1/d$ for this case, the current induced in G_2 by each electron in the gap is $\Delta I = eu/d$. If the area of the grids is A, the volume of the gap is Ad and there are $\rho \frac{Ad}{e}$ electrons in the gap at any instant. The total induced current at any instant is thus

$$I = \rho \frac{Ad}{e} \frac{eu}{d}.$$

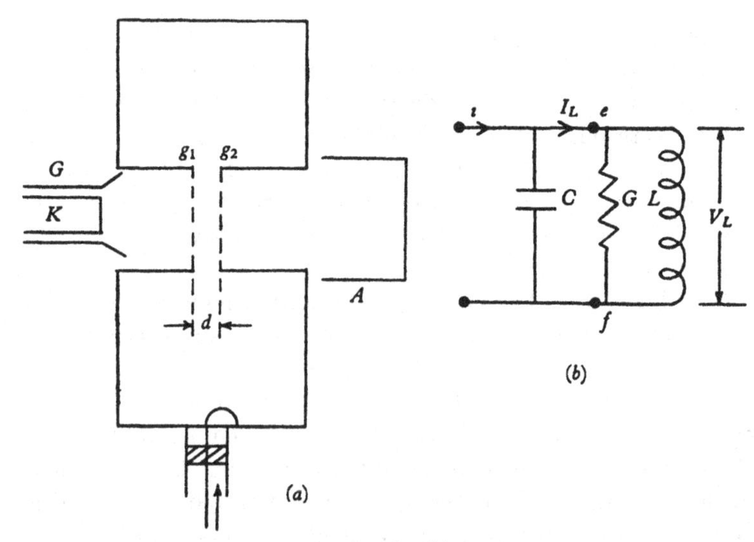

Fig. 7.2. A reflex klystron.

But $\qquad \rho u =$ conduction current density,

$$= i/A.$$

Therefore $\qquad I = i,$

$$= i_0(1+\alpha \sin \omega t).$$

The amplitude of the h.f. current circulating in the external circuit is thus

$$I_{\text{h.f.}} = \alpha i_0.$$

We can look at this result in a slightly different way. Fig. 7.2 *b* shows an equivalent circuit for the gap and external circuit. If the gap alone is considered, it is equivalent to putting an open circuit at e, f. The open-circuit voltage is then $V_0 = -ji/\omega C$. When the inductance and capacitance are connected at e, f, the voltage and current are V_L, I_L, so that the driving current must equal I_L plus the current in C. Therefore

$$i = I_L + j\omega C V_L.$$

The equivalent circuit of the gap alone is thus the circuit seen looking back into the terminals of e, f, i.e. a capacitance across a constant current generator of infinite internal impedance and the current is equal to the h.f. component of the beam current. We shall see later on that the impedance of the generator is not, in general, infinite, but depends on the transit angle across the gap. Practical valves usually operate with dimensions such that the impedance falls as the transit angle increases, but this need not invariably be the case. When the generator impedance is not infinite, an extra admittance Y_B is connected across C, and the driving current must maintain the voltage V_L across this. In this case the current in the external circuit is

$$I_L = i - V_L Y_g,$$

where

$$Y_g = \text{gap admittance,}$$

$$= Y_B + j\omega C.$$

The problem of calculating Y_B is discussed in § 7.3.

These relations are very important, and will be used continuously in the chapters on velocity modulated valves and ultra-high frequency triodes.

7.2. Current flow at very high frequencies

Let us now put these ideas on a firmer theoretical basis, as is done by Jen,† and, with even greater generality, by Gabor.‡ We consider a closed region of space bounded by a number of electrodes n held at potentials $\Phi_1, \Phi_2, \ldots, \Phi_k$, and with charges

† Jen, loc. cit.

‡ Gabor, *J. Instn. Elect. Engrs.* **91**, pt. III (1944), 128.

$q_1, q_2, \ldots, q_k$. Inside the volume there is a space charge whose density at any point is $\rho(x, y, z, t)$ and the electrons travel with velocities $\mathbf{u}(x, y, z, t)$. $\mathbf{u}$ refers to the normal vector in the outward direction as usual. Green's first identity† can be used to find the relation between the field and the charges. The vector identity is

$$\int_V (\nabla\phi)^2 dV + \int_V \phi \nabla^2 \phi dV = \int_S \phi \frac{\partial \phi}{\partial n} dA. \tag{6}$$

Now
$$\mathbf{E} = -\nabla\phi,$$

and
$$\nabla^2\phi = -\nabla . \mathbf{E} = -\rho/\epsilon_0,$$

therefore
$$\epsilon_0 \int_V \mathbf{E}^2 dV - \int_V \phi\rho dV = \sum_{k=1}^{n} \Phi_k q_k. \tag{7}$$

The right-hand side results from the application of Gauss's theorem over the electrode surfaces. Eqn. (7) is an energy equation, and we are interested in power relations, so we differentiate to obtain

$$\sum_{k=1}^{n} (\Phi k \dot{q}_k + \dot{\Phi} k q_k) + \int_V (\dot{\rho}\phi + \dot{\phi}\rho) dV = 2\epsilon_0 \int_V \mathbf{E} . \dot{\mathbf{E}} dV. \tag{8}$$

We can find expressions for $\dot{\rho}\phi + \dot{\phi}\rho$ as follows.

The equation of continuity is

$$\dot{\rho} + \nabla . \rho\mathbf{u} = 0,$$

therefore
$$\dot{\rho}\phi + \phi\nabla . \rho\mathbf{u} = 0.$$

But
$$\nabla . \phi\rho\mathbf{u} = \rho\mathbf{u} . \nabla\phi + \phi\nabla . \rho\mathbf{u},$$

therefore
$$\int_V \dot{\rho}\phi dV = -\int_V (\nabla . \phi\rho\mathbf{u} - \rho\mathbf{u} \operatorname{grad} \phi) dV,$$

$$= -\int_S \phi\rho\mathbf{u} dA - \int_V \rho\mathbf{u} . \mathbf{E} dV,$$

$$= -\sum_{k=1}^{n} \Phi_k \int_S (\rho\mathbf{u}) n dA_k - \int_V \rho\mathbf{u} . \mathbf{E} dV. \tag{9}$$

† Stratton, *Electromagnetic Theory*, McGraw Hill, 1941, p. 165.

$$\int_V \phi\rho dV = \epsilon_0 \int_V \phi\nabla.\mathbf{E}dV,$$

$$= \epsilon_0 \int_V (\nabla.\dot{\phi}\mathbf{E} - \mathbf{E}.\nabla\dot{\phi})dV,$$

$$= \epsilon_0 \int_S \phi\mathbf{E}ds + \epsilon_0 \int_V \mathbf{E}.\dot{\mathbf{E}}dV,$$

$$= -\sum_{k=1}^{n} \Phi_k q_k + \epsilon_0 \int_V \mathbf{E}.\dot{\mathbf{E}}dV. \tag{10}$$

Using these results in eqn. (8) we obtain

$$\sum_{k=1}^{n} \Phi_k[\dot{q}_k - \int_S (\rho\mathbf{u})_n dA_k] = \int_V (\rho\mathbf{u} + \epsilon_0 \dot{\mathbf{E}})\mathbf{E}dV. \tag{11}$$

If i_k = conduction current to the kth electrode from the external connexion, we can apply the equation of continuity to the kth electrode and obtain

$$\dot{q}_k = i_k + \int_S (\rho\mathbf{u})_n dS_k; \tag{12}$$

or, in words, the time rate of change of surface charge equals the conduction current plus the incident space current.

Using eqn. (12) in (11) we finally obtain

$$\sum_{k=1}^{n} \Phi_k . i_k = \int_V (\rho\mathbf{u} + \epsilon_0 \dot{\mathbf{E}})\mathbf{E}dV. \tag{13}$$

It will be seen that Maxwell's total current has reappeared in eqn. (13).

Eqn. (13) can be given a more significant form by inserting the expressions for the kinetic energy of the electrons and the energy stored in the electrostatic field, viz.

$$\dot{T} = \int_V \rho\mathbf{u}\mathbf{E}dV,$$

$$\dot{U} = \epsilon_0 \int_V \mathbf{E}\dot{\mathbf{E}}dV.$$

Then $$\sum_{k=1}^{n} \Phi_k i_k = \frac{d}{dt}(T+U). \tag{14}$$

Thus we see (*a*) that the rate of change of the total energy is equal to the sum of the instantaneous powers on all the electrodes; (*b*) that the electrodes are all coupled together by means of the field and the space current; (*c*) that the current to an electrode depends only on the total current, and not on either the conduction current or the displacement current alone.

Let us now use eqn. (14) to derive Ramo's theorem. We divide the potential into two parts, one depending on the electrode potentials, and the other on the space charge, i.e. $\phi = \phi_1 + \phi_2$, with

$$\nabla^2\phi_1 = 0, \quad (\phi_1)_k = \Phi_k,$$

$$\nabla^2\phi_2 = -\rho/\epsilon_0, \quad (\phi_2)_k = 0 \ (k = 1, 2, \ldots, n).$$

Then
$$\sum_{k=1}^{N} \Phi_k i_k = \int_V (\rho\mathbf{u} + \epsilon_0 \dot{\mathbf{E}})\mathbf{E}_1 \, dV,$$

$$0 = \int_V (\rho\mathbf{u} + \epsilon_0 \dot{\mathbf{E}})\mathbf{E}_2 \, dV.$$

For the quasi-stationary case where $\dot{\mathbf{E}} = 0$, we have

$$\sum_{k=1}^{N} \Phi_k i_k = \int_V \rho \mathbf{u} \mathbf{E}_1 \, dV. \tag{15}$$

We now apply this to a parallel-plane diode with the cathode at zero potential and the anode at unit potential. For a single electron we can integrate over a small sphere containing the electron to obtain

$$i_k = e\mathbf{u}E_k. \tag{16}$$

Eqn. (16) is identical with eqn. (15) and therefore Ramo's equation is proved.

Another useful relation which can be derived is an equation for the power balance. Let p equal the period of the oscillation, then the average of a function $\phi(t)$ over a single cycle is

$$\overline{\phi(t)} = \frac{1}{p}\int_0^p \phi(t)\,dt,$$

and the average of a periodic quantity is zero.

Now E is periodic, and therefore $\overline{\mathbf{E}}$ and $\overline{d\mathbf{u}/dt}$ are zero. From eqn. (14) we have

$$\Sigma \overline{\Phi_k i_k} = \int_V \overline{\rho \mathbf{u} \mathbf{E}}\, dV. \tag{17}$$

We note that, if Φ_k is a periodic function of t, the product $\overline{\Phi_k i_k}$ need not be zero even if $\bar{i}_k = 0$, and so there may be an interchange of energy between the field and the electrons, even if the electrons are not collected by the electrodes.

In the above development we have not taken account of radiation effects, and we also assume that h.f. magnetic fields are negligibly small. In a later paper Jen† removes the first restriction while the work of Gabor‡ eliminates the second.

7.3. Power relations in gaps

As a first application of these relations to a practical problem, let us consider the simplest problem of ultra-high-frequency electronics, the power interchange between an initially uniform beam and a gap across which a small h.f. voltage V_1 is developed. In this context a small h.f. voltage is defined as one for which the ratio V_1/V_0, usually denoted by the symbol α and called the 'depth of modulation', obeys the condition $\alpha < 0{\cdot}5$. Fig. 7.2 shows the arrangement; the gap is to be considered as bounded by ideal grids which constrain the field without intercepting any electrons. By Ramo's theorem each electron induces a current pulse $(e/d)u$ into the circuit, across which a voltage $V_1 \sin \omega t$ is maintained by an external generator. The current at entry is

$$I_0 = \rho_0 . u_0, \quad u_0 = \sqrt{\left(2\frac{e}{m}\right)} V_0^{\frac{1}{2}}.$$

Again using Ramo's theorem, the current at the time t due to the charge entering the gap in the interval dt_0 between times t_0 and $t_0 + dt_0$ is

$$di = \frac{I_0}{d} u\, dt_0.$$

The total current at time t is the integral of this over all the electrons in the gap, i.e. over $t_0 = t - \tau$ to $t_0 = t$ when τ is the

† Jen, *Proc. Inst. Radio Engrs., N.Y.* **29** (1941), 464.

‡ Gabor, loc. cit.

transit time of those electrons which reach g_2 at t:

$$i = \frac{I_0}{d}\int_{t-\tau}^{t} u\,dt_0. \tag{18}$$

We must now determine u and τ by dynamical methods. The equation of motion is

$$m\ddot{z} = \frac{eV_1}{d}\sin\omega t.$$

Carrying out two integrations and inserting the initial conditions that $\dot{z} = u_0$ at $t = t_0$ and $z = 0$ at $t = t_0$, we obtain

$$\dot{z} = u_0 + \frac{eV_1}{md\omega}(\cos\omega t_0 - \cos\omega t), \tag{19}$$

$$z = \left(u_0 + \frac{eV_1}{md\omega}\cos\omega t_0\right)\left(t - t_0\right) + \frac{eV_1}{md\omega^2}(\sin\omega t_0 - \sin\omega t). \tag{20}$$

Let us now introduce some notation. We put $V_1/V_0 = \alpha$ and $\omega d/u_0 = \phi_0$. ϕ_0 is the transit angle in the gap if the h.f. field is removed. Remembering that $u_0^2 = 2\frac{e}{m}V_0$, we have

$$\frac{eV_1}{md\omega} = \frac{\alpha u_0}{2\phi_0},$$

and $\quad z = u_0\left(1 + \frac{\alpha}{2\phi_0}\cos\omega t_0\right)\left(t - t_0\right) + \frac{\alpha u_0}{2\phi_0\,\omega}(\sin\omega t_0 - \sin\omega t).$

First we determine τ by using the condition that $t_0 = t - \tau$ for the electrons reaching g_2 at any instant. To obtain an explicit relationship we put

$$\tau = \frac{d}{u_0} + \frac{\delta}{\omega}, \tag{21}$$

where δ is a small correction factor. We approximate by putting sin, cos $(\omega t - \phi - \delta)$ equal to sin, cos $(\omega t - \phi)$ and neglecting terms involving the product $\alpha\delta$. Then

$$d = d + \frac{u_0\delta}{\omega} + \frac{\alpha d}{2\phi_0}\cos(\omega t - \phi_0) + \frac{\alpha u_0}{2\phi_0\omega}[\sin(\omega t - \phi_0) - \sin\omega t], \tag{22}$$

whence

$$\delta = \frac{\alpha}{2\phi_0}[\sin\omega t(1 - \cos\phi_0 - \phi_0\sin\phi_0) + \cos\omega t(\sin\phi_0 - \phi_0\cos\phi_0)] \tag{23}$$

From eqns. (18) and (19)

$$i = \frac{I_0}{d}\left[u_0 t_0 + \frac{\alpha u_0}{2\phi_0}\left(\frac{\sin\omega t_0}{\omega} - t_0\cos\omega t\right)\right]_{t-\tau}^{t}$$

$$= \frac{I_0}{d}\left[u_0\tau + \frac{\alpha u_0}{2\phi_0}\left(\frac{\sin\omega t - \sin\omega(t-\tau)}{\omega} - \tau\cos\omega t.\right)\right].$$

Using the value of τ from eqns. (21) and (23) we have

$$\frac{i}{I_0}\left[1+\alpha_0\left\{\left(\frac{1-\cos\phi_0}{\phi_0^2} - \frac{\sin\phi_0}{2\phi_0}\right)\sin\omega t + \left(\frac{\sin\phi_0}{\phi_0^2} - \frac{1+\cos\phi_0}{2\phi_0}\right)\cos\omega t\right\}\right]. \tag{24}$$

The h.f. current therefore has components both in phase and in quadrature with the driving voltage $V_1\sin\omega t$. The power given up to the beam by the field is clearly

$$P_b = \frac{V_1^2}{2R_0}\left[\frac{1-\cos\phi_0}{\phi_0^2} - \frac{\sin\phi_0}{2\phi_0}\right], \tag{25}$$

where R_0 = d.c. beam impedance,

as can be proved by evaluating the average of $iV_1\sin\omega t$ over a single cycle. In later parts of the book we shall refer to a quantity η_D, the diode efficiency of the gap. η_D is defined as P_b divided by I_0V_0, i.e. it is the ratio of the h.f. power gained by the beam to the d.c. power carried by the beam. For this case

$$\eta_D = \frac{\alpha^2}{2}\left[\frac{1-\cos\phi_0}{\phi_0^2} - \frac{\sin\phi_0}{2\phi_0}\right] \tag{26}$$

$$= \frac{\alpha^2}{2}\left[\frac{\phi_0^2}{24} - \frac{\phi_0^4}{360} + 0(\phi_0^6)\right]. \tag{27}$$

Another useful way of looking at the system consisting of the beam and the gap is to consider that the beam is represented by an admittance $G+jB$ across the terminals of the gap:

$$G = G_0\left(\frac{1-\cos\phi_0}{\phi_0^2} - \frac{\sin\phi_0}{2\phi_0}\right), \tag{28}$$

$$B = G_0\left(\frac{\sin\phi_0}{\phi_0^2} - \frac{1+\cos\phi_0}{2\phi_0}\right), \tag{29}$$

where $$G_0 = \frac{1}{R_0} = \frac{I_0}{V_0}. \tag{30}$$

In fig. 7.3, the conductance and susceptance are shown as junctions of ϕ_0.

The case of the parallel-plane diode can be handled in exactly the same way. This has been done by many authors, among

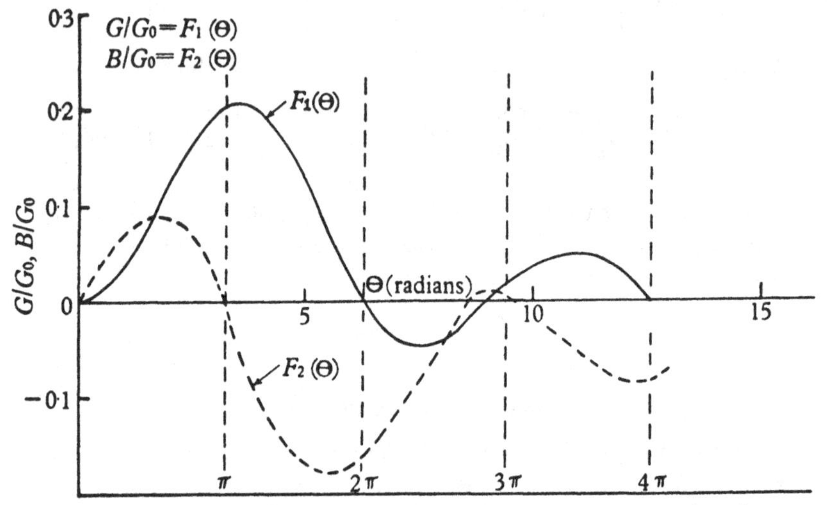

Fig. 7.3. Conductance and susceptance of simple gap as a function of frequency.

others by Benham,† Müller,‡ Llewellyn§ and North‖. The derivation by North uses exactly the method used above. The result, subject to certain assumptions which we shall discuss in a moment, is

$$Z = r_\rho + jx_p = \frac{12r_c}{\phi_0^4}[2(1-\cos\phi_0) - \phi_0 \sin\phi_0$$

$$-j\{\phi_0(1+\cos\phi_0) - 2\sin\phi_0.\}]. \qquad (31)$$

Eqn. (31) represents the beam alone. The ordinary low-frequency capacity of the gap can be included in this case, because the capacity and the current bear an obvious relationship in the planar diode. The free-space capacity can easily be shown to contribute an extra $2r_c/\phi_0$ to the reactive part.

† Benham, *Phil. Mag.* **5** (1928), 641 and **11** (1931), 457.
‡ Müller, *Hochfrequenztech. u. Elektroakust.* **41** (1933), 156.
§ Llewellyn, *Electron Inertia Effects*, Cambridge University Press, 1941.
‖ North, *Proc. Inst. Radio Engrs., N.Y.* **24** (1936), 108.

Including this and expanding the trigonometric functions, we obtain

$$Z = r_c\left[\left\{1-\frac{\phi_0^2}{15}+\ldots\right\}-j\left\{\frac{3\phi_0}{10}-\frac{\phi_0^3}{84}+\ldots\right\}\right] \tag{32}$$

In these expressions r_c is the static slope resistance of the diode

$$r_c = -\left.\frac{\partial V_a}{\partial I_0}\right|_{Z=d} = -\frac{2V_a}{3I_0} = -\frac{2{\cdot}85\times 10^5 d^2}{V_a^{\frac{1}{2}}} = -\frac{378\times 10^3 d^{\frac{4}{3}}}{I_0^{\frac{1}{3}}}. \tag{33}$$

r_c is measured in ohms/unit area. $\phi_0 = \omega\tau$ as usual. For the space-charge limited diode, eqn. (85) of Chapter 6 gives

$$\tau = \frac{5{\cdot}07d}{10^6 V_a^{\frac{1}{2}}}. \tag{34}$$

The assumptions made in deriving the diode expressions are the following:

(*a*) The electrons in the cathode-anode space only travel in one direction, i.e. towards the anode. This means that the analysis only applies strictly under those field conditions for which the potential gradient is zero at the cathode surface. The loading effects of electrons going to the space-charge minimum and returning are not considered and, in fact, the necessary mathematical equipment for doing this has not been developed.

(*b*) In spite of (*a*) it is assumed that space charge is complete and that as V_A swings up the cathode current increases according to the space-charge law.

(*c*) Initial velocities are neglected except that a correction term may be applied to V_A to take the *mean* initial velocity into account.

The agreement between experiment and theory is fairly good in spite of the approximations, as is shown by the performance of the diode oscillator of Llewellyn and Bowen.†

Eqn. (31) can be inverted to obtain the value of the admittance; the low-frequency value is

$$Y_d = g_c\left[1+j\frac{3\phi_0}{10}\right]. \tag{35}$$

† Llewellyn and Bowen, *Bell Syst. Tech. J.* **18** (1939), 280.

Eqn. (35) is the admittance of a resistance shunted by a condenser. It is easy to show that the condenser has six-tenths of the capacitance of the diode electrodes when the cathode is cold, i.e. the presence of the electrons reduces the cold capacitance to 60 per cent of the original value.

Fig. 7.4 shows the values of r_p and x_p as a function of ϕ_0 as tabulated by Llewellyn.†

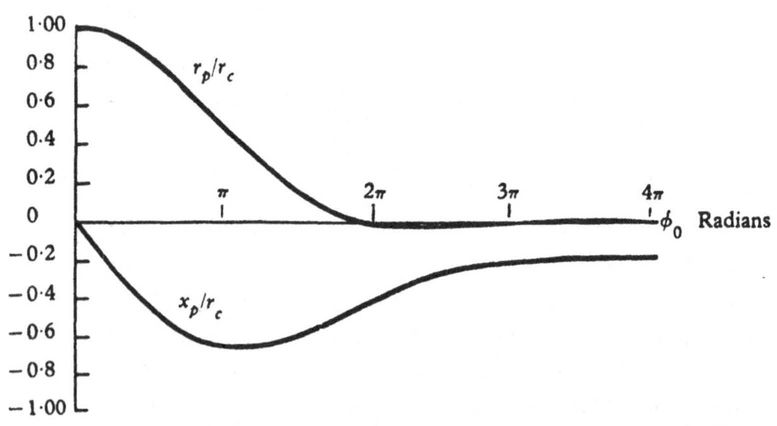

Fig. 7.4. Internal impedance of diode as a function of transit angle, θ.

The above results, besides being applicable to the diode, can also be used to discuss the cathode-grid, grid-screen, screen-anode, etc., spaces in parallel-plane multi-electrode valves. Before we discuss these applications we shall briefly indicate some further methods of handling the transit time problem.

7.4. Further methods of treating transit time phenomena

The method which was used by Llewellyn to derive the results given in the last section, is based on the following reasoning.

The total current per unit area is

$$\mathbf{I} = \rho\mathbf{u} + \epsilon_0 \frac{d\mathbf{E}}{dt}, \tag{36}$$

† Llewellyn, *Electron Inertia Effects*, Cambridge University Press, 1941.

and $$m\ddot{z} = e\mathbf{E}. \tag{37}$$

The divergence relation is

$$\nabla . \epsilon_0 \mathbf{E} = \rho,$$

which for parallel planes is

$$\epsilon_0 \frac{\partial \mathbf{E}}{\partial z} = \rho. \tag{38}$$

Putting (38) into (36) we obtain

$$\mathbf{I} = \epsilon_0 \left(\frac{\partial \mathbf{E}}{\partial z} \mathbf{u} + \frac{\partial \mathbf{E}}{\partial t} \right). \tag{39}$$

Now $$\mathbf{E} = f(z, t),$$

so $$\frac{d\mathbf{E}}{dt} = \frac{\partial \mathbf{E}}{\partial z} \frac{dz}{dt} + \frac{\partial \mathbf{E}}{\partial t}; \tag{40}$$

therefore $$\mathbf{I} = \epsilon_0 \frac{d\mathbf{E}}{dt}. \tag{41}$$

Eqn. (41) is to be interpreted as meaning that the rate of change of field must be evaluated at each instant on the path of a reference electron. If now we use (37) to substitute for $\mathbf{E}$ we obtain as the final result

$$\dddot{z} = \frac{e\mathbf{I}}{m\epsilon_0}, \tag{42}$$

or, in words, the total current is directly proportional to the rate of change of acceleration evaluated at each instant on the path of the reference electron. I is not a function of z but, in general, is a function of t.

To make use of eqn. (42) Llewellyn writes

$$\dddot{z} = K + \phi'''(t), \quad K = \text{const.}$$

Integrating, we obtain successively

$$\ddot{z} = K(t - t_0) + \phi''(t) - \phi''(t_0) + a_0 + \alpha(t_0), \tag{43}$$

where

$$a_0 + \alpha(t_0) = \text{acceleration at the plane of origin,}$$

$$\dot{z} = \tfrac{1}{2}K(t-t_0)^2 + \phi'(t) - \phi'(t_0) - (t-t_0)\phi''(t_0) + (t-t_0)[a_0+\alpha(t_0)] + u_0 + v(t_0) \quad (44)$$

where $u_0 + v(t_0)$ is the velocity at the plane of origin, and finally

$$z = \tfrac{1}{6}K(t-t_0)^3 + \phi(t) - \phi(t_0) - (t-t_0)\phi'(t_0) - \tfrac{1}{2}(t-t_0)^2\phi''(t_0) + \tfrac{1}{2}(t-t_0)^2[a_0+\alpha(t_0)] + (t-t_0)[u_0+v(t_0)], \quad (45)$$

as z is assumed zero at the origin, time $= t_0$.

This theory is only applicable to single-velocity streams. If the a.c. velocity distribution is such that electrons can overtake one another it is inapplicable. This follows because the total current is defined as a single-valued function of z and t. In cases where a large d.c. component of velocity or acceleration exists, this condition is almost automatically fulfilled.

Llewellyn makes use of eqn. (45) by inserting the value of the transit time when there are no a.c. fields present, i.e.

$$z = \tfrac{1}{6}K(t-t_0)^3 + \tfrac{1}{2}a_0(t-t_0)^2 + u_0(t-t_0),$$

$$= \tfrac{1}{6}K\tau^3 + \tfrac{1}{2}a_0\tau^2 + u_0\tau. \quad (46)$$

When the a.c. fields *are* present a correction factor is applied by writing $t - t_0 = \tau + \delta$.

Functions of $(t-t_0)$ and t_0 can now be expanded in series, i.e.

$$\left.\begin{aligned} f(t-t_0) &= f(\tau) + f'(\tau)\delta + (1/2!)f''(\tau)\delta^2, \text{ etc.} \\ f(t_0) &= f(t-\tau_0) + f'(t-\tau)\delta + (1/2!)f''(t-\tau)\delta^2, \text{ etc.} \end{aligned}\right\} \quad (47)$$

Now let $$\delta = \delta_1 + \delta_2 + \delta_3 + \ldots, \text{ etc.} \quad (48)$$

Eqns. (46), (47) and (48) can be substituted back into eqn. (45) to give expressions for the deltas. This enables solutions for the alternating quantities to be found.

Another method, due to Benham,† may be called the current compression method. A derivation is given by Llewellyn,‡ but here we use a simple physical argument to make the relation plausible.

† Benham, *Nature, Lond.*, **139** (April 1937).
‡ Llewellyn, op. cit.

Consider an electron beam passing between four planes A, B, C, D (see fig. 7.5). Consider the motion of those electrons which passed between the planes A and B in a time interval Δt_1 (A and B are considered to be very close together, but may be widely separated from C and D). These electrons arrive in the space CD during a time interval Δt_2, then, since current is conserved,

$$\left.\begin{aligned} i_1\,\Delta t_1 &= i_2\,\Delta t_2,\\ i_2 &= i_1\frac{\Delta t_1}{\Delta t_2},\end{aligned}\right\} \tag{49}$$

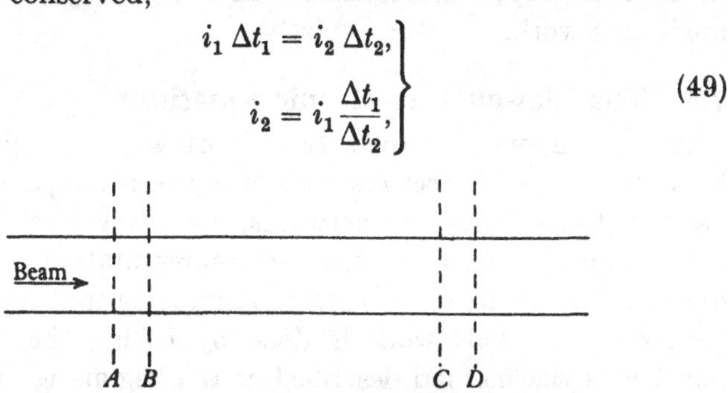

Fig. 7.5. The basis of the klystron.

and in the limit
$$i_2 = i_1\frac{dt_1}{dt_2}. \tag{50}$$

In particular, if the current density is constant at the plane A

$$i_2 = I_0\frac{dt_1}{dt_2}, \tag{51}$$

where I_0 is the d.c. beam current. The application of this relation to the evaluation of the input admittance of a parallel-plane gap, as given by eqns. (24) and (25), is given in detail by Beck.† This method of calculating radio-frequency current is of great importance, because it leads to the simplest method of treating velocity modulation tubes and travelling wave tubes.

The last method of calculating the interchange of energy between a beam and a field applies in the case of electron devices where the h.f. fields are confined to the inside of resonators. The radio-frequency currents flow only on the inside surfaces of the resonators, and the power input to the device is therefore constant in time. Eqn. (14) therefore shows

† Beck, *Velocity-modulated Thermionic Tubes*, Cambridge University Press, 1948.

that the change of kinetic energy of the electrons is the negative of the change in field energy. It is usually fairly simple to evaluate the gain or loss in kinetic energy by kinetic methods. The total gain or loss of kinetic energy is then obtained by averaging over all phases of entry to the system. The decrease or increase in the field energy is then known. This method can be used for graphical determinations of efficiency as well as for analytical work.

7.5. The Llewellyn electronic equations

In two important papers Llewellyn† and his collaborators have studied the correct methods of representing parallel-plane vacuum tubes by active networks, when the frequencies are sufficiently high that the ordinary representation in terms of dynamic impedance and constant input and output capacitances breaks down. This work is done by taking the first-order solution of the method described at the beginning of the last section, i.e. putting

$$\frac{eI}{m\epsilon_0} = K + J_1 e^{pt},$$

where

$$p = \alpha + j\omega.$$

The equations of the last section are then solved for δ_1. This gives the first-order component of the acceleration which in turn allows an expression for the voltage difference between the two planes to be derived.‡ The result is

$$\left.\begin{aligned} V_b - V_a &= AI + Bq_a + Cv_a, \\ q_b &= DI + Eq_a + Fv_a, \\ v_b &= GI + Hq_a + Kv_a, \end{aligned}\right\} \tag{52}$$

where V_b = alternating voltage at plane B,
V_a = alternating voltage at plane A,
I = alternating current,
q = alternating conduction current,
v = alternating velocity,

† Llewellyn and Peterson, *Proc. Inst. Radio Engrs., N.Y.* **32** (1944), 144. Peterson, *Proc. Inst. Radio Engrs., N.Y.* **35** (1947), 1264.

‡ For the details, see Llewellyn, *Electronic Inertia Effects*, chaps. 2 and 4.

A–K = electronic coefficients, depending on the extent of the space charge between the planes.

The necessary relations for making use of this system of equations are given in Table 2.

As an example of the use of these equations, we apply them to a space-charge limited diode. Since we are not considering the effect of initial velocities, but only the velocity variations due to applied fields, both u_a and v_a can be put equal to zero.

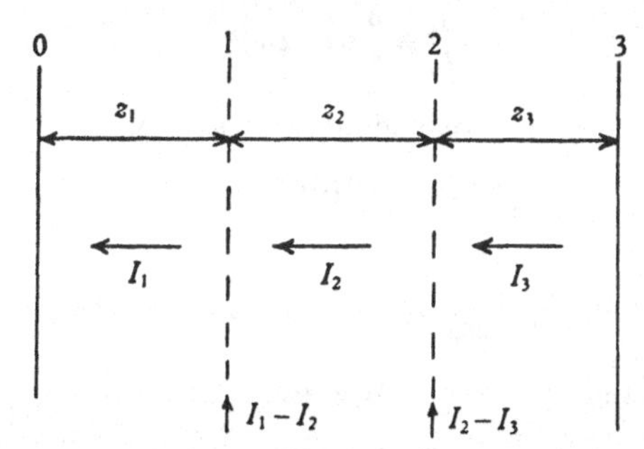

Fig. 7.6. Notation for the Llewellyn electronics equations.

$u_a = 0$ means that $B = 0$, so the first of eqns. (52) reads

$$V_b - V_a = AI. \tag{53}$$

Thus A is the h.f. impedance of the diode.

Now
$$(u_a+u_b)^2 = 2\eta(\sqrt{V_{Da}}+\sqrt{V_{Db}})^2, \tag{54}$$

and
$$\frac{\eta}{\epsilon_0} I_0 = \frac{2(u_a+u_b)}{\tau^2}, \tag{55}$$

the last equation being Child's law rewritten in terms of τ. Using these equations A, the impedance, can be reduced to

$$Z_d = \frac{2}{3}\frac{(\sqrt{V_{Da}}+\sqrt{V_{Db}})^2}{I_0}\left[\frac{2}{\beta}+\frac{12S}{\beta^4}\right]$$
$$= r_c\left[\frac{2}{\beta}+\frac{12S}{\beta^4}\right]. \tag{56}$$

In the last equation we have put $V_{Da} = 0$ which is the case for an emitting cathode, and which is necessary for $u_a = 0$. It is easily shown that eqn. (56) reduces to the value of impedance, in terms of the transit angle, which was given in § 7.2.

Let us now consider a multi-electrode tube, fig. 7.6. We

TABLE 2. CONSTANTS FOR LLEWELLYN'S ELECTRONIC EQUATIONS

Space-charge factor $\quad \zeta = 3\left(1-\frac{\tau_0}{\tau}\right) \quad (0<\zeta<1),$

where τ_0 = transit time, zero space charge, τ = transit time, with actual value of space charge. ζ is calculated from

$$\frac{I_0}{I_m} = \frac{9}{4}\zeta(1-\zeta/3)^2,$$

where $\quad I_m$ = current given by Child's law

$$= \frac{2{\cdot}336\times 10^{-6}(\sqrt{V_{Da}}+\sqrt{V_{Db}})^3}{s^2}.$$

$$\eta = 1{\cdot}77\times 10^{11}, \quad \epsilon_0 = \frac{1}{36\pi\times 10^9}, \quad \eta/\epsilon_0 \doteqdot 2\times 10^{26}, \quad \beta = j\phi, \quad \phi = \omega\tau.$$

Potential variation $\quad \eta V_D = \frac{1}{2}u^2.$

Distance $\quad z = (1-\zeta/3)(u_a+u_b)\tau/2.$

Current density $\quad (\eta/\epsilon_0)I_0 = (u_a+u_b)\frac{2\zeta}{\tau^2}.$

$$P = 1-e^{-\beta}-\beta e^{-\beta} = \frac{\beta^2}{2}-\frac{\beta^3}{3}+\frac{\beta^4}{8}\cdots$$

$$Q = 1-e^{-\beta} = \beta-\frac{\beta^2}{2}+\frac{\beta^3}{6}-\frac{\beta^4}{24}\cdots$$

$$S = 2-2e^{-\beta}-\beta e^{-\beta}-\beta = -\frac{\beta^3}{6}+\frac{\beta^4}{12}-\frac{\beta^5}{40}+\frac{\beta^6}{180}\cdots$$

$$A = \frac{1}{\epsilon_0}(u_a+u_b)\frac{\tau^2}{2}\frac{1}{\beta}\left[1-\frac{\zeta}{3}\left(1-\frac{12S}{\beta^3}\right)\right].$$

$$B = \frac{1}{\epsilon_0}\frac{\tau^2}{\beta^3}[u_a(P-\beta Q)-u_b P+\zeta(u_a+u_b)P].$$

$$C = -\frac{2\zeta}{\eta}(u_a+u_b)\frac{P}{\beta^2}.$$

$$D = 2\zeta\left(\frac{u_a+u_b}{u_b}\right)\frac{P}{\beta^2}.$$

$$E = \frac{1}{u_b}[u_b-\zeta(u_a+u_b)]e^{-\beta}.$$

$$F = \frac{\epsilon_0}{\eta} \frac{2\zeta}{\tau^2} \left(\frac{u_a + u_b}{u_b} \right) \beta \, e^{-\beta}.$$

$$G = -\frac{\eta}{\epsilon_0} \frac{\tau^2}{\beta^3} \frac{1}{u_b} [u_b(P - \beta Q) - u_a P + \zeta(u_a + u_b)P].$$

$$H = -\frac{\eta}{\epsilon_0} \frac{\tau^2}{2} \left(\frac{u_a + u_b}{u_b} \right) (1 - \zeta) \frac{e^{-\beta}}{\beta}.$$

$$K = \frac{1}{u_b} [u_a - \zeta(u_a + u_b)] \, e^{-\beta}.$$

Space-charge limitation	Zero space charge
$\zeta = 1$	$\zeta = 0$
$A = \frac{1}{\epsilon_0} (u_a + u_b) \frac{\tau^2}{3\beta} \left(1 + \frac{6S}{\beta^3} \right)$	$A = \frac{1}{\epsilon_0} (u_a + u_b) \frac{\tau^2}{2} \frac{1}{\beta}$
$B = \frac{1}{\epsilon_0} \frac{\tau^2}{\beta^3} u_a (2P - \beta Q)$	$B = \frac{1}{\epsilon_0} \frac{\tau^2}{\beta^2} [u_a (P - \beta Q) - u_b P]$
$C = -\frac{2}{\eta} (u_a + u_b) \frac{P}{\beta^2}$	$C = 0$
$D = 2 \left(\frac{u_a + u_b}{u_b} \right) \frac{P}{\beta^2}$	$D = 0$
$E = -\frac{u_a}{u_b} e^{-\beta}$	$E = e^{-\beta}$
$F = \frac{\epsilon_0}{\eta} \frac{2}{\tau^2} \left(\frac{u_a + u_b}{u_b} \right) \beta \, e^{-\beta}$	$F = 0$
$G = -\frac{\eta}{\epsilon_0} \frac{\tau^2}{\beta^3} (2P - \beta Q)$	$G = -\frac{\eta}{\epsilon_0} \frac{\tau^2}{\beta^3} \frac{1}{u_b} [u_b (P - \beta Q) - u_a P]$
$H = 0$	$H = -\frac{\eta}{\epsilon_0} \frac{\tau^2}{2} \left(\frac{u_a + u_b}{u_b} \right) \frac{e^{-\beta}}{\beta}$
$K = -e^{-\beta}$	$K = \frac{u_a}{u_b} e^{-\beta}$

assume that electrons are emitted from O which is at zero potential. Now, for region (1),

$$V_1 = A_1 I_1, \quad q_1 = D_1 I_1, \quad v_1 = G_1 I_1,$$

all the other terms having disappeared because of the initial conditions.

For region (2)

$$\left.\begin{aligned} V_2 - V_1 &= A_2 I_2 + B_2 \alpha_1 q_1 + C_2 v_1, \\ q_2 &= D_2 I_2 + E_2 \alpha_1 q_1 + F_2 v_1, \\ v_2 &= G_2 I_2 + H_2 \alpha_1 q_1 + K_2 v_1, \end{aligned}\right\} \tag{57}$$

where α_1 = the proportion of current intercepted by the grid.

Substituting for q_1 and v_1 we get

$$\left.\begin{aligned} I_2 &= (V_2 - V_1)\frac{1}{A_2} - \frac{V_1}{A_1 A_2}(B_2 \alpha_1 D_1 + C_2 G_1), \\ q_2 &= (V_2 - V_1)\frac{D_2}{A_2} + \frac{V_1}{A_1 A_2}[A_2(D_1 \alpha_1 E_2 + G_1 F_2) \\ &\qquad - D_2(D_1 \alpha_1 B_2 + G_1 C_2)], \\ v_2 &= (V_2 - V_1)\frac{G_2}{A_2} + \frac{V_1}{A_1 A_2}[A_2(D_1 \alpha_1 H_2 + G_1 K_2) \\ &\qquad - G_2(D_1 \alpha_1 B_2 + G_1 C_2)]. \end{aligned}\right\} \tag{58}$$

The process can be repeated for regions (3) and (4).

We can then write

$$\left.\begin{aligned} I_1 &= V_1 y_{11}, \\ I_2 &= (V_2 - V_1) y_{22} - V_1 y_{12}, \\ I_3 &= (V_3 - V_2) y_{33} - (V_2 - V_1) y_{23} - V_1 y_{13}, \end{aligned}\right\} \tag{59}$$

and so on, the admittances being defined by

$$\left.\begin{aligned} y_{11} &= \frac{1}{A_1}, \quad y_{22} = \frac{1}{A_2}, \quad y_{33} = \frac{1}{A_3}, \quad \text{etc.}, \\ y_{21} &= \frac{1}{A_1 A_2}(B_1 \alpha_1 D_2 + C_2 G_1), \\ y_{23} &= \frac{1}{A_2 A_3}(B_3 \alpha_2 D_2 + G_2 C_3), \\ y_{13} &= \frac{1}{A_1 A_2 A_3}[A_2\{\alpha_2 B_3(D_1 \alpha_1 E_2 + G_1 F_2) + C_3(D_1 \alpha_1 \\ &\qquad \times H_2 + G_1 K_2)\} - \{\alpha_2 B_2 D_2(D_1 \alpha_1 B_2 \\ &\qquad + G_1 C_2) + C_3 G_2(D_1 \alpha_1 B_2 + G_1 C_2)\}]. \end{aligned}\right\} \tag{60}$$

It should be remembered that direct voltages used in evaluating the necessary quantities are the equivalent voltages at the planes of the grids.

Llewellyn and Peterson† give many examples of the use of these equations to derive networks for the representation of tube performance. Fig. 7.7 shows a typical example, that of a

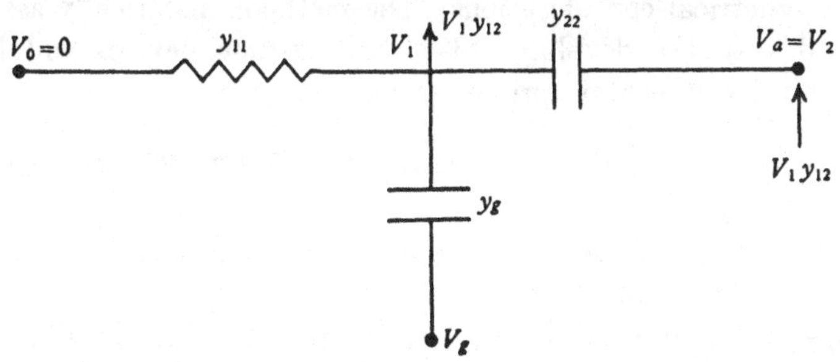

Fig. 7.7. A network for the negative grid triode.

triode. Here y_{11} is the input admittance and y_{22} the output admittance, mainly the plate-grid capacitance. $y_g = \mu C_{ag}$ and y_{12} is the transadmittance, shown plotted in Chapter 8. It should, perhaps, be emphasized that fig. 7.7 is only intended to represent the electron stream and the active part of the electrodes. Lead inductance and capacitances are not included. These quantities are, naturally, of importance to the circuit engineer, but to us they represent effects which are not fundamental in the operation of the tube. Moreover, modern constructional techniques tend to eliminate unwanted reactances.

It is worth noting that the Llewellyn analysis for the input admittance of a diode, applied to the cathode-grid region of the triode, yields a value having the same frequency dependance as that due to Ferris,‡ whose result was

$$G_{in} = kg_m f^2 \tau^2, \tag{61}$$

where k = constant, g_m = mutual conductance. This agrees with the Llewellyn analysis for small transit angles.

The use of Llewellyn's equations is not, of course, confined to conventional gridded valves. The equations are applicable

† Llewellyn and Peterson, loc. cit.
‡ Ferris, loc. cit.

to velocity-modulated tubes, as long as the electrode system can be considered as plane parallel, and Llewellyn, in the paper cited above, gives a derivation of the transconductance of a two-resonator klystron which agrees with the value calculated by other methods. However, dynamical methods are generally used instead, because they are not limited so much by geometrical considerations. The methods historically associated with the development of transit-time devices will be described in the appropriate section.

7.6. Remarks on the inadequacy of single stream electronics

Eqn. (42) is only true for a single-velocity stream of electrons, since, if electrons of several velocities or with a significant range of velocities are under consideration, it is not possible to pass from the partial differential to the total differential as was done in eqns. (39) and (40). If a range of velocities must be considered

$$i = \int_{u_1}^{u_2} \rho u \, du,$$

and eqn. (36) has to be replaced by

$$I = \int_{u_1}^{u_2} \rho u \, du + \frac{\epsilon_0 m}{e} \ddot{z}. \tag{62}$$

Now a practical way in which such a situation could arise is that a cross-section of the beam at the plane of observation contains groups of electrons accelerated by different h.f. fields, as could happen in a magnetron. Then the velocity limits will depend on the acceleration of the electrons. But this is precisely what we wish to calculate, so eqn. (62) is an integral equation. In small-signal theory the difficulty can be overcome, but in large-signal theory, such as is required to give an adequate picture of magnetron operation where the output efficiency may be as high as 70 per cent, this is not the case. Adequate mathematical techniques for attacking this problem have not been devised, and as a result a satisfactory magnetron theory is lacking. Another instance of the failure of single-stream techniques is in the analysis of transit-time effects in the cathode-potential minimum space of the triode, where the range of velocities is big in comparison with the mean velocity.

Some progress in these matters has been made by the use of semi-graphical, semi-analytical methods by Wang† and by Gundlach.‡ The American General Electric Company have made studies by the differential analyser, which give some insight into the problems of triodes.

7.7. Scaling

A technique which is of the utmost importance in practice is that in which tube designs which are successful in one frequency band are scaled so that a similar performance is obtained in a higher frequency band. It should be clear from the earlier sections of this chapter that the requirements for similarity are twofold. First, the transit angles must be the same in the two tubes, and secondly, the space-charge conditions must be identical.

Consider two tubes, one with linear dimensions expressed in units of l_1 working at a frequency f_1, and the other with dimensions l_2 and frequency f_2. Let $l_1/l_2 = n$, $f_2/f_1 = m$, i.e. n and m will usually be >1.

The generalized transit angle is $\omega l/u$, so for similarity

$$\frac{\omega_1 l_1}{u_1} = \frac{\omega_2 l_2}{u_2},$$

therefore

$$\frac{u_2}{u_1} = \sqrt{\frac{V_2}{V_1}} = \frac{\omega_2 l_2}{\omega_1 l_1} = \frac{m}{n},$$

therefore

$$\frac{V_2}{V_1} = \left(\frac{m}{n}\right)^2. \tag{63}$$

The fields scale as V/l, i.e. as m^2/n. Then, using eqn. (41) and the relation $dt_1 = m dt_2$, the current density scales as

$$\frac{J_2}{J_1} = \frac{dE_2/dt_2}{dE_1/dt_1} = \frac{m^3}{n}. \tag{64}$$

Scale factors for other quantities are easily deduced, and their values are tabulated in Table 3. If the linear dimensions are scaled in direct proportion to the wave-length scale factor,

† Wang, *Proc. Inst. Radio Engrs., N.Y.* **29** (1941), 200.

‡ Gundlach, *Funk. u. Ton*, no. 8 (Aug. 1948), 407; no. 9 (Sept. 1948), 454; no. 10 (Oct. 1948), 516.

TABLE 8. SCALE FACTORS

Quantity	Scale factor	Complete scaling $m = n$
Voltage	$(m/n)^2$	1
Field	(m^2/n)	m
Current density	(m^3/n)	m^2
Current	(m^3/n^3)	1
Inductance	n^{-1}	m^{-1}
Capacitance	n^{-1}	m^{-1}
Power	$(m/n)^5$	1
Power density	m^5/n^3	m^2
Conductance	m/n	1

the values of the second column apply. From the practical point of view the current density and power density columns are very interesting. The current density and power density in complete scaling go up as the square of the frequency ratio. Since there are definite physical limits to the values of these quantities which can be obtained with present-day materials, the clear indication is that the frequency of operation cannot be increased without limit. We shall discuss scaling again when we consider specific ultra-high-frequency tube types. More complete and elaborate discussions are given by Lehmann,† Raymond,‡ and Martinot-Lagarde.§

† Lehmann, *Proc. Inst. Radio Engrs., N.Y.* **33** (1945), 663; *Onde élect.* **26** (1946), 175.
‡ Raymond, *Onde élect.* **27** (1947), 209.
§ Martinot-Lagarde, *Onde élect.* **28** (1948), 440.

Chapter 8

FLUCTUATION NOISE IN VALVES

To complete our survey of the general mathematical theory of valves we must consider the production of unwanted signals. The basic experimental fact is that if one applies a small signal through a variable attenuator to the input of a high-gain amplifier and listens to the output on telephones or observes it on a cathode-ray tube, it is found that the wanted signal is eventually lost in a high-pitched crackling hiss or, in the second case, in a thick mass of signals of random amplitude and position. If the input of the amplifier is short-circuited the unwanted signals persist though they are usually much reduced in amplitude. The origin of the signals is therefore partially in the amplifier and partially in the input circuit. The noise in the input circuit is known as thermal, Johnson or resistance noise, while the noise in the amplifier, which, as we shall see, is mostly due to the current in the first valve, is called shot noise. In radar slang the noise signal as seen on a cathode-ray tube screen is called 'grass', a graphic and useful word.

Although we are here mainly concerned with the phenomena inside valves, it is perhaps somewhat easier to understand the reasons for thermal noise in resistors, so we shall discuss these first. In Chapter 1 we pointed out that the electric current in a metallic conductor is due to the difference between the number of electrons drifting to the right through a fixed plane and the number drifting to the left. When no current flows enormous numbers of electrons drift with thermal energies in each direction every second, but, taken over a long time such as a second, the numbers moving in each direction are equal. Taken over a very short time, so short that only a few electrons can drift in each direction, the numbers may be far from equal. Thus each microscopic instant of time contributes a minute current pulse in one direction or the other. On the average the current is zero, but it is always fluctuating by a small amount around the mean value. This fluctuation in current manifests itself as a voltage across the terminals of the resistance, and it is this voltage which is amplified to produce 'grass'. The value of the noise voltage can be deduced from thermodynamics or from quantum statistics. The original derivation by Johnson† and Nyquist‡ used the first discipline. The result is

$$\overline{e^2} = 4kTR\Delta f, \tag{1}$$

where $\overline{e^2}$ = mean square noise voltage,

T = temperature of resistor,

R = resistance,

k = Boltzmann's constant, (e.v.)

Δf = band-width of amplifier (cycles).

Numerically

$$e_{rms} \doteqdot 1{\cdot}26\sqrt{(R\Delta f)} \times 10^{-10} \text{ volts}, \tag{2}$$

if the resistor is at room temperature.

We now introduce the concept of 'available noise power'.§ The resistor R is a noise generator of internal resistance R and open-circuit e.m.f. e_{rms}. The maximum power is taken from it when the load resistance is also R. The open-circuit e.m.f. is then halved and the power output is e_0^2/R. But $e_0^2 = \frac{1}{4}\overline{e^2}$. Thus the available noise power is

$$\frac{\overline{e^2}}{4R} = kT\,\Delta f. \tag{3}$$

This concept is of great importance as it sets up a standard of merit. If the amplifier has a power gain S and the noise power in the output with matched input is $SkT\Delta f$, then the amplifier is perfect, since all the noise comes from the input circuit. Normally the noise output will be $nSkT\Delta f$, i.e. the amplifier contributes $(n-1)$ times the noise of the input circuit. This amplifier is then said to have a noise figure of n, which is

† Johnson, *Phys. Rev.* **32** (1928), 97.
‡ Nyquist, *Phys. Rev.* **32** (1928), 110.
§ Friis, *Proc. Inst. Radio Engrs., N.Y.* **32** (1944), 419.

frequently expressed in decibels. The noise figure can be measured by matching a signal generator to the input and measuring the output with the signal generator switched off. The generator is then switched on and the generator output adjusted until the amplifier output power is doubled. The signal generator output is then $nkT\Delta f$, so that if T and Δf are also measured n can be calculated. The theory of the measurement is therefore very simple. The practical technique is elaborate if accurate results are to be obtained. The design of the final detector to give a linear power indication demands careful consideration.

Inside the valve essentially the same phenomenon takes place. The number of electrons incident on the emitting surface from the interior of the emitter is constant when averaged over a macroscopic time interval, but on a microscopic time scale the number fluctuates violently. The energy distribution of these electrons is determined by the appropriate statistics, Fermi-Dirac or Maxwell-Boltzmann, but the times of arrival at the emitting surface are random. The emission therefore shows the same random structure if sampled in very short time intervals. The current through the valve therefore shows the same sort of fluctuation as the resistor e.m.f. If the emission is temperature limited,

$$\overline{i^2} = 2eI\Delta f, \tag{4}$$

where $\overline{i^2}$ = mean square noise current, e = electronic charge, I = electron current.

Other types of valve noise are met with in practice, but we shall have little to say about them since they are evitable, at least theoretically. An exception is flicker noise, which we describe later. These types of noise are:

(1) noise due to the pressure of gas,

(2) noise due to vibration of the electrodes (microphonicity),

(3) noise due to poor contacts.

Returning now to the question of shot noise it is interesting to note that eqn. (4) can be established to within a constant multiplier by very simple reasoning. Suppose we count the number of electrons passing into the anode in short equal intervals Δt. The number can be written as $N_{\Delta t}$. If N is the number in a long time T, the expected number in Δt is $N\frac{\Delta t}{T}$. The fluctuation is then

$$u = N_{\Delta t} - N\frac{\Delta t}{T}. \tag{5}$$

The probability of a particular electron crossing in the time Δt is simply $\Delta t/T = p$, and of not crossing is $1-\Delta t/T = q$.

Now the standard deviation of a normal distribution is

$$\sigma = \sqrt{(Npq)} = \sqrt{\left\{N\frac{\Delta t}{T}\left(1-\frac{\Delta t}{T}\right)\right\}} \doteqdot \sqrt{\left(N\frac{\Delta t}{T}\right)}.$$

Thus the average fluctuation

$$\bar{u} = \sqrt{N\frac{\Delta t}{T}}.$$

The mean current fluctuation is then

$$i = \frac{e\bar{u}}{\Delta t},$$

or

$$\overline{i^2} = \frac{e^2 N}{\Delta t T} = \frac{eI}{\Delta t}. \tag{6}$$

We can now ascribe a meaning to Δt by the following reasoning. If a pulse of length Δt is to pass through a low-pass filter without much distortion, the filter band-width must equal $1/\Delta t$ at least. Therefore an amplifier of band-width Δf will analyse the fluctuations over the time $\Delta t = 1/\Delta f$. Eqn. (6) can therefore be rewritten

$$\overline{i^2} = eI\Delta f, \tag{7}$$

which agrees with eqn. (4) apart from the factor 2. Naturally this only indicates the general correctness of eqn. (4).

To conclude these introductory remarks, let us discuss briefly two questions which have caused a great deal of trouble in the theory of noise. First, is there any physical distinction between thermal noise and shot noise? Secondly, is it correct to use Fourier series to calculate noise spectra, since the use of Fourier series presupposes that the phenomenon to be analysed is repetitive?

The answer to the first question is that fundamentally thermal noise and shot noise are due to the same thing, the fact that electron motions in solids are not regular but random, if the interval of observation is taken to be short enough. Thermal noise is essentially the noise observed from a system in thermal equilibrium, while shot noise is the noise from a dynamic system to which the concepts of thermal equilibrium cannot be applied. It is usually obvious which theoretical technique will be most directly applicable, but there is no real physical difference between the two 'sorts' of noise. The correctness of this attitude is shown by the fact that Johnson's thermal noise formula can be derived in the following different ways:

(1) by classical statistical mechanics (thermodynamics),

(2) by quantum statistical mechanics,

(3) by a detailed application of the Lorentz theory of metallic conduction,

(4) by the quantum theory of metallic conduction.

The last two methods are the only ones bringing the actual motions into the picture, and many authors state that they lack generality as compared with the statistical methods. This is true in a way, but it does not mean that the approach from electron theory is less correct. Conversely, there have been many attempts† to show directly that shot noise is the thermal noise in the valve conductance and, from what has been said above, it is obvious that this is logically possible; though it is doubtful whether any practical advantages result, as some of the steps of the argument are very obscure.

The second question can also now be answered with some confidence. The argument that, since noise is non-repetitive, Fourier series cannot be used to describe noise spectra is correct but irrelevant. Since all measuring instruments have only a limited pass band, one is only concerned with investigating spectra in this band, and it turns out that it is immaterial

† Particularly by D. A. Bell, *J. Instn. Elect. Engrs.* **93**, pt. 3 (1946), 37, and earlier papers.

whether one associates a spectrum with the noise impulses or with the effect they produce in the circuit. The difficulty that, if the noise from a resistance or a valve is integrated over the whole range of frequencies from zero to infinity, it would have an infinite value is eliminated by quantum-mechanical and transit-time treatments. These show that eqn. (1) and eqn. (4), respectively, are only low-frequency approximations to the correct results; the difficulty is not therefore introduced by any question about the validity of Fourier representations.†

8.1. Thermal noise in resistors

We first give the derivation of eqn. (1) due to Nyquist. Referring to fig. 8.1 we consider a long, lossless transmission

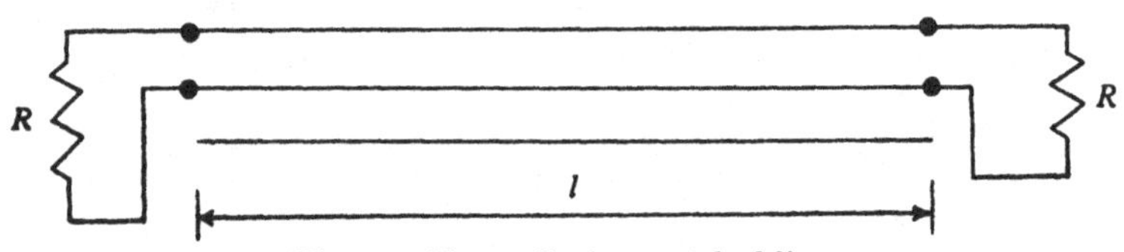

Fig. 8.1. Non-radiating matched line.

line (which we have drawn as a coaxial to emphasize the fact that it does not radiate), terminated at both ends in resistors R so that the energy reaching either end is absorbed or reflected back down the line. The line is in thermal equilibrium. The length of the line is l. A short-circuited line forms a resonator of gravest frequency

$$f_1 = \frac{c}{2l}, \tag{8}$$

and in addition all the harmonics of f_1 are possible resonances of the system. Moreover, there is an electric and a magnetic field to be considered, so the system has two degrees of freedom for each resonance. Thus the number of degrees of freedom in the frequency interval $f, f+\Delta f$ is

$$dN_f = \frac{2\Delta f}{f_1} = \frac{4l}{c}\Delta f. \tag{9}$$

† A stimulating discussion of the basis of noise calculations which takes a viewpoint opposed to that given here, is N. R. Campbell's *Monograph on Noise*, G.E.C. Research Labs., 1942.

Now it is a well-known result of classical statistical mechanics that energy equal to $\frac{1}{2}kT$ must be associated with each degree of freedom. The average electromagnetic energy in the range Δf is then

$$dU_f = \frac{2l}{c} kT \Delta f. \tag{10}$$

Now, if the resistances R are equal to the characteristic impedance of the line, i.e. $R = \sqrt{(L/C)}$, there will be no reflexion from them and the line will act exactly like an infinite line short-circuited at the ends. Thus eqn. (10) will also apply to the line with matched terminations. Since the system is in equilibrium each resistor must be responsible for $\frac{1}{2}dU_f$. Also it takes l/c seconds for the wave to traverse the line, so the power produced by each resistor is

$$\frac{c\,dU_f}{2l} = kT \Delta f. \tag{11}$$

But eqn. (11) is just the noise power available from the resistor R, and therefore eqns. (1) and (3) have been verified.

Bernamont† gives several other methods of deriving these equations, the most important of which are those based on electron theory, of which there are three, two based on quantum theory and one on the classical theory.

8.2. Some circuit aspects of thermal noise

Although circuit considerations are not strictly within the bounds of this work, it would be unrealistic not to say anything about methods of handling the thermal noise in circuit calculations. A resistor can be represented from the noise viewpoint by either of the equivalents shown in fig. 8.2, where $G = 1/R$ is the conductance.

The next question which arises is what happens when two resistors are connected in parallel? It is clear that the noise currents in the two resistors are quite uncorrelated, i.e. molecular phenomena in one resistor produce no correlated effect in the other. In these circumstances the r.m.s. noise currents simply add. If the resistors are R_1 and R_2, conductances G_1 and

† Bernamont, *Ann. Phys., Paris*, 11ème série, 7 (1937), 71.

G_2, the r.m.s. noise current is

$$i = \sqrt{\{4kT\,\Delta f(G_1+G_2)\}}$$
$$= \sqrt{\left\{4kT\,\Delta f\left(\frac{R_1 R_2}{R_1+R_2}\right)\right\}}. \tag{12}$$

Eqn. (12) is exactly the value obtained for the noise from a resistance equal to R_1, R_2 in parallel, as it should be.

When the noise-producing resistor forms part of a reactive network it is usual to resort to formulæ given by Johnson,† i.e.

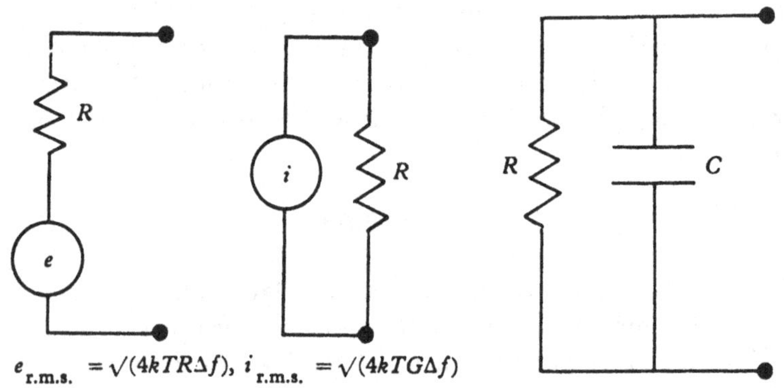

Fig. 8.2. Equivalent noise representations for a resistor.

Fig. 8.3. Noise producing circuit.

$$\overline{i^2} = \frac{2kT}{\pi}\int_0^\infty R(\omega)\,Y^2(\omega)\,d\omega, \tag{13}$$

$$\overline{v^2} = \frac{2kT}{\pi}\int_0^\infty R(\omega)\,G^2(\omega)\,d\omega. \tag{14}$$

Here $R(\omega)$ is the real part of the impedance of the noisy network and $Y(\omega)$ is the transfer admittance of the amplifier in which the noise is measured. $G(\omega)$ is the gain. As an application of eqn. (13), consider fig. 8.3:

$$R(\omega) = \mathfrak{R}_e\left(\frac{R\dfrac{1}{j\omega C}}{R+j\omega C}\right),$$
$$= \frac{R}{1+\omega^2 C^2 R^2}.$$

† Johnson, loc. cit.

Then
$$\overline{v^2} = \frac{2kT}{\pi}\int_0^\infty \frac{R}{1+\omega^2C^2R^2}\,d\omega,$$
$$= \frac{kT}{C},$$
or
$$\tfrac{1}{2}C\overline{v^2} = \tfrac{1}{2}kT. \tag{15}$$

It is interesting to note that, since $\frac{1}{2}C\overline{v^2}$ is the energy stored in the condenser, eqn. (15) is another way of stating the 'equipartition of energy' theorem of statistical mechanics which was used in Nyquist's derivation of eqn. (1). Furthermore, it is interesting to note that, though the integration is carried to infinity, the result is finite. This is because the high-frequency components of the noise are short-circuited by the condenser so that the result is independent of their values. Since all electrical measuring apparatus can be represented as a capacitative reactance if the frequency is made high enough, this means that in practice electrical measurements are not capable of observing the quantum phenomena of resistor noise which only become measurable at optical frequencies.

To conclude this section we note that it has been shown theoretically and experimentally by Ornstein† and his collaborators that, if the various resistances of a circuit are at different temperatures, the noise can be computed by using the foregoing formulæ with an effective temperature. This effective temperature is given by
$$T_e = \frac{\sum_n R_n T_n}{\sum_n R_n}. \tag{16}$$

8.3. The shot noise in temperature-limited diodes

The possible effects of the random emission of electrons in the production of noise in amplifiers were first discussed by Schottky,‡ although similar fluctuations in the rate of emission of α-particles had been studied much earlier by Schweidler§ and Campbell‖. The mathematical methods introduced by the latter

† Ornstein, *Z. Phys.* **41** (1927), 848; *Proc. Roy. Soc.* A, **115** (1927), 391.
‡ Schottky, *Ann. Phys., Lpz.* **57** (1918), 541.
§ Schweidler, *Phys. Z.* **11** (1910), 225 and 614.
‖ Campbell, *Proc. Camb. Phil. Soc.* **15** (1909), 117.

author are nowadays almost universally used for shot-noise calculations.

The problem of this section is to calculate the current fluctuation in the resistance R in fig. 8.4 when a known mean current I_0 flows through the temperature-limited diode. The frequency characteristic of the measuring amplifier is assumed to be known. Now, the mean current I_0 is due to the passage of

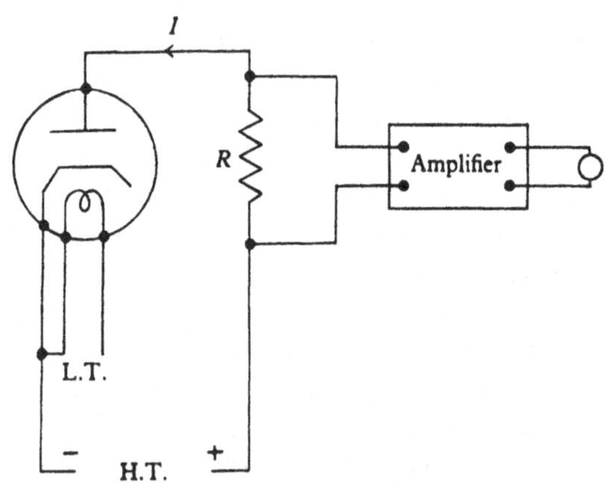

Fig. 8.4. Measurement of shot noise.

a very large number of electrons. If n_e = number of electrons flowing from cathode to anode in 1 sec.,

$$I_0 = n_e e.$$

Owing to the fact that the electron emission is random in time, the number of electrons in equal time intervals $\Delta t_0, \Delta t_1, \ldots, \Delta t_n$ at different times is not equal to the average number in Δt, i.e.

$$n_0 \Delta t_0 \neq n_1 \Delta t_1 \neq n_2 \Delta t_2 \neq \ldots \neq n_n \Delta t_n,$$

although $$\Delta t_0 = \Delta t_1 = \ldots = \Delta t_n.$$

We have to make some assumption about the probability that an event, in this case the traversal of the cathode-anode space by an electron, will occur in a given time interval. The natural assumption is that of equal probabilities, i.e. that if the average number of electrons flowing is n_e the probability is $n_e \delta t$, where δt is chosen so that $n_e \delta t \ll 1$. It is also assumed that the events are independent of one another. It is shown in text-books on statistics that these assumptions lead to Poisson's law of small probabilities, which states that if the number of electrons flowing in equal intervals t is measured for very many such intervals and $p(M)$ is defined as the number of trials giving M electrons divided by the total number of trials, evaluated in the limit as the number of trials $\rightarrow \infty$,

i.e. $$p(M) = \underset{N\rightarrow\infty}{Lt} \frac{\text{No. of trials yielding } M \text{ electrons}}{N = \text{no. of trials (total)}},$$

then $$p(M) = \frac{(n_e t)^M}{M!} \exp(-n_e t). \tag{17}$$

We shall only need one property of the Poisson distribution, that it is indistinguishable from the normal distribution when $n_e t$ is large. For a normal distribution it is well known that the standard deviation is equal to the square root of the mean rate or

$$\sigma^2 = \overline{(n-n_e)^2} = \overline{n^2 - n_e^2} = n_e = \frac{I_0}{e}. \tag{18}$$

Eqn. (18) tells us the expected fluctuation from the mean value $n.$; we now have to convert this into a voltage or current fluctuation in the resistance R.

This is done by applying Campbell's† theorem which states that if the arrival of each particle produces an effect $F(t)$ in the output circuit the average value of the output will be

$$\overline{I_{(t)}} = n_e \int_{-\infty}^{+\infty} F(t)\,dt, \tag{19}$$

and the mean square fluctuation will be

$$\overline{(I_{(t)} - \overline{I_{(t)}})^2} = n_e \int_{-\infty}^{+\infty} F^2(t)\,dt. \tag{20}$$

The next step which must be taken to make eqn. (20) useful in circuit analysis is to convert it from a time function to a

† Proofs of Campbell's theorem are given (*a*) in his *Monograph on Noise*; (*b*) by Whittaker, *Proc. Camb. Phil. Soc.* **33** (1937), 451; **34** (1938), 158; (*c*) by Rice, *Bell Syst. Tech. J.* **23** (1944), 282; (*d*) by Rowland, *Proc. Camb. Phil. Soc.* **32** (1936), 580; **34** (1938), 329.

frequency function, since our ordinary methods of circuit theory are based on frequency and not on time. This is done in the usual manner by forming the Fourier transform. We know that if a current pulse is given in the form $i = F(t)$, we can write the Fourier transform pair:

$$F(t) = \frac{1}{\sqrt{(2\pi)}}\int_{-\infty}^{+\infty} G(\omega)e^{j\omega t}d\omega, \tag{21}$$

$$G(\omega) = \frac{1}{\sqrt{(2\pi)}}\int_{-\infty}^{+\infty} F(t)e^{-j\omega t}dt. \tag{22}$$

To interpret eqn. (20) we need the F transform of $F^2(t)$. This can be obtained by the use of Parseval's theorem.† The result is

$$\int_{-\infty}^{+\infty} F^2(t)dt = \int_{-\infty}^{+\infty} G(\omega)G^*(\omega)d\omega,$$

where $G^*(\omega)$ = complex conjugate of $G(\omega)$.

We note that $G(\omega)G^*(\omega)$ must be an even function, so we can now write, using eqn. (20),

$$\overline{i^2} = \overline{(I_{(t)} - \overline{I_{(t)}})^2} = \frac{2I_0}{e}\int_0^{\infty} G(\omega)G^*(\omega)d\omega. \tag{23}$$

We next have to determine $F(t)$ which in turn allows us to evaluate $G(\omega)$. For the temperature-limited planar diode $F(t)$ is given by eqn. (4 a) of the last chapter; it is $F(t) = \dfrac{2et}{\tau^2}$.

Therefore $$G(\omega) = \frac{1}{\sqrt{(2\pi)}}\frac{2e}{\tau^2}\int_0^{\tau} te^{-j\omega t}dt,$$

$$= \frac{2e}{\sqrt{(2\pi)}}\frac{1}{\theta^2}\int_0^{\theta} z\,e^{-jz}dz,$$

$$= \frac{2e}{\sqrt{(2\pi)}}\frac{1}{\theta^2}[(\theta\sin\theta + \cos\theta - 1) - j(\sin\theta - \theta\cos\theta)],$$

where $\theta = \omega\tau$ = transit angle.

† See Titchmarsh, *Introduction to Theory of Fourier Integrals*, Oxford University Press, 1937, chaps. II and III.

Then

$$G(\omega)G^*(\omega) = \frac{2e^2}{\pi}\frac{1}{\theta^4}[(\theta\sin\theta+\cos\theta-1)^2+(\sin\theta-\theta\cos\theta)^2],$$

$$= \frac{2e^2}{\pi}\frac{1}{\theta^4}[\theta^2+2(1-\cos\theta-\theta\sin\theta)].$$

Let us now consider that the amplifier shown in fig. 8.4 has a very small band-width, small enough for $G(\omega)G^*(\omega)$ to be considered constant in the band. Eqn. (23) can then be re-

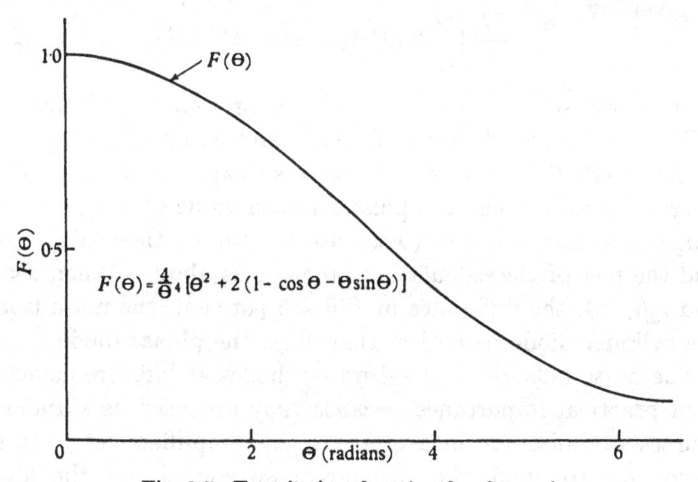

Fig. 8.5. Transit time function for shot noise.

placed by

$$\overline{i^2} = \frac{4eI_0}{\pi\theta^4}[\theta^2+2(1-\cos\theta-\theta\sin\theta)]\Delta\omega$$

$$= \frac{8eI_0}{\theta^4}[\theta^2+2(1-\cos\theta-\theta\sin\theta)]\Delta f. \tag{24}$$

Eqn. (24) gives the fluctuation current which would be observed in the range of frequencies $f, f+\Delta f$ when θ is given the value $2\pi f\tau$. Eqn. (24) can be written

$$\overline{i^2} = 2eI\,\Delta f F(\theta), \tag{25}$$

$$F(\theta) = \frac{4}{\theta^4}[\theta^2+2(1-\cos\theta-\theta\sin\theta)]. \tag{26}$$

As $\theta\to 0$, $F(\theta)\to 1$ so that eqn. (26) shows the behaviour we should expect at the low frequencies as it then reduces to eqn. (4). Eqn. (26) is plotted in fig. 8.5.

If the noise current in the resistor is to be measured in an amplifier with a very wide band-width, so that it is impossible to consider it constant over the whole band, it is easy to integrate eqn. (23) graphically between the upper and lower cut-off frequencies of the amplifier. In general, the amplifier gain will be represented by some function of ω, $A(\omega)$ say. If this is the case, the fluctuation current read on a true square-law instrument, thermocouple or thermistor in the amplifier output will be given by

$$\overline{i^2} = \frac{2I_0}{e}\int_{\omega_1}^{\omega_2} G(\omega)G^*(\omega)A(\omega)A^*(\omega)d\omega, \qquad (27)$$

which again will usually have to be integrated graphically.

The case of the cylinder diode has been treated by an approximation method by Fraser.† He derives an approximate analytic expression for the current pulse for each value of r_a/r_c; e.g. for $r_a/r_c = 4$ he finds $i = 2e/\tau^2(0{\cdot}8t - 0{\cdot}43t^2)$. $G(\omega)$ is then calculated and the rest of the calculation proceeds as above. When $\theta < 2$ and $r_a/r_c < 4$, the difference in $F(\theta) < 5$ per cent, the noise from the cylinder diode being less than from the planar diode.

The noise behaviour of saturated diodes at high frequencies is of practical importance because they are used as standard sources of noise for measurements on amplifiers, etc. It is impossible to make the spacings close enough for the low-frequency analysis to apply, and therefore accurate theoretical knowledge of the behaviour for large-transit angles is necessary.

The analysis of shot noise has been carried out using a different mathematical technique by Blanc-Lapierre,‡ who bases his work on the use of correlation functions. A similar method is used by Schremp in his interesting chapter on noise in *Vacuum Tube Amplifiers*.§

8.4. The reduction of shot noise by space charge

If a diode is set up in the circuit of fig. 8.4 and the anode current held constant while the heater power is reduced, the

† Fraser, *Wireless Engr.* **26** (April 1949), 129–32.
‡ Blanc-Lapierre, *Rev. sci., Paris,* **84** (1946), 75.
§ Vol. **18** of M.I.T. series on radar, ed. Valley and Wallman, McGraw Hill, 1948.

noise output increases until the point is reached at which the valve is temperature-limited. The noise then remains constant. We require to give a theoretical analysis of the effect of space charge in reducing shot noise and to deduce some quantitative results for the magnitude of the effect.

It was at first thought that the reduction in shot noise when a diode becomes space-charge limited was simply due to the reduced effect of a given fluctuation in the saturation current. It is an obvious experimental fact that the anode current in a space-charge limited diode is practically independent of I_s when $I_s/I_a \gg 1$. In fact, $I_s/I_a > 3$ is a condition sufficient to ensure that I_a only varies very slowly with I_s. Using the results of § 6.7 and § 6.8, it is easy to show that

$$\frac{\Delta I_s}{I_s} \doteqdot \frac{\Delta I_a}{I_a}\left(1+\frac{2V_a}{3kT}\right), \tag{28}$$

and the term in the bracket, very approximately $8V_a$, is in all normal circumstances a large number.

If we write the following expression for the noise current in the space-charge limited case

$$\overline{i_a^2} = 2eI_a\Gamma^2\Delta f, \tag{29}$$

where Γ^2 = space-charge noise reduction factor, then by using eqn. (28) with eqn. (29) we can obtain an expression for

$$\Gamma^2 = \frac{I_s}{I_a}\left(1+\frac{2V_a}{3kT}\right)^2. \tag{30}$$

This expression for Γ^2 is in complete disagreement with experiment, as it forecasts values of Γ^2 much smaller than those which are observed. Eqn. (28) is correct to a high degree of approximation and cannot be the source of the error. Clearly, we must re-examine the physics of the problem.

In fig. 8.6 we show a plot of the number of electrons with energies in the interval V, $V+\Delta V$ for a Maxwellian distribution. The lower line represents the number when the saturation current is equal to I_s and the upper the case when the saturation current changes to a higher value, $I_s+\Delta I_s$. Initially, when the saturation current is I_s, the potential minimum in the space-charge limited diode is at V_m in fig. 8.7 and $I_a = I_s \exp(V_m/kT)$. In fig. 8.6, therefore, the area under the lower curve to the right of the line AB represents the anode current. When I_s

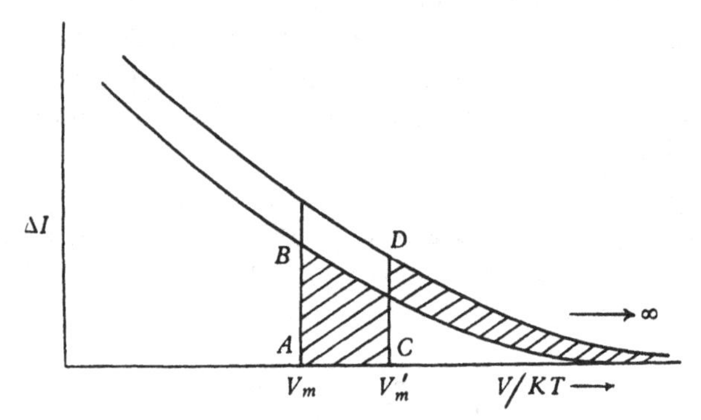

Fig. 8.6. Effect of change of emission on V_m.

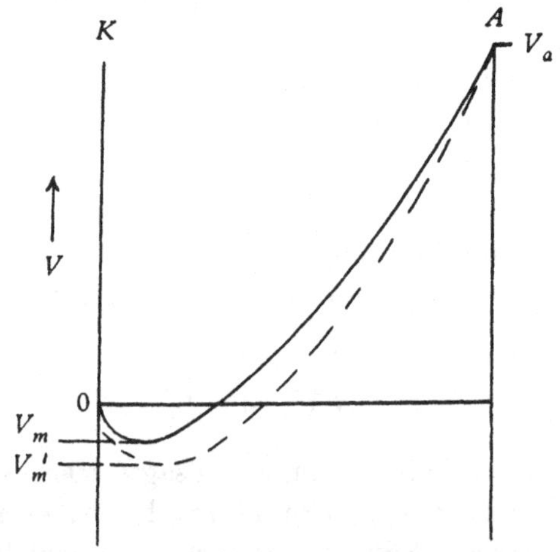

Fig. 8.7. Potential profiles as emission changes.

increases to $I_s + \Delta I_s$, a new equilibrium is established with a deeper potential minimum and a practically unchanged anode current. The new anode current is the area under the upper curve to the right of CD. Since the anode-current is nearly constant the two cross-hatched areas must be approximately equal, and, in fact, the fluctuation in V_m is just sufficient to maintain this near equality. It is clear that if V_m were held constant as I_s increased, as is the case in the diode under retarding field conditions, the fluctuation in the anode current would be equal to the anode current times the relative change in I_s, and no smoothing by the space charge can result. This can be verified experimentally. In the case of space-charge limitation another effect must be looked for. Fig. 8.7 gives the clue. To the right of the potential minimum, the potential is of the form $c(V_a - V_m)^{\frac{3}{4}}$ when I_s flows, and $c(V_a - V_m')^{\frac{3}{4}}$ when $I_s + \Delta I_s$ flows, the constants being the same because the shift of the minimum position is negligible. Thus when $I_s + \Delta I_s$ flows all the electrons at any point in the space to the right of the minimum are moving faster than were the electrons at the same point when I_s alone flowed. The currents induced in the anode are therefore greater and, although the number of electrons decreases, a larger anode current is produced when the emission fluctuates positively. In addition, there is a small effect due to the currents induced by the electrons which only move to the minimum and back. Their effect is known to be only about 10 per cent of that due to the electrons reaching the anode. The noise in space-charge limited diodes is therefore primarily due to velocity variations in the potential minimum-anode space.

Several authors have given the detailed calculations necessary to evaluate the space-charge smoothing factor, the analyses of North, Schottky and Spenke, and Rack† being the best known. These analyses are very elaborate, and we shall only indicate the basic ideas of Spenke and North's calculation here. They start from the Maxwell distribution in the form

$$dI_s = \frac{I_s}{kT} \exp\left[-V_s/kT\right] dV_s, \tag{31}$$

where V_s equals the volt equivalent of the electrons' normal velocity component, together with the relation

$$I_a/I_s = \exp\left(V_m/kT\right).$$

† North, *R.C.A. Rev.* **4** (1940), 441. Schottky and Spenke, *Wiss. Veröff. Siemens-Konz.* **16** (1937), 1 and 19. Rack, *Bell Syst. Tech. J.* **17** (1938), 592.

Introducing the notation

$$\lambda = \frac{V_s+V_m}{kT}, \tag{32}$$

we obtain the following expression for the current due to electrons with velocities between λ and $\lambda+d\lambda$,

$$dI_s = I_a e^{-\lambda} d\lambda. \tag{33}$$

This current exhibits true shot-noise fluctuations

$$\overline{(dI_s)^2} = 2eI_a e^{-\lambda} d\lambda \Delta f. \tag{34}$$

They next consider that every variation in the emission is linearly reduced at the anode by a factor $\gamma(\lambda)$ depending on the emission velocity. Therefore

$$\overline{(di_a)^2} = 2e I_a \Delta f[\gamma^2(\lambda) e^{-\lambda} d\lambda],$$

and

$$\overline{i_a^2} = 2e I_a \Delta f \int_{V_m/kT}^{\infty} \gamma^2(\lambda) e^{-\lambda} d\lambda, \tag{35}$$

Therefore $\Gamma^2 = \Gamma_\alpha^2 + \Gamma_\beta^2$

$$= \int_{V_m/kT}^{0} \gamma^2(\lambda) e^{-\lambda} d\lambda + \int_0^{\infty} \gamma^2(\lambda) e^{-\lambda} d\lambda. \tag{36}$$

For the electrons which reach the anode (β group),

$$\gamma(\lambda) = 1 + \frac{\hat{I} - I_a}{i_s(\lambda)} \quad (0 \leqslant \lambda \leqslant \infty).$$

For the α group, $\gamma(\lambda) = \dfrac{\hat{I} - I_a}{i_s(\lambda)}, \quad \dfrac{V_m}{kT} \leqslant \lambda \leqslant 0.$

Here $i_s(\lambda)$ = an increase in the emission with velocity λ, I_a = mean anode current, $\hat{I}$ = anode current when $i_s(\lambda)$ flows in addition to I_s, but not including $i_s(\lambda)$. Thus $\gamma(\lambda)$ measures the change in anode current for a change in the number of emitted electrons in a specified velocity class. $e^{-\lambda}$ is a weighting factor belonging to the distribution. The evaluation of the integrals in eqn. (36) depends on the use of Langmuir's development for the effect of initial velocities, and it has to be carried out by numerical methods.

Rack† gives two methods for the calculation, the first of which is a detailed application of the Langmuir theory assuming that there is an instantaneous deviation from the mean rate of emission. His second method uses the Llewellyn analysis of transit time electronics and is therefore much closer to the spirit of this book. The analysis consists of the application of the Llewellyn electronics equations to the space between potential minimum and anode. The origin is taken very slightly beyond the potential minimum so that the electron motion is in one direction only, but the d.c. acceleration can still be taken as zero.

Eqn. (52) of Chapter 7 gives us the relation

$$V_1 = AI + Bq_a + Cv_a, \tag{37}$$

where V_1 = a.c. voltage between the minimum and the anode and the other symbols have their earlier meaning.

Now we have already shown that q_a, the fluctuating conduction current density at the minimum, is negligibly small (see eqn. (28)). Also, we already know from the example in Chapter 7 that A is simply the h.f. impedance of the diode. The term Cv_a, where it will be remembered that v_a is the fluctuating component of velocity at the minimum, is new.

Thus
$$V_1 = AI + Cv_a,$$

and
$$Cv_a = -\frac{I_0 v_a}{\omega^2 \epsilon_0}(j\theta\,\mathrm{e}^{-j\theta} + \mathrm{e}^{-j\theta} - 1). \tag{38}$$

We can look on eqn. (38) as a generator of e.m.f. equal to Cv_a located inside the diode. The mean-square noise voltage is then

$$\overline{V_n^2} = \frac{I_0^2 \overline{v_a^2}}{\omega^4 \epsilon_0^2}\left| j\theta\,\mathrm{e}^{-j\theta} + \mathrm{e}^{-j\theta} - 1 \right|^2. \tag{39}$$

We must next evaluate $\overline{v_a^2}$ for those electrons which cross the minimum, remembering that $\overline{v_a}$ is the fluctuation velocity, which the electron stream passing the minimum must have in order to behave as does the real stream with fluctuating mean velocity.

† Rack, loc. cit.

By definition the mean velocity for the β electrons is

$$\bar{u} = \frac{\int_{V_m}^{\infty} u_m \, dI_{(E)}}{\int_{V_m}^{\infty} dI_{(E)}}, \tag{40}$$

where $$u_m = \left((V - V_m)\frac{2kT}{m}\right)^{\frac{1}{2}}$$

= velocity after passing the minimum of an electron which initially had an energy with volt equivalent V.

Now, the number of electrons in each group $\Delta I_{(E)}$ is subject to shot fluctuations, and not only $\Delta I_{(E)}$ changes but also the lower limit of the integration. Carrying out the variation, we get

$$\bar{u} + \delta\bar{u} = \frac{\int_{V_m}^{\infty} u_m \, dI_{(E)} + u_m \, \delta I_{(E)} - \bar{u}\left(\frac{dI}{dE}\right) V_m \, \delta V_m}{\int_{V_m}^{\infty} dI_{(E)} + \delta I_{(E)} - \left(\frac{dI}{dE}\right) V_m \, \delta V_m}. \tag{41}$$

Therefore

$$\delta\bar{u} = \frac{\int_{V_m}^{\infty} u_m \, dI_{(E)} - \bar{u}\int_{V_m}^{\infty} dI_{(E)} + (u_m - \bar{u})\, \delta I_{(E)}}{\int_{V_m}^{\infty} dI_{(E)} + \delta I_{(E)} - \left(\frac{dI}{dE}\right) V_m \, \delta V_m}. \tag{42}$$

We now approximate by neglecting the variations in the denominator, which amounts to assuming that the number of electrons does not change. Then, using eqn. (40), the first two terms cancel and

$$\delta\bar{u} = \left(\frac{u_m - u}{I}\right) \delta I_{(E)},$$

$$\overline{\delta\bar{u}^2} = \overline{v_a^2} = \frac{\overline{(u_m - u)^2}\, \overline{\delta I_{(E)}^2}}{I^2}, \tag{43}$$

and since $\delta I_{(E)}$ is a pure shot effect, we can put

$$\overline{\delta I_{(E)}^2} = 2eI_0 \Delta f,$$

therefore $$\overline{v_a^2} = \frac{2e}{I_0}(\overline{u^2} - \overline{u_m}^2)\Delta f. \tag{44}$$

The square of the mean velocity and the mean square velocity are readily evaluated from the properties of the Maxwellian distribution† to be

$$\overline{u_m}^2 = \frac{\pi kT}{2m}, \tag{45}$$

and

$$\overline{u^2} = \frac{2kT}{m}; \tag{46}$$

therefore

$$\overline{v_a^2} = \frac{4kT\,e}{I_0 m}\left(1-\frac{\pi}{4}\right)\Delta f. \tag{47}$$

Inserting eqn. (47) into eqn. (39) we have

$$\overline{V_n^2} = \frac{4kTI_0}{\omega^4\epsilon_0^2}\frac{e}{m}\left(1-\frac{\pi}{4}\right)\left|j\theta\,\mathrm{e}^{-j\theta}+\mathrm{e}^{-j\theta}-1\right|^2\Delta f,$$

but using eqns. (53) and (54) of Chapter 7 we can show that the slope resistance of the diode (r_c) is given by

$$\frac{eI_0}{m\epsilon_0^2} = \frac{12r_c}{\tau^4} = \frac{12r_c\omega^4}{\theta^4}.$$

Therefore

$$\overline{V_n^2} = 4kT\,r_c\frac{12}{\theta^4}\left(1-\frac{\pi}{4}\right)\Delta f\,[(\cos\theta+\theta\sin\theta-1)^2+(\sin\theta-\theta\cos\theta)^2]$$

$$= 4kT\,r_c\left(1-\frac{\pi}{4}\right)\Delta f\frac{12}{\theta^4}\,[\theta^2+2(1-\cos\theta-\theta\sin\theta)]. \tag{48}$$

At low frequencies

$$\theta\to 0,\quad \theta^2+2(1-\cos\theta-\theta\sin\theta)\to\frac{3\theta^4}{12},$$

and the low-frequency value of the noise voltage is

$$\overline{V_n^2} = 4kT(0{\cdot}644r_c)\Delta f. \tag{49}$$

† The integrals to be evaluated are

$$\overline{u_m} = \frac{\int_0^\infty u^2\exp(-mu^2/2kT)du}{\int_0^\infty u\exp(-mu^2/2kT)du} = \sqrt{\frac{\pi kT}{2m}},$$

and

$$\overline{u^2} = \frac{\int_0^\infty u^3\exp(-mu^2/2kT)du}{\int_0^\infty u\exp(-mu^2/2kT)du} = \frac{2kT}{m}.$$

At high frequencies we can use $F(\theta)$ as defined by eqn. (26) and write

$$\overline{V_n^2} = 4kT(0{\cdot}644r_c)F(\theta)\Delta f. \tag{50}$$

It will at once be observed that eqns. (49) and (50) are in the form of thermal noise equations, and they are often interpreted by saying that the space-charge limited diode behaves as a resistance equal to the slope resistance at a temperature equal to 0·644 times the cathode temperature. The noise-voltage formulation is much more convenient in circuit calculations than the current formulation of eqns. (29) and (35)† because the voltage is independent of the space-charge conditions in the valve. We may look on the potential minimum as a region where density modulation is converted to velocity modulation. It should also be noted that the frequency dependence of eqn. (50) is that of shot noise and not of thermal noise. It is, however, correct to say that the action of the space-charge minimum in a diode is to convert current modulation of the emission current into velocity modulation of the anode current. This leads us to the interesting question of whether the smoothing action of space charge persists even when the beam is made very long, i.e. when the distance from minimum to anode is very great. This may happen in, for example, a klystron amplifier where the distance from the cathode to the first gap (buncher) may be quite large and the distance from cathode to second gap (catcher) will be very large. This question has not been definitely decided experimentally; but, theoretically, it seems that the velocity modulation must produce current modulation, so that noise currents depending on the transit angle from the observation plane to the minimum will be found. A theoretical paper by MacDonald‡ may be consulted, and the elements of another approach are given later in the book. Recent work is listed on pp. 559–65.

† This can be used to show that $\Gamma^2 \doteqdot 9\dfrac{(1-\frac{1}{4}\pi)}{\eta_a}$, where η_a is the eta parameter of Langmuir's diode theory, evaluated for the anode side of the minimum, i.e.

$$\eta_a = e(V_a - V_m)/kT.$$

‡ MacDonald, *Phil. Mag.* **40** (1949), 561. Ibid. **41** (1950), 863.

8.5. Partition noise

The last question on shot noise which we shall answer in this chapter is the determination of noise currents and voltages when the electron current from a cathode is divided between a plurality of collecting electrodes. The behaviour to be expected depends very much on the circumstances of the current division, e.g. if the valve consists of a cathode and two large anodes

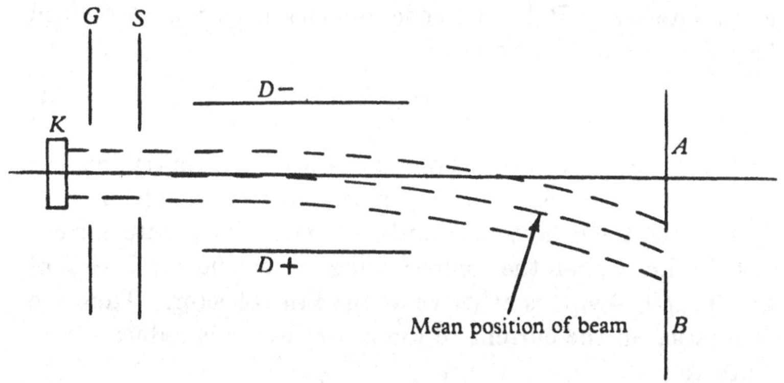

Fig. 8.8. A deflexion tube.

disposed symmetrically with respect to it, the noise in the anode currents will be

$$\overline{i_{a_1}^2} = 2eI_{a_1}\Gamma_1^2\Delta f,$$
$$\overline{i_{a_2}^2} = 2eI_{a_2}\Gamma_2^2\Delta f,$$

just as would be expected. If, however, there is any mechanism by which the compensating effect of space charge can be partially annulled, we speak of the extra noise introduced as partition noise. Examples of annulment of the smoothing are the following. If the current to the collectors is determined by velocity sorting, for instance, in a deflexion valve or in a diode in the retarding field regime, then if the extra noise current in a microscopic time interval has an energy such that it goes to collector A, the compensating current which is returned to the cathode will be taken from collector B (fig. 8.8). Thus, although the total beam current remains constant, the sum of the noise currents in the collector leads is greater than the space-charge smoothed shot noise in this current. In the simplest case, the retarding field diode, only electrons with energies greater than that equivalent to the anode voltage can reach the anode. The smoothing effect of space charge is therefore entirely absent, since the distribution of the electrons is fixed once and for all and the noise is pure shot noise, i.e.

$$\overline{i^2} = 2eI_r\Delta f.$$

With these ideas we can now calculate the partition noise in general terms.† If the cathode supplies a current I divided between n collectors we have

$$I = \sum_1^n I_n. \tag{51}$$

We suppose that there is a unit fluctuation of initial current to the nth electrode. The compensating current will be divided up between the collectors according to the steady-state current distribution. Then the compensating current flow is $1-\gamma_{(\beta)}$ of which $I_n/I(1-\gamma_{(\beta)})$ is effective at the nth collector. Thus the fluctuation in the current to the nth collector is reduced from unity to

$$\left[1-\frac{I_n}{I}(1-\gamma_{(\beta)})\right].$$

If pure shot variations in I_n are considered and the result is averaged over the emission velocities which are greater than u_m,

$$\overline{i_n^2} = \left[1-\frac{2I_n}{I}(1-\overline{\gamma_{(\beta)}})+\left(\frac{I_n}{I}\right)^2(1-2\overline{\gamma_{(\beta)}}+\overline{\gamma_{(\beta)}{}^2})\right]I_n\,2e\Delta f. \tag{52}$$

Next we consider the effect of unit fluctuation in the current to all the other electrodes, i.e. in $I-I_n$. The compensating current to n is $-\frac{I_n}{I}(1-\gamma_{(\beta)})$ and the corresponding noise is

$$\overline{i_n^2} = \left[\left(\frac{I_n}{I}\right)^2(1-2\overline{\gamma_{(\beta)}}+\overline{\gamma_{(\beta)}{}^2})\right](I-I_n)\,2e\Delta f. \tag{53}$$

The last contribution is that due to the α electrons, i.e. to shot noise in (I_s-I). This contributes

$$\overline{i_n^2} = \left[\left(\frac{I_n}{I}\right)^2\overline{\gamma_{(\alpha)}{}^2}\right]I\,2e\Delta f. \tag{54}$$

† North, *R.C.A. Rev.* **4** (1940), 244.

The sum of eqns. (52), (53) and (54) is then the total noise in I_n, therefore

$$\overline{i_n^2} = \left[1 - \frac{I_n}{I}(1-\Gamma^2)\right] I_n 2e\Delta f. \tag{55}$$

The fluctuation current in the cathode lead is

$$\overline{i^2} = 2eI\Gamma^2\Delta f,$$

therefore

$$\frac{\overline{i_n^2}}{\overline{i^2}} = \frac{\left[1 - \frac{I_n}{I}(1-\Gamma^2)\right]}{\Gamma^2}\frac{I_n}{I}, \tag{56}$$

instead of I_n/I as would be expected from straightforward current division. It will be observed that the factor is fairly small when I_n is a large proportion of I but is large for small I_n.

Thus if $\quad I_n/I = 0{\cdot}9, \quad \Gamma^2 = 0{\cdot}1, \quad \dfrac{\overline{i_n^2}}{\overline{i^2}} = \dfrac{1{\cdot}9 I_n}{I},$

but if $\quad \dfrac{I_n}{I} = 0{\cdot}2, \quad \dfrac{\overline{i_n^2}}{\overline{i^2}} = 8{\cdot}2\dfrac{I_n}{I}.$

Eqn. (55), however, shows that the noise is always less than pure shot noise in the current.

To conclude our study of partition noise we wish to emphasize that, when high-frequency effects are considered, it is not enough simply to consider variations in the number of electrons reaching an electrode. The induced current has also to be considered, so that if, for instance, a high initial-velocity electron takes a different path through a grid or through a resonator from that taken by a lower initial-energy particle, partition noise arises from this effect even if no electrons reach either grid or resonator. In this case partition noise can occur in a negative grid triode and some calculations have been made by Bell.† The effect is clearly important in velocity-modulation tubes of the gridless type, i.e. when the current induced by an axial electron is much less than the current induced by an electron at the edge of the gap.

8.6. Flicker noise

To conclude this chapter we shall say a little about a third type of noise which is only evident at low frequencies, i.e.

† Bell, *Wireless Engr.* **25** (1948), 294.

below a few thousand cycles. If the noise output per unit-frequency interval be measured for a temperature-limited diode it is found to increase when the mean frequency is below 3000–5000 cycles per second. Below this frequency the mean-square noise varies approximately as $\omega^{-1\cdot 25}$ (Johnson).† The

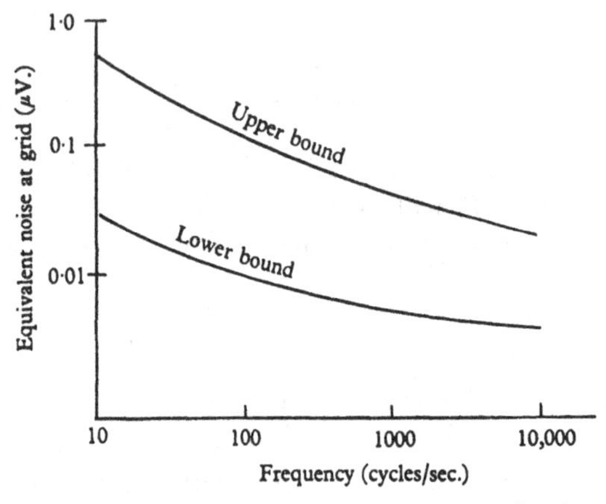

Fig. 8.9. Behaviour of flicker noise with increasing frequency.

effect is much larger in oxide cathode valves than in tubes with thoriated tungsten emitters, which in turn are more noisy than pure tungsten. In fact, some investigators say that pure tungsten (and presumably other pure metal) emitters do not show the effect at all. This noise, in excess of temperature-limited shot noise, is called flicker noise. It is of great practical importance because it sets a limit to the signal/noise ratio of low-frequency amplifiers, for the effect persists under space-charge limited conditions though it is much reduced. To give an idea of the magnitude of the effect, modern amplifying valves worked under normal (space-charge) conditions produce flicker noise which is equivalent to a grid signal of between 0·05 and 0·20 μV., when the noise is measured in a 4~ band (General Radio Harmonic Analyser) centred at 30~. Fig. 8.9 illustrates the form of the curve of noise versus frequency and indicates the boundaries between which most valves fall.

† Johnson, *Phys. Rev.* **26** (1925), 71.

Schottky† gave the first theory of the effect. He put forward the hypothesis that the flicker noise was due to changes in the emission from relatively large patches of the surface. Using the older picture of emission from oxide cathodes this would mean that barium would diffuse on to the surface of the cathode from the interior, and would lower the work function of a patch of the surface. Since each adsorbed barium ion only has a restricted life-time on the surface, the patch would gradually evaporate, but in this time other patches would have become active so that the average emission over a very long time would remain constant. This picture is immediately applicable to thoriated tungsten, for which it was originally developed. Schottky found theoretically

$$\overline{\Delta f^2} = \frac{kJ^2}{\omega^2}, \tag{57}$$

where $\overline{\Delta f^2}$ = mean square flicker current, k = constant, J = current density, $\omega = 2\pi f$; whereas the empirical data were of the form

$$\overline{\Delta f^2} = \frac{kJ^{2\cdot 3}}{\omega^{1\cdot 25}}. \tag{58}$$

The theory therefore is not in good agreement with the measurements. More recently MacFarlane‡ has given a treatment based on a diffusion-conduction process bringing barium to the surface and calculating the life-times of whole patches rather than of single ions, following Spoull's§ work on emission decay. MacFarlane's result is

$$\overline{\Delta f^2} = \frac{kJ^{x+1}}{\omega^x}, \quad (1 < x < 2). \tag{59}$$

Since the theory of emission decay put forward by Spoull is almost certainly incorrect (although it can be rewritten in terms of poisoning of the cathode by oxygen—a much more acceptable picture), the validity of eqn. (59) is debatable, and our knowledge, both theoretical and experimental, of the flicker effect leaves much to be desired.

† Schottky, *Phys. Rev.* **28** (1926), 74.
‡ *Proc. Phys. Soc.* **59** (1947), 366.
§ Spoull, *Phys. Rev.* **67** (1945) 166.

PART III

TYPES OF VALVE

Chapter 9

TRIODES FOR LOW AND MEDIUM FREQUENCIES

In the third part of this book we seek to apply the physical and mathematical theories of the first and second parts to the study of specific valve types actually used in modern electronic engineering. As usual we start from the simplest types and work towards the more complicated. The simplest type of valve which is capable of the function of amplification is the triode or three-electrode valve in which a grid is inserted between the anode and cathode. The grid may be a mesh or a helix or a number of parallel rods or in general any metallic structure which is permeable by electrons, and it is used to control the potential of the cathode-anode space which in turn controls the current from the cathode to the anode. Let us first consider the experimental characteristics of such a valve. Characteristics with constant cathode temperature can be taken (*a*) with constant grid voltage, (*b*) with constant anode voltage. These characteristics are shown in figs. 9.1 and 9.2. Early studies by van der Bijl† showed that, in the operating regions, there is a relation of the form

$$I_T = I_a + I_g = G(\mu V_g + V_a)^n, \quad (1)$$

between the currents and voltages. Here I_T = total space current, I_a, V_a = anode current and voltage, I_g, V_g = grid current and voltage, G, n, μ = constants whose significance will be discussed. For receiving valves and linear amplification in general V_g is negative so that $I_g = 0$ and the relation reduces to

$$I_a = G(\mu V_g + V_a)^n. \quad (2)$$

We shall later give theoretical explanations of these empirical formulæ which allow us to calculate the constants from the geometrical form of the valve. Before we proceed to the

† Van der Bijl, *Thermionic Vacuum Tubes*, McGraw Hill, 1920.

theoretical development we shall give a brief discussion of the way in which the valve is reduced to an active network for use in circuit calculations. The following parameters are introduced:

(1) The amplification factor† $= \left(\frac{\partial V_a}{\partial V_g}\right)$, I_a const. (symbol μ).

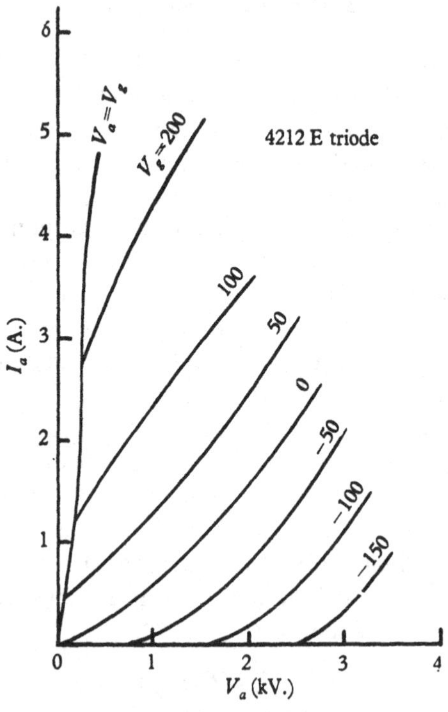

Fig. 9.1. I_a, V_a characteristics for a triode.

(2) The mutual conductance or slope $= \left(\frac{\partial I_a}{\partial V_g}\right)$, V_a const. (symbol g_m).

(3) The dynamic resistance $= \left(\frac{\partial V_a}{\partial I_a}\right)$, V_g const. (symbol r_a).

Clearly these are related by the expression

$$\mu = g_m r_a. \tag{3}$$

† In Continental works, particularly in German, use is made of $\frac{\delta V_g}{\delta V_a} = \frac{1}{\mu}$. This quantity is given the symbol D from the German *Durchgriff*.

In circuit theory the triode is represented either as a series (constant voltage) generator or as a parallel (constant current)

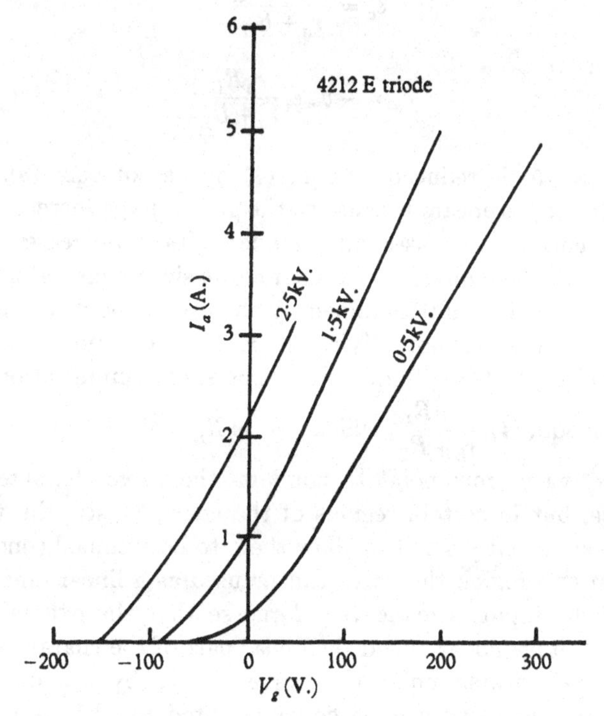

Fig. 9.2. I_a, V_g characteristics for a triode.

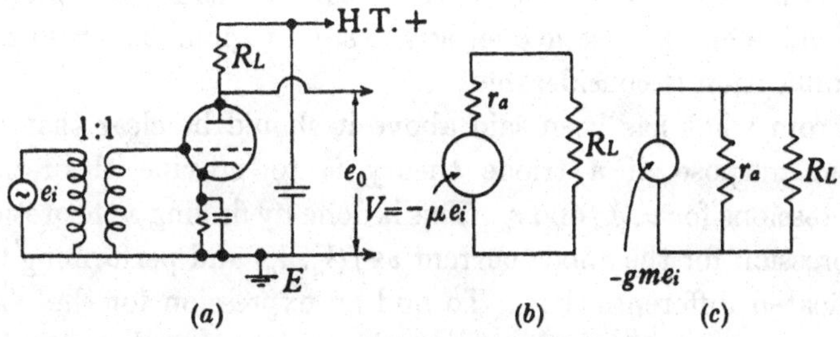

Fig. 9.3. Triode connected as amplifier and equivalent circuits.

generator. Fig. 9.3 shows an amplifying stage together with the constant-voltage and constant-current equivalents. If

e_o = a.c. voltage across R_L we have for (*b*)

$$e_o = \frac{-\mu e_i R_L}{r_a + R_L}, \tag{4}$$

and for (*c*)

$$e_o = -g_m e_i \frac{r_a R_L}{r_a + R_L}; \tag{5}$$

and eqn. (5) is reduced to eqn. (4) by use of eqn. (3). The negative sign appears because as the grid voltage increases, the anode current increases and anode voltage decreases. The validity of the equivalent circuit follows since eqns. (4) and (5) agree with the results obtained by the use of the definitions of μ, g_m, r_a. A single amplifying stage therefore produces 180° phase change at low frequencies. The stage amplification e_o/e_i is, from eqn. (4), $\dfrac{-\mu R_L}{r_a + R_L}$, which $\rightarrow \mu$ as $R_L \rightarrow \infty$. The parameters μ, g_m, r_a vary from point to point on the valve characteristic surface, but in certain regions of the characteristic the variation is sufficiently small to allow them to be assumed constant. Within this region the valve can be used as a linear amplifier, but if the input is made very large so that the excursion in anode voltage goes beyond the linear part of the characteristic, or if the working point is incorrectly set by the d.c. bias voltages, the valve cannot be represented by the equivalent circuits above. The resulting distortion is usually calculated by power-series expansions which can be made to represent the valve characteristic over any desired region. The resulting complication is considerable.

From what has been said above it should be clear that the main purpose of a triode theory is to provide theoretical expressions for μ, g_m and r_a. This is done by finding a theoretical expression for the anode current as $f(V_a, V_g)$ and performing the indicated differentiations. To find an expression for the plate current we have to find out how the grid and anode potentials control the potential and field inside the valve. Once this has been done the anode current can be calculated in a fairly simple manner.

9.1. The electrostatic problem in a planar triode

We first make a few qualitative points. Consider the cathode-grid space when no current flows. The potential of a point in this space depends on the grid potential, taking $V_c = 0$, and, to a lesser extent, on the anode potential. The anode potential plays less part because (*a*) it is farther away and (*b*) it is partially screened by the grid. This can be put in another way: the potential at a point in the valve is the potential due to superposing the potentials due to grid alone and anode

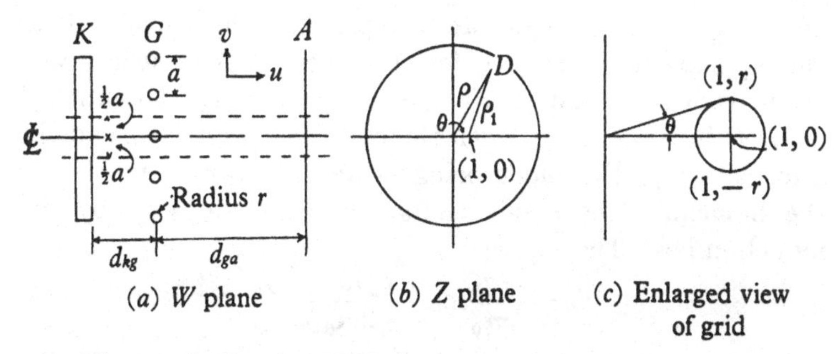

Fig. 9.4. Section through planar triode and transformed section.

alone minus the effect due to the image of the grid charge in the anode. The closer the mesh of the grid, the more completely it will determine the potential of the cathode-grid region. The electrostatic problem thus posed can be solved in several different ways, but it is most commonly solved by the use of conformal transformations which are used to convert the real shape of the valve or part of the valve into another shape in which the potential can be deduced by simple means.

As a first example let us consider the parallel-plane triode of fig. 9.4, where the dimensions are marked. If we now restrict ourselves to the region between the dotted lines with a grid wire on the centre line, we find that it can be transformed into a circular grid-wire at (1, 0) by the following transformation, with the centre of the grid wire as origin:

$$W = u+jv = \frac{a}{2\pi}\ln Z = \frac{a}{2\pi}\ln \rho\, e^{j\theta}. \tag{6}$$

Solving for the real and imaginary parts we have

$$u = \frac{a}{2\pi}\ln\rho, \quad v = \frac{a\theta}{2\pi}, \tag{7}$$

$$\rho = e^{2\pi u/a}, \quad \theta = 2\pi v/a. \tag{8}$$

At the grid wire centre $(u, v) = (0, 0)$, so the Z plane coordinates are $(1, 0)$ as stated. The cathode is at $-d_{kg}$, i.e.

$$\rho_k = \exp\left(-\frac{2\pi d_{kg}}{a}\right),$$

and if $d_{kg} \geqslant a$ the cathode can be represented by a point. Thus this approximation is good for close-meshed grids spaced well away from the cathode. Now let us evaluate the potential at point D in the Z plane when the cathode charge is q_k and the grid charge q_g, the anode being assumed at zero potential for the moment. The anode charge is then $-(q_k+q_g)$ by the induction law. Then

$$V_D = \frac{-q_k}{2\pi\epsilon_0}\ln\rho - \frac{q_g\ln\rho_1}{2\pi\epsilon_0} + C, \tag{9}$$

by the application of Laplace's equation in cylindrical co-ordinates.

But
$$\rho_1^2 = \rho^2 + 1 - 2\rho\cos\theta,$$
therefore

$$V_D = -\frac{q_k}{4\pi\epsilon_0}\ln\rho^2 - \frac{q_g}{4\pi\epsilon_0}\ln(\rho^2+1-2\rho\cos\theta) + C. \tag{10}$$

We next have to evaluate the constant C. At the anode $V_D = V_A$ and $u = d_{ga}$; also if $d_{ga} \geqslant a$,

$$\ln(\rho^2+1-2\rho\cos\theta) \doteqdot \ln\rho^2 = \frac{4\pi d_{ga}}{a},$$

therefore
$$V_A = -\frac{q_k+q_g}{a\epsilon_0}d_{ga} + C. \tag{11}$$

At the cathode $V_D = V_C$ and $u = -d_{kg}$; also, from the assumption made about d_{kg},

$$\ln(\rho^2+1-2\rho\cos\theta) \doteqdot \ln 1 = 0.$$

Therefore
$$V_C = \frac{q_k}{a\epsilon_0}d_{kg} + C. \tag{12}$$

At the grid surface $V_D = V_g$, $u = r$ and $\rho = \exp\left(\frac{2\pi r}{a}\right)$ which is of the form $1+\delta$ where δ is a small number, when practical values of r are inserted. When this is the case

$$\rho^2+1-2\rho\cos\theta = 2(1+\delta)(1-\cos\theta),$$
$$= (1+\delta)(2\sin\tfrac{1}{2}\theta)^2,$$

therefore $$\ln(\rho^2+1-2\rho\cos\theta) = 2\ln(2\sin\tfrac{1}{2}\theta),$$

and $$\ln\rho^2 \doteqdot 2\ln(1+\delta) = 0,$$

$$V_g = \frac{-q_g}{2\pi\epsilon_0}\ln\left(2\sin\frac{\pi r}{a}\right)+C. \tag{13}$$

If we now take V_c to be the zero potential we find from (12)

$$C = -\frac{q_k}{a\epsilon_0}d_{kg}.$$

Then

$$\left.\begin{aligned} V_A &= \frac{-(q_k+q_g)d_{ga}-q_k d_{kg}}{a\epsilon_0}, \\ V_g &= \frac{-aq_g\ln(2\sin\pi r/a)-2\pi q_k d_{kg}}{2\pi a\epsilon_0}. \end{aligned}\right\} \tag{14}$$

Solving these for the unknown charges we get

$$\left.\begin{aligned} q_k &= \frac{-a\epsilon_0(V_a+mV_g)}{d_{ga}+d_{kg}(1+m)}, \\ q_g &= \frac{a\epsilon_0 m(d_{ga}+d_{kg})(V_g-d_{kg}V_a)}{d_{ga}\{d_{ga}+d_{kg}(1+m)\}}, \end{aligned}\right\} \tag{15}$$

where $$m = \frac{-2\pi d_{ga}}{a\ln(2\sin\pi r/a)}. \tag{16}$$

From the first of eqns. (15) we see that $q_k = 0$ if $V_a = -mV_g$. But if $q_k = 0$ the field at the cathode surface is also zero by Gauss's theorem, so the condition $V_a = -mV_g$ is the condition that the field due to the anode evaluated at the cathode surface is equal and opposite to the field due to the grid. Thus the grid potential is m times as effective as the anode in determining the off-cathode field, or

$$\mu = \left(\frac{\partial V_a}{\partial V_g}\right) = -m. \tag{17}$$

Eqn. (16) therefore gives us a theoretical expression for the amplification factor of the parallel-plane triode. This μ is usually called the electrostatic μ because it is evaluated from the geometry of the cold tube making no allowance for the existence of the potential minimum, variations in field along the cathode surface and other departures from the assumed conditions. In addition, it should be remembered that some fairly vigorous approximations have been made in deriving eqn. (16).

Let us discuss eqn. (16) for a moment. μ increases as d_{ga} increases, it is inversely proportional to a, and it increases as r increases up to the maximum value of r (approx. $a/10$), for which the analysis is valid. It may seem strange that d_{kg} does not appear in the expression for μ, but a little thought shows that μ really depends on the relative field strengths and not on the actual potentials, so that d_{kg} need not be present. It is also interesting to note that, if all the dimensions, including the grid-wire radius, are scaled by the same factor, μ remains unchanged, which is, of course, a general property of the electrostatic field.

We have now given a first approximation to the solution of the potential problem in a planar triode; for the potential at any point can be calculated from eqn. (10) together with eqn. (15) and the value of C. In fact, however, we are not interested in the potential except for the values in the immediate vicinity of the cathode which determines the emission current. Fig. 9.5 is a sketch of the equipotentials with a slightly negative grid and highly positive anode. No allowance has been made for space charge.

9.2. The electrostatic field in a cylindrical triode

Until recently most small receiver triodes have approximated fairly closely to cylindrical form rather than to the planar structure discussed above, and we shall now indicate the derivation of the relations corresponding to eqns. (14), (15) and (16).

The appropriate transformation is (see fig. 9.6)

$$W = R_g . Z^{1/N}, \tag{18}$$

where R_g = the mean radius of the grid wire, N = number of

Fig. 9.5. Equipotentials in a planar triode.

grid wires. The sector of angle $2\pi/N$ is transformed into the Z-plane representation of fig. 9.4 by this transformation. The defining relations are $\rho = \left(\frac{R}{R_g}\right)^N$, $\theta = N\phi$. By repeating the analysis of § 9.1 we find

$$\mu = \frac{N \ln (R_a/R_g)}{\ln [2 \sin (Nr/2R_g)]}. \tag{19}$$

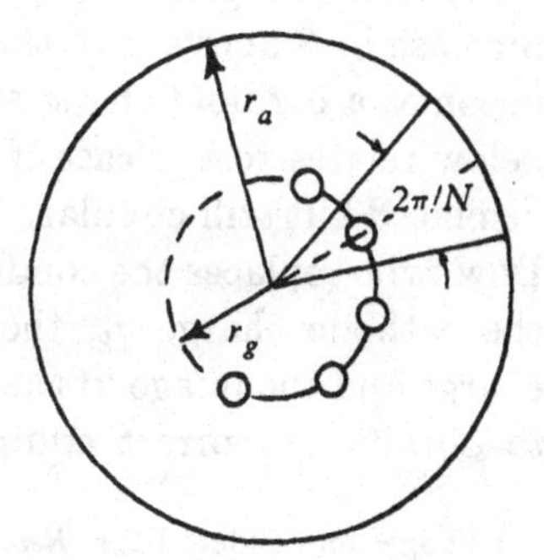

Fig. 9.6. Cylindrical triode.

9.3. More exact treatment

The formulæ given in the last two sections are only very approximate.

They apply fairly well to valves in which μ is small or in which the grid wires are very thin and the gridcathode spacing large. Few modern valves, and those of only small interest, fulfil these conditions. Valves for ultra-high-frequency work and even valves for use as radar intermediate-frequency amplifiers and for television and f.m. radio-frequency work depart very widely from the assumptions made. In this section we give an account of some better approximations for the planar case. The treatment is based on those of Vodges and Elder† and of Dow.‡ The main approximation made in § 9.1 is in the derivation of eqn. (13), where it is assumed that the transformed grid wire is circular and has its centre at (1, 0).

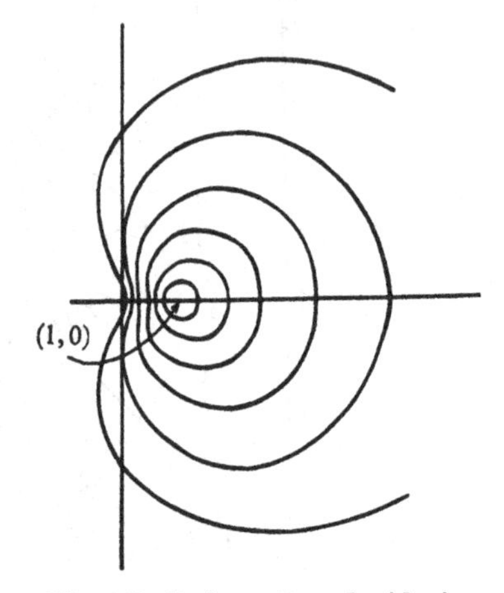

Fig. 9.7. Deformation of grid wire as shielding factor increases.

To discuss the situation it is convenient to introduce a new parameter which is the ratio of grid-wire diameter to grid-wire spacing, i.e. it is the geometrical stopping factor of the grid. This is called the screening factor S and $S = 2r/a$. As S increases the centre of the transformed grid wire first moves away from the point (1, 0) towards the anode, and with further increase in S above 0·16 the grid wire ceases to be circular and becomes a cardioid shape as sketched in fig. 9.7. The analysis below relates to the case if $S < 0{\cdot}16$, i.e. grid wire transform is displaced but still circular. The simplest development is that of Dow, who replaces the conductors by a set of three line charges, the cathode charge q_k, the image in the grid of the cathode charge and the image in the grid of the anode charge. Actually, to obtain the correct equipotentials at the anode, the anode

† Vodges and Elder, *Phys. Rev.* **24** (1924), 683.
‡ W. G. Dow, *Fundamentals of Engineering Electronics*, John Wiley, 1937.

image charge should be located at the centre of the system (fig. 9.8); but since the radius of the transformed anode is very large, it is a very good approximation to locate the charge at the centre of the transformed grid. The transformed grid radius and centre coordinate (since the grid is assumed circular) are given by

$$R = \tfrac{1}{2}(e^{\pi S} - e^{-\pi S}) = \sinh \pi S, \tag{20}$$

$$\rho_g = \tfrac{1}{2}(e^{\pi S} + e^{-\pi S}) = \cosh \pi S. \tag{21}$$

The image of the cathode charge q_k can be located by use of the fact that the image of a charge distant a from the centre of a circular conductor radius R is at b where $ab = R^2$. $-q_k$ is therefore located at

$$\rho^1 = \rho_g - b = (\rho_g^2 - R^2)/\rho_g.$$

Using eqns. (20) and (21) we get $\rho^1 = 1/\rho_g$. The potential at point $D(\rho, \theta)$ is then

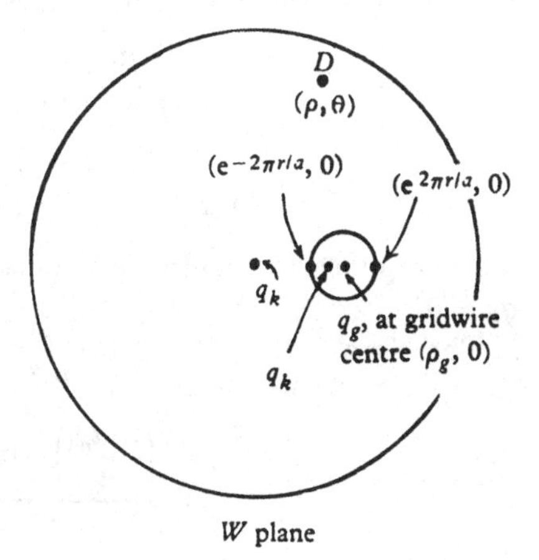

Fig. 9.8. Transformed section of cylinder triode.

$$V_D = -\frac{1}{4\pi\epsilon_0}\left[q_k \ln \rho^2 - q_k \ln\left(\rho^2 + \frac{1}{\rho_g^2} - \frac{2\rho}{\rho_g}\cos\theta\right) - q_a \ln(\rho^2 + \rho_g^2 - 2\rho\rho_g\cos\theta) + C\right]. \tag{22}$$

C can be evaluated by inserting the value of ρ at the cathode surface and equating the resultant potential to zero. If the cathode transforms to a circle of radius ρ_k we get, for $\theta = 0$,

$$C = -q_k \ln \rho_k^2 + q_k \ln\left(\rho_k + \frac{1}{\rho_g}\right)^2 + q_a \ln(\rho_k + \rho_g)^2. \tag{23}$$

We next have to determine the charges by inserting the grid and anode potentials V_g and V_a. At the grid we can put $\rho = \rho_g - R$, also $\rho_g^2 = 1 + R^2$ and $\rho_k \ll \rho_g$ or ρ^1. Then, inserting

(23) in (22), and using these facts, we get

$$V_g = \frac{1}{2\pi\epsilon_0}\left[q_k \ln(R\rho_k) + q_a \ln\left(\frac{R}{\rho_g}\right)\right]. \tag{24}$$

To find V_a, we put $\rho = \rho_a$, $\rho_k \ll 1$, $\rho_a \gg 1$ and obtain

$$V_a = \frac{1}{2\pi\epsilon_0}\left[q_k \ln(\rho_k \rho_g) + q_a \ln\left(\frac{\rho_a}{\rho_g}\right)\right]. \tag{25}$$

Solving eqns. (24) and (25) for q_k and q_a, we get

$$q_k = 2\pi\epsilon_0\left[\frac{V_g \ln(\rho_a/\rho_g) - V_a \ln(R/\rho_g)}{\ln(R\rho_k)\ln(\rho_a/\rho_g) - \ln(R/\rho_g)\ln(\rho_a\rho_g)}\right], \tag{26}$$

$$q_a = 2\pi\epsilon_0\left[\frac{V_a \ln(R\rho_k) - V_g \ln(\rho_k\rho_c)}{\ln(R\rho_k)\ln(\rho_a/\rho_g) - \ln(R/\rho_g)\ln(\rho_a\rho_g)}\right]. \tag{27}$$

The amplification factor is evaluated as in § 9.1 by putting $q_k = 0$.

Therefore
$$\mu = -\ln\left(\frac{\rho_a}{\rho_g}\right)\Big/\ln\left(\frac{R}{\rho_g}\right), \tag{28}$$

$$= \frac{\left(\dfrac{2\pi d_{ga}}{a}\right) - \ln\cosh \pi S}{\ln\coth \pi S} \quad (S < 0{\cdot}16). \tag{29}$$

For values of $S > 0{\cdot}16$ eqn. (29) gives too low a value for μ. Since μ is only a function of d_{ga}/a and πS, it is a simple matter to prepare a chart giving μ. For fixed S, $\mu \propto d_{ga}/a$.

The derivation of analogous results for the cylindrical triode is left as an exercise for the reader. The transformed co-ordinates of the points where the grid wire cuts $\theta = 0$ are expanded in series to obtain expressions similar to (20) and (21). The result is

$$\mu = \frac{N\ln(R_a/R_g) - \ln\cosh\pi S}{\ln\coth\pi S}, \tag{30}$$

where
$$S = \frac{Nr}{\pi R_g}. \tag{31}$$

In spite of the fact that eqns. (28) and (29) are not sufficiently accurate to serve as the basis for the study of modern close-spaced valves, we shall leave the consideration of μ for the moment and turn our attention to the evaluation of the anode current in the triode.

9.4. Current flow and mutual conductance in the triode

To calculate the current recourse is usually made to the device of the 'equivalent diode', i.e. the diode which has the same current ~ voltage characteristic as the triode under discussion. Eqn. (2) of this chapter shows that, for negative V_g,

$$I_a = G(\mu V_g + V_a)^n,$$

or more correctly

$$I_a = G\{\mu(V_g + V_c) + V_a\}^n, \tag{32}$$

where V_c = contact p.d. between cathode and grid. Clearly if $n = \frac{3}{2}$ a diode spacing can be found for which G = diode constant$/d^2$, and we can replace the triode by a diode with the voltage $\{\mu(V_g + V_c) + V_a\}$ on its anode and the calculated spacing d. It is a matter of experimental fact that $n \doteqdot \frac{3}{2}$ over wide regions of the triode characteristic, and from the theoretical point of view we should expect the current from the cathode to depend only on the field at the cathode surface in the absence of space charge.

The usual treatment, as given by Dow,† for instance, is to evaluate the potential gradient at the cathode surface and to equate this to the potential gradient in a diode whose anode voltage is the same as the voltage effective at the grid plane of the triode, i.e. $V_D = V_g + V_c + V_a/\mu$. This is a more convenient choice than the use of the voltage effective at the anode plane of the triode $V_{a\,\text{eff.}} = V_a + \mu(V_g + V_c) = \mu V_{g\,\text{eff.}}$. The potential gradient is seen from Gauss's law to be $-q_k/a\epsilon_0$ for the planar triode, since the cathode charge is q_k for each transformed section, i.e. q_k/a per unit length. Eqn. (26) gives q_k. For the diode the

† Op. cit. Another treatment often used is that of Tellegen, *Physica*, 5e (1925), 301. For planar valves his result is

$$I = 2{\cdot}336 \times 10^{-6} \frac{\left(Vg + \dfrac{V_a}{\mu}\right)^{\frac{3}{2}}}{\left(1 + \dfrac{1}{\mu} + \dfrac{4}{3\mu}\dfrac{d_{ga}}{d_{gk}}\right)^{\frac{3}{2}}}.$$

The derivation of this diode is not purely electrostatic as it is in our expressions. This expression reappears in Chapter 13.

gradient is $1/d(V_g + V_a/\mu)$ if we now include V_c in V_g. Equating these expressions gives

$$d = \frac{a}{2\pi}\left[\frac{\ln(\rho_k \rho_g)\ln(R/\rho_g)}{\ln(\rho_a/\rho_g)} - \ln(\rho_k R)\right],$$

and, after reduction,

$$d \doteqdot d_{kg}\left[1 + \frac{d_{kg} + d_{ga}}{\mu d_{kg}}\right]. \tag{33}$$

This expression has the unfortunate property that $d \to \infty$ as $\mu \to 0$, so that the current $\to 0$. A triode with a low μ is clearly very nearly equivalent to the diode with the same cathode-anode spacing, so d should $\to d_{kg} + d_{ga}$ as $\mu \to 0$.

Walker† has shown that a way can be found out of this difficulty by imposing the further condition that $\partial E_k/\partial V_k$, the change in cathode field with cathode potential, should be the same for the triode and for the equivalent diode. Eqns. (22) and (23) show that if we take the same arbitrary value for V_k eqn. (26) holds if we replace V_g by $(V_g - V_k)$ and V_a by $(V_a - V_k)$. Then, from eqn. (26),

$$E_k = \frac{2\pi}{a}\left[\frac{\mu(V_g - V_k) + (V_a - V_k)}{D}\right], \tag{34}$$

$$\frac{\partial E_k}{\partial V_k} = -\frac{2\pi}{aD}(1+\mu), \tag{35}$$

where D = denominator of eqn. (26).

For the diode

$$E_k = \frac{1}{d}(V_D - V_k), \tag{36}$$

$$\frac{\partial E_k}{\partial V_k} = -\frac{1}{d}. \tag{37}$$

From eqns. (35) and (37)

$$d = \frac{aD}{2\pi}\frac{1}{1+\mu} \tag{38}$$

$$\doteqdot d_{kg}\left[1 + \frac{d_{ga}}{(1+\mu)d_{kg}}\right]. \tag{39}$$

† Walker, *Wireless Engr.* **24** (1947), 5.

Putting $V_k = 0$ and using eqns. (34) and (36) we obtain

$$V_D = \frac{V_a + \mu V_g}{1+\mu}, \tag{40}$$

V_D is therefore slightly different from the effective grid-plane voltage. It can be seen that d as given by eqn. (39) behaves correctly as $\mu \to 0$, but is indistinguishable from eqn. (33) for large μ.

The triode current will then be equal to the current in the equivalent diode, which is

$$I = \frac{2{\cdot}336 \times 10^{-6}\, V_D^{3/2}}{d^2}$$

$$= \frac{2{\cdot}336(1+\mu)^{1/2}(V_a+\mu V_g)^{3/2} \times 10^{-6}}{[d_{kg}(1+\mu)+d_{ga}]^2}. \tag{41}$$

The mutual conductance of the triode can now be found by differentiating eqn. (41) with respect to V_g. The result is, for unit cathode area,

$$g_m = \frac{3{\cdot}504 \times 10^{-6}\, \mu(1+\mu)^{1/2}(V_a+\mu V_g)^{1/2}}{[d_{kg}(1+\mu)+d_{ga}]^2}. \tag{42}$$

To obtain r_a we differentiate eqn. (41) with respect to V_a and take the reciprocal. The result is

$$r_a = \frac{2{\cdot}855 \times 10^5 [d_{kg}(1+\mu)+d_{ga}]^2}{(1+\mu)^{1/2}(V_a+\mu V_g)^{1/2}} = \frac{\mu}{g_m}. \tag{43}$$

The last expression agrees with eqn. (3), as it should.

Comparing eqns. (42) and (41) we see that g_m is directly proportional to $I^{1/3}$, which is well borne out by experiment. If we consider variations in tube geometry which leave μ invariant g_m is mainly controlled by d_{kg}^{-2}, so eqn. (42) brings out the very important role of the grid-cathode distance in determining the mutual conductance. For high-frequency work, where the quality of a valve is measured by its gain and band-width product, a high g_m is always desirable. In addition, we have already seen that the input admittance is smaller for small values of d_{kg}. Thus it is doubly important to obtain the minimum d_{kg} in valves for ultra-high-frequency work, and the design of these tubes is nowadays reduced to the mechanical problem of designing a grid for which small and constant values of d_{kg} can be obtained in assembly and maintained during life.

Eqns. (41) and (42) are not restricted in validity to the region in which the expressions for μ developed in § 9.3 are valid. They are valid whenever the field at the cathode surface can be obtained as a linear function of the applied voltages. The restriction is in fact quite different from that operating in the analysis of § 9.3, because the value of the screening factor is not important in determining the linearity of the cathode field; what is important is the value of d_{kg}/a, the ratio of grid-cathode spacing to grid-wire spacing. If this ratio is small, less than 0·6 say, the field on the cathode surface is much greater at a point midway between the wires than it is under grid wire, so that the current density varies along the cathode surface. This phenomenon is known as *Inselbildung* or shadow formation, and it is of great importance in modern ultra-high-frequency valves. The effect will be discussed in some detail later on.

We next consider the case of the cylindrical triode. Since there are N grid sectors, the total cathode charge is Nq_k, i.e. N times the relation corresponding to (26). The field is $Nq_k/2\pi\epsilon_0 R_k$, i.e.

$$E_k = -\frac{N}{R_k}\left[\frac{\mu(V_g - V_k) + (V_a - V_k)}{D}\right], \tag{44}$$

here $D = [\ln(R.\rho_k) \ln(\rho_a/\rho_g) - \ln(R/\rho_g).\ln(\rho_k.\rho_g)]$,

therefore
$$\frac{\partial E_k}{\partial V_k} = \frac{N}{DR_k}[\mu + 1]; \tag{45}$$

for the diode
$$E_{kD} = -\frac{(V_D - V_k)}{R_k \ln(R_d/R_k)}, \tag{46}$$

therefore
$$\frac{\partial E_k}{\partial V_k} = \frac{1}{R_k \ln(R_d/R_k)}. \tag{47}$$

From these we get $V_D = \dfrac{\mu V_g + V_a}{\mu + 1}$ as before, and

$$\ln\frac{R_d}{R_k}=\frac{D}{N}\frac{1}{\mu+1}$$

$$=\frac{-1}{(\mu+1)N}[\mu\ln(R\,.\,\rho_k)+\ln(\rho_k\,.\,\rho_g)]$$

$$=\frac{-1}{(\mu+1)N}[(1+\mu)\ln\rho_k+\mu\ln R+\ln\rho_g]$$

$$=\frac{-1}{(\mu+1)N}\left[(1+\mu)N\ln\frac{R_k}{R_g}+\mu\ln\sinh\pi S+\ln\cosh\pi S\right]. \quad (48)$$

We can now use eqn. (30) to eliminate $\ln\sinh\pi S$, i.e.

$$\mu\ln\sinh\pi S=(1+\mu)\ln\cosh\pi S-N\ln R_a/R_g.$$

Then

$$\ln\frac{R_d}{R_k}=\frac{1}{\mu+1}\left[(1+\mu)\ln\left(\frac{R_g}{R_k}\right)+\ln\left(\frac{R_a}{R_g}\right)+\left(\frac{\mu+2}{N}\right)\ln\cosh\pi S\right]. \quad (49)$$

For $S<0{\cdot}1$ the last term is negligible and we get

$$\ln\frac{R_d}{R_k}=\frac{\mu}{\mu+1}\left[1+\frac{1}{\mu}+\frac{1}{\mu}\frac{\ln R_a/R_g}{\ln R_g/R_k}\right]\ln\frac{R_g}{R_k}. \quad (50)$$

The current in the equivalent diode is then

$$I=\frac{2{\cdot}336\times10^{-6}\,V_D^{3/2}\,A}{R_k\,.\,R_d\,.\,\beta^2},$$

and

$$g_m=\frac{\partial I}{\partial V_g}=\frac{\partial I}{\partial V_D}\frac{\partial V_D}{\partial V_g}=\frac{3{\cdot}504\times10^{-6}\,AV_D^{1/2}}{R_kR_g\beta^2}\frac{\mu}{1+\mu}$$

$$=\frac{3{\cdot}504\times10^{-6}\mu}{R_kR_d\beta^2}\frac{(\mu V_g+V_a)^{1/2}}{(1+\mu)^{3/2}}A. \quad (51)$$

Thus to find the current and mutual conductance, eqn. (50) is used to determine R_d which is then inserted into (51). Similarly, we obtain for r_a

$$r_a=\frac{2{\cdot}855\times10^5\,R_kR_d\beta^2(1+\mu)^{3/2}}{(\mu V_g+V_a)^{1/2}\,A}. \quad (52)$$

These formulæ are not so convenient in form as those for the planar diode. A point which is worth noting is that eqn. (50) shows the equivalent diode radius to be somewhat greater than the grid radius even for the limit of large μ. This is in line with physical expectation because the field in a cylindrical system is mainly determined by the cathode radius, i.e. if the grid-plane equivalent voltage is used, we should expect the equivalent diode radius to be very nearly equal to the grid radius.

9.5. Application to real triodes

In general, triodes used in receiver valves at frequencies below about 600 Mc./sec. are neither planar nor cylindrical in

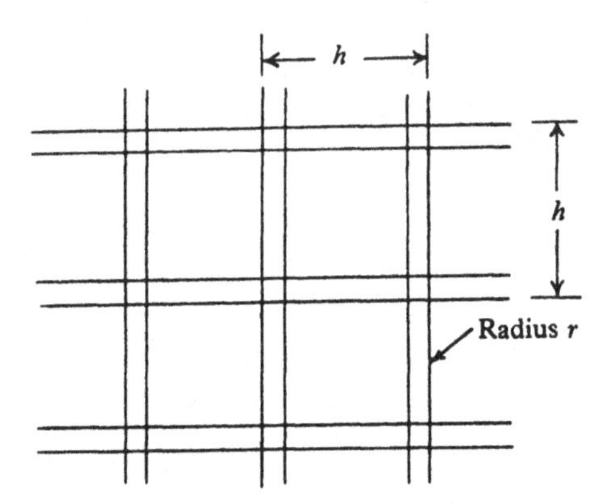

Fig. 9.9. A square meshed grid.

geometry but somewhere between the two. An additional departure from the assumed condition is that the grid in cylindrical structures is usually a helix instead of the parallel-wire structure assumed. The thick wires on which the grid is supported cause beam formation which also plays a part in the operation.

First, we consider departures of the grid from the assumed form. It is legitimate to use the formulæ already developed if the screening fraction is used to represent the opacity of the grid. Kusunose† redefines N in the cylindrical case to mean the active length of grid wire per unit axial length. Thus for a square mesh of wires of diameter r at centres h (fig. 9.9),

$$N = \frac{4\pi R_g}{h}\left(1 - \frac{r}{h}\right), \tag{53}$$

† Kusunose, *Proc. Inst. Radio Engrs., N.Y.* **17** (1929), 1706.

and $$\pi S = \frac{Nr}{R_g}. \tag{54}$$

Similar formulæ can be derived for other cases.

Jervis† has shown that the geometrical difference can be taken care of by the use of the following expression:

$$\mu = \mu_p - K(\mu_p - \mu_c), \tag{55}$$

where K = constant, depending on the geometry, μ_p = planar μ, μ_c = cylindrical μ. K varies from 0 for a planar triode to 1 for a cylinder triode. Unfortunately, the structures for which he tabulated K correspond to rather old types, but some intelligent guesswork will usually give a reasonable result.

In practice it is usual to determine the design of a receiver valve empirically. A structure is designed which fulfils the various conditions imposed by the requirements of mass production, i.e. the structure must be rigid, of fairly simple shape, and so on. Sample valves are made up and their characteristics determined. The characteristics can be modified in the required direction; the two most useful variables are the grid-wire diameter and winding pitch, as these are easier to alter than the cathode-grid spacing. In the case of high-power transmitter valves, it is usually not possible to go to the expense of making a fairly large number of experimental valves to determine the parameters, but, on the other hand, such valves tend to conform more closely to the theoretical assumptions and therefore the design formulæ apply fairly accurately. These remarks are only meant to apply to valves intended for fairly low-frequency operation. For frequencies above about 30 Mc./sec. considerably more care with the preliminary design is necessary.

It is useful to have a relation for μ which is valid for small changes, such as the adjustment of tube geometry after the first approximation to the correct geometry has been determined. Such a formula is the following:

$$\mu = \text{const.}\, N^2 rD, \tag{56}$$

where N = number of grid wires per unit length, r = grid wire radius, $D = d_{ga}$ for a planar triode and $R_g \ln R_a/R_g$ for a cylinder

† Jervis, *Electronics*, **12** (1939), 1945.

triode. Naturally this expression can only be accurate for quite small variations, but it is a useful aid in rapid computation.

9.6. The electrostatic theory of triodes for high frequencies

The theory developed up to now has been restricted to cases in which the screening factor is less than 0·16 and in which $d_{kg} > a$, i.e. *Inselbildung* has not been considered. Moreover, the effect of initial velocities has not been brought into the analysis. In valves meant for use at 30 Mc./sec. and above, all these points must be considered. The restriction in magnitude of screening factor is perhaps the least irksome condition because the wire diameter decreases as d_{kg} decreases, because one naturally requires a grid with a fairly high transparency if good use is to be made of the cathode. However, this process is limited by manufacturing considerations, as very fine wires are very difficult to handle and special grid-winding techniques must be adopted. The Bell Laboratories have been able to use 0·0003 in. tungsten wire for grids for valves at 4000 Mc./sec., but ordinary techniques are capable of handling only 0·001 in. wires or thicker. The effects of *Inselbildung* and initial velocity may be important in valves in which the screening factor is quite low.

An important simplification in the work results from the fact that nearly all ultra-high-frequency valves approximate quite closely to a true planar geometry. The well-known 'Lighthouse' valve in which a circular disk cathode is placed close to a planar grid wound round a thin perforated metal strip is a case in point. The grid wires are soldered to one side of the strip after winding and removed from the other. The grid is then soldered to a metal disk which is sealed through the glass envelope to form the grid lead. The cathode and anode subassemblies are next sealed in parallel to the grid. Clearly such a tube is a very good approximation to the assumed shape. Even in valves of more normal external appearance, such as the miniature 6 AK5 pentode, the cathode-grid system approximates fairly well to planar form. We shall, therefore, confine ourselves to the study of the planar geometry from now on.

There is a large literature devoted to improving the accuracy of μ calculations. Ollendorf† produced some formulæ which are useful for large S by using the conformal transformation $W = \ln \sin Z$. Herne‡ transformed a section containing a quadrant of one grid wire by the Schwartz-Christoffel transformation. His results agree well with those of Ollendorf, whose expressions are simpler. We quote Ollendorf's third approximation, valid up to $S \doteqdot 0{\cdot}5$:

$$\mu = \frac{\dfrac{2\pi d_{ga}}{a} - \dfrac{1}{2}\dfrac{(\pi S)^2}{1+\frac{1}{2}(\pi S)^2}}{-\ln(\pi S) + \dfrac{\frac{1}{4}(\pi S)^2}{1+\frac{1}{12}(\pi S)^2} - \dfrac{\frac{1}{228}(\pi S)^4}{1+\frac{1}{210}(\pi S)^4}}. \tag{57}$$

It is interesting to note that the result of § 9.3 is about 10 per cent lower than the value of eqn. (56) for $S = 0{\cdot}3$. Fremlin,§ in a paper which in addition gives another approach to the equivalent diode problem, has given formulæ for μ valid for $S < 0{\cdot}1$, $a/d_{kg} > 2{\cdot}5$, i.e. conditions in which considerable *Inselbildung* would be expected.

The most powerful method of attacking the problem is due to Noether|| and Rosenhead and Daymond.¶ We shall give a brief account of their method here. The basic physical idea is to replace the valve electrodes by the images produced by the grid charge in anode and cathode, and the images produced by these first images and so on. Consider fig. 9.10 which shows the first few sets of images. The grid charge imaged in the cathode is $-q_k$ at d_{kg} behind the cathode. The grid charge imaged in the anode is $-q_k$ at d_{ga} behind the anode. The image in the anode of the charge $-q_k$, d_{kg} behind the cathode, is $+q_k$ at $d_{ga}+2d_{gk}$ behind the anode and so on. The valve can therefore be replaced by these charges and the condition that the cathode and anode planes must be equipotentials.

† Ollendorf, *Elektrotech. u. Maschinenb.* **52** (1934), 585.

‡ Herne. *Wireless Engr.* **21** (1944), 59.

§ Fremlin, *Phil. Mag.* **27** (1939), 709, reprinted in *Electy. Commun.* **18** (1939), 33. Fremlin, Hall and Shatford, *Electy. Commun.* **23** (1946), 426.

|| Noether, Reimann and Weber, *Differentialgleichungen*, F. Vieweg, Braunschweig, (1935), vol. 2.

¶ Rosenhead and Daymond, *Proc. Roy. Soc.* **161** (1937), 382.

The potential problem can then be solved in terms of elliptic functions, which are doubly periodic and can be adjusted so that the poles and zeros of the functions coincide with the charges. If the grid wires are sufficiently small they can be fitted to the equipotentials near the poles and zeros as these are circular. If we take the origin at the centre of a grid wire

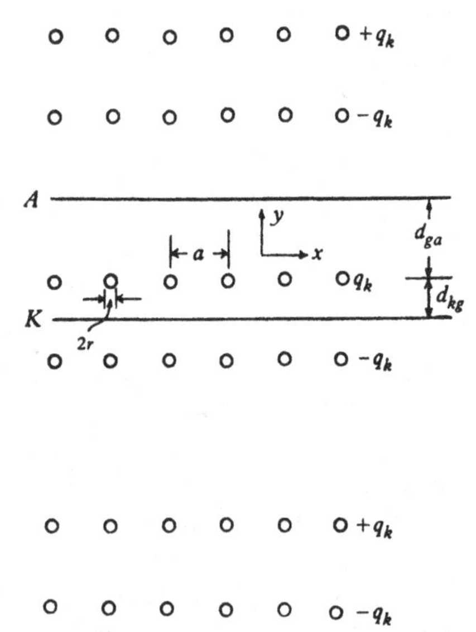

Fig. 9.10. Image representation of planar triode.

the positive charges are at $z = ma+j\,2n(d_{ga}+d_{gk})$ and the negative charges at $z = ma-j\,2[d_{gk}-n(d_{ga}+d_{gk})]$; or, if we write $d_{ga}+d_{gk} = d$ for convenience,

$$\text{positive charges, } z = ma+j\,2nd,$$

$$\text{negative charges, } z = ma-j\,2(d_{gk}-nd). \qquad (58)$$

Now, we know that in the immediate vicinity of the positive-line charges the potential varies as $\ln(z-z_p)$, where z_p is the coordinate of the pth positive charge, and near the negative charges as $\ln(z-z_p)^{-1}$, so the potential function we require is of the form $W = C\ln[f(z)+\text{const.}]$. $f(z)$ has to have poles at the points given by eqn. (58). This can be done by making

$f(z)$ the quotient of two θ functions. It can then be shown without much difficulty† that the complete expression for V is

$$V(x,y) = \frac{V_g}{A}\left[\frac{2\pi}{a}d_{ga}(y-d_{ga})+df(x,y)\right]$$
$$+\frac{V_a}{A}\left[B(y+d_{gk})-\frac{2\pi}{a}d_{ga}y-d_{gk}f(x,y)\right], \qquad (59)$$

where

$$f(x,y) = \ln\left|\frac{\theta_1\left[\frac{\pi}{a}(x+jy-2jd_{ga})\right]}{\theta_1\left[\frac{\pi}{a}(x+jy)\right]}\right|, \qquad (60)$$

$$A = dB-\frac{2\pi}{a}d_{ga}^2, \qquad (61)$$

$$B = \ln\left|\frac{a\theta_1\left(\frac{2\pi}{a}d_{ga}\right)}{\pi\tau\theta_1'(0)}\right|. \qquad (62)$$

Once V has been determined it is easy to find μ, the electric field and the current or current density. V, as given by eqn. (59), is the real part of $W = V+jU$. Then

$$W' = \frac{\partial V}{\partial x}-j\frac{\partial V}{\partial y} = -E_x+jE_y \quad \text{(Cauchy-Riemann equation).}$$

W is easily constructed by putting $-jz$ wherever y occurs (eqn. (59) is real and $z = x+jy$ therefore $R_e(jz) = -y$) and using the complex forms of eqns. (60) and (62) instead of their moduli. On taking the derivative we require the following relation from elliptic function theory:

$$\frac{\theta_1'(z-\pi\tau)}{\theta_1(z-\pi\tau)} = \frac{\theta_1'(z)}{\theta_1(z)}+2j.$$

† Bennett and Peterson, *Bell Syst. Tech. J.* **28** (1949), 303. Elliptic functions are dealt with in Whittaker and Watson, *Modern Analysis*, Cambridge University Press, and Neville, *Introduction to the Jacobian Elliptic Functions*, Oxford University Press, 1944. The notation used is that of Whittaker and Watson, i.e.

$$\theta_1(z)=2\sum_{n=0}^{\infty}(-1)\exp[j(n+\tfrac{1}{2})^2\tau]\sin(2n+1)z, \quad \tau=2j\frac{d}{a}, \quad \theta_1'=\frac{d\theta_1}{dz}.$$

Then, at the cathode,

$$F'(x-jd_{gk}) = \frac{2\pi j}{a}[1+C(x)] \tag{63}$$

$$= \text{complex form of eqn. (60)},$$

where
$$C(x) = I_m \frac{\theta_1'[\pi/a(x+jd_{gk})]}{\theta_1[\pi/a(x+jd_{gk})]}. \tag{64}$$

Since $E_x = 0$ on the cathode, i.e. at $y = -d_{gk}$, we get

$$\frac{a}{2\pi}AE_y = [d_{gk}+d\,C(x)]V_g + \left[d_{ga}-d_{gk}-\left(\frac{aB}{2\pi}\right)-d_{gk}C(x)\right]V_a, \tag{65}$$

and
$$\mu = \frac{\partial E_y/\partial V_g}{\partial E_y/\partial V_a} = \frac{d_{gk}+d\,C(x)}{d_{ga}-d_{gk}-(aB/2\pi)-d_{gk}C(x)}. \tag{66}$$

Eqn. (66) reduces to a form closely analogous to eqn. (16) when the distances d_{gk} and d_{ga} are large.

Eqn. (66) is superior to all the other results† for μ which we have derived, in that it shows clearly that μ varies along the cathode surface, being greatest immediately behind a grid wire and least at the midpoint between two grid wires. Fig. 9.11 shows the variation in μ for a fixed value of d_{ga}, a, and different values of d_{gk}. It can be seen that for $d_{gk}/a < 0{\cdot}66$ μ varies very considerably along the cathode.

If eqn. (66) is solved for $C(x)$ and the result substituted in the expression for the field on the cathode, we get

$$-E_y = \frac{V_g + V_a/\mu(x)}{d_{gk}+d/\mu(x)}. \tag{67}$$

We write $\mu(x)$ to emphasize the fact that eqn. (67) depends on x. Fig. 9.12 shows this function for the same values as we used in fig. 9.11. The field changes sign in some cases, so that the cathode does not emit at all in certain regions if the grid-cathode spacing is too small. This condition should be avoided if best use of the cathode is to be made. Eqn. (67) can be used to obtain an equivalent diode in several ways. The most obvious is to put the equivalent grid-plane voltage $V_g + V_a/\mu$ at

† For a still more accurate expression with two line charges per grid wire, see Strutt, *Schweiz. Bautztg.* **67** (1949), 36.

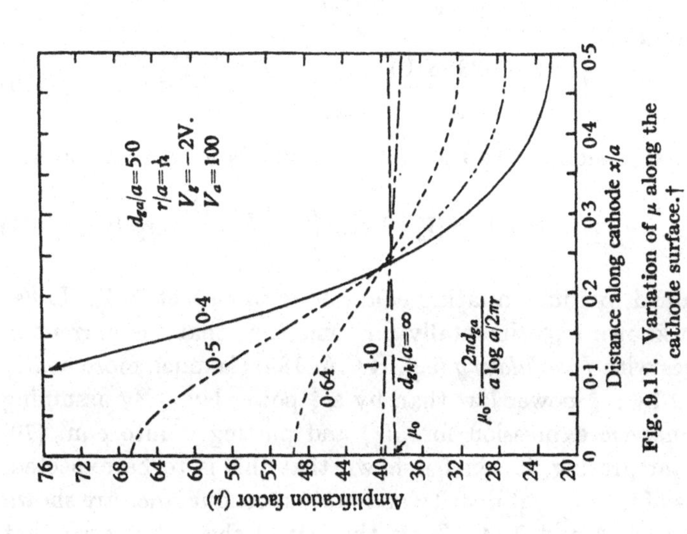

Fig. 9.11. Variation of μ along the cathode surface.†

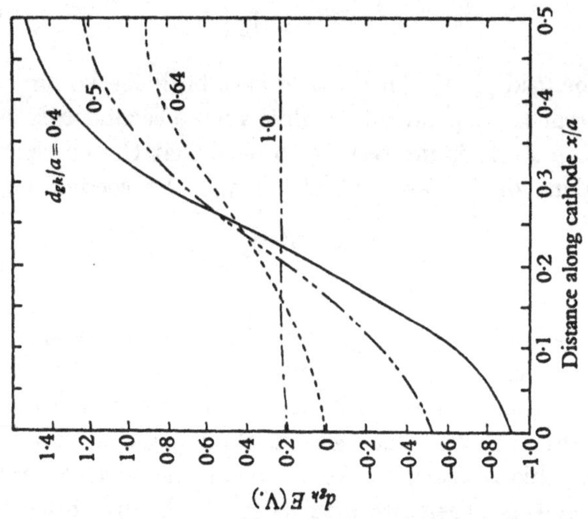

Fig. 9.12. Variation of field along the cathode surface. Dimensions and voltages as in Fig. 9.11. E = volts/unit of distance.

† Figs. 9.11—9.14 are reproduced by courtesy of *Bell Syst. Tech. J.*

distance $d_{gk}+d/\mu(x)$. Another is to take a voltage

$$V_e = V_g + V_a/\mu \frac{1}{1+d/\mu d_{gk}} \tag{68}$$

acting at the grid plane. In the first case both the voltage and the diode spacing depend on x, while in the second case only V_e depends on x. It is interesting to note that the currents in the equivalent diodes are not the same, the second choice leading to

$$J = \frac{2{\cdot}336 \times 10^{-6} V_e^{\frac{3}{2}}}{d_{gk}^{\;2}}, \tag{69}$$

which is $[1+d/\mu d_{gk}]^{\frac{1}{2}}$ times greater than the first choice. If $\mu d_{gk} \gg d$ the result is the same, but this is not invariably the case. On the whole it seems better to adopt the second choice, as this agrees with the analysis we shall make at high frequencies. If this is done, the total current is found by integrating eqn. (69) between two grid wires. If the dimensions are such that the emission is cut off at a point x_0, where

$$V_g + V_a/\mu(x_0) = 0,$$

the result is

$$I = \frac{2 \times 2{\cdot}336 \times 10^{-6}}{d_{gk}^{\;2}} \int_{x_0}^{a/2} \left[\frac{(\mu V_g + V_a)}{(\mu + d/d_{gk})}\right]^{\frac{3}{2}} dx. \tag{70}$$

The expression giving the mutual conductance per unit area is

$$g_m = 3{\cdot}51 \times 10^{-6} (V_g + V_a/\mu)^{\frac{1}{2}} \left(\frac{d_{gk}}{d_{gk}+d/\mu}\right)^{\frac{3}{2}} \text{ amp./V.} \tag{71}$$

obtained by differentiating eqn. (69) with respect to V_g. It has been known experimentally for some time that the current in valves with *Inselbildung* (e.g. R.C.A. 1851) is much more nearly given by a $\frac{5}{2}$ power law than by a $\frac{3}{2}$ power law. By assuming an analytic expression for $\mu(x)$ and putting it into eqn. (70) and integrating, it can be shown that this is to be expected. Plots of the current density and mutual conductance are shown in figs. 9.13 and 9.14. From the last of these it is seen that part of the cathode makes no contribution whatever to the conductance when the *Inselbildung* is extreme. Clearly it is of

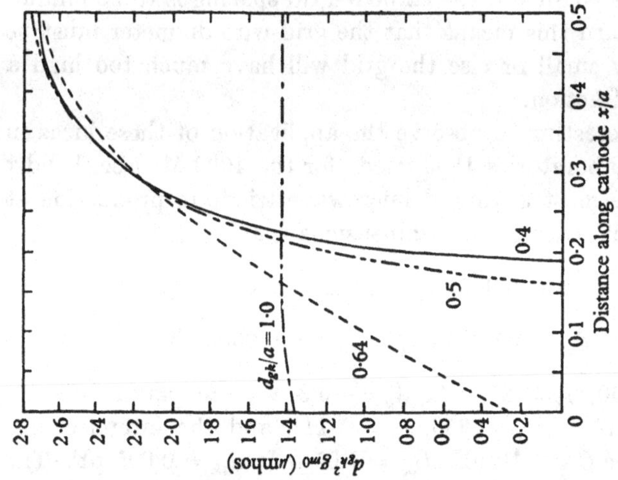

Fig. 9.14. Variation of g_m along the cathode surface. Dimensions and voltages as in Fig. 9.11. g_{mo} = transconductance per unit cathode surface area.

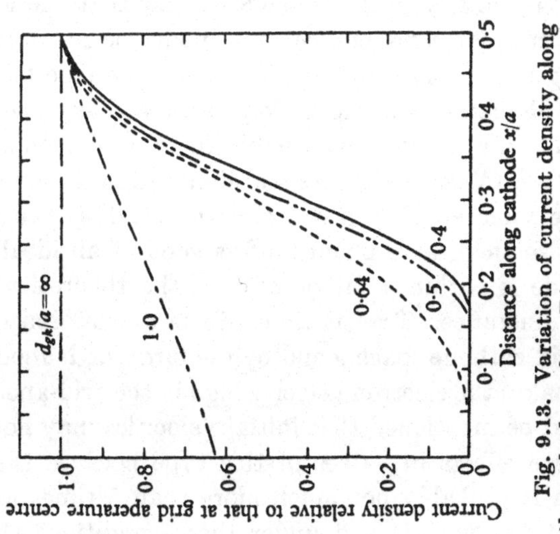

Fig. 9.13. Variation of current density along the cathode surface. Dimensions and voltages as in Fig. 9.11.

no value to diminish d_{gk} much below the spacing between grid-wire centres. Thus the pitch of the grid winding must be made as small as possible if the cathode-grid spacing is to be minimized. In turn this means that the grid-wire diameter must be made very small or else the grid will have much too high a screening fraction.

It is interesting to observe the application of these ideas in the Bell Laboratories 1553 triode for the 4000 Mc./sec. band.† This is the most advanced microwave triode in production at the time of writing. The dimensions are

$$a = 0{\cdot}001 \text{ in.,} \qquad d_{ga} = 0{\cdot}010 \text{ in.,}$$

$$d_{gk} = 0{\cdot}0006 \text{ in.,} \qquad 2r = 0{\cdot}00033 \text{ in.}$$

At $V_a = 250$, $I_a = 25$ mA., $V_g = -0{\cdot}3$ V., the parameters are $G_m = 50$ mA./V., $\mu = 350$, $r_a = 7\ k\Omega$, and the interelectrode capacitance $C_{gk} = 10$ pF., $C_{ga} = 1{\cdot}05$ pF., $C_{ak} = 0{\cdot}005$ pF. The current density (presumably the average current density) is 180 mA./cm.^{2}, which is a good deal higher than is used in more orthodox valves. The ratio of mutual conductance to anode current is 2 mA./V./mA. In § 6.8 we saw that the theoretical limit of mutual conductance in a diode, or in a triode with perfect grid, was $11{,}600/T$ mA./V./mA. The true temperature of an oxide cathode would be kept at less than 1160° K. for a reasonable life, so it is reasonable to take Morton's figure of 10 mA./V./mA. as the theoretical limit. He states that diodes with the same spacing gave twice the conductance per milliampere, so the grid is about half as good as an ideal grid and the triode is within a factor of 5 of the theoretical limit of transconductance. Transit time effects are minimized by the small grid-cathode spacing and by the fairly high anode voltage which makes the electron velocity high in the grid-anode region. It should be mentioned that initial velocities may not produce very large effects in valves of this type because the cathode emission is probably not much more than 5 times as large as the working current, and under these conditions the virtual cathode is very close to the cathode surface.

† Morton, *Bell Lab. Rec.* **27** (1947), 166.

9.7. Interelectrode capacitance in triodes

Fig. 9.15 shows the arrangement of the interelectrode capacitance network representing the triode at low and medium frequencies.

If the cathode potential is taken as zero and the grid and anode potentials as V_g and V_a respectively we have

$$-q_k = V_g C_{gk} + V_a C_{ak}, \quad (72)$$

$$q_a = (V_a - V_g)C_{ga} + V_a C_{ak},$$

$$= -V_g C_{ga} + V_a (C_{ga} + C_{ak}). \quad (73)$$

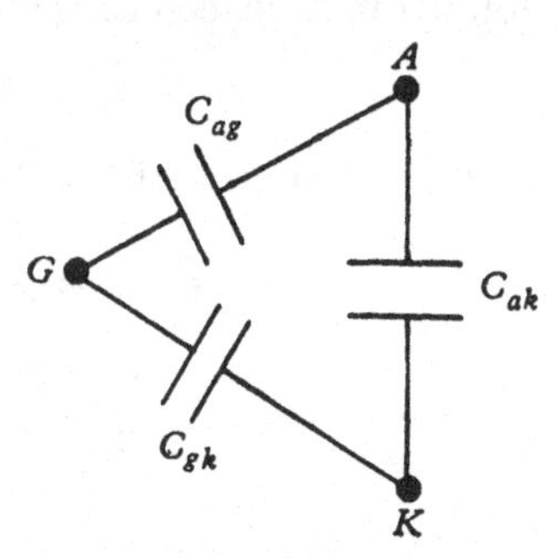

Fig. 9.15. Triode as a capacitance Δ.

These expressions are analogous to eqns. (26) and (27) if

$$\mu = C_{gk}/C_{ak}. \quad (74)$$

Eqn. (74) only refers to the active portion of the electrodes, i.e. to that portion which intersects the electron stream. For most valves of conventional pinch construction a large part, or even a major part, of the interelectrode capacitance is located outside the electron stream, so that if μ, C_{gk} and C_{ak} are measured under specific operating conditions eqn. (74) will not hold. However, for modern planar microwave triodes the electrode capacitance is largely due to that portion of the system in which the beam flows, and for such valves eqn. (74) may be a better approximation. Even for these the agreement is not good. For the Bell 1553 $C_{gk}/C_{ak} = 2000$ but $\mu = 350$. For the CV257, a larger valve used up to 1000 Mc./sec., $C_{gk}/C_{ak} = 14/0{\cdot}3 = 47$ and $\mu = 23$. In such tubes the additional stray capacitance is mostly located in the cathode circuit. In conventional tubes the reverse may be the case.

We shall see in a later chapter that the network of fig. 9.15 must be modified at frequencies for which transit-time effects become appreciable.

9.8. Noise in triodes

We conclude our study of low-frequency effects in triodes by considering the shot noise produced by the electron beam. At low frequencies we can make use of the fact that the current induced in the grid circuit is negligibly small in comparison with the conduction current, to treat the conduction current as the total current. We restrict ourselves to the case of space-charge limited currents which is the only one of practical importance. In § 8.4 we showed that the mean square noise voltage in a space-charge limited diode was (eqn. (49))

$$\overline{V_n^2} = 0{\cdot}644\,.\,4kTr_c\Delta f, \tag{75}$$

where r_c = slope resistance of diode, T = cathode temperature.

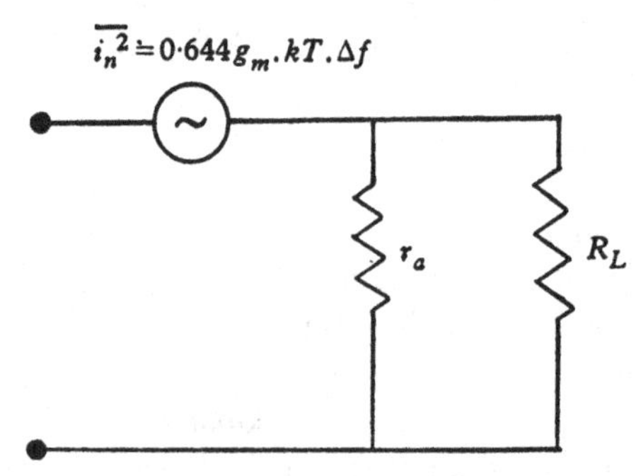

Fig. 9.16. Triode noise equivalent for short-circuited input.

We can apply this equation to give the noise voltage developed across the diode equivalent to our triode. This voltage can then be considered as the shot voltage acting between cathode and grid of the triode, i.e. the noise current in the plate circuit is g_m times the voltage defined by eqn. (75) or

$$\overline{i_n^2} = 0{\cdot}644 g_m^2\,.\,4kTr_c\Delta f.$$

But $g_m = 1/r_c$, by definition. Therefore

$$\overline{i_n^2} = 0{\cdot}644 g_m\,.\,4kT\Delta f. \tag{76}$$

Some writers put $g_m = \sigma/r_c (0{\cdot}5 < \sigma < 1{\cdot}0)$ which has the advantage of allowing one to obtain precise agreement with measured results for a given tube type. Fig. 9.16 shows the noise equivalent circuit of the low-frequency triode. Sometimes it is more convenient in circuit analysis to replace the noise-current generator in the plate circuit by a fictitious noise resistor in the grid circuit. If this resistor is $R_{eq.}$ we have

$$g_m^2 . kT_A R_{eq.} \Delta f = 0{\cdot}644 g_m . 4kT \Delta f$$

where T_A = ambient temperature, therefore

$$R_{eq.} = \frac{0{\cdot}644T}{g_m T_A}.$$

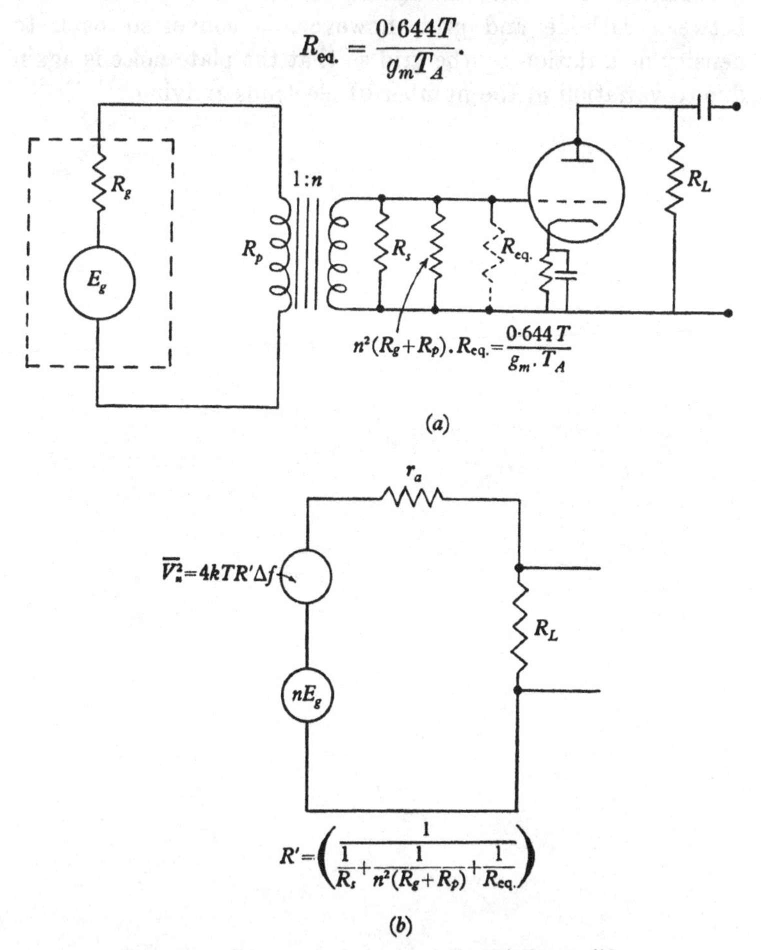

Fig. 9.17. Noise equivalent circuit for triode amplifier.

This device is convenient when there are several real noise-producing resistors in the grid circuit, since they can be lumped together with the noise-equivalent resistor to form a single noise source. The equivalent circuit for this case is shown in fig. 9.17.

The physical basis for our picture of noise production in the triode is as follows. The fluctuations in cathode current are smoothed out by the virtual cathode which changes density modulation to velocity modulation. The velocity modulation between cathode and grid, however, is converted back to density modulation by the grid so that the plate noise is again due to variation in the number of electrons arriving.

Chapter 10

MULTIGRID RECEIVING VALVES

This chapter deals with receiving valves or low-power valves which include more than one grid. Extra grids have been introduced from time to time for various purposes. Historically the first large-scale use of multigrid valves was in the form of h.f. amplifier tetrodes. Somewhat later pentodes were introduced as high-power gain audio-output amplifiers. The next stage of development introduced a variety of multigrid valves all designed for use in super-heterodyne mixer stages. To-day the field is more or less reduced to pentode radio-frequency amplifiers, mixers with several grids, and output amplifiers, which may be either pentodes or the so-called beam tetrodes.

10.1. Screen-grid tetrodes

In the screen-grid tetrode a second grid is introduced between the control grid and anode. This 'screen' grid is biased to a positive d.c. potential, but is held at earth potential as far as radio-frequency currents are concerned, by a large by-pass capacitance. The screen grid produces two desirable effects:

(*a*) the interelectrode capacitance between anode and grid is much reduced, thereby eliminating the necessity of neutralizing the valve;

(*b*) the anode resistance of the valve is very much increased because the current is determined by the instantaneous grid potential and the (fixed) screen potential instead of by the grid and anode potentials, varying in opposite directions.

Since g_m is determined mainly by the grid-cathode system, property (*b*) really means that the screen grid has also a high μ, from eqn. (3), Chapter 9.

Fig. 10.1 shows some typical characteristics of a screen-grid tetrode. Both screen and anode take current, the sum of I_s and I_a being the total space current in the valve. It is rather difficult to obtain a reliable estimate of the real electrode currents from the measured values of I_s and I_a because of secondary emission. When the anode potential is well above the screen potential, secondary electrons emitted from the screen move to the anode so the observed screen current $I_{so} = I_{st}(1-\delta)$, where δ is the secondary emission coefficient of the screen. The

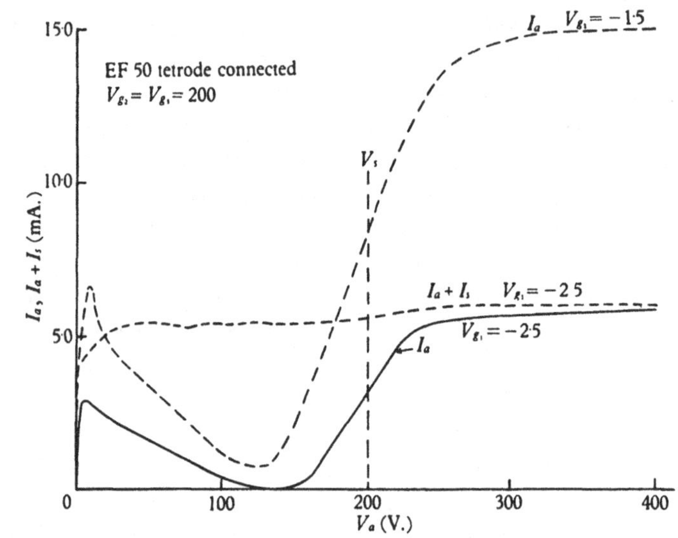

Fig. 10.1. I_a, V_a characteristic of tetrode.

observed anode current is $I_{ao} = I_a + \delta I_{st}$. On the other hand, when the anode is below the screen potential, secondary electrons from the anode go to the screen. The anode current then decreases while the screen current rises. δ is, as we saw in Chapter 2, a function of voltage and may be considered zero for voltages below about 20 V., rising from there to a maximum at a few hundred volts. Therefore the anode current behaves normally below about 20 V., increasing as the anode potential increases. Secondary emission then sets in and the anode current starts to *decrease* as the voltage increases up to a value 10–12 V. above the screen potential. At this point secondaries can no longer reach the screen, or, more correctly, only a small percentage can reach the screen, so that the anode current rises sharply to the value corresponding roughly with the space current minus the geometrical interception current to the screen. Beyond this point the anode current increases with voltage and the anode draws secondary current from the screen as we have already said. This reasoning accounts for the general shape of the characteristics and, in particular, it shows why the anode-current characteristic has a negative resistance region around the point for which $V_a = V_s$. This negative resistance region was made use of in dynatron oscillators, but in linear amplifiers it is a great disadvantage, since the input voltage swing must be kept small enough to ensure that the valve is never operated in the region of anomalous anode current. This means that the anode swing must be limited to quite a small percentage of the anode voltage, perhaps 20–25 per cent, so that the screen-grid tetrode is useless as an output amplifier where anode efficiency is clearly desirable. For this reason screen grids were used only as radio-frequency amplifiers where the anode swing is always small in comparison with the d.c. potential and efficiency is of no importance.

It should be clear from the above that the dimensions of the screen grid itself must be arrived at by a compromise between the requirement of good screening between grid and anode and the requirement that it should not take too high a proportion of the total space current. In early valves the screen current was fairly large, but in modern tetrodes, especially beam tetrodes, screen current is reduced by arranging the screen wires in the shadow of the grid wires, so that the grid prevents electrons from reaching the screen directly. This is of obvious importance in output tetrodes, where the screen current is merely wasted, while in radio-frequency tetrodes low screen current is important because of the consequent reduction in partition noise.

To discuss the theory of the tetrode a little further we introduce the following notation, grids being numbered outwards from the cathode:

$$\mu_a = -\left(\frac{\partial V_a}{\partial V_1}\right)_{I_a=\text{const.}}, \quad \mu_2 = -\left(\frac{\partial V_2}{\partial V_1}\right)_{I_a=\text{const.}}.$$

Then, empirically, we can show that

$$I_t = I_s + I_a = G\left(V_1 + \frac{V_2}{\mu_2} + \frac{V_a}{\mu_a}\right)^n. \tag{1}$$

We are now faced with the problem of determining the fields in the valve in order to determine the μ's. This can be done by the method of conformal transformation applied in Chapter 9. The work is much simplified if the grids are assumed to be alined, and this will be done below. Rosenhead and Daymond† deal with the following particular cases for

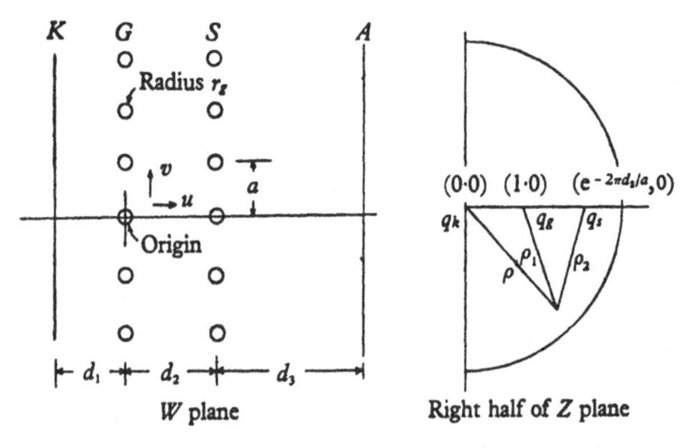

Fig. 10.2. Section of planar tetrode together with its transform.

planar and cylindrical valves: (1) screen pitch twice the grid pitch but wires alined, (2) screen pitch twice the grid pitch, screen wires half-way between grid wires. However, the new problem is really that of determining the μ of the screen, because the cathode-grid-screen may be treated as a triode for the determination of the grid μ. Since the distances grid-screen and grid-anode are usually considerably bigger than the screen pitch, we may determine its μ by the simplest methods. We shall only consider the planar case for which the transformations are those of § 9.1. Fig. 10.2 shows the W plane. Proceeding exactly as in § 9.1, we evaluate the potential at the arbitrary

† Rosenhead and Daymond, *Proc. Roy. Soc.* **161** (1937), 382.

point D,

$$V_D = -\frac{q_k}{2\pi\epsilon_0}\ln\rho - \frac{q_g}{2\pi\epsilon_0}\ln\rho_1 - \frac{q_s}{2\pi\epsilon_0}\ln\rho_2 + C. \tag{2}$$

Substituting for ρ_1 and ρ_2 in terms of ρ and the abscissae we get

$$V_D = -\frac{q_k}{2\pi\epsilon_0}\ln\rho - \frac{q_g}{2\pi\epsilon_0}\ln(\rho^2+1-2\rho\cos\theta) - \frac{q_s}{2\pi\epsilon_0}$$
$$\times\ln\left[\rho_1^2\exp\left(\frac{4\pi d_2}{a}\right) - 2\rho\exp\left(\frac{2\pi d_2}{a}\right)\cos\theta\right] + C.$$

Transforming by the use of

$$\rho = \exp\left(\frac{2\pi u}{a}\right), \quad \theta = \frac{2\pi v}{a},$$

$$V_D = -\frac{uq_k}{a\epsilon_0} - \frac{q_g}{a\epsilon_0}\ln\left[1+\exp\left(\frac{4\pi u}{a}\right) - 2\exp\left(\frac{2\pi u}{a}\right)\cos\frac{2\pi v}{a}\right]$$
$$-\frac{q_s}{2\pi\epsilon_0}\ln\left[\exp\left(\frac{4\pi(u+d_2)}{a}\right) - 2\exp\left(\frac{2\pi}{a}(u+d_2)\right)\cos\frac{2\pi v}{a}\right] + C. \tag{3}$$

C is found by putting $u = -d_1$, $v = 0$, $V_D = V_K = 0$. Since we are assuming $d_1 > a$, we can approximate to obtain

$$C = \frac{1}{a\epsilon_0}(-q_k d_1 + q_s d_2). \tag{4}$$

We can now write down the grid, screen and anode potentials. For V_1, we have $u = 0$, $v = r_g$, $r_g < a/20$, therefore

$$V_1 = -\frac{d_1 q_k}{a\epsilon_0} - \frac{q_g}{2\pi\epsilon_0}\ln\frac{2\pi r_g}{a}. \tag{5}$$

For V_2, $u = d_2$, $v = r_s$, $d_2 > a$, therefore

$$V_2 = -\frac{d_2+d_1}{a\epsilon_0}q_k - \frac{d_2}{a\epsilon_0}q_g - \frac{q_s}{2\pi\epsilon_0}\ln\frac{2\pi r_s}{a}. \tag{6}$$

For V_a, $u = d_2+d_3$, $v = 0$, $d_2+d_3 \gg 2a$, therefore

$$V_a = -\frac{d_1+d_2+d_3}{a\epsilon_0}q_k - \frac{d_2+d_3}{a\epsilon_0}q_g - \frac{d_3 q_s}{a\epsilon_0}, \tag{7}$$

Eqns. (5), (6) and (7) can be solved for q_k. The result is

$$q_k = \frac{1}{\|q\|\, a^2 \epsilon_0^2} \left\{ \left[d_2\, d_3 - \frac{a}{2\pi} (d_2 + d_3) \ln \frac{2\pi r_s}{a} \right] V_1 - \frac{aV_2}{2\pi} \right.$$

$$\left. \times \ln\left(\frac{2\pi r_g}{a}\right) d_3 - V_a \left(\frac{a}{2\pi}\right)^2 \ln\left(\frac{2\pi r_a}{a}\right) \ln \frac{2\pi r_s}{a} \right\}, \quad (8)$$

where $\| q \|$ is determinant of the q's in eqns. (5), (6) and (7). Eqn. (8) can be used to evaluate the μ's in the way that was used in § 9.1:

$$\mu_2 = \frac{d_2 d_3 - (a/2\pi)(d_2 + d_3) \ln (2\pi r_s/a)}{(ad_3/2\pi) \ln (2\pi r_s/a)}; \quad (9)$$

$$\mu_a = \frac{d_2 d_3 - (a/2\pi)(d_2 + d_3) \ln (2\pi r_s/a)}{(a/2\pi)^2 \ln (2\pi r_c/a) \ln (2\pi r_s/a)}. \quad (10)$$

Also

$$\mu_a = \mu_2 \frac{2\pi}{a} \frac{1}{\ln (2\pi r_s/a)}. \quad (11)$$

So we see that $\mu_a \gg \mu_2$ or, in other words, the screen is far more effective in determining the field in the neighbourhood of the cathode than is the anode. It must be remembered that eqns. (9) and (10) are not generally valid because they refer to alined grids only. In other cases a useful approximation is to calculate the μ's of the two triode systems: cathode-grid-screen, grid-screen-anode. μ_a is then given by the product of these μ's and μ_2 is approximately the μ of the first triode.

10.2. Pentodes. Characteristics and gain × band-width product

In the pentode, a third grid is inserted near to the anode. It is usually of much more open mesh than the control or screen grids, and it is held near cathode potential. The function of this suppressor grid is to prevent secondary electrons emitted from the anode from reaching the screen, even when the anode potential is considerably lower than the screen potential. The grid at cathode potential produces a strong field at the anode surface which retards secondaries, turns them round and thus returns them to the anode. On the other hand, the suppressor must have a fairly open mesh because if it is too closely wound, the primary electron current will be unable to reach the anode. However, the field around the suppressor wires is of such a shape that it tends to focus the primary electrons through the openings in the suppressor, and it proves possible to make a suppressor which effectively eliminates secondaries without altering the primary current. Fig. 10.3 shows a typical pentode characteristic. The dotted curve illustrates the effect of strapping the suppressor to the anode, i.e.

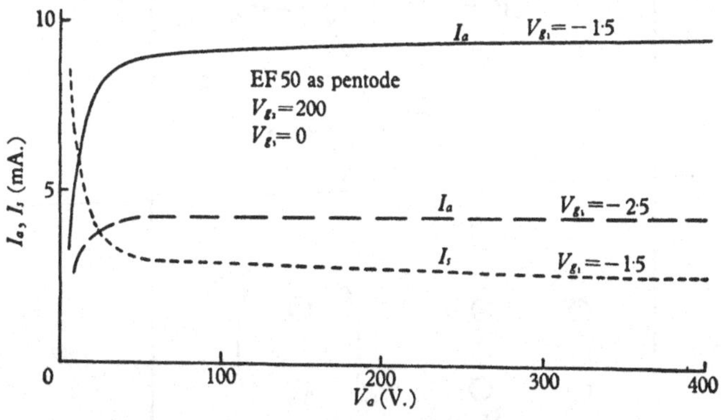

Fig. 10.3. I_a, V_a characteristic of pentode.

converting the valve to a screen-grid tetrode. The suppressor also has the effect of reducing still further the effectiveness of the anode in determining the off-cathode field, so that μ_a is even larger than in a screen-grid valve. Of course, r_a increases to keep $\mu = g_m r_a$. Fig. 10.4 is a sketch of part of the electric field in the screen-anode region which is given to illustrate the focusing action of the suppressor wires. A particular case has been chosen in which the fields on either side of the suppressor are nearly equal. It can be seen that all the electrons whose paths lie between AA' will penetrate the suppressor and those between AB will be returned. These returned electrons may either be captured by the screen, thus increasing the screen current, or enter the grid-screen region from which they will eventually return. On their second journey they may be collected by the screen, or they may reach the anode, or they may be reflected by the suppressor to start the cycle once again. For this reason even if the grid and screen are perfectly alined, a pentode takes screen current or, what is more important, even if secondary emission is neglected the screen current is higher than simple geometrical considerations would lead one to believe.

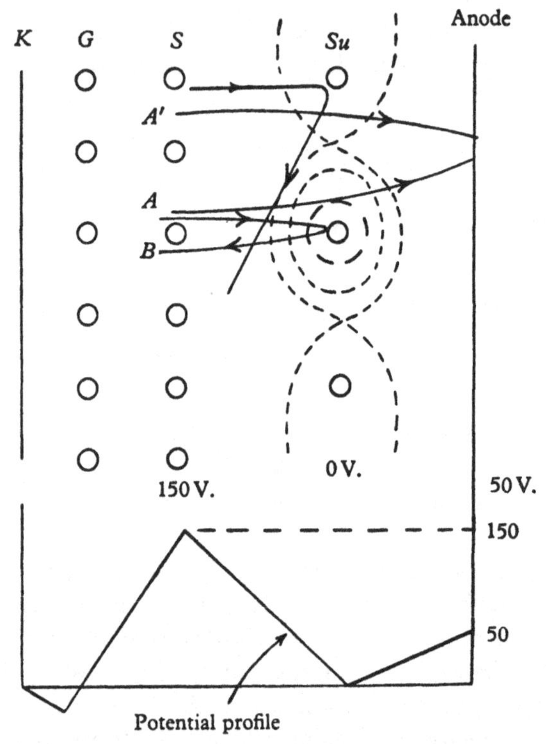

Fig. 10.4. Focusing action of suppressor grid.

It will be clear from the above that the exact theory of the pentode is very much more complicated than that of the triode. In practice, however, most of the theoretical difficulties can be avoided by one means or another. As an example, if the suppressor is of fine wire and has an open mesh and is located close enough to the anode, it can be ignored except in its primary role of secondary emission eliminator. We proceed then to develop a very simplified pentode theory which applies only to low-power types and not to transmitter valves, without modification.

Before we do this, we shall very briefly discuss the circuit aspects of pentode valves, and in particular introduce the important concept of gain × band-width product. Pentodes are usually used with all electrodes except grid and anode at radio-frequency ground potential. This is done by earthing the screen and cathode and suppressor through capacitances having reactances negligibly small in comparison to the d.c. resistances. For some special purposes the screen and cathode are connected to ground through a network of more complex characteristics, but we are not concerned with such cases. Since the pentode has a very high internal impedance, the constant current generator is a better representation than the constant voltage generator. The basic amplifying stage, at medium frequencies, is shown in fig. 10.5, together with the equivalent circuit. At low frequencies Z_L is replaced by a pure resistance R_L, and the gain is given by the triode expression eqn. (5) of Chapter 9,

$$-g_m \frac{r_a R_L}{r_a + R_L} = -g_m \frac{R_L}{1+(R_L/r_a)}. \tag{12}$$

For the pentode, r_a may be considerably greater than R_L, so the stage gain is approximately $A = g_m R_L$.

At resonance the parallel circuit of fig. 10.5 behaves like a pure resistance whose value depends on the following factors:

(1) the anode resistor R_A,

(2) the losses in L and C,

(3) the resistive component of the grid input impedance of the next stage.

This resistance can be measured by finding the two frequencies at which the response of the stage has dropped to 0·707 of the value at resonance, at $f = f_0$ say. If these frequencies are Δf cycles apart we have

$$Q = \frac{f_0}{\Delta f} = \omega_0 CR,$$

or

$$R = \frac{1}{2\pi C\, \Delta f}. \tag{13}$$

Then the stage gain at resonance is $g_m R$, i.e.

$$A = \frac{g_m}{2\pi C\,\Delta f}. \tag{14}$$

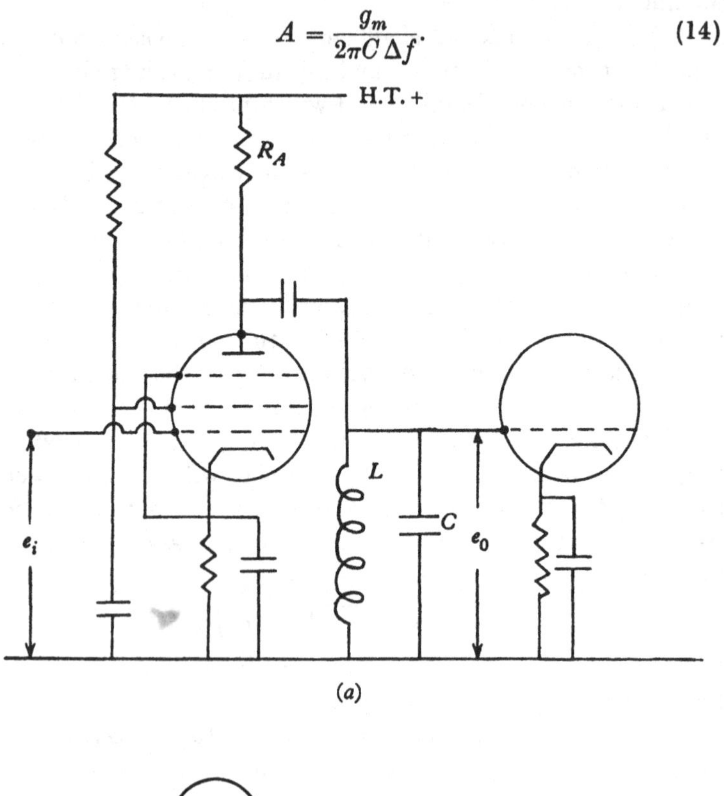

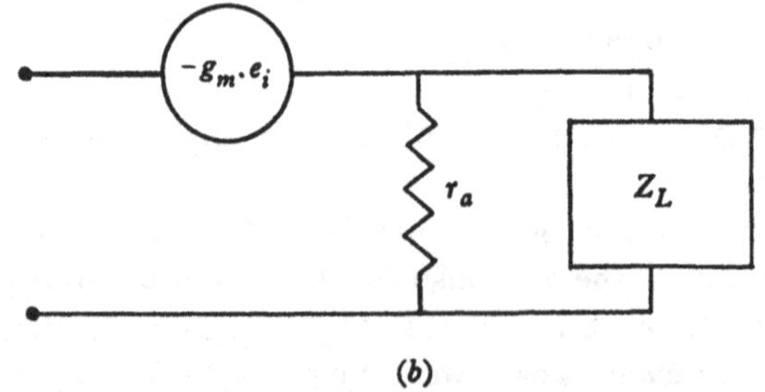

Fig. 10.5. (*a*) Pentode amplifying stage and (*b*) equivalent circuit.

Now let us consider the maximum value of eqn. (14) which can be obtained from a given valve at a prescribed band-width. Following the usual communication practice, the band-width is defined as the number of cycles between the half-power points. But the half-power points are the 0·707 voltage points, so Δf as used to measure Q is the band-width as usually defined. Thus maximizing the gain × band-width product is seen from eqn. (14) to be equivalent to minimizing C. But the minimum value which C can take is $C_{out}(1)+C_{in}(2)$, i.e. the sum of the output capacity of valve 1 and the input capacity of valve 2 with no additional tuning condenser. If a chain of similar valves is considered $C_{min} = C_{in}+C_{out} = C_T$ say. Thus the gain × band-width product of a single tuned stage is $g_m/2\pi C_T$, and the best valve is the one which has the biggest value of g_m/C_T and not the one with the biggest value of g_m. In practice, the wiring will contribute a little extra capacitance, so this value cannot quite be reached. The important thing is that the gain × band-width product gives a valuable figure of merit which can be used to compare valves of widely different characteristics. We should remark that when modern pentodes are used as broad-band amplifiers ($\Delta f > 1$ Mc./sec.), it is a very close approximation to ignore the shunting effect of r_a. As a typical example, we consider a miniature valve used in radar intermediate frequency strips, the 6AK5. For this tube $g_m = 5·0$ mA./V., $C_{in} = 3·9$ pF., $C_{out} = 2·85$ pF. Gain × Δf Mc./sec. = 117. A similar British valve, CV 138, has $g_m = 7·65$ mA./V., $C_{in} = 7·5$ pF., $C_{out} = 3·2$ pF. Gain × Δf Mc./sec. = 114. The figures given relate to the cold tube. The 'hot' input capacity is 20–30 per cent greater than the figure quoted, so that a gain × band-width product of about 80 is a more realistic figure for use in circuit calculation. The best commercially available pentode of this class is, at the time of writing, the Western Electric 404A with a 'hot' $g_m.\Delta f$ of approx. 120. It does not seem likely that this figure will be exceeded by a large amount, using more or less conventional receiving tube structure.

Consider that a CV 138 is to be used as a radio-frequency amplifier in a television receiver. The band-width required is 4 Mc./sec. The stage gain then cannot exceed 20. Thus $g_m.R_L = 20$. Therefore

$$R_L = \frac{20}{7·65}\,k\Omega \doteqdot 2·60k\Omega.$$

Since $r_a = 1\ M\Omega$ for the CV 138 our previous remark on the insignificance of R_L/r_a compared with unity is substantiated.

10.3. Pentodes. Mutual conductance and capacitance

Now that we have emphasized the importance of capacity, particularly C_{in}, in the design of pentodes, we can consider the theoretical design in a little more detail. The simplest procedure is to consider that the cathode, grid and screen form a triode whose μ and g_m can be calculated from the formulæ of the last chapter. The screen grid intercepts a certain proportion of the current arriving at the screen plane so that the anode current is less than the current in the triode. If the anode current $= MI_c$, where I_c = cathode current, the mutual conductance of a planar pentode is M times the mutual conductance of the planar triode defined above. That is, from eqn. (42) of Chapter 9,

$$g_m = 3{\cdot}504 \times 10^{-6} \frac{AM\mu_{12}(1+\mu_{12})^{\frac{1}{2}}(V_S+\mu_{12}V_g)^{\frac{1}{2}}}{[d_1(1+\mu_{12})+d_2]^2}, \tag{15}$$

where A = cathode area,

$\mu_{12} = \mu$ of screen grid through control grid,

d_1 = spacing between cathode and control grid (cm.),

d_2 = spacing between control grid and screen (cm.),

V_S = screen potential,

M = proportion of current intercepted by screen.

A first approximation to M is obviously given by calculating the optical transparency of the screen. This will invariably underestimate the current taken by the screen, which will be increased by reflexion of electrons from the suppressor and by electron optical deviations in the control grid region which can cause large departures from ideal linear flow. Values of two to five times the calculated interception are met, but it should be possible to keep between two and three, by using fine wire for the screen and alining the grids. The screen current should be kept low to reduce the wastage of cathode current, to reduce the screen dissipation and to reduce partition noise.

By using eqn. (41) of Chapter 9 to give I_c, and inserting I_o for the cathode current density, we can show that $g_m \propto I_o^{\frac{1}{3}}$, as was the case for the triode. The next step is to obtain an approximation for the total capacity, which is the sum of the capacitances measured between grid and earth with the other electrodes earthed. This is done by assuming that the electrodes form a parallel-plate condenser whose area is the active area of the structure. Then

$$C_{\text{tot.}} = \frac{1{\cdot}11 \times 10^{-12}}{4\pi}\left(\frac{a_1}{d_1} + \frac{a_2}{d_2} + \frac{a_4}{d_4}\right), \tag{16}$$

where a_1, a_2, a_4 = grid, screen and anode active areas (sq. cm.), and d_1, d_2, d_4 = spacings in order previously defined. The screen-suppressor dimension does not appear because both these electrodes are earthed in each measurement. Eqn. (16) is only a rough approximation, but its use, together with eqn. (15), allows the relative importance of the various parameters in the valve design to be correctly assessed. These two equations will obviously overestimate the gain × band-width product because the stray capacitances between the leads, fringing effects and so on have been neglected.

A more important source of error in modern close-spaced tubes is the effect of *Inselbildung* which appears in the following manner. In deriving eqn. (42) from eqn. (41) in Chapter 9 it was assumed in the differentiation that μ was independent of V_g. The consideration of the variation of μ along the cathode surface when $d_1 < a$, presented at the end of Chapter 9, shows that this is not the case, because when *Inselbildung* is present, the effect of driving the grid more negative is not to reduce the emission from the part of the cathode which is emitting, but to reduce the potential in the region immediately behind the grid wires which is already negative enough to cut off the emission. The apparent μ thus decreases for negative swings and increases for positive swings, resulting in a reduction in mutual conductance. Let us calculate the magnitude of the effect. From eqn. (41) of Chapter 9,

$$I_c = \frac{2{\cdot}336 \times 10^{-6} A (1+\mu_{12})^{\frac{1}{2}} (V_S + \mu_{12} V_g)^{\frac{3}{2}}}{[(1+\mu_{12})d_1 + d_2]^2}.$$

For large μ_{12},

$$I_c \doteqdot \frac{2{\cdot}336 \times 10^{-6} A (V_g + (V_S/\mu_{12})^{\frac{3}{2}}}{d_1^2 [1 + (d_2/\mu_{12} d_1)]^2}. \tag{17}$$

Differentiating eqn. (17) with respect to V_g but neglecting the variation of μ_{12} in the denominator, which is justified as $(d_2/\mu_{12}d_1) \ll 1$, we obtain

$$g_m = \frac{3}{2}\frac{2{\cdot}336\times10^{-6}A[V_g+(V_S/\mu_{12})]^{\frac{1}{2}}}{d_1^2[1+(d_2/d_1\mu_{12})^2]}\left[1-\frac{V_S}{\mu_{12}{}^2}\frac{d\mu_{12}}{dV_g}\right] \tag{18}$$

$$= \frac{3}{2}\frac{(2{\cdot}336\times10^{-6}A)^{\frac{2}{3}}I_c^{\frac{1}{3}}}{d_1^{\frac{4}{3}}[1+(d_2/d_1\mu_{12})]^{\frac{2}{3}}}\left[1-\frac{V_S}{\mu_{12}{}^2}\frac{d\mu_{12}}{dV_g}\right]. \tag{19}$$

Since $C_{tot.}$ is substantially a constant $\times\, d_1^{-1}$ for a high slope valve (cf. eqn. (16)) the functional form of g_m/C is

$$\frac{g_m}{C} = K\sqrt[3]{(I_c/d_1)}\left[1-\frac{V_S}{\mu_{12}{}^2}\frac{d\mu_{12}}{dV_g}\right]. \tag{20}$$

The gain $\times$ band-width product is thus proportional to $(I_o/d_1)^{\frac{1}{3}}$ and independent of the cathode area. The factor in the square bracket is not easy to evaluate theoretically. It is known that it decreases rapidly when $a/d_1 \doteqdot 1{\cdot}0$. Eqn. (20) shows that it is very difficult to obtain increases in gain $\times$ band-width product by increasing I_o or decreasing d_1 because a 2 : 1 increase in g_m/C involves an eightfold increase in I_o/d_1. For modern oxide cathodes at 1050° K. the limit of I_o is a few hundred mA./cm.^{2}. In ordinary receiving tube structures it is impractical to make $d < 0{\cdot}003$ in., mainly because of manufacturing tolerances which would lead to enormous variations between the characteristics of a series of valves. The grid pitch should be, as stated, approximately the same as d_1. The grid-wire diameter should be as small as possible so as to make use of as much of the cathode area as possible and because thick grid wires help to provoke *Inselbildung* and μ variation. These processes are carried to the limit in the Western Electric 404A† which has $d_1 = 0{\cdot}0025$ in., $a = 0{\cdot}0025$ in., $2r = 0{\cdot}0003$ in., i.e. the grid transparency is about 87·5 per cent. The tungsten grid is wound to about half its breaking tension on a rigid molybdenum frame and is then brazed in position. By this means the difficulties of conventional grid-winding techniques are overcome, at the cost of considerable manufacturing complication. It is

† Ford, *Bell Lab. Rec.* **27** (1949), 59.

interesting to note that even this tube, which has $g_m = 12{\cdot}5$ mA./V. at 13·0 mA. anode current, does not approach the theoretical figure of I/kT within a factor of 10.

Considerations such as those of the last paragraph lead one to believe that the limit of performance for gridded pentodes has nearly been reached, and that any progress made in this field will be at the cost of extreme manufacturing difficulty. It is a problem of considerable importance that this limitation should be overcome, since fields of great practical importance such as radar with 0·1–0·2 μsec. pulses and microwave f.m. and p.c.m. communications demand amplifiers with band-widths of between 10 and 20 Mc./sec. For special measurements band-widths of 200 Mc./sec. are necessary. Using present valves, amplifiers for such purposes can only be made by using large numbers of valves with consequent expense, likelihood of valve failure and large power drain. Improved circuits have somewhat better gain $\times$ band-width than the simple network discussed, but all depend on the ratio of g_m/C. The latest of these to be put into practice was originally described by Percival,† and it is worth a brief description here because analogous schemes are used in ultra-high-frequency technique.

The principle is to build the valves into two transmission lines which have equal propagation velocities. The capacitances forming the shunt elements of the input line are the input capacitances of the valves, and the output capacitances, with additional loading capacity if necessary, fulfil the same function for the output line. Fig. 10.6 shows the general arrangement. Since the time of propagation per section is the same in both input and output lines, the current pulse $g_m e_i$ in the anode circuit of the second valve caused by e_0 at the grid, arrives at the same instant as the pulse $g_m e_i$ from the anode of the first valve and the pulses add. An amplified wave thus propagates to the left in the anode line. The wave to the right, on the other hand, is cancelled out if the circuit constants are accurately maintained. The amplifier thus behaves as a single valve with mutual conductance ng_m, where n = number of valves. A

† Percival, B.P. 460,562 (1937).

paper by Ginzton *et al.*† should be consulted for practical details.

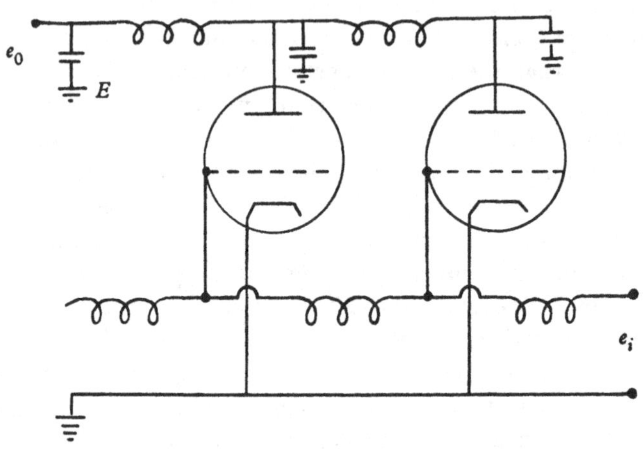

Fig. 10.6. Schematic distributed amplifier.

10.4. Pentodes. Remarks on the suppressor grid

As yet we have said very little about the suppressor grid except to mention the main function of preventing secondaries from reaching the screen. From what has been said it is clear that the suppressor grid must be of very open pitch, since otherwise it will return many electrons by electrostatic repulsion which is just as objectionable as secondary emission from some points of view. On the other hand, if the pitch is too great, the suppressor field will only be effective over part of the anode. We have already seen that if d_4, the suppressor-anode spacing, is less than p_3 the suppressor pitch, the field on the anode will be non-uniform since the problem is just the same as that encountered in the grid-cathode region. Thus the limiting value of d_4/p_3 is approximately 1. Another point that is important is that $C_{\text{out}} \propto d_4^{-1}$, so that d_4 has a certain effect on the gain × band-width product. Since $d_4 \gg d_1$ usually, the variation with d_4 is generally small. Finally, the distance d_3, the screen-suppressor clearance, must be sufficiently large to make the effect of the suppressor evaluated inside the screen rather

† Ginzton *et al.*, *Proc. Inst. Radio Engrs., N.Y.* **36** (1948), 956.

small. This is not difficult to do, since the suppressor pitch is much greater than the screen pitch, and $d_3 > d_2$ is probably a sufficient condition in most cases.

10.5. Noise in pentodes

From our earlier results we can very easily deduce the noise current in the anode circuit of a pentode. The result is the same as that for a triode, but due allowance must be made for the effect of partition noise which arises because of the division of current between anode and screen. The simplest way of making the calculation is to write the expression for the noise component of the cathode current from eqn. (29) of Chapter 8 as

$$\overline{i_c^2} = 2\Gamma^2 e I_c \Delta f. \tag{21}$$

We then use eqn. (56) of Chapter 8 to calculate the effect of partition noise. Inserting eqn. (21) into eqn. (56) we get

$$\overline{i_a^2} = 2eI_a\Delta f\left(1 - \frac{I_a}{I_c}(1-\Gamma^2)\right) = 2eI_a\Delta f\left(\frac{I_c - I_a + I_a\Gamma^2}{I_c}\right).$$

But $I_c - I_a$ must equal I_{g_2}, therefore

$$\overline{i_a^2} = 2eI_a\Delta f\left(\frac{I_{g_2} + I_a\Gamma^2}{I_c}\right). \tag{22}$$

If $I_g = 0{\cdot}2I_c$, $I_a = 0{\cdot}8I_c$, and $\Gamma^2 = 0{\cdot}2$, which are reasonable figures for modern valves, the factor in the brackets is 1·8 times as large as the mean square noise in the triode with the same slope and anode current. The difference is large enough to be noticeable in amplifiers designed for minimum noise figure at intermediate frequencies, and so there has been a tendency in recent years to use neutralized triodes instead of pentodes for the input stages of such amplifiers. An equivalent noise resistance can be deduced as in the last chapter. The result is

$$R_{\text{eq.}} = \frac{0{\cdot}644T}{g_m T_A}\frac{I_a}{I_c}\left(1 + \frac{I_{g_2}}{\Gamma^2 I_a}\right). \tag{23}$$

A little manipulation yields

$$R_{\text{eq.}} \doteqdot \frac{0{\cdot}644T}{g_m T_A}\frac{I_a}{I_c}\left(1 + \frac{8I_{g_2}}{g_m}\right). \tag{24}$$

Returning to eqn. (22) we notice that it is important to keep the screen current low if partition noise is to be minimized. Since the screen must also be a good electrostatic shield or, in other words, must have a fairly high μ, it is necessary to use thin wire for the winding. The screen current can be reduced by alining the screen wires directly behind the grid wires and, particularly in larger valves, by electron optical methods, i.e. by designing the valve so that the elementary beams leaving the cathode come to a focus at or near the screen plane as shown in fig. 10.7. For more information on this question the

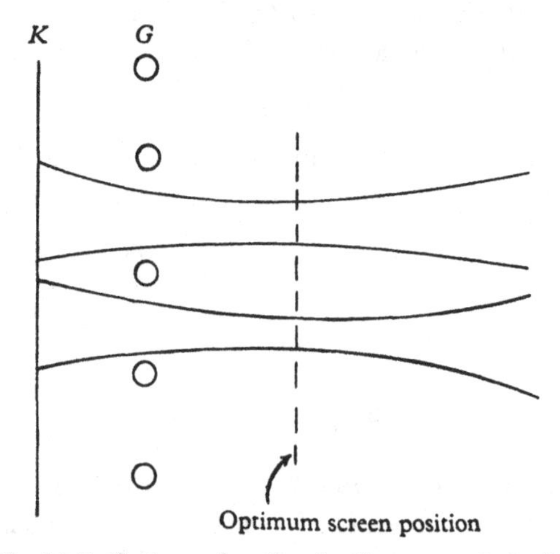

Fig. 10.7. Optimum location for the screen electrode.

reader is referred to a paper by Bull.†

These remarks on noise conclude our present study of the pentode. It has been tacitly assumed throughout that there will be no effects due to transit time in the frequency region we have been considering. This is not so, and, as is well known, modern high-slope pentodes have input impedances between 2 and 10 kΩ at frequencies around 50 Mc./sec. Since one obvious way of minimizing transit time effects is to minimize d_1, the means used to obtain high slope also serve the second purpose, and none of our conclusions on the design of pentodes

† Bull, *J. Instn. Elect. Engrs.*, **92**, pt. III (1945), 86; see also Jonker, *Philips Res. Rep.* **4** (1949), 357.

will have to be modified. In these circumstances it is logical to defer the question of transit time ultra-high-frequency effects until we consider the whole question in relation to u.h.f. valves in a later chapter.

10.6. Beam tetrodes

In the middle thirties it was discovered that tetrodes could be given pentode characteristics if special design conditions

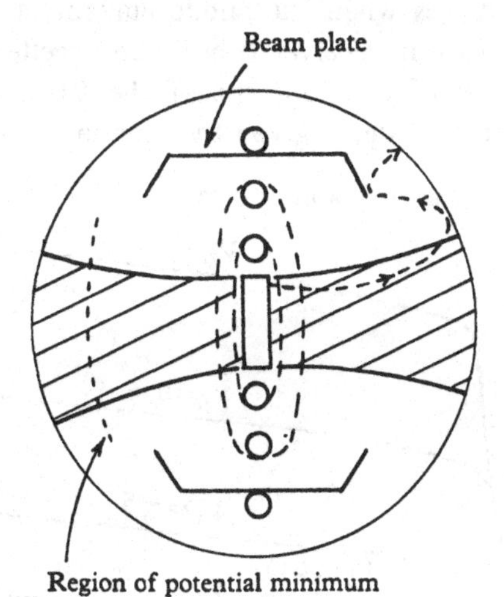

Fig. 10.8. Section through beam tetrode.

were employed. The cathode current was formed into a beam and the anode was located a long way behind the screen. Under these conditions the space charge can be made to lower the potential in the screen-anode space so that an electric field acts towards the anode, causing secondary electrons to be returned to the anode. A section through a valve of this type is shown in fig. 10.8. The cathode, grid and screen are conventional in design. Outside the screen two beam-forming plates are located. These are at cathode potential. The anode is a considerable distance away from the screen. The main function of the beam-forming plates is not, in fact, to form the beam. This is done mostly by the thick grid support wires at the sides of the cathode. The beam plates prevent secondary electrons, which are not emitted in the direction of the potential minimum, from reaching the screen. The dotted trajectory indicates an electron which grazes the anode, producing a secondary moving in the forward direction along the second part of the dotted trajectory. If it were not for the beam plates this would reach the screen, but as it is, it is returned to the anode. The anode must be a considerable distance away from the screen if the potential minimum is to be deep enough to repel secondaries when the anode current is a few tens of mA., as is the case in receiver tubes. An excellent description of the experimental development of the 6L6, an early and successful beam tetrode, is given by Schade.† Fig. 10.9 shows

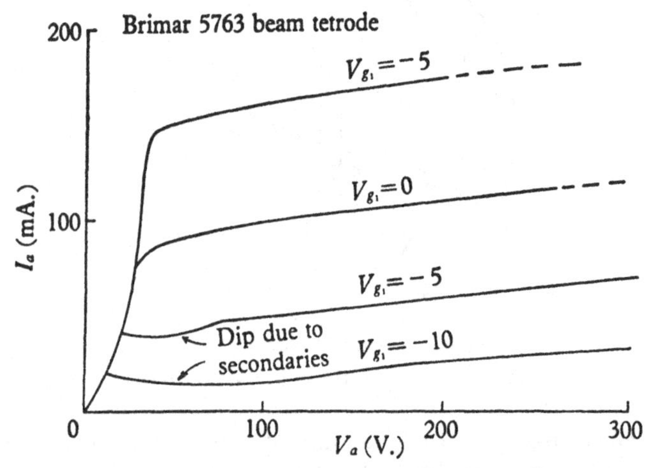

Fig. 10.9. I_a, V_a characteristic of beam tetrode.

an I_a, V_a family for a similar but more recent valve.

The phenomena observed when electron beams are injected into a space at high velocity have previously been discussed in § 6.9. This is precisely the situation found in the beam tetrode, so that discussion can be carried over to our present work. The most important questions to be answered are the extent to which a given current will depress the potential between two given planes, the converse question of how far apart the planes must be if a given current is to produce a given depression,

† Schade, *Proc. Inst. Radio Engrs., N.Y.* **26** (1938), 137.

and lastly the value of V_a at which the sharp knee in the I_a, V_a characteristic will occur. The answer to the last question has been given already in § 6.9. Here we must calculate the potential depression when current flows between two planes. We only consider the case of planes which extend to infinity in the direction perpendicular to the beam; we also assume that the electrons travel normally to the planes, without electron optical deviations. This is not the case in real valves, and it is not difficult to obtain better agreement between theory and experiment if reasonable allowance is made for such deviation, but the difficulty of giving a general theory far outweighs the putative advantages of such a theory.

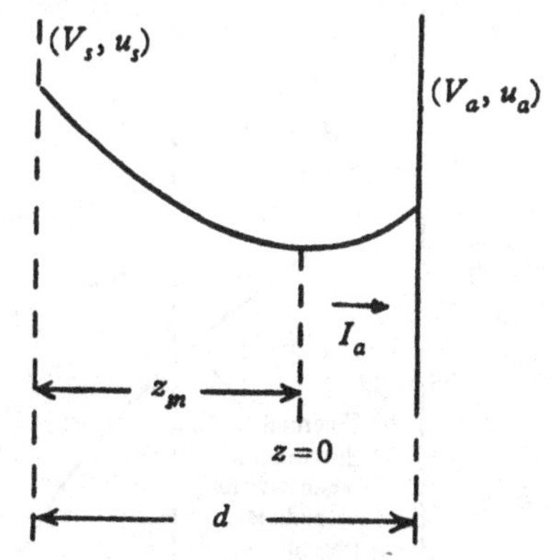

Fig. 10.10. Screen-anode potential variation in a beam tetrode.

The notation is given in fig. 10.10; for convenience we take the z origin at the potential minimum and calculate the potential variation by means of Poisson's equation. The details are given in § 6.1. The first integration gives

$$\left(\frac{dV}{dz}\right)^2 = \frac{4J_a}{\epsilon_0}\left(\frac{m}{2e}\right)^{\frac{1}{2}} V^{\frac{1}{2}} + K_1.$$

But $dV/dz = 0$ at $z = 0$ where $V = V_m$, therefore

$$\frac{dV}{dz} = \left(\frac{4J_a}{\epsilon_0}\right)^{\frac{1}{2}} \left(\frac{m}{2e}\right)^{\frac{1}{4}} [V^{\frac{1}{2}} - V_m^{\frac{1}{2}}]^{\frac{1}{2}},$$

and
$$\left(\frac{4J_a}{\epsilon_0}\right)^{\frac{1}{2}} \left(\frac{m}{2e}\right)^{\frac{1}{4}} z = \int \frac{dV}{(V^{\frac{1}{2}} - V_m^{\frac{1}{2}})^{\frac{1}{2}}}.$$

This can be integrated by substituting $\zeta = V^{\frac{1}{2}} - V_m^{\frac{1}{2}}$. The integral is

$$4/3(V^{\frac{1}{2}} - V_m^{\frac{1}{2}})(V^{\frac{1}{2}} + 2V_m).$$

Squaring and reducing we find

$$V^{\frac{3}{2}} + 3V_m^{\frac{1}{2}}V - 4V_m^{\frac{3}{2}} = \frac{9J_a}{\epsilon_0}\left(\frac{m}{2e}\right)^{\frac{1}{2}} z^2. \tag{25}$$

The distance from the screen to the minimum is given by

$$V_S^{3/2} + 3\ V_m^{1/2} V_S - 4V_m^{3/2} = \frac{9J_a}{\epsilon_0}\left(\frac{m}{2e}\right)^{1/2} z_m^2. \tag{26}$$

The constant $\dfrac{9}{\epsilon_0}\left(\dfrac{m}{2e}\right)^{1/2}$ is the reciprocal of the diode constant D. Bull† has given a neat method for dealing with eqns. (25) and (26). Dividing by $V^{3/2}$ we get

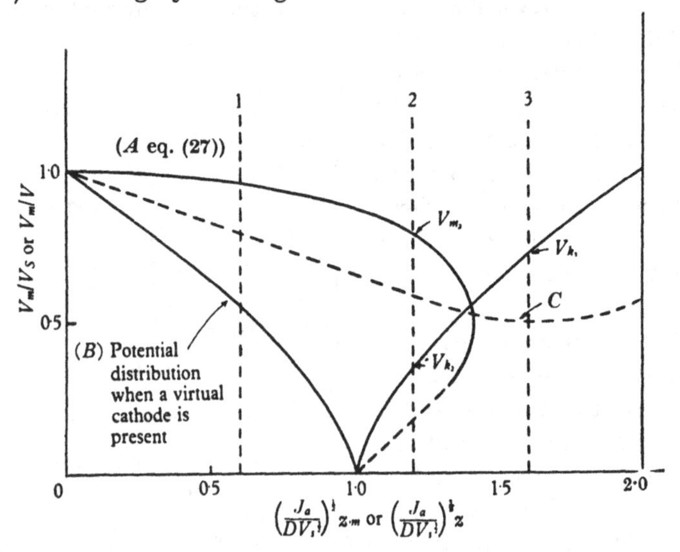

Fig. 10.11. Universal current voltage characteristic.

$$1 + 3\left(\frac{V_m}{V}\right)^{1/2} - 4\left(\frac{V_m}{V}\right)^{3/2} = \frac{J_a z^2}{DV^{3/2}},$$

or

$$\left[1 + 3\left(\frac{V_m}{V}\right)^{1/2} - 4\left(\frac{V_m}{V}\right)^{3/2}\right]^{1/2}$$

$$= \left(\frac{J_a}{DV^{3/2}}\right)^{1/2} z, \tag{27}$$

with a similar expression connecting V_m, V_s and z_m. If we now plot $(1+3y-4y^3)^{1/2}$ and measure z in units of $\left(\dfrac{J_a}{DV_S^{3/2}}\right)^{1/2}$, we obtain a universal curve giving the location of the potential minimum. In fig. 10.11 this curve has been plotted, together with the

† Bull, *J. Instn. Elect. Engrs.*, **95**, pt. III (1948), 17.

$V^{\frac{3}{2}}$ power-law curve which represents the potential variation when a virtual cathode is formed. From the results of § 6.9, $\hat{J}_u = DV_S^{\frac{3}{2}}/b^2$ when the virtual cathode is just formed. But the virtual cathode is also a potential minimum, so $b = z_m$ for $V_m = 0$ and the coordinates of this point are $y = 0$, $z_m = 1$.

We can now see what happens when the anode voltage varies. Three positions for the anode have been marked in fig. 10.11, and we consider these in turn. If the anode is in position 1, the injected current is not high enough to form a virtual cathode and all the current reaches the anode for any voltage above zero. The anode voltage is the lowest potential in the system for all values below that at which the plane 1 intersects curve A, i.e. at $(V_m/V_S)^{\frac{1}{2}} \doteqdot 0{\cdot}96$. For higher anode voltages a minimum is present. When the anode is at plane 2, the anode current varies as $\dfrac{DV_a^{\frac{3}{2}}}{(d-b)^2}$ for the range $0 < V_a < V_{k_2}$. For voltages above V_{k_2}, the knee voltage, all the current reaches the anode. The anode potential is the lowest potential in the system, when $V_{k_2} < V_a < V_m$, so that between these values secondary emission is not suppressed. For voltages above V_{m_2} a minimum is formed. In case 3, the current varies as $\dfrac{DV_a^{\frac{3}{2}}}{(d-b)^2}$, $0 < V_a < V_{k_3}$; but for all voltages above V_{k_3} the secondary emission will be suppressed. It is clear that the anode plane must be beyond the plane of intersection of the curves A and B if the secondary suppression is to be effective over the largest possible range of voltage. On the other hand, the knee voltage must not be too high, for an output valve must operate over a large range of anode voltages if it is to give good efficiency. Moreover, as the injected current density varies with the grid voltage at constant screen voltage, the size of the distance unit varies or, what is the same thing, the position of the anode plane in fig. 10.11 varies. The position of the virtual cathode, of course, remains fixed at unity. If the injected current decreases the anode plane moves farther away. Thus the knee voltage is smaller for smaller injected currents.

The choice of anode plane is thus a compromise between the conflicting requirements that the knee voltage must not be too high at the maximum current, and that the suppression of secondary emission should be effective at low current densities. Two circumstances help to ease the problem in practice; first, the beam plates tend to depress the potential in the screen-anode space even when no space charge is present so that the knee forms at lower voltages, and secondly, the beam is not plane parallel but to some extent focused so that the current density varies along the beam. If the density is maximum at about the centre of the screen-anode space, this will also tend to promote the formation of a virtual cathode. It is, however, difficult or impossible to suppress secondaries at low currents as the 5763 characteristics show.

It may be inquired why a potential curve of the type C in fig. 10.11 is not permitted. This curve is characterized by the fact that all the current would reach an anode at plane 3 for voltages below V_{k_3}. Bull, in the paper cited, has shown that curves of this type can be excluded by the principle of least action.

It will be seen that the explanation of the operation of beam tetrodes given above excludes the possibility of hysteresis loops in the characteristics, which are indicated by the theories given by Salzburg and Haeff, Fay *et al.* and Walker, for instance. Such hysteresis loops are sometimes observed in valves, but their occurrence is due to Barkhausen-Kurz oscillations as can be proved by the use of an h.f. search receiver. The frequencies are often high enough to involve oscillations on the internal structure of the valve, and these cannot be eliminated by decoupling the external leads. This has led some observers to attribute the disturbances to the necessities of the valve, which is not the case.

10.7. Maximum current flow between planes

We now take up a question which is not logically in place here. This is the question of maximum current flow between planes at different potentials neither being zero. This question was not included in the discussion of space charge because it will be solved by the Llewellyn equations which we had not then discussed. The problem has some bearing on the development of the last section, and in fact provides an alternative approach to that work. We shall not use this approach† here, but we shall obtain an expression for the current flow. The development depends on the fact that at low frequencies the conduction current is equal to the total current. We can thus write the integral of eqn. (42) of § 7.4 for this case as

$$a(z) = \frac{eJt}{m\epsilon_0} + a(0), \tag{28}$$

$$u(z) = \frac{eJt^2}{2m\epsilon_0} + a(0)t + u(0), \tag{29}$$

$$z = \frac{eJt^3}{6m\epsilon_0} + \frac{a(0)t^2}{2} + u(0)t. \tag{30}$$

If τ = transit time from 0 to d, we use eqns. (29) and (30) to get

$$d = -\frac{eJ\tau^3}{12m\epsilon_0} + \left(\frac{u(d)+u(0)}{2}\right)\tau. \tag{31}$$

Eqn. (31) is to be regarded as a cubic equation giving τ in terms of known quantities. By differentiation of eqn. (31) it is easily shown that J has a maximum value at $\tau = 3d/[u(d)+u(0)]$. The corresponding value of J is

$$\frac{eJ}{m\epsilon_0} = \frac{2[u(d)+u(0)]^3}{9d^2}. \tag{32}$$

Now $$u(d) = \left(2\frac{e}{m}\right)^{\frac{1}{2}} V_d^{\frac{1}{2}}, \quad u(0) = 2\left(\frac{e}{m}\right)^{\frac{1}{2}} V_0^{\frac{1}{2}},$$

therefore $$\hat{J} = \frac{4\epsilon_0(2e/m)^{\frac{1}{2}}}{9d^2}[V_d^{\frac{1}{2}} + V_0^{\frac{1}{2}}]^3$$

$$= \frac{D}{d^2}[V_d^{\frac{1}{2}} + V_0^{\frac{1}{2}}]^3 \tag{33}$$

Eqn. (33) is the generalization of eqn. (6) of § 6.1, for the case when neither plane is at zero potential. It is interesting to note that the transit time for this current density, $\tau = 3d/[u(d)+u(0)]$,

† Knipp, *M.I.T. Rad. Lab. Rep.* 22 March 1944, p. 534,.

is $3/2\tau_0$, where τ_0 is the transit time for zero space charge, also in agreement with what we have found before.

10.8. Valves for frequency conversion

To conclude this chapter we shall give some very brief descriptions of valves used for frequency conversion. These are called convertors, when the same valve serves as local oscillator and frequency changer combined, and mixers, when the signal is mixed with the output from a separate local oscillator. Essentially, both types consist of valves having two control grids working at approximately cathode potential. One control grid is used to control the anode current at local oscillator frequency, and the other is used to control the anode current at signal frequency. The mixer is probably the best technical solution to the conversion problem, but it is infrequently used in broadcast reception because of the extra valve required. The main problems of frequency changers is exactly the same as that met in the design of pentodes and beam power tubes, i.e. the valve must have a high mutual conductance (this is necessary since the mutual conductance is directly proportional to the conversion conductance). Other problems, however, arise. The two most important are the arrangement of the valve so that there is minimum undesired internal coupling due to electrons repelled from the region of the second grid and the question of minimum noise. Any back-coupling produces undesired shifts in the oscillator frequency, a very objectionable phenomenon when the intermediate frequency is only a small percentage of the radio-frequency. Frequency changers are usually noisy, since there is much more partition noise than in a pentode and also because the returned electrons, which may take very long and complicated paths, produce transit effects at much lower frequencies than is the case, for example, in a pentode.

To understand the operation of frequency changers we assume that the local oscillator voltage is much bigger than the signal, which is always true in practical systems. Consider for a moment the schematic valve of fig. 10.12, where the local oscillator and signal are applied to two grids inserted one after the other in the A–K space. Since the g_m of the triode formed by the two grids and the anode varies as the third power of the injected current, we have

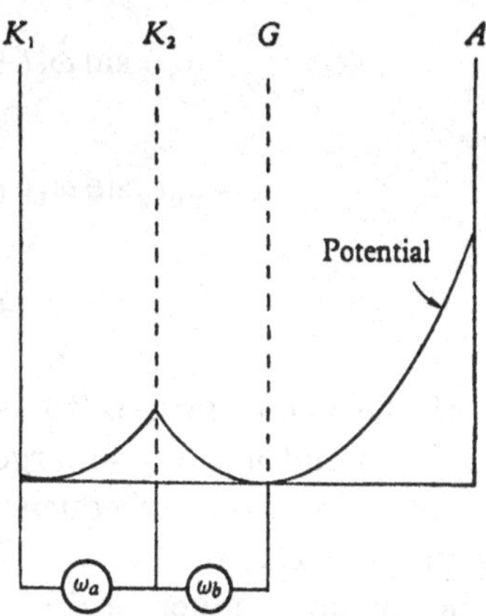

Fig. 10.12. Potential distribution in elementary mixer.

$$g_{m_2} \propto k i_1^{\frac{1}{3}}$$

But $$i_1 \propto V_g^{\frac{3}{2}},$$

therefore $$g_{m_2} \propto V_g^{\frac{1}{2}}. \qquad (34)$$

This functional relationship is approximately true for real valves as the characteristics of fig. 10.14 indicate. We now make a Fourier expansion of g_{m_2} in terms of the oscillator frequency ω_0, i.e.

$$g_m = a_0 + a_1 \cos \omega_0 t + a_2 \cos 2\omega_0 t, \text{ etc.}$$

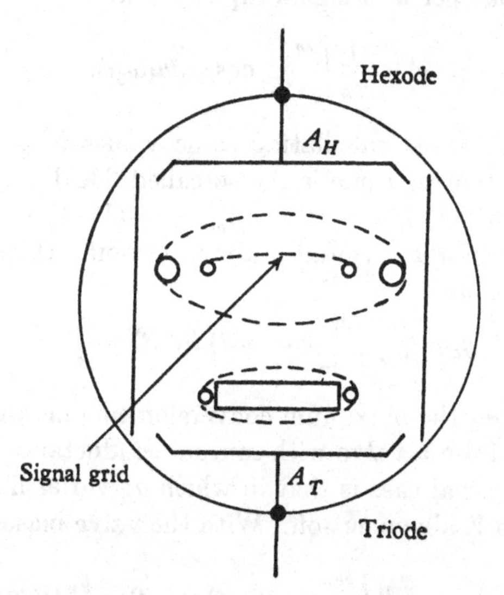

Fig. 10.13. Section through triode hexode convertor.

It is not necessary to consider sin terms, since g_{m_2} is a single-valued function of the oscillator voltage which varies as $\cos \omega_0 t$. If now a small signal $e_2 \sin \omega_1 t$ is applied to the second grid, the a.c. anode current is

$$
\begin{aligned}
i_a &= g_{m_2} e_2 \sin \omega_1 t \\
&= a_0 e_2 \sin \omega_1 t + e_2 \sum_1^\infty a_n \sin \omega_1 t \cos n\omega_0 t \\
&= a_0 e_2 \sin \omega_1 t + \frac{e_2}{2} \sum_1^\infty a_n \sin (\omega_1 + n\omega_0) t \\
&\qquad + \frac{e_2}{2} \sum_1^\infty a_n \sin (\omega_1 - n\omega_0) t. \qquad (35)
\end{aligned}
$$

If the anode load is tuned to the intermediate frequency $(\omega_1 - \omega_0)$ and has a Q so large that it has negligible impedance to any of the other frequencies generated, the output voltage is proportional to $\frac{1}{2}(e\ a_1)$. The quantity $\frac{1}{2}a$ is defined as the 'conversion conductance' of the valve, and is denoted by the symbol g_c. By its use we can calculate the intermediate frequency output per unit signal input. Clearly

$$g_c = \frac{1}{2\pi} \int_0^{2\pi} g_{m_2} \cos \omega_0 t d(\omega_0 t).$$

It is worth while calculating some values of g_c for special cases. An interesting one is the so-called 'ideal' mixer characteristic for which $i_a = 0$, $V_g < -V_{g_1}$, $i_a = kV_g$, $V_g > -V_{g_1}$ i.e. g_{m_2} is a constant for all voltages above cut-off. If the valve is biased to cut-off

$$g_c = \frac{g_{m_2}}{2\pi} \int_{-\frac{1}{2}\pi}^{\frac{1}{2}\pi} \cos (\omega_0 t) d(\omega_0 t) = \frac{g_{m_2}}{\pi}. \qquad (36)$$

Eqn. (36) gives the maximum conversion conductance that can be obtained from a valve with mutual conductance g_{m_2}.

Another special case is that in which g_m varies linearly from 0 to g_{m_2} with V_g above cut-off. With the valve biased to cut-off

$$g_c = \frac{g_{m_2}}{2\pi} \int_{-\frac{1}{2}\pi}^{\frac{1}{2}\pi} \cos^2 (\omega_0 t) d(\omega_0 t) = \frac{g_{m_2}}{4}. \qquad (37)$$

Here g_{m_2} is the value of the mutual conductance at the point where the local oscillator grid is driven to its most positive value. The conversion conductance is thus a function of the local oscillator amplitude, as in practice, whereas in the ideal mixer g_c is independent of the local oscillator amplitude.

In general, of course, g_{m_1} is not a simple analytic function of the local oscillator voltage. It is possible to determine g_c by the usual graphical methods of Fourier analysis if a curve of $g_m \sim V_g$ is provided. Alternatively, it is easy to determine g_c experimentally by accurate measurements at broadcast frequencies.

The considerations on mixers which we have dealt with above are common to all types. We now pass to the qualitative consideration of the specialized types used to-day.

10.9. Triode hexode and triode heptode

These two types are very commonly used in modern broadcast receivers to-day. A common type (6K8) of triode hexode structure is shown diagrammatically in fig. 10.13. Typical characteristics are shown in figs. 10.14 and 10.15. One side of the cathode provides the electron stream for the oscillator triode, and the other for the hexode. The grid is common to both structures so that the local oscillator voltage is injected into the hexode by the first grid. The signal grid of the hexode is surrounded by a positive screen grid which has the double function of providing the positive field on the hexode cathode and of isolating the signal grid from the oscillator grid. Since the signal grid is biased negatively, there is a danger that electrons will be repelled from its vicinity, particularly when the local oscillator swings to high positive values and the injected current is therefore large. Any returned electrons are collected by the screen and only a few return to the oscillator grid circuit where they produce only a negligible change in the electronic susceptance across the oscillator circuit. The other side of the screen grid acts as a screen electrode in the signal grid-anode part of the hexode. In the triode heptode, a suppressor grid is located between screen and anode to prevent secondary emission and to increase the anode impedance.

Another type of triode hexode has been developed and is mainly used in Europe.† In this the triode structure is below

† Shelton, *Wireless Engr.* **35** (1934), 283.

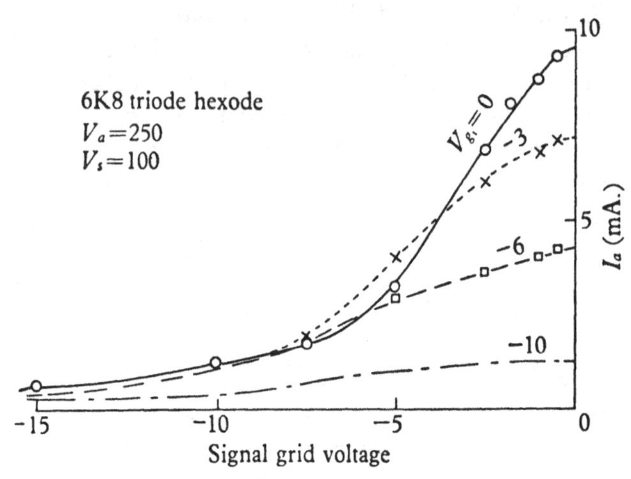

Figs. 10.14. Triode hexode characteristics.

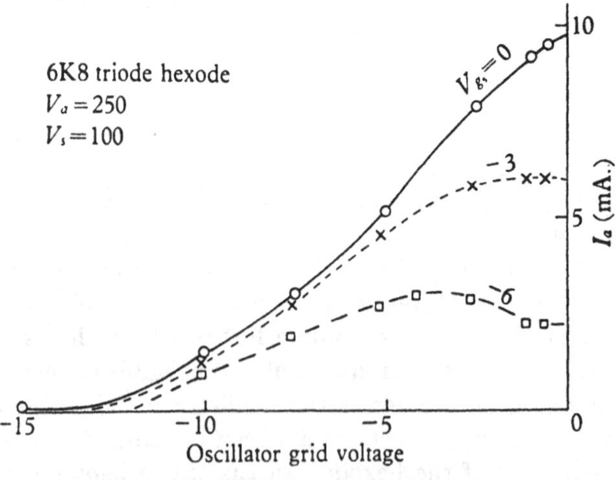

Fig. 10.15. Triode hexode characteristics.

the hexode structure, both using the same cathode. This has the advantage that there is no direct electronic interaction between the hexode and the triode, but the interaction due to changes of load on the oscillator still exists. A further possibility which is often used with this last structure is to connect the local oscillator to the third grid and the signal to the inner grid. This tends to reduce the interaction between oscillator and signal circuits still more and also tends to give a very broad maximum to the curve of $g_c \sim V_g$. On the other hand, the big swing of the local oscillator repels many electrons into the signal grid region of the hexode and these long transit electrons damp and detune the signal circuit. The signal-grid current may be high for the same reason. These disadvantages are overcome by designing the valve on electron optical principles so that the returned electrons are deflected on to a part of the screen which is made solid. This technique however, is usually applied to pentagrid convertors and not to triode hexodes.

10.10. The pentagrid

Early types of pentagrid simply consisted of a cathode together with five concentric grids and an anode. The cathode,

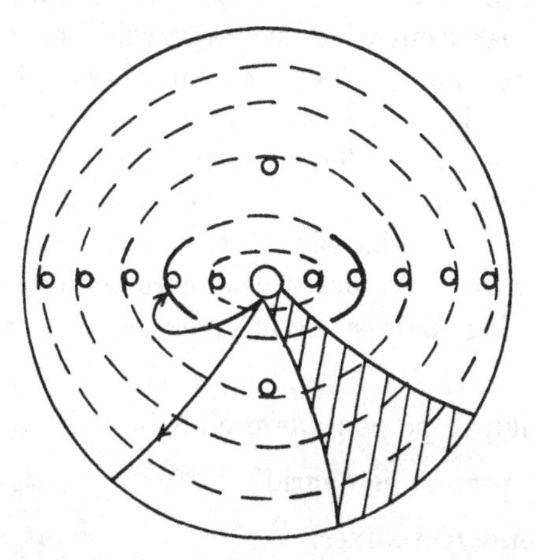

Fig. 10.16. Section through modern type of pentagrid.

together with g_1 and g_2 formed a triode which was connected as the local oscillator. g_3 was the signal grid, g_4 another screen and g_5 a suppressor grid. These early pentagrids suffered from a great deal of interaction between the signal and local oscillator circuits, because the whole local oscillator current is common to both circuits. Using the ideas mentioned at the end of the last section, it has been possible to produce a much more satisfactory valve. A cross-section is shown in fig. 10.16. The electron flow is confined to a beam in each quadrant by means of the thick, negative, support wires of g_1 and g_3. The screen g_2 is made solid in the region of the support wires so that electrons returned from the region of g_3 are collected on g_2 and cannot penetrate into the triode. An example of this construction is the 6SA7.

10.11. The 6L7 pentagrid mixer

In the pentagrid mixer, the local oscillation is generated by a separate valve. The signal is applied to g_1 and the oscillator voltage to g_3; g_2 and g_4 form a screen around the oscillator grid. The geometry is nearly the same as that in the original pentagrids, i.e. a circular cathode with five concentric grids and an anode. The development and application of this valve are described in some detail by Nesslage, Herold and Harris.†

It will be clear from what has been said above that all frequency convertors are noisy. For this reason simpler devices are used at ultra-high-frequency where the signal/noise ratio is of extreme importance. The use of convertor valves is thus limited to frequencies below about 60–100 Mc./sec., and only the more recent glass-based valves will reach frequencies as high as 100 Mc./sec. An idea of the increased noisiness is given by the following figures giving the equivalent grid noise resistance:

6AC7	High slope pentode used as mixer,	$R_n \doteqdot 20\text{k}\Omega$;
6SA7	Improved pentagrid,	$R_n \doteqdot 200\text{k}\Omega$;
6J7	Pentagrid mixer,	$R_n \doteqdot 200\text{k}\Omega$.

Since valves are too noisy, mixing at microwave frequency is performed by silicon crystals, at least in all low-level applications. These crystals show a conversion loss rather than a gain,

† Nesslage *et al.*, *Proc. Inst. Radio Engrs.*, *N.Y.* **24** (1936), 207.

but, in spite of this, the noise figure of a good 10 cm. crystal is of the order 7–10 db., better than any thermionic mixer yet made.

The reader is referred to a paper by Herold† for an interesting general account of convertor development and a full bibliography. A book by M. J. O. Strutt‡ also contains much interesting material particularly on effects at moderately high frequencies.

† Herold, *Proc. Inst. Radio Engrs., N.Y.* **30** (1942), 84.
‡ Strutt, *Mehrgitterelektronenröhren*, **2** (1938), sects. 11 and 13. J. Springer.

Chapter 11

TRANSMITTER VALVES

Transmitter valves are usually divided into three categories, according to the method by which the heat is removed from the anode. The categories are:

(1) Radiation-cooled anodes in glass bulbs. The biggest valves in this class have about 1 kW. anode dissipation.

(2) Cooled anode valves with forced air-cooled anodes.

(3) Cooled anode valves with water-cooled anodes.

A subcategory of (1) exists; it consists of valves with radiation cooled anodes in silica bulbs instead of glass. Silica is much more transparent to red and infra-red radiation than are even hard glasses, and it will also stand operation at a very high temperature because its softening point is very high (approx. 1500° C.). For these reasons silica envelope valves can be used with much higher anode dissipations than can glass valves, and one triode dissipating 4 kW. is made. Silica valves were developed by the Admiralty Signal Establishment for use in Naval transmitters where their ease of replacement and low overall weight and space requirements (absence of cooling blowers, etc.) is of prime importance. They have not seen much application outside the Service as yet, but they would seem to have advantages as valves for mobile or portable radio-frequency heating units, etc.

Progress in the design of forced air-cooled valves has been fairly extensive in recent years, and the design of coolers and circulating systems has been much improved so that air cooling is now used for much bigger valves than was recently the case. Valves of as high as 25–30 kW. anode dissipation are now air-cooled, and one may suppose that gradually the limit will be forced upwards. Water-cooling must be used for the largest valves and this will continue to be the case, since there are physical limits to the dissipation per unit area which can be handled using air as a cooling fluid.

In this chapter we propose to discuss some of the questions which are peculiar to transmitter valves as distinct from receiver tubes. Some of these questions relate to the electronic theory of the valves, and some to questions of mechanical and thermal design. From the electronic point of view, transmitter valves have the peculiarity that they often have to function with the grid driven highly positive to the cathode. Previously, we have assumed that the grid is always sufficiently negative not to acquire any electrons. Our earlier results allow us to calculate the total current drawn from the cathode; we have now to find how this current divides between grid and anode when the former is positive. Before we take up this question, however, it is necessary to say something about the electrostatic theory of valves with filamentary cathodes and the design of tungsten filaments.

11.1. Characteristics of valves with filamentary cathodes

Many transmitter valves, particularly the higher powered ones used at broadcast and short-wave frequencies, have filamentary cathodes. Typical geometries are indicated in figs. 11.1 and 11.2. Fig. 11.1 represents a typical 200–250 W.

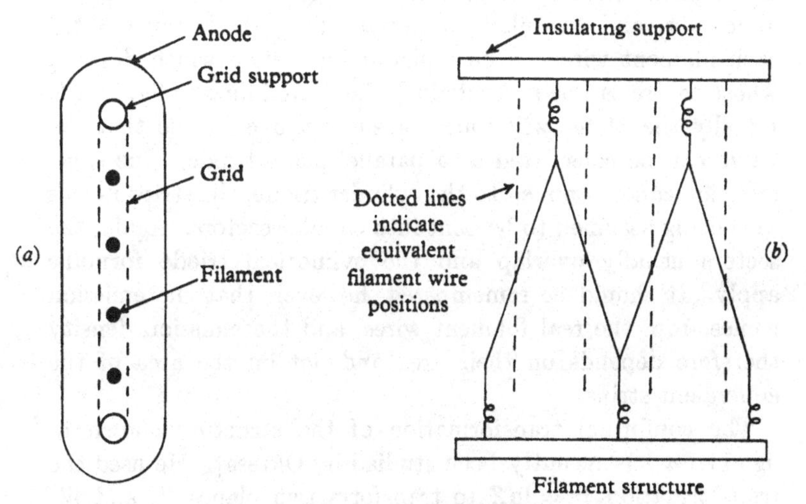

Fig. 11.1. Filamentary cathode triode.

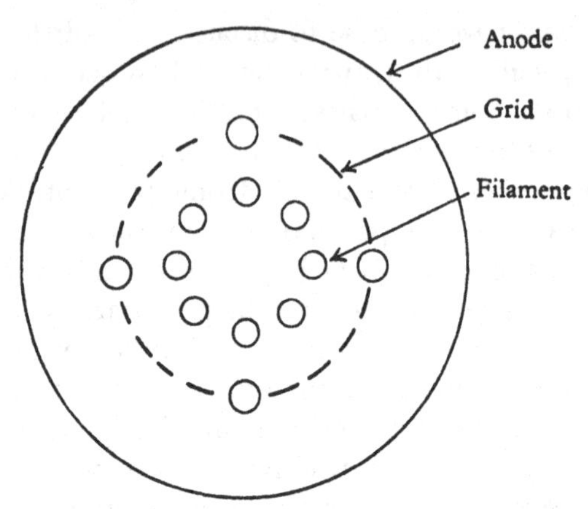

Fig. 11.2. Cylindrical filamentary cathode triode.

radiation-cooled triode with a sprung W-shaped filament. Fig. 11.2 represents the cylindrical structure typical of high-power cooled anode valves. The question obviously arises how are we to calculate the valve parameters. A useful rule, whose origin seems lost in the past, is to replace the filament wires by plates of width equal to the filament diameter plus twice the filament grid spacing. Thus in fig. 11.1 if the filament diam. $= d_f$ and the filament surface to grid clearance $= d_{fg}$, each filament wire can be replaced by a strip width $d_f + 2d_{fg}$ whose centre can be located along the dotted lines in fig. 11.1 *b*. Usually the plates will touch or slightly overlap, so that the valve can be considered as a parallel plane triode. The same criterion can be applied to the cylinder triode, the strips in this case being assumed to be bent into circular sectors. Again, the sectors usually overlap and the cylindrical triode formulæ apply. It should be remembered, however, that the emission comes from the real filament wires, and the emission density therefore depends on their area and not on the area of the equivalent strips.

The conformal transformation of the structure shown in fig. 11.1 *a* has recently been studied by Okress†. He used the transformation $W = \ln Z$ to transform two planes W_1 and W_2,

† Okress, *J. Appl. Phys.* **20** (1949), 850.

W_1 having the centre of one of the filament strands as origin and W_2 having the centre of one of the grid wires as origin. The potential at any point in the valve is then the sum of the potentials in the Z_1, Z_2 planes. The resulting expressions can only be stated in simple form when the grid-filament spacing is large compared to the pitch. For this case μ turns out to be the same as given in eqn. (29) of Chapter 9, and therefore the remarks made above are verified. Particular cases which do not admit of this restriction can be studied by the full expressions given in the original paper.

If it is necessary to know the μ of a structure which either does not fulfil the requirements of the approximate solution, or is of an unusual geometry, μ can be determined experimentally either in a scale model valve or in the electrolytic trough.

The most obvious way in which filamentary cathodes differ from unipotential cathodes is that there is a voltage drop along the filament so that if the high-tension negative is returned to the negative side of the filament, the anode current per unit length of filament is always less than for the corresponding unipotential cathode, although the two are equal in the limit of very high voltages. Let the filament length $= l$ and the filament voltage $= V_f$. If the effective anode-plane voltage is $V_a + \mu V_g = V_a'$, eqn. (41) of Chapter 9 shows that $I = GV_a'^{3/2}$, where $G =$ perveance of the tube. For the filamentary cathode

$$V'_{a(x)} = V_a' - \frac{xV_f}{l}(1+\mu),$$

therefore
$$I = \frac{G}{l}\int_0^l \left(V_a' - (1+\mu)\frac{xV_f}{l}\right)^{3/2} dx$$

$$\doteqdot \frac{GV_a'^{3/2}}{l}\int_0^l \left(1-(1+\mu)\frac{3x V_f}{2lV_a'}\right)dx, \quad \text{for } (1+\mu)\frac{V_f}{V_a} \ll 1,$$

$$= GV_a'^{3/2}\left[1-(1+\mu)\frac{3V_f}{4V_a'}\right]. \tag{1}$$

Eqn. (1) is a very close approximation to the exact result for $(1+\mu)V_a/V_f > 4$. For $(1+\mu)V_a/V_f > 15$, the correction is negligible.

11.2. The electrical design of tungsten filaments

As we have said earlier in the book, tungsten is the most generally satisfactory of the pure metal emitters. To recapitulate the advantages, we can say that the high melting-point and low vapour pressure allow very substantial emission densities to be drawn in spite of the high work function. All the oxides of tungsten are volatile so that the emitter is not permanently damaged by a sudden burst of gas, such as may occur in a flash arc, whereas the oxides of molybdenum and tantalum are much more difficult to dislodge. Tantalum, moreover, acts as a getter to some gases in specified temperature regions and for this reason is unsuitable for valve production. Finally, tungsten sheet has recently become available, so that one of the limitations to its use has been removed.

The saturated emission and total thermal radiation from tungsten have been studied over a very wide range of temperatures. Perhaps the most widely used results are those of Jones and Langmuir† with corrections due to Forsythe and Worthing.‡ In fig. 11.3 these values have been plotted against I/T for the small range of temperatures used in the production of valves. The emission and total radiation are given for a surface area of 0·25 cm.^{2}, so as to shift the total radiation curve to a more open region of the log scale.

It has been found empirically that a tungsten filament breaks when the diameter of the smallest section has been reduced to approximately 10 per cent below the original value at the same point. Since the rate of evaporation of tungsten has also been carefully measured as a function of temperature, this allows one to calculate the life of a filament assuming that the valve fails for no other reason beforehand. Let the original wire radius in cm. $= r_o$ and the final radius $= r_1$ $(= 0{\cdot}90 r_o)$, d = density of tungsten and $g(T)$ = rate of evaporation in gm./cm.^{2}/sec. Then

$$\text{Weight of tungsten evaporated} = \pi dl(r_o^2 - r_1^2),$$

$$\text{Mean surface area} = \frac{\pi l}{2}(r_o + r_1);$$

† Jones and Langmuir, *Gen. Elect. Rev.* **30** (1927), 310, 354, 408.
‡ Forsythe and Worthing, *Astrophys. J.* **61** (1925), 146.

therefore

$$\text{weight evaporated/unit surface area} = d(r_o - r_1),$$
$$= 0{\cdot}10dr_o,$$

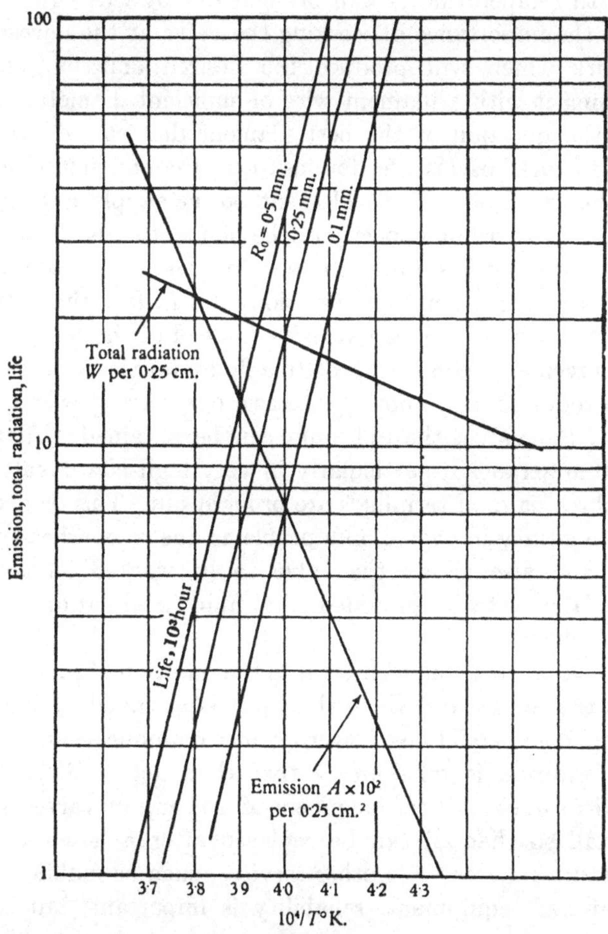

Fig. 11.3. Emission, life and total radiation data for tungsten.

therefore
$$\text{life} = \frac{0{\cdot}10dr_o}{3600g(T)} \text{ hours.}$$

The density is 19·4 gm./c.c., therefore

$$\text{life} = \frac{5{\cdot}4 \times 10^{-4} r_o}{g(T)}. \tag{2}$$

Thus the life is directly proportional to the original diameter. Fig. 11.3 shows values computed from eqn. (2) for three values of r_0 using Zwikker's† values of $g(T)$. The rapid variation of life with temperature is well brought out by this figure. This shows the importance of working the valve at the lowest temperature which will produce the desired emission, and of designing it with a filament wire of sufficient diameter. However, the question of the best filament diameter depends on other factors besides life, for instance, the emission efficiency and the economics of the filament power supply and also on the current-carrying capacity of the filament seals. Taking the first question, it is simple to determine the number of filament watts required per mA. of emission, and in fact this quantity mA./W. is often used as a variable instead of the temperature. For a given emission this leads to a figure for the total filament power required, and knowing the cost of electric power and the cost of the valve, the optimum can be obtained. Filaments are often arranged, particularly in very high-power valves, so that three pairs of terminals are brought out. This device is of obvious utility in overcoming problems due to overheating the seals, and also allows the valve to be worked on a three-phase a.c. supply, provided the hum modulation can be tolerated.

Another factor which enters into the question of the desirable life is the service for which the valve is intended. For broadcasting, commercial and Government communication, etc., the main criterion is reliability rather than actual life, i.e. the spread of lives between members of a group of valves should be small so that all can be replaced after n hours with no premature failures. For other services, such as mobile marine and aircraft equipment, reliability is important, but weight and size more so, and a slightly less conservative filament design may be better. In general, large European transmitting valves are designed for a maximum life of the order of 10^5 hours, i.e. this would be their life if there were no failures

† Zwikker, Dissertation, Amsterdam, 1925. A paper by Aboville, *Rev. gén. élect.* **37** (1935), 161, gives experimental results for thin ($<74\mu$) wires. The lives are shorter by factors of 3–5 than those given.

except for filament burn-out. American valves are usually designed for a somewhat shorter life and a lower first cost.

A few words on the question of ratings and operation of tungsten filaments will not be out of place. It is general to rate a tungsten filament so that the maximum anode current is 90 per cent of the saturated emission from the filament. Owing to the manufacturing tolerances in filament wire diameter, resistivity, etc., the emission varies from valve to valve, so the operating filament voltage and current are marked on all big valves. These values are determined by measurement of the emission characteristics of the valve. Several techniques are common; pulse-testing to full emission may be used or Davisson power emission charts can be used to plot the saturated emission at low-filament powers and the resulting straight line extrapolated to find the power giving 111 per cent of the maximum anode current. Normally these marked volts will differ from the mean volts for a group of valves by only a small percentage, but the departure may be sufficient to lower life or performance seriously if the marked volts are not adhered to. It is also important to work big valves at constant voltage; for, as the filament diameter decreases owing to evaporation, the resistance rises. If the valve is worked at constant current, the voltage will have to be increased throughout life, increasing the power input. As the surface area is also diminishing, the temperature will rise, since more watts have to be radiated by a smaller total surface, and the life will be curtailed. If the voltage is held constant, the power input decreases throughout life and the power per unit area remains much more nearly constant. The power per unit area actually decreases with life, and it is possible to predetermine values of hot filament resistance at which the filament voltage should be increased by a stated percentage. Operation on such a schedule reduces the spread of lives very much.†

This concludes our survey of the general design of tungsten filaments. It would be very desirable to give a similar general account of the properties of thoriated tungsten. Unfortunately, there are not sufficient data to do this at present. All one can

† Bell, Davies and Gossling, *J. Instn. Elect. Engrs.* **83** (1938), 176.

say is that the temperature of operation is so much lower than those we have considered above, that there is no question of the tungsten evaporating. Failures are due to poisoning of the surface, to crystal growth in the tungsten and consequent fracture by vibrations, etc. There is some evidence that the best modern thoriated filaments have very long lives, equal to the figure of 10^5 hours quoted above, but this figure is not consistently obtained and a considerable factor of safety is allowed in the emission, the saturated emission being initially 5–10 times the peak current.

For the sake of completeness, it should be mentioned that some use has been made of filaments consisting of a layer of thoria deposited by spraying or cataphoresis on a molybdenum or tantalum strip. Sometimes the core is covered with a sintered layer of powdered metal to obtain better adhesion. The properties of such composite filaments are very similar to those of the sintered thoria cathodes which were briefly discussed in Chapter 1. These filaments have been studied fairly widely in research laboratories, but the only commercial example of their use is, at present, the R.C.A. 8D21, a double tetrode for ultra-high-frequency use, which includes several novel design features.

11.3. Current to a positive grid

We now take up the question of current flow to a positive grid. This is a problem of considerable complexity because not only the primary current, but also the secondary electron current, is involved. When the grid is positive but the anode is more positive than the grid, the measured grid current is made up of the proportion of the space current going direct to the grid minus the secondary emission from the grid. The secondary emission current is itself not connected with the primary grid current in any simple way, since a large range of angles of incidence is possible and the grid surface can hardly be a homogeneous secondary emitter. When the grid is more positive than the anode, the measured grid current is the sum of the primary grid current and the secondary emission from the anode. In both cases matters are complicated still further by thermionic emission from the grid and by the fact that small numbers of electrons are very strongly deflected by the grid wires, and these may not have sufficient forward energy to reach the anode when it is approximately at grid potential. Fortunately, the conditions when the grid potential is higher than that of the anode are of little practical importance for in even the hardest driven class C oscillators, it is rare for V_a to be less than $0{\cdot}8\ V_g$. We therefore restrict ourselves to voltage ratios V_a/V_g less than $0{\cdot}8$ in what follows.

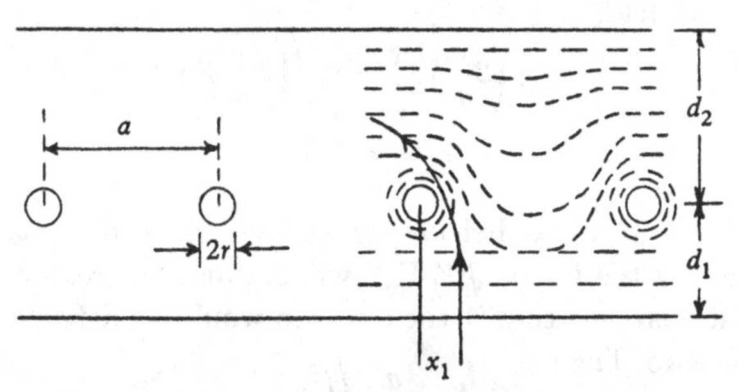

Fig. 11.4. Electron deflexion in positive grid triode.

It may be asked what use there is in developing a theory which from the outset cannot hope to include all the relevant phenomena. The answer is that, although we can only calculate the grid current under very idealized conditions, i.e. secondary emission minimized, grid dissipation lowered enough for thermionic emission to be neglected, the calculation for the grid-power dissipation is much more accurate. This is because grid emission and secondary emission remove only very small amounts of power, as the mean energy loss per electron is very small in either case. Thus if we can calculate the primary grid current as a function of grid and anode voltages, we shall have a good idea of the grid dissipation even if the measured grid current is widely different from our forecast.

The first approximation to the current division is given in the form due to Tellegen.† Fig. 11.4 shows the notation for the problem and indicates the equipotential plot which, in the grid-cathode region, can be assumed to be a system of parallel planes perturbed into circles in the immediate vicinity of the

† Tellegen, *Physica*, **6** (1926), 113.

grid. Consider the path of an electron which leaves the cathode at a point distance x_1 from the grid-wire centre line and just grazes the grid wire after deflexion in the central field. We can assume that this path is tangential to the central field at x_1, and the velocity at entry to the central field is that due to the effective grid potential V_{eg}. The electron finally leaves the central field by grazing the grid at radius r with velocity corresponding to V_g. Since angular momentum is conserved in the central field,

$$x_1\sqrt{\left(2\frac{e}{m}V_{eg}\right)} = r\sqrt{\left(2\frac{e}{m}V_g\right)},$$

or

$$x_1 = r\sqrt{\frac{V_g}{V_{eg}}}. \tag{3}$$

Thus the grid wires behave as though their diameter were increased by the factor $\sqrt{(V_g/V_{eg})}$, which must be greater than 1, since if it were less than 1 the electron would be deflected away from the wire. Then

$$\frac{I_k}{I_g} = \frac{a}{2r}\sqrt{\left(\frac{V_{eg}}{V_g}\right)},$$

$$\frac{I_a}{I_g} = \frac{I_k - I_g}{I_g} = \left[\frac{a}{2r}\sqrt{\left(\frac{V_{eg}}{V_g}\right)} - 1\right]. \tag{4}$$

In Tellegen's theory

$$V_{eg} = \frac{V_a + \mu V_g}{1 + \mu + d_2/d_1}. \tag{5}$$

Eqn. (5) does not reduce to $V_g + V_a/\mu$ for very large μ, because d_2/d_1 increases with μ. We can rewrite (5) as follows:

$$V_{eg} = V_g\left(\frac{1 + V_a/\mu V_g}{1 + 1/\mu(1 + d_2/d_1)}\right), \tag{6}$$

i.e. $V_{eg}/V_g < 1$ when $V_a/V_g < 1 + d_2/d_1$. For many valves this limit lies in the range 4–6. Beyond this value of V_a/V_g the grid current should drop sharply as electrons will be deflected away from the wires instead of towards them. For values of $V_a/\mu V_g$ less than 0·50 we can expand the square roots in eqn. (4) by the binomial theorem to obtain

$$\frac{I_a}{I_g} \doteqdot \frac{a}{2r}\left(1 + \frac{V_a}{2\mu V_g}\right)\left\{1 - \frac{1}{2\mu}\left(1 + \frac{d_2}{d_1}\right)\right\} - 1. \tag{7}$$

Eqn. (7) neglects the effects of space charge and initial velocities entirely and also assumes that the grid-cathode spacing is large enough in comparison with the grid pitch to eliminate *Inselbildung*.

More recently Jonker and Tellegen† have given a more accurate form of eqn. (7) by using a more accurate representation of the grid-cathode field. The result is

$$\frac{I_a}{I_g} = \frac{a}{2r}\sqrt{\left(\frac{V_{eg}}{V_g}\right)}\left\{\frac{2V_{eg}\ln(a/2\pi r)}{2V_{eg}\ln(a/2\pi r)+(V_g-V_{eg})}\right\}-1. \tag{8}$$

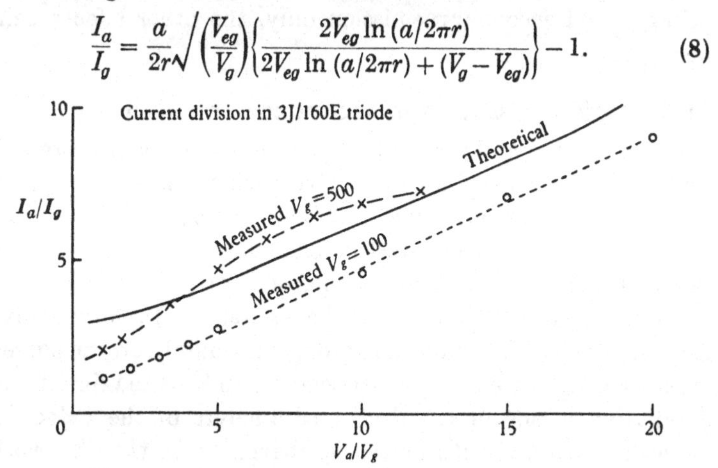

Fig. 11.5. Current distribution in positive grid triode.

The grid current according to eqn. (8) is higher than the value given by eqn. (7) when

$$\frac{V_a}{V_g} < \frac{1}{\mu}\left(1+\frac{d_2}{d_1}\right),$$

and smaller when

$$\frac{V_a}{V_g} > \frac{1}{\mu}\left(1+\frac{d_2}{d_1}\right).$$

Eqn. (8) was checked experimentally on a rubber sheet model and gave good agreement with the observed results, but, of course, this does not relate very directly to the situation in an actual valve. It must also be remembered that eqns. (7) and (8) do not include current taken to the grid support wires, which, for mechanical reasons, are much thicker than the grid wires and in some arrangements may make an appreciable contribution to the grid current.

Fig. 11.5 shows experimental values of I_a/I_g plotted against V_a/V_g with the curve deduced from the valve geometry using

† Jonker and Tellegen, *Philips Res. Rep.* **1** (1945), 13.

eqn. (8) for comparison. The agreement is only fair, as is perhaps to be expected in view of the complexity of the actual situation in a valve, where magnetic effects, finite emission velocities, etc., must be considered. In the curves shown, that for $V_g = 100$ will include effects due to these causes and secondary emission, while that for $V_g = 500$ includes effects due to primary and secondary emission only, the other effects being negligible.

11.4. Grid emission and secondary emission

Grid emission and secondary emission from the grid are both undesirable effects which the valve designer is at some pains to minimize. Grid emission is reduced (*a*) by maintaining the grid temperature at the lowest possible value, and (*b*) by making the grid of high work function material or by coating it with some material which reduces the emission. In large valves used under class C conditions (*a*) depends mainly on the current division as discussed in the last section, and the maximum grid dissipation is one of the design parameters of the valve. In valves not worked with positively driven grids, the grid works at a temperature depending on the valve geometry and the anode and cathode temperatures.

In high-power valves with tungsten filaments the grid is commonly made of molybdenum wire, but the problem of grid emission is more acute in lower powered valves with thoriated tungsten or oxide-coated cathodes, since in both these cases there is some evaporation of active material from the cathode on to the grid with a resulting increase in grid emission during life. In these cases grids are often coated to reduce emission. The best coating seems to depend on the nature of the emitter. With oxide cathodes gold-plated grids† seem to be fairly effective. In this case the barium evaporated from the cathode diffuses into the gold instead of remaining on the surface. The emission, however, is not permanently eliminated as the gold eventually becomes saturated with barium and the emission rises to the value obtained with non-plated wires. A useful life can be obtained before the emission rises. Soft

† Fay, *Bell Syst. Tech. J.* **26** (1947), 818.

iron wires are also found to have low emission but make it very difficult to obtain high cathode activities.

For valves with thoriated filaments, zirconium is a useful grid coating, or sometimes zirconium and carbon are used together.

Grid emission is undesirable for several reasons, first because of the damping imposed on the input circuits, and secondly because of the tendency to cause the grid current to change sign with resultant blocking. Lastly, in ultra-high-frequency

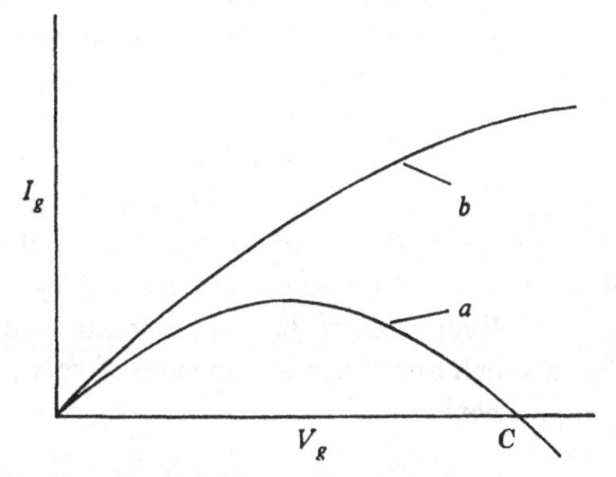

Fig. 11.6. Good and bad grid emission characteristics.

valves the electrons emitted from the grid may cause appreciable damping in the output circuit.

Secondary emission from the grid is not quite so serious a problem as grid emission simply because it can only occur when the grid is driven positive. However, it is in these circumstances that blocking can occur. If the total grid current in a valve is a function of the grid voltage of the sort (*a*) shown in fig. 11.6, the voltage drop in the bias resistor of the usual self-biased class C oscillator will reverse sign at *C* and the grid will not only be driven positive by the drive voltage, but will also bias itself positively, the anode current will rise disastrously and a high-powered valve can easily destroy itself. The danger can be overcome by connecting a diode clamp across the input circuit, thereby preventing the grid from going more than a predetermined amount positive, but it is clearly preferable to design the valve so that the characteristic is of the type indicated at *b* in fig. 11.6, where blocking cannot happen.

On the other hand, a modicum of secondary emission reduces the grid current, and therefore that portion of the driving power which is lost in the grid circuit, so that valves with a suitable value of secondary emission are slightly more efficient than those without. The tendency in modern valves is to reduce secondary emission as much as possible and to put up with the loss of efficiency. Some of the grid coatings described above, particularly the zirconium-carbon combinations, have low values of secondary emission, probably because they are extremely rough on a microscopic scale, and trap secondaries as we have described in Chapter 2.

These brief remarks have served merely to indicate the nature of the problems discussed. In many valves it is a matter of the greatest technical difficulty to find acceptable methods of reducing grid emission and secondary emission to tolerable values. Every design has to be treated on its merits and procedures which are of use in one type of valve are useless or deleterious in another.

11.5. Measurement of current distribution by the method of de la Sablonière

A useful method for measuring the current distribution in valves, which makes allowance for the effect of secondary emission but not for the effects of space charge, grid emission or the effects of initial velocities, has been described by de la Sablonière.† The method has recently been re-examined and improved by Hamaker,‡ and it is the latter account we follow here.

The valve characteristic is divided into two regions, the first in which $V_a > V_g > 0$, and thesecond in which $V_g > V_a > 0$.

Case 1. $V_a > V_g > 0$.

Let the primary grid current be given by

$$I_{gp} = p_g I_k, \quad p_g = f(V_g/V_a) = f(u). \tag{9}$$

† de la Sablonière, *Hochfrequenztech. u. Elektroakust.* **41** (1933), 195.
‡ Hamaker, *Appl. Sci. Res.* B, no. 2 (1948), 77.

Each electron produces δ_g secondaries, but only a fraction s_g of these reach the anode. The secondary current leaving the grid is thus $I_{gs} = s_g \delta_g I_{gp}$, and the ratio of measured grid current to cathode current is

$$y_g = \frac{I_g}{I_k} p_g(1 - \delta_g s_g). \tag{10}$$

Next, the reasonable assumption is made that $\delta_g = \delta(V_g)$ and $s_g = s_g(V_g/V_a) = s_g(u)$. Thus

$$y_g(u, V_g) = p_g(u)\{1 - \delta_g(V_g) s_g(u)\}. \tag{11}$$

Case 2. $V_g > V_a > 0$.

An exactly similar argument yields the relation

$$y_a(u, V_a) = p_a(u)\{1 - \delta_a(V_a) s_a(u)\}, \tag{12}$$

where the various functions are now defined for the anode. Eqns. (11) and (12) are not the most general forms of the relationship we are seeking to establish, as can be seen if we put

$$\delta(v) = \alpha\delta'(v) + \beta. \tag{13}$$

where v = generalized voltage.

Then

$$y(u, v) = p(u)\{1 - \beta s(u)\}\left[1 - s'(v)\frac{\alpha s(u)}{1 - \beta s(u)}\right]. \tag{14}$$

And the expressions

$$\left.\begin{aligned} p'(u) &= p(u)\{1 - \beta s(u)\}, \\ \delta'(v) &= \frac{1}{\alpha}\delta(v) - \frac{\beta}{\alpha}, \\ s'(u) &= \frac{\alpha s(u)}{1 - \beta s(u)}, \end{aligned}\right\} \tag{15}$$

will also satisfy (10), α and β being arbitrary constants. Eqns. (15) can be shown to be the most general forms which satisfy eqns. (11) or (12). Therefore we must determine α and β in addition to the functions. We now return to the question of determining the functions.

Determination of $p(u)$. For a fixed value, u_0 say, of u, eqns. (11) and (12) show that $y(u_0, v)$ is linear in $\delta(v)$. Thus two curves

$y(u_0, v)$, $y(u_1, v)$ are linear functions of one another. If then $y(u, v)$ is measured for a series of values of v with fixed values of u as parameter, we can test eqns. (11) and (12) by taking the values for one particular run $y(u_0, v)$ as abscissa, and plotting the values of the other runs $y(u_1, v)\ldots y(u_n, v)$ as ordinates and the result should be a series of straight lines (see fig. 11.7). If the lines are not straight, the conditions assumed are being

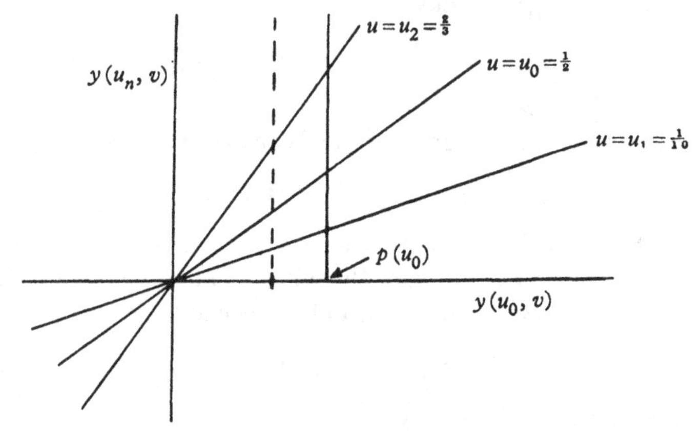

Fig. 11.7. Illustrating the graphical determination of positive grid characteristics.

violated in some way. Now, eliminating δ between two members of eqn. (11), say $y(u_0, V_g)$, $y(u_1, V_g)$, we find that

$$\frac{y(u_1, V_g)}{p(u_1)s(u_1)} - \frac{1}{s(u_1)} = \frac{y(u_0, V_g)}{p(u_0)s(u_0)} - \frac{1}{s(u_0)}. \tag{16}$$

Eqn. (16) is obviously satisfied if we put $y(u_1, V_g) = p(u_1)$, $y(u_0, V_g) = p(u_0)$. Thus we get a series of values of $p(u_0)$ by erecting a vertical at u_0 on fig. 11.7 and reading off the intersections with the straight lines. Other values of $p(u_0)$ give different sets of p values. These sets can be plotted, and the values obtained for each region must form a continuous curve, since, clearly,

$$p_g = 1 - p_a.$$

When this has been done, we have to choose between the curves thus determined. A possible set of curves is indicated in fig. 11.8. One of these curves has to be chosen, and in the original method of de la Sablonière this was done by requiring a smooth junction. Hamaker, however, makes the more usable assumption that, at the u value for which the effective grid plane potential is equal to the grid wire potential, the p value is the ratio $2r/a$ as defined in § 11.3. Thus one curve can be chosen and this is the one corresponding to eqn. (15), i.e. β is now fixed.

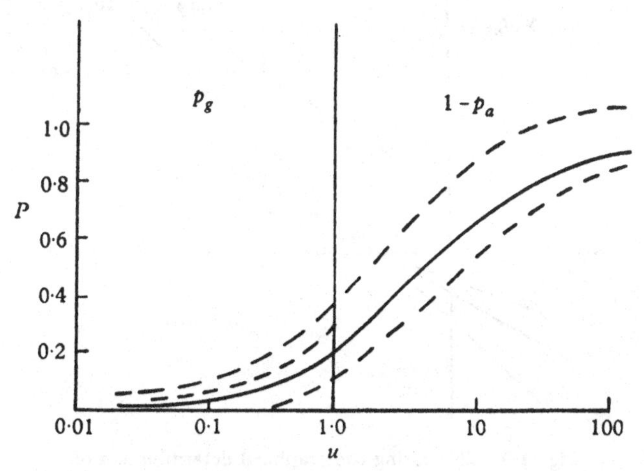

Fig. 11.8 Illustrating the graphical determination of positive grid characteristics.

Now to determine $s(u)$ we first recall that $\lim_{u\to 0} s_g(u) \to 1$, $\lim_{u\to\infty} s_a(u) \to 1$, since in these cases all the secondaries are collected by the other electrode. It is easy to see what values of u are good enough approximations by the following reasoning. Solve eqn. (16) for the value of $y(u_0, v)$ which makes $y(u_1, v) = 0$, i.e. the point at which the curve cuts the abscissa. The result is

$$j_0 = y(u_0, v)_{u_1=0} = p(u_0)\,\frac{s(u_1) - s(u_0)}{s(u_1)}. \tag{17}$$

If $s(u) = 1$, j_0 is independent of u and all the u curves will cut the axis in a single point. This point is

$$j_0' = p(u_0)\,[1 - s(u_0)]. \tag{18}$$

Then from eqns. (17) and (18)

$$s(u) = \frac{p(u_0) - j_0'}{p(u_0) - j_0}. \tag{19}$$

Eqn. (19) allows us to determine $s(u)$ for each of the curves. In fig. 11.9 we have plotted three curves for the case $s_g \to 1$ and

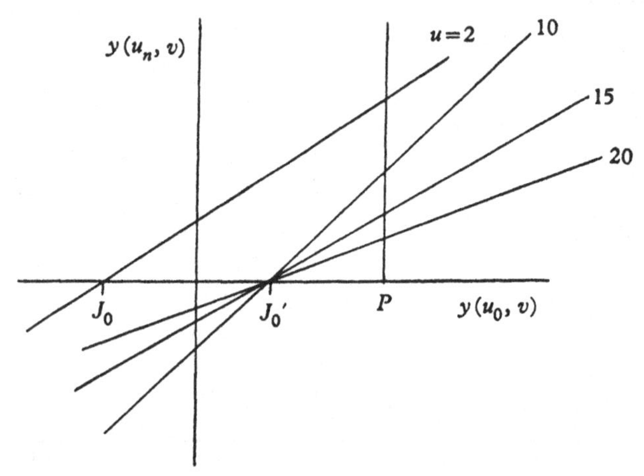

Fig 11.9. Illustrating the graphical determination of positive grid characteristics.

a curve for which s_g is some other value. From eqn. (19) $s_g = PJ_0'/PJ_0$, where P is the value corresponding to the correct $p(u)$ already determined. The determination of s_g fixes α.

Finally, δ can be determined from eqns. (11) and (12),

$$\delta(v) = \frac{p(u) - y(u, v)}{p(u)s(u)}. \tag{20}$$

For each value eqn. (20) can be solved for $\delta(v)$, since $p(u)$ and $s(u)$ have been determined. Hamaker gives a step further and sums over the different values of u, getting

$$\delta(v) = \frac{\sum\limits_n p(u_n) - \sum\limits_n y(u_n, v)}{\sum\limits_n p(u_n)s(u_n)}. \tag{21}$$

This completes the determination of the current division and secondary emission factors for the tube. The reader should refer to the original paper for an extensive discussion of some of the cases in which the method fails.

11.6. Cooling by radiation

Valves of up to several hundred watts anode dissipation are cooled simply by radiation. The factors involved are the anode temperature and material and the diathermancy of the bulb material. According to the Stefan-Boltzmann law, the power dissipated by unit area of an anode at temperature T° K. is

$$P = \epsilon_T S(T^4 - T_a^4), \tag{22}$$

where ϵ_T = total emissivity of anode material,

S = Stefan's constant

$= 5{\cdot}73 \times 10^{-8}$ W. metre^2 per degree^4,

T_a = ambient temperature ° K.

Eq. (22) can usually be written as $P = \epsilon_T S T^4$, since T_a is small compared with T for the applications we are interested in. The anode must dissipate a proportion of the filament or cathode heating power, in addition to the anode-current dissipation. The proportion depends on the tube geometry, but for most purposes it is not worth while to evaluate the exact amount, and the whole heating power may be added to the anode dissipation. The total emissivity ϵ_T is a function of temperature, and varies considerably from metal to metal; it also depends on the degree of polish and other surface phenomena. For these reasons it is better to quote the experimentally determined value of the total radiated power as a function of temperature. Table 4 gives values for commonly used anode materials.

In the case of triode and other gridded valves, the maximum permissible anode temperature is determined by the vapour pressure of the anode material, the softening temperature of the bulb and the transparency of the glass to red and infra-red radiation. In the case of rectifiers, however, the maximum anode temperature depends on the thermionic emission from the anode material since there must be an inappreciable reverse current when the voltage acts from cathode to anode. Rectifier anodes are therefore generously designed so that their temperature never exceeds a few hundred degrees C.

The transparency of glass in the infra-red varies very much with composition, and there are special glasses which have low infra-red absorption. However, the glasses generally used for vacuum valve envelopes are not transparent in the far infra-red region. As a typical example† we may take Wembley soda-lime glass, used for receiver and other low-power valves. For

TABLE 4. RADIATION FROM ANODE MATERIALS

Material	Temperature °K. (units W./cm.²)					
Ni	800	1000	1200	1400		
	0·25	0·90	2·0	5·0		
Mo	1000	1200	1400	1600	1800	2000
	0·6	1·4	3·2	6·2	11·3	19·5
Ta	1600	1800	2000	2200	2400	
	7·0	13·0	21·5	34·0	51·0	

1 mm. thickness this glass transmits 90 per cent of the incident radiation at all wave-lengths in the visible spectrum and up to 2 μ but the absorption then increases until only 20 per cent transmission is measured at 4·5 μ. These figures include the reflexion loss at the two glass-air surfaces. Harder glasses, i.e. containing more silica, have higher transmissions at long wave-lengths, but the difference is not especially marked. Now according to Wien's law, a black body at temperature T° K. radiates maximum energy at a wave-length given by

$$\lambda = \frac{0{\cdot}288}{T}\,\text{cm.} \tag{23}$$

A carbon anode is a fair approximation to a black body, and eqn. (23) shows that at 1000° K. the maximum of the spectral

† A very small percentage of iron as an impurity increases the absorption in glasses very rapidly, so that considerable differences may be observed between batches of glass, particularly if sand from different deposits is used in their manufacture.

distribution of energy is at 2·88 μ, a wave-length for which the absorption is increasing fairly fast. Molybdenum or tantalum anodes can run hotter than carbon because their vapour pressure is lower, and we see that this is advantageous also from the viewpoint of reducing the bulb absorption.

In practice, the valve manufacturer generally has a table giving the power which can be radiated through the walls of the standard bulbs used, a maximum bulb temperature of 80–90° C. being a common basis for the rating. The considerations discussed above are useful if it is necessary to design a valve for some special purpose which involves exceeding normal ratings. Such circumstances often arise in designing valves for the 100–500 Mc./sec. region where size is important, but it is not really necessary to adopt the methods of construction used at higher frequencies. In these circumstances it is becoming common practice to use molybdenum or tantalum anodes run at high temperatures and coated with zirconium to act as a getter.

11.7. Air blast cooling

We next consider the question of cooling by means of a blast of air passed over the external anode. The surface area of the anode is normally increased by soldering it into a system of cooling vanes, frequently extending radially from the anode. In addition to increasing the surface area, the fins break up the air flow and help to maintain turbulence. In the theory of heat transmission it is well known that the heat transfer between, say, a metal plate and a stream of air is much greater when the air velocity is sufficiently great to make the flow turbulent, since when the flow is viscous a thin film of static air remains in contact with the metal surface and prevents the cool incoming stream from reaching it. For this reason air coolers should be operated in such conditions that turbulent flow is assured. The criterion for the existence of turbulent flow is the dimensionless constant, known as Reynold's number, which should exceed a certain value

$$R = \left(\frac{\rho D V}{\mu}\right) > 4000, \tag{24}$$

where V = velocity of cooling fluid,

ρ = density of coolant,

$D = 4 \times$ hydraulic mean radius. This term is borrowed from hydraulics, and is used to denote the quotient of section area and perimeter, therefore

$$D = \frac{4 \times \text{section area}}{\text{perimeter}},$$

μ = coefficient of viscosity.

Fig. 11.10 shows two common types of longitudinal flow cooler: (*a*) is a cooler with milled parallel-sided slots, and (*b*) is a fabricated cooler with radial parallel-sided vanes. Let the minimum fin thickness equal δ, and the minimum fin spacing equal s. The length of the fins is l. The outer radius of the fins is R_f, and the outer radius of the core is R_c. Then the number of fins is

$$n = \frac{2\pi R_c}{(\delta + s)}.$$

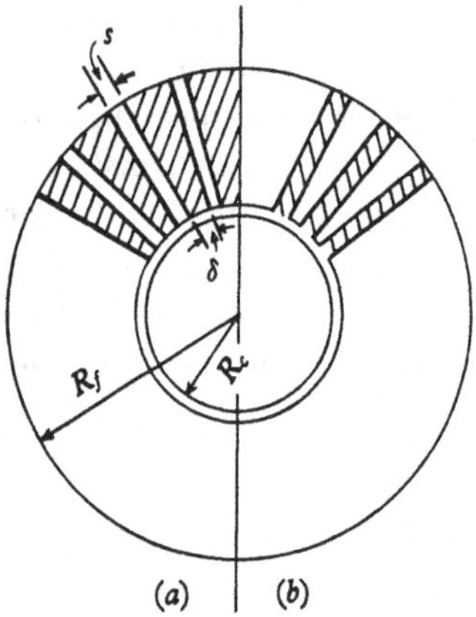

Fig. 11.10. Axial flow air blast cooler sections.

The total surface area is

$$S = 2(R_f - R_c)\frac{2\pi R_c}{(\delta + s)}.l. \tag{25}$$

Differentiating with respect to R_c and maximizing, we find

$$R_f = 2R_c,$$

therefore

$$S_{\text{max.}} = \frac{4\pi R_c^2 l}{(\delta + s)}. \tag{26}$$

We can now readily evaluate $D_{\text{max.}}$ for the shapes (*a*) and (*b*) in fig. 11.10.

(*a*)

$$D_a = 4s(R_f - R_c)\left(\frac{2\pi R_c}{\delta + s}\right)\frac{l}{S}$$

$$= 2s. \tag{27}$$

(*b*)

$$D_b = 4\left\{3\pi R_c^2 - \frac{2\pi R_c}{\delta + s}\,\delta\,.\,R_c\right\}\frac{l}{S}$$

$$= 3s + \delta. \tag{28}$$

These shapes are particularly simple; other shapes, particularly those in which the direction of flow changes abruptly, are not so easy, but several examples can be found in books on hydraulics and aerodynamics.

Having defined the Reynold's number and the quantities used in its calculation, we can return to the question of heat transfer, using the assumption that flow is turbulent throughout the work. The operation of the cooler can be determined by equating two expressions, one giving the power dissipated in terms of the rise in temperature of the air, and the other the rate of heat transfer through unit cooler surface. Since the process takes place at substantially constant pressure, the first expression is simply

$$P = 4{\cdot}2C_p q \Delta T \text{ W.}, \tag{29}$$

where C_p = specific heat at constant pressure (cal./gm./° C.),†
$= 0{\cdot}241$;
q = quantity of air passing cooler (gm./sec.);
ΔT = rise in air temperature.

The second expression is

$$P = 4{\cdot}2St\delta T \text{ W.}, \tag{30}$$

where t = rate of heat transfer from cooler to air per unit surface, per degree C.,
δT = average temperature difference between cooler and air.

For turbulent flow,

$$t = 0{\cdot}024\frac{k}{D}R^{0{\cdot}8}\left(\frac{C_p\,.\,\mu}{k}\right)^{0{\cdot}4}. \tag{31}$$

The only new quantity here is k, the thermal conductivity of the coolant. We can rewrite t as

$$0{\cdot}024A\frac{V^{0{\cdot}8}}{D^{0{\cdot}2}}, \tag{32}$$

where A depends only on the physical constants of the coolant. In the case of air, a great simplification results from the fact that A turns out to be independent of temperature over the

† We use c.g.s. units throughout as the constants used in heat transmission are not normally tabulated in m.k.s. units.

range from -20 to $+100°$ C., the constant value being $A = 0{\cdot}0215 \times 10^{-2}$ cal./sec.$^{0{\cdot}2}$/cm.$^{2{\cdot}6}$/° C. Then, for air,

$$t = 5{\cdot}16 \times 10^{-6} \frac{V^{0{\cdot}8}}{D^{0{\cdot}2}}. \tag{33}$$

Eqn. (33) shows that the important variable is V because the variation with D is extremely slow. Finally, there is a relation between ΔT and δT. Let T_1 = air-inlet temperature, T_a = temperature of fins, then

$$\delta T = T_a - \left(T_1 + \frac{\Delta T}{2}\right). \tag{34}$$

We have now obtained all the necessary relations for the cooler design. The best way of handling them depends on what information is given. We can certainly treat P, T_1 and T_a as known. T_a is fixed by two considerations: (*a*) that the maximum permissible temperature for the surface of the anode exposed to the vacuum shall not be exceeded, (*b*) that the solder used to join the cooler and the anode shall not melt. The second is to-day the limiting factor. Typical solders are pure tin, m.p. $\doteqdot 230°$ C., and pure cadmium, m.p. $\doteqdot 312°$ C. A technique of handling the equations is then to assume a value of ΔT and solve eqn. (29) which becomes

$$q \doteqdot \frac{P}{\Delta T}. \tag{35}$$

Eqn. (34) then yields δT. Reynold's number can be put equal to some value well above the turbulent limit (6000–10,000). Now, by definition,

$$q = \frac{DSV\rho}{4l} = \frac{\mu RS}{4l},$$

therefore

$$S = \frac{4lq}{\mu R}. \tag{36}$$

Eqn. (36) fixes S since q is known, R has been fixed, μ is a constant, and l depends only on the anode length. Eqns. (29) and (30) together with (35) yield

$$t = \frac{C_p P}{S\,\delta T}. \tag{37}$$

Eqn. (37) can then be solved for V, assuming $D = 1$. It may happen that the value of S obtained by this procedure is greater than $S_{max.}$ as determined by mechanical considerations. If this is so, ΔT or R may be increased, thereby reducing S and increasing V. Also, having determined V, the value of D resulting from the expression $q = \frac{DSV\rho}{4l}$ should be checked to ensure that $D^{0 \cdot 2}$ is not, in fact, very different from 1.

By taking a series of trial values it is thus possible to determine the best compromise between the cooler surface area and the quantity of air required, the cooler cost and weight being balanced against the blower cost and weight.

In the treatment given above it has been implicitly assumed that the cooler fins are thick enough and made of material of sufficiently high thermal conductivity to ensure that the temperature is uniform over the fin surface. Such a design would actually be uneconomic, and in practice there exists a radial temperature drop in the fins and a longitudinal temperature drop due to the fact that the anode is heated only over the central portion, since the active part of the structure does not extend to the ends of the anode. These questions are discussed in two papers by Mouromtseff.†

Another topic which has not been discussed is the pressure required to force the air through the cooler, which is determined by the ordinary methods of hydraulics. The pressure required varies as V^2 and the power as V^3.

It will have been noticed that the limitation of anode temperature to the melting-point of the solder very materially increases the amount of cooling air or surface required. In recent years a good deal of work has been done on methods of improving the joint between cooler and anode, since valves are normally processed at temperatures high enough to allow them to be worked at considerably above the 230–250° C. figure.

In a cooler described by Prévost *et al.*,‡ axial flow is dispensed with in favour of transverse flow from both ends of a diameter. Transverse flow requires less pressure than axial

† Mouromtseff, *J. Appl. Phys.* **12** (1941), 491; *Proc. Inst. Radio Engrs., N.Y.* **30** (1942), 190.

‡ Prévost, Boissière and Loukovski, *Ann. Radioélect.* **4** (1949), 138.

flow, particularly in high-power valves with long anodes. The fins, of rather a complicated shape, are so arranged in pairs that the flow between the nth and the $n+1$th fin is a mirror image of the flow between the $n-1$th and the nth, and thus regions of still air are eliminated. The fins are attached to the anode by spraying metal from a Schoop gun (the process by which receiving valves used to be metallized) over the joint. The sprayed metal can be operated up to 400° C. Results given for a 14 kW. anode dissipation show that the maximum temperature at the anode-fin joint was 300° C. and the temperature at the outside edge of the fin 140° C. The air flow required was 2·28 m.^{3}/min./kW., which is comparable with figures for axial flow coolers (2–4 m.^{3}/min./kW. or 70–140 ft.^{3}), but the pressure required was less than $\frac{1}{2}$ in. instead of 1–2 in. of water. Other improved methods of air cooling are described by de Brey and Rinia,† whose coolers required only 0·8–1·0 m.^{3}/min./kW. but at a rather higher pressure than the Société Française Radioélectrique use.

The physical constants of air are given in Table 5.

TABLE 5. PHYSICAL PROPERTIES OF DRY AIR
(c.g.s. units)

Property	Temperature (°C.)					
	0	20	40	60	80	100
Density (ρ)‡	0·001293	0·001205	0·001130	0·001063	0·001010	0·000947
Viscosity (μ) $\times 10^4$	1·73	1·83	1·93	2·02	2·12	2·21
Specific heat (C_P)	—	0·2417	—	—	—	0·2366
Thermal conductivity (k) $\times 10^5$	5·77	—	—	—	—	7·55

11.8. Water cooling

The theory of water cooling very closely parallels the last section. The most important difference is that A for water is

† De Brey and Rinia, *Philips Tech. Rev.* **9** (1947), 171.

‡ The density figures are for a pressure of 760 mm. Hg. They should be multiplied by $P/760$ to obtain the density at another pressure P.

not constant with temperature and has a much greater value. For water A varies almost linearly from 0·09 at 0° C. to 0·20 at 75° C. This fact shows that it is not always preferable to use the lowest possible input temperature, since the heat transfer may be so much reduced that the water may boil at outlet. Another important question is the maximum permissible water temperature. Since the water is fed to the valve through long pipes which act as electrical insulation, a considerable pressure gradient exists, and the boiling-point at the mean pressure in the water jacket may be used instead of the N.T.P. boiling-point. Experimentally it has been found that the anode temperature can somewhat exceed this value, without the water boiling, probably because of the formation of a thin layer of steam on the anode. It is usual not to allow the water to boil, so as to avoid scale and cavitation of the anode; but with especially pure water supplies, this precaution may not be necessary. Knowing the inlet and outlet temperatures, the rate of flow necessary to remove the heat can be calculated. The value of A, at the mean water temperature, gives the constant in t (eqn. (33)) so that V can be determined. A check must then be made ensure that the flow is turbulent. The physical constants are given in Table 6.

TABLE 6. PHYSICAL PROPERTIES OF WATER
(c.g.s. units)

Property	Temperature (°C.)					
	0	20	40	60	80	100
Density (ρ)	0·9987	0·9982	0·9922	0·9832	0·9718	0·9584
Viscosity (μ)	0·0179	0·0101	0·0066	0·0047	0·0036	0·0028
Specific heat (C_p)	1·0094	1·000	0·9982	1·000	1·0033	1·0074
		10		50	80	
Thermal conductivity (k)		$14{\cdot}7\times10^{-4}$		$15{\cdot}4\times10^{-4}$	$16{\cdot}0\times10^{-4}$	

Water coolers are very often arranged as sketched in fig. 11.11, and for this form the flow takes place through an area

πD_{ad}, the wetted perimeter is $2\pi D_a$, therefore $D = 2d$. It is worth noting that since the area is directly proportional to d, the velocity is inversely proportional to d and Reynold's number is therefore independent of d. If it is desired to increase Reynold's number, the rate of flow must be increased. Baffles and other mechanical devices are often used to increase the turbulence in the jacket.

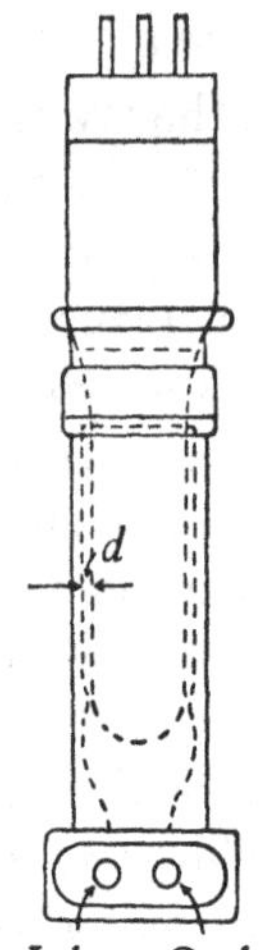

Fig. 11.11. Typical water cooling jacket.

An effect which has not been mentioned so far can cause cooling troubles especially in those stages, such as the linear modulator, where the grid is always held negative. The current is then focused by the grid wires and the anode is much hotter half-way between the centres of the projection of the grid wires on the anode than it is under the wires. In valves where the grid is driven positive, the beams vary in width at the oscillation frequency and the anode is more evenly heated. Some calculations on this effect have been made by Mouromtseff.†

11.9. Mechanical design of filaments

To continue our survey of special questions which arise in transmitter tubes, we propose to discuss the electromechanical design of the filaments, i.e. the problem of designing a filament which will withstand the magnetic forces due to the passage of large currents, and the problem of overcoming the expansion of the wire. The latter is a considerable problem, since the expansion of a 50 cm. tungsten wire between 0 and 2500° K. is about 4 mm. In modern designs the filament is usually allowed to hang vertically downwards with the bottom end free to move, thus taking up the expansion. In other designs, the filament is often held in tension by springs, but this necessitates locating insulators at parts of the tube where they may become too hot or may have films of tungsten or copper deposited on them. For these reasons the free-hanging filament is now used

† Mouromtseff, *Communications*, **20**, (1938), 11.

for the larger valves, although it necessitates the use of a special mounting to ensure that the tube is vertical. In spite of this, many excellent designs having sprung filaments are used, particularly in shorter valves designed for ultra-high-frequency.

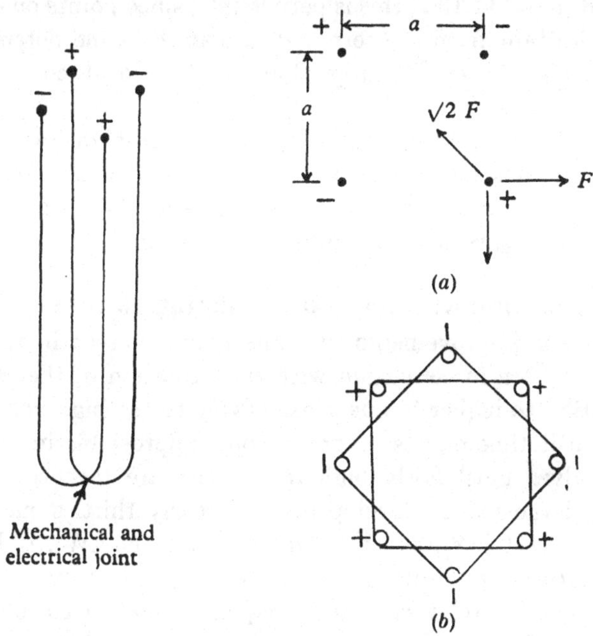

Figs. 11.12, 11.13. Illustrating the mechanical design of filaments.

In a free-hanging filament, the individual wires are hairpin-shaped loops; and because the wires are closely matched in length and diameter, the potential at the end of each loop is the same, so that all the loops can be rigidly joined mechanically at the free end. When this is done, the fixed ends of the wires are connected so that the current flows in opposite directions in adjacent wires. Fig. 11.12 shows the arrangement of a four-wire filament, and fig. 11.13 *a* represents the magnetic forces acting on a wire. If the wires are spaced at a distance a and the currents are in opposite directions, the repulsive force per unit length is

$$F = \frac{2\mu_0 I^2}{a},$$

and in the square formation of fig. 11.13 *a* the net force on each wire is $F/\sqrt{2}$ directed diagonally outwards. As more pairs of wires are added, the resultant force always remains outwardly directed. The filament is prevented from bowing outwards by lateral wires joined to all the wires of the same polarity as indicated in fig. 11.13 *b*; this is permissible, since points on each wire equidistant from the terminals are at the same potential. The laterals are made fairly thin so that local cooling is avoided.

In the much shorter valves used for ultra-high-frequency operation it has become common practice to make the filament in the form of a multistart helix, as this gives a simple and rigid structure without additional bracing.

11.10. Lead inductances and circulating currents

The major improvements in transmitting valves in recent years have been in connexion with the extension of the range of operation to higher frequencies. Owing to the high voltages used, transit time effects do not become appreciable in transmitter valves until fairly high frequencies are reached. For instance, large valves developed in the early thirties such as the CAT 14 (150 kW., 20 kV.) and the 4030C (65 kW., 16 kV.) will operate on full ratings up to 20 Mc./sec. or over. The development of television and frequency-modulated broadcasting as well as the extension of aircraft communication and similar special communication facilities at much higher frequencies however, has produced the need for high powers at these frequencies. The earlier transmitter tube designs turned out to be defective, not so much because the electronic performance was at fault, as because the high-frequency currents could not reach the regions where electron flow took place, and because the heavy currents flowing in the grid leads, due to the interelectrode capacitance, overheated the seals. For these reasons modern valves for high frequencies are designed so as to be short and compact. The interelectrode capacitances are minimized and the lead inductances reduced. Here we can only indicate the main ideas behind this work; the reader who wishes to study the historical details of the development is referred to an excellent series of papers from the research laboratories of the General Electric Co.†

First we consider the lead inductance. The inductances of a straight wire, of length 1 cm. and diameter d cm., is

$$L = 2l\left(\ln\frac{4l}{d} - 1 + \frac{d}{2l}\right) \times 10^{-9} H. \tag{38}$$

or

$$X = \frac{\pi l}{250}\left(\ln\frac{4l}{d} - 1 + \frac{d}{2l}\right) \Omega/\text{Mc.} \tag{39}$$

The inductance can be reduced by reducing the length and increasing the diameter, or by joining leads in parallel. It is impossible to make leads very great in diameter for reasons of convenience and because of the limitations introduced by glass-working technique, so the second alternative has been used.

Grids were first mounted on two or more leads instead of one, but it was soon realized that it was preferable to pass to the limit and make the grid lead in the form of a ring seal, either the feather-edged Houskeeper seal in copper or a seal using one of the newer matched sealing alloys. Such seals can be made of very considerable diameter, up to at least 6 in. if necessary, and they therefore also solve the problem of providing a means of dealing with the large capacitance currents, which, at the frequencies under discussion, are confined to the surface of the metal. The following example indicates the magnitude of these currents. A particular 10 kW. dissipation valve for use up to 100 Mc./sec. gives a voltage gain of about 10. The anode voltage is 10 kV. so the r.m.s. grid voltage may be as high as 900 V. The grid cathode capacitance is 40 pF. The circulating current may therefore be as high as 25 A. The skin depth in copper is

$$2{\cdot}57 \times 10^{-3} f^{-\frac{1}{2}} \text{ in.} \tag{40}$$

and the surface resistivity R is

$$2{\cdot}61 \times 10^{-7} f^{\frac{1}{2}}\, \Omega/\text{unit square.} \tag{41}$$

† Picken, *J. Instn. Elect. Engrs.* **65** (1927), 791; Le Rossignol and Hall, *G.E.C. Jl.* **7** (1936), 176; Bell, Davies and Gossling, loc. cit.; Bell and Davies, *G.E.C. Jl.* **16** (1949), 138. Chevigny, *Proc. Inst. Radio Engrs., N.Y.* **31** (1943), 331, gives an account of the development of a 165 kW. dissipation valve useful up to 25 Mc./sec. This valve is known in England as the 3 Q 331 E (S.T.C.).

For other materials (40) has to be multiplied by $1/\sqrt{(\mu\sigma)}$ (μ = permeability, σ = conductivity) and (41) has to be multiplied by $\sqrt{(\mu/\sigma)}$. At $10^8 \sim$, $R = 2{\cdot}61 \times 10^{-3}$ Ω/unit square, and a seal, glassed over a length of 5 mm. and 10 cm. in diameter, would have a resistance of $\dfrac{2{\cdot}61 \times 10^{-3}}{2 \times 10\pi} = 4{\cdot}3 \times 10^{-5}\Omega$. Even with 25 A. flowing, the heat generated in the seal is not important. However, the calculation has assumed that the seal is made to copper instead of to copper oxide, which is probably the case. The resistance may be increased considerably above the calculated value for this reason; but the seal will stand a good deal more than $2{\cdot}5 \times 10^{-2}$ W. However, if the diameter were reduced by a factor of say 10, and the poor conductivity of the oxide considered, the power dissipated in the seal could easily reach a dangerous figure.

It may be noted in passing that sealing alloys, such as kovar, are magnetic and therefore their skin resistance is very much greater than that of copper. At very high frequencies this is very disadvantageous, and the trouble has to be overcome by plating the sealing alloy with copper or silver. In spite of this, the increased simplification in structure and glass working obtained by the use of these alloys makes their use worth while.

To conclude this section we must say a little about interelectrode capacitances. Since the transit angles must be kept below certain values, it is inevitable that the electrode spacings become closer as the frequency rises. The resulting increase in interelectrode capacitance is somewhat reduced by the fact that the electrode system must be made shorter as it should not exceed about $\frac{1}{8}\lambda$ in any dimension. From eqn. (85) of Chapter 6 it follows that the transit time $\propto (d/I)^{\frac{1}{3}}$, so that, if the voltage is fixed, the only other way of improving matters is to increase I. This in turn means that the cathode must be worked at a higher average emission density. In general, both d and I are varied so that the shorter the wave-length the smaller the valve, the higher the cathode loading and the higher the electrode capacitance per unit length. The effects of the increased capacitance are partially compensated for by building the valve into a coaxial transmission line or a cavity resonator, these structures being desirable also for other reasons.†

It is tempting at this point to enter into a discussion of the application of the general theory of dynamic similarity to valve design. Some brief remarks were made on this at the end of Chapter 7, but a much more general theory has been worked out. By means of this theory it is easy to find the necessary relations to allow one to scale a valve maintaining any desired parameter constant, e.g. a valve can be scaled from one wavelength to another n times as great keeping the current density constant. In this case, lengths scale as n^3, voltages as n^{-4}, currents as n^{-6} and powers as n^{-10}. The last factor shows only too clearly how important the cathode current density is in ultra-high-frequency valves; for it means that if we determine the frequency at which a valve with a certain definite cathode loading will just give a defined minimum output, a valve scaled to operate at twice the frequency will only give $\frac{1}{2^{10}}$ as much output. Other types of scaling may be more appropriate for specified purposes. This theory of similarity allows one to set useful targets for the development of high-frequency valves by scaling the dimensions of valves known to be efficient at lower frequencies. The reader is referred to the papers of Lehmann, Raymond and Martinot-Lagarde for full details.‡

11.11. Transmitting tetrodes and pentodes

We have confined the discussion in this chapter to triodes. Tetrodes and pentodes are used for transmission at low frequencies but mostly at rather low powers. The advantages of multigrid valves are their increased power gain and the fact that neutralization is either easier or, at lower frequencies, unnecessary. On the other hand, the limiting power output is determined by the screen dissipation, since it is not possible to

† The design and construction of valves for use in coaxial line circuits is discussed by, *inter alia*, Bell and Davies, loc. cit.; Frankel *et al.*, *Proc. Inst. Radio Engrs., N.Y.* **34** (1946), 986; Bennett *et al.*, *Proc. Inst. Radio Engrs., N.Y.* **36** (1948), 1296.

‡ Lehmann, *Onde élect.* **230** (1946), 175. Raymond, *Onde élect.* **242** (1947), 209. Martinot-Lagarde, *Onde élect.* **28** (1948), 440.

cool the screen as efficiently as can be done with an external anode. Valves with a few tens of kW. output are possible. As the frequency increases, multigrid valves have more attraction and will possibly become standard in the 100–600 Mc./sec. frequency range. At still higher frequencies constructional difficulties influence one in favour of the triode once more. The arguments in favour of tetrodes at frequencies up to this limit are given by Wagener.†

† Wagener, *Proc. Inst. Radio Engrs., N.Y.* **36** (1948), 611.

Chapter 12

VELOCITY-MODULATED VALVES

By the middle of the thirties it had become clear that there were very grave practical difficulties in using conventional gridded valves at very high frequencies. In Chapter 7 we saw that the input impedance of a diode decreases approximately as the square of the frequency for a fixed spacing. This is also true for the cathode-grid circuit of a triode. Two possibilities exist for overcoming this effect. First, the spacing can be made as small as possible, and, secondly, the voltage used may be increased. Since the velocity varies as $V^{\frac{1}{2}}$ the first is more efficacious and, in addition, does not introduce the problems of high-power dissipation inherent in high-voltage operation. Small, close-spaced valves were therefore made and operated down to about 40 cm. wave-length. However, the pressure to develop radar systems made the second approach of interest too, because pulsed working overcame the power dissipation troubles, and special valves for pulsed operation were also designed and used in many early radar systems. From both classes of valve the trend of results was such that it was unreasonable to expect useful valve characteristics in the region of 10 cm. wave-length, which was the target for aerial-scanning systems. For these reasons many other types of valve were investigated, for instance, deflexion valves, inductive output tubes, and split-anode magnetrons. Of the several types worked on, only two are of any practical interest now, velocity-modulation valves and cavity magnetrons. In this chapter we give an account of the operation and design of velocity-modulated tubes. This is discussed before we give any details of the behaviour of gridded valves at high frequencies because the physical principles used in velocity modulation are operative in all tubes used at very high frequencies and, moreover, are exhibited in their simplest and most obvious form. It is therefore preferable to depart from the conventional order of treatment by considering velocity modulation first.

The principles of velocity modulation have been dealt with in three recent books: in an elementary manner by Harrison†, more deeply by Hamilton, Kuper and Knipp‡, and by Beck§. In view of the existence of these sources, we confine ourselves to the physical principles, a few of the more important mathematical results, and some new results on frequency pulling, etc., not included in our earlier book.

12.1. The physical principles of velocity modulation

The basic principle of velocity modulation is to abandon any attempt to modulate the current density in the valve directly by the application of the h.f. voltage to a grid; instead, the velocities of the electrons are modulated in the input region of the tube, say between the limits $u_0 \pm u_1$. The electrons then enter a space which is free of h.f. fields but not necessarily free of direct fields. In this space the faster electrons catch up the slower ones in front of them and the beam, instead of having a constant number of electrons in every cross-section, begins to show variations of density. At some points in space the values of electron density show maxima, and if a resonant circuit is located at these points it will be driven by the 'bunches' of electrons. Let us perform an order of magnitude calculation. If the period of the radio-frequency voltage is $\tau_0 = 1/f_0$ and the distance between the centres of the two h.f. fields is s_0, the time interval between the instant at which the fastest electron velocity $u_0 + u_1$, leaves the first h.f. field and the instant at which the slowest electron $u_0 - u_1$ leaves, is $\frac{1}{2}\tau_0$. If the second field is to be at the point where the fastest electron catches up with the slowest, we have

$$\frac{s_0}{u_0\left(1+\dfrac{u_1}{u_0}\right)} + \frac{\tau_0}{2} = \frac{s_0}{u_0\left(1-\dfrac{u_1}{u_0}\right)},$$

therefore
$$s_0 = \frac{u_0 \tau_0}{4} \frac{u_0}{u_1}, \tag{1}$$

† Harrison, *Klystron Tubes*, McGraw Hill, 1947.
‡ Hamilton *et al.*, M.I.T. series, vol. 7.
§ Beck, *Velocity-modulated Thermionic Tubes*, Cambridge University Press, 1948.

if $u_1/u_0 \ll 1$. If the modulating voltage is triangular in form, eqn. (1), with τ_0 redefined, will give the condition not only that the fastest and slowest electrons arrive at the second field at the same instant, but that all electrons will arrive simultaneously. For this case, therefore, it gives the correct position for maximum current. For a sinusoidal modulation it is only very approximate. To summarize, a velocity-modulated tube

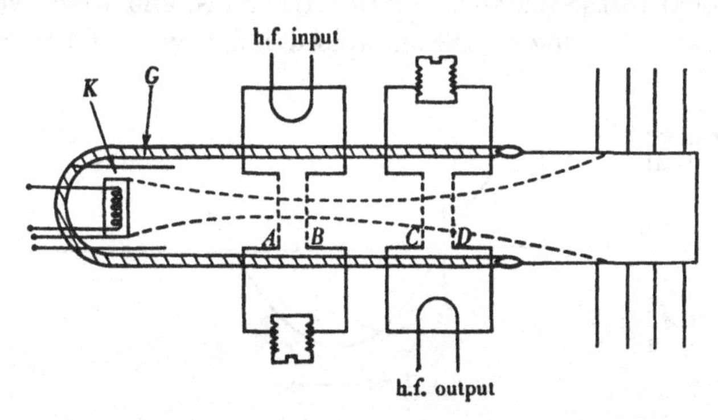

Fig. 12.1. The schematic diagram of a klystron amplifier.

consists of (*a*) a region in which the electron velocity is varied in accordance with the h.f. input wave, (*b*) a space free of h.f. field in which the electrons catch one another and the h.f. current density therefore increases from zero at the input, and (*c*) a second region of h.f. field where the electron bunches give up their energy to maintain oscillations in the output circuit. Let us now consider the simplest application of these ideas, the two-resonator klystron amplifier (fig. 12.1).

The electrons are focused through the valve by means of an electrostatic electron gun, which must be designed to produce as much current as possible at the working voltage. The h.f. circuits are both cavity resonators through which the beam flows by means of apertures closed by grids which are permeable to electrons but not to the h.f. field. A cavity resonator may be looked on as a high Q resonant circuit in which the h.f. field is confined to the interior of the metallic boundaries. The grids serve to complete these boundaries. The cavities are provided with tuning slugs, which alter their volume and thereby allow the circuits to be tuned to the same wavelength. After traversing the system, the electrons are removed by the collector which is designed to dissipate the necessary amount of energy without overheating. The action is as follows. The electrons enter the h.f. field at plane A and leave it at plane B. Between A and B they experience a force which is proportional to the voltage across the circuit, inversely proportional to the distance between the grids, and which varies in sense according to the impressed h.f. wave. If the input

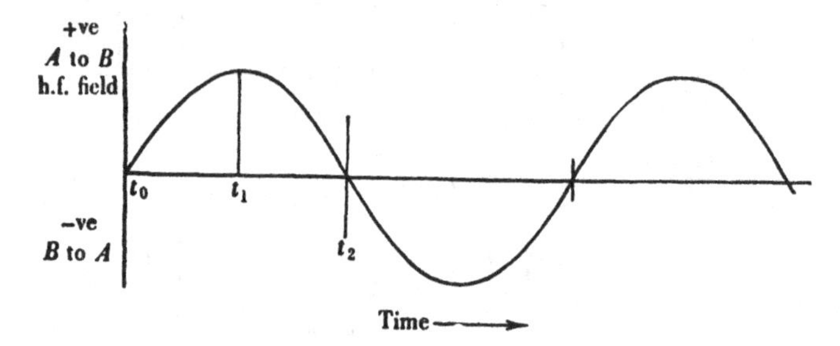

Fig. 12.2. The h.f. field in the modulating gap.

modulation is sinusoidal, the force varies sinusoidally and is accelerating in one half-cycle and decelerating in the other. In the space between B and C there is no h.f. field acting, so that each electron has a constant velocity, that at which it left plane B, and in this region the faster electrons catch up the slower ones. Consider now that the modulating voltage is as shown in fig. 12.2. The electrons passing the gap at t_0 and t_2 experience no force, and therefore travel on with velocity u_0. They can thus be used to define moving planes of reference. All the electrons passing through the gap in the half-cycle before $t = t_0$ were slowed down, and all those in the half-cycle after t_0 are accelerated. The electrons in the half-cycle after t_0 catch up those in the previous half-cycle, and the electron concentration around the $t = t_0$ reference plane increases. Consideration of the $t = t_2$ reference plane shows that the concentration there decreases. Thus the operation of the field-free space, usually called the drift space, is to form the initially uniform beam into a series of bunches of electrons, one bunch for every cycle of the modulating wave. The bunches form around the electrons originally entering the field at instants when the field was changing from positive to negative. The beam, now modulated in current density, next enters the second resonator, i.e. the space *CD* (see fig. 12.1). Here the bunches of electrons induce an electric field which adjusts its phase so as to extract the maximum amount of energy from each bunch. That is, the second resonator field goes through its maximum negative value at the instant the electron bunch traverses *CD*. The electrons in the bunch are slowed down by this field, and the total energy lost by the beam is, of course, equal to the total energy gained by the field. Since the Q of the output resonator is high, oscillations persist for far longer than one cycle of the driving wave and therefore build up to a constant value determined by the resonator impedance and the coupling to the external load. The two-resonator klystron is therefore capable of amplifying when conditions are such that the power fed to the load is larger than the input power, which in the first approximation is just the power required to maintain the given amplitude of oscillation in the high Q input resonator. The energy which provides the output power comes from the d.c. supply and the drift region acts as a device for converting d.c. energy to radio-frequency energy. The maximum voltage amplitude which can occur at the second resonator is approximately the amplitude required to bring the slowest electron to rest. If the amplitude increases beyond this point, some electrons will be turned back and they will reduce the available output. The remaining d.c. power has to be dissipated at the collector, and it will be clear that the electrons reaching the collector will have a range of energies. Theoretically at the maximum amplitude it would be possible to collect all the electrons on a collector at cathode potential, but in practice this is not possible and the electrons must be collected at a fairly high potential. In fact, they are usually collected at resonator potential although a gain in overall efficiency is obtained by working at a lower potential. It is usual to term the first, input, resonator the buncher, and the second, output, resonator the catcher.

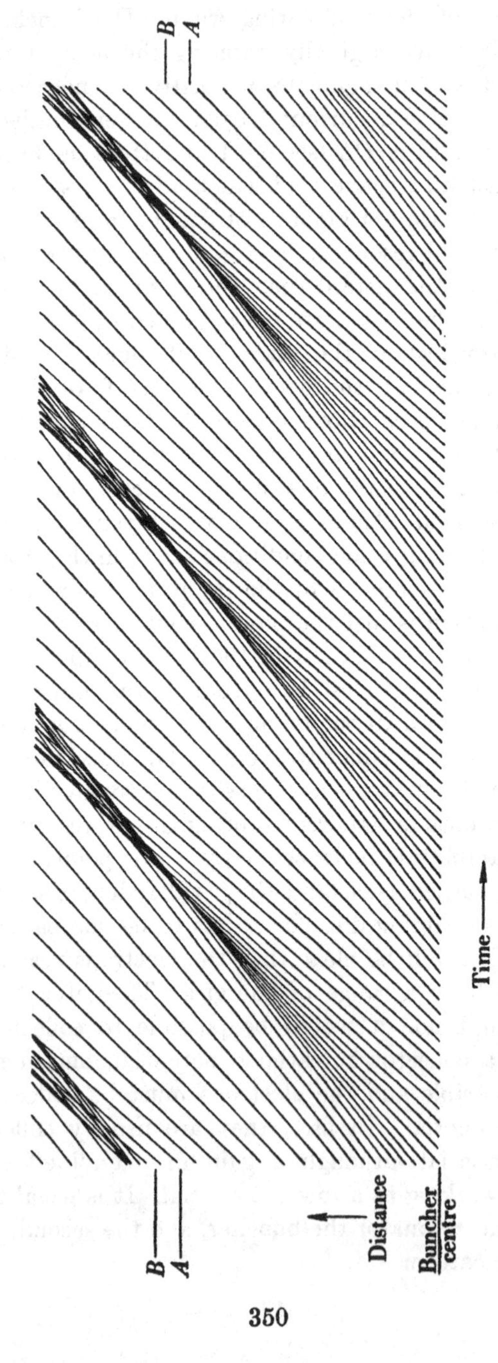

Fig. 12.3. Applegate diagram, showing electron position as a function of distance and time.

The process of bunching is well illustrated by fig. 12.3, an Applegate diagram. This is a distance-time plot for twenty-four electrons in each cycle. The slope of the line representing each electron is equal to the velocity at exit from the buncher, and the diagram shows clearly the way in which the electrons bunch together in between the lines marked A and B. Beyond B the electrons diverge once more and the radio-frequency current density decreases again; but if the diagram were extended, it would show further regions of bunching. Owing to the action of space-charge forces, neglected in the above account, these further maxima will not be as large as the first maximum and are never used in practical tubes.

The two-resonator klystron amplifier described above can be made to oscillate by feeding power back from the catcher to the buncher, the conditions for oscillation being the usual ones that the loop gain $\geqslant 1$ and the total phase shift be $2n\pi$.

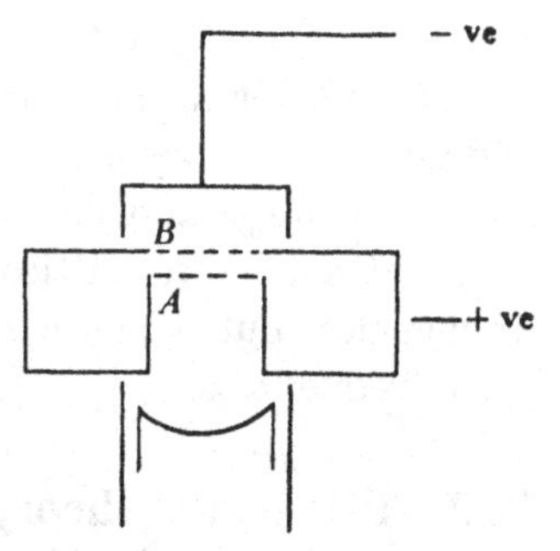

Fig. 12.4. Schematic diagram of a reflex klystron.

Perhaps the most important type of velocity-modulated tube to-day is the single-resonator reflex klystron in which the electrons pass through a single resonator, are reflected by a negative electron mirror or reflector and pass back through the same resonator a second time. During the first transit they are velocity-modulated, and, by the time they return, the velocity modulation has been converted into current modulation. so that the returning beam drives the resonator and the system is a self-oscillator. The arrangement of a reflex klystron is shown diagrammatically in fig. 12.4. The advantages of the reflex are as follows: (1) there is only a single resonator to tune, (2) the manufacture is cheaper and simpler than that of the two resonator tube, and (3) automatic frequency control can easily be applied by varying the reflector voltage. We shall deal with (3) in detail later. For these reasons all the low-power klystrons used as local oscillators in radar and similar microwave receivers are of the reflex type. The reflex is a natural choice as a transmitter in microwave frequency-modulated systems, because of the ease with which considerable frequency deviations can be obtained with only minute modulating power.

The reflex can be shown to behave in exactly the same way as a unity-coupled two-resonator klystron if the reflector field is a linear function of distance from grid B. If the reflector field is non-linear, an additional and valuable modification is introduced but no essential difference results.

We shall next give some mathematical results on the klystron. These are based on the well-known fact that one can only measure power at microwave frequencies and not current or voltage. The theory is therefore based on observable quantities, the d.c. voltage and current, the h.f. power, the resonator Q, etc., and not on quantities which are usually invoked at lower frequencies, but which cannot be directly measured at microwave frequencies.

12.2. First-order theory of velocity modulation

We now derive a very simple theory of velocity-modulated devices originally due to Webster,† which neglects the power consumed in bunching or debunching the electron beam. Some general assumptions are made which will persist throughout the rest of the work; other assumptions are made which will be discarded later. The general assumptions are:

(1) Relativity effects can be neglected. This confines us to the study of accelerating voltages below, say, 15 kV.

(2) Electrons travel in straight lines parallel to the z axis unless they have been deflected by an h.f. field.

(3) Secondary emission is neglected.

Special assumptions used for the moment are that the modulating voltage is much smaller than the direct voltage, and that the grids constrain the field to a plane parallel form. The equation of motion of an electron between grids A and B

† Webster, *J. Appl. Phys.* **10** (1939), 501.

of fig. 12.5, which shows the notation used in this section, is

$$m\ddot{z} = \frac{eV_1}{d}\sin\omega t, \tag{2}$$

therefore
$$\dot{z} = \frac{eV_1}{md\omega}(\cos\omega t_0 - \cos\omega t) + u_0, \tag{3}$$

if we use the initial condition $\dot{z} = u_0$ at $t = t_0$, where t_0 = time of entry into the field. Using the first special assumption

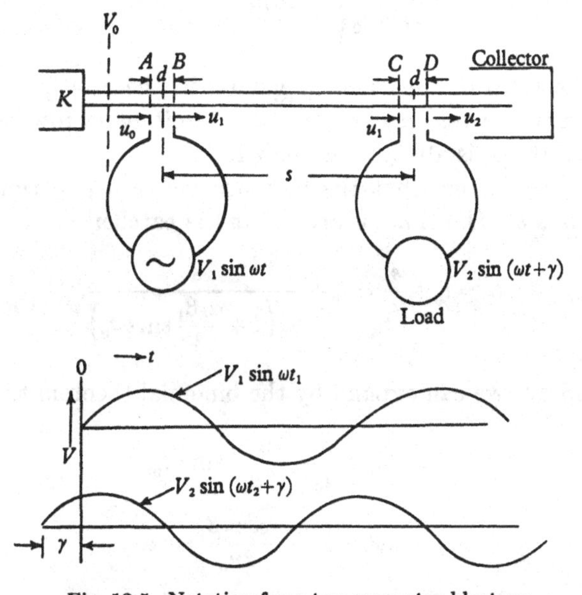

Fig. 12.5. Notation for a two resonator klystron.

$V_1/V_0 \ll 1$, we can write t_1, the time of exit from the field, as $t_1 = t_0 + \dfrac{d}{u_0}$. The velocity at exit is then u_1, defined by

$$u_1 = u_0 + \frac{eV_1}{mu_0}\frac{\sin\frac{1}{2}\phi}{\frac{1}{2}\phi}\sin(\omega t_0 + \tfrac{1}{2}\phi), \tag{4}$$

where $\phi = \omega d/u_0$, the transit angle across the gap. If we now shift the time origin to the gap centre we get

$$u_1 = u_0 + \frac{e\beta_1 V_1}{mu_0}\sin\omega t_0, \tag{5}$$

where β_1 = gap coupling coefficient

$$= \frac{\sin \frac{1}{2}\phi}{\frac{1}{2}\phi}.$$

Also $$u_1 = u_0\left(1+\frac{e\beta_1 V_1}{mu_0^2}\sin\omega t_0\right) \quad \text{and} \quad u_0^2 = 2\frac{e}{m}V_0.$$

If we put $V_1/V_0 = \alpha_1$, the depth of modulation, we finally obtain

$$u_1 = u_0\left(1+\frac{\alpha_1\beta_1}{2}\sin\omega t_0\right). \tag{6}$$

β_1 is the ratio of the velocity gained in the real gap with V_1 across it to the velocity gained in an infinitely narrow gap with V_1 across it. β_1 is always less than 1.

We now assume that the distance between gap centres in fig. 12.5 is s. The time of arrival at the catcher centre is then

$$t_2 = t_0+\frac{s}{u_1} = t_0+\frac{s}{u_0\left(1+\frac{\alpha_1\beta_1}{2}\sin\omega t_0\right)}. \tag{7}$$

Since $\alpha\beta \ll 1$, we can expand by the binomial theorem to obtain

$$t_2 = t_0+\frac{s}{u_0}-\frac{s\alpha_1\beta_1}{2u_0}\sin\omega t_0,$$

or $$\omega t_2 = \omega t_0+\frac{\omega s}{u_0}-\frac{\omega s\,\alpha_1\beta_1}{2u_0}\sin\omega t_0. \tag{8}$$

Now $\omega s/u_0$ is the transit angle of an unmodulated electron velocity u_0 through the drift space. We can write S for this angle. Lengths multiplied by ω/u_0 are often called 'normalized' lengths. Then

$$\omega t_2 - S = \omega t_0 - \frac{S\alpha_1\beta_1}{2}\sin\omega t_0. \tag{9}$$

Now by eqns. (47) and (48) of Chapter 7, we can use continuity of current to show that

$$i_0\,dt_0 = i_2\,dt_2. \tag{10}$$

Differentiating (8) to obtain dt_2/dt_0 we get

$$\frac{dt_2}{dt_0} = 1-\frac{S\alpha_1\beta_1}{2}\cos\omega t_0,$$

and
$$i_2 = \frac{i_0}{1-\frac{S\alpha_1\beta_1}{2}\cos\omega t_0}. \tag{11}$$

But i_0 is, in this case, simply the d.c. beam current I_0; therefore

$$i_2 = \frac{I_0}{1-\frac{S\alpha_1\beta_1}{2}\cos\omega t_0}. \tag{12}$$

We next expand i_2 as a Fourier series in $(\omega t_2 - S)$, using eqns. (9) and (10), and

$$i_2 = \tfrac{1}{2}a_0 + a_1\cos(\omega t_2 - S) + a_2\cos 2(\omega t_2 - S) + b_1\sin(\omega t_2 - S) + b_2\sin 2(\omega t_2 - S).$$

Using eqn. (9) we can easily determine the coefficients in the usual way:

$$a_0 = 2I_0,$$

$$a_n = \frac{I_0}{\pi}\int_{-\pi}^{\pi}\cos n\left(\omega t_0 - \frac{S\alpha_1\beta_1}{2}\sin\omega t_0\right)d(\omega t_0)$$

$$= 2I_0 J_n\left(n\frac{S\alpha_1\beta_1}{2}\right),$$

$$b_n = \frac{I_0}{\pi}\int_{-\pi}^{\pi}\sin n\left(\omega t_0 - \frac{S\alpha_1\beta_1}{2}\sin\omega t_0\right)d(\omega t_0) = 0.$$

Finally, then,

$$i_2 = I_0\left[1 + 2\sum_{n=1}^{\infty} J_n\left(n\frac{S\alpha_1\beta_1}{2}\right)\cos n(\omega t_2 - S)\right]. \tag{13}$$

There are some questions of validity which we have not raised in discussing the derivation of eqn. (13); e.g. eqn. (12) shows that i_2 has an infinite singularity for $\frac{1}{2}S\alpha_1\beta_1 = 1$ at $\omega t_0 = 0, 2n\pi$. Hamilton† gives a full discussion of these points, demonstrating that eqn. (13) is correct for all values of $\frac{1}{2}S\alpha_1\beta_1$.

The next step is to calculate the radio-frequency power produced in the catcher resonator, which in an ordinary amplifier is tuned to the fundamental, $n = 1$. We assume that the output voltage is $V_2\sin(\omega t_2 + \gamma)$. Using Ramo's theorem the power

† Hamilton, op. cit. p. 207.

extracted from the beam is minus the time average of the product, $i_2 . V_2$, i.e.

$$P_b = -\frac{2I_0}{2\pi}\int_0^{2\pi} \beta_2 V_2 \sin(\omega t_2+\gamma) J_1\left(\frac{S\alpha_1\beta_1}{2}\right)\cos(\omega t_2 - S)\, d(\omega t_2)$$

$$= -\beta_2 V_2 I_0 J_1\left(\frac{S\alpha_1\beta_1}{2}\right)\sin(S+\gamma). \tag{14}$$

In the case of an amplifier the output voltage adjusts its phase to extract maximum power from the beam and

$$S+\gamma = (4n+3)\pi/2 \quad (n = 0,\ 1,\ 2,\ \text{etc.}).$$

Therefore $$\sin(S+\gamma) = -1.$$

The efficiency is $P_b \div I_0 V_0$ as $I_0 V_0$ = d.c. input power, therefore

$$\eta = \beta_2 \alpha_2 J_1\left(\frac{S\alpha_1\beta_1}{2}\right). \tag{15}$$

And, if no electrons are turned back, $\beta_2 \alpha_2 \leqslant 1$; thus

$$\eta \leqslant J_1\left(\frac{S\alpha_1\beta_1}{2}\right). \tag{16}$$

The maximum value of $J_1(\theta)$ is 0·584 at $\theta = 1{\cdot}84$, so the maximum conversion efficiency which can be obtained from a conventional klystron in which no electrons are turned back is 58·4 per cent. By the conversion efficiency we mean the efficiency of conversion of d.c. energy, taken from the h.t. supply, into radio-frequency energy. Not all this radio-frequency energy can be taken out to appear across the load, since some energy is lost in maintaining the voltage V_2 across the output grids. This power appears as heat in the walls of the resonator. Thus the useful power output is always less than the radio-frequency power generated and, as we shall see later, is sometimes very much less than the generated power. If we write η_C for the conversion efficiency and η_L for the circuit efficiency, the overall efficiency is

$$\eta_0 = \eta_C \eta_L. \tag{17}$$

To summarize the results of this section we have calculated the velocity modulation in the beam after it has passed through the buncher gap and the result is given by eqn. (6). In the drift space of length S, in which the transit angle of an unmodulated electron is $S = \omega s/u_0$ radians, the velocity modulation changes to a conduction current density modulation, given by eqn. (13). The passage of the bunched beam through the catcher then leads to the power interchange given by eqn. (14). The operation of any velocity-modulated device can be calculated by splitting the electron motion into these three parts, even if they are not all physically distinct, as is the case, for instance, in the reflex klystron.

12.3. Phase conditions and electron beam transconductance

We digress for a moment to discuss the operation of velocity-modulated devices in terms of transconductance, an aspect which is of importance because of the analogy with ordinary valve theory at low frequencies.

In the derivation of the output power given in the last section we saw that if the input voltage was of the form $V_1 \sin \omega t_2$ and the output of the form $V_2 \sin(\omega t_2 + \gamma)$, i.e. if the output voltage leads the input by γ, the power gained by the catcher resonator is a maximum when $S + \gamma = (4n+3)\pi/2$. Two special cases are of interest:

(*a*) $\gamma = 0$, resonators in phase, $S = (4n+3)\pi/2$,

(*b*) $\gamma = \pi$, resonators in anti-phase, $S = (4n+1)\pi/2$.

If we consider the two-resonator klystron oscillator for a moment we see that, if the coupling device used to feed power back from the output to the input is such that the resonators oscillate in phase, oscillations will only be possible for those values of anode voltage which make

$$S = \frac{\omega s}{u_0} = \frac{\omega s}{\left(2\frac{e}{m}\right)^{\frac{1}{2}} V_0^{\frac{1}{2}}} = (4n+3)\pi/2. \qquad (18)$$

Similarly, the antiphase condition is

$$S = (4n+1)\pi/2. \qquad (19)$$

In general the coupling device will have an electrical length ϕ which will enter into the phase condition; but the important point is that oscillation is only possible for voltages fairly close to a series of well-defined values depending on the tube geometry. All transit-time oscillators show this behaviour.

It is convenient to base our terminology for these modes on the number of quarter wave-lengths in the transit angle between gap centres. For example, if $n = 0$ in eqn. (18) we should call the oscillation the 'mode 3 oscillation', while eqn. (19) with $n = 2$ is the 'mode 9 oscillation'. This nomenclature eliminates any difficulties connected with the definition of n or γ. There are two special types of double-gap transit-time oscillator which make use of resonators in which the fields are either in antiphase or in phase. In the coaxial line of *Heil* tube the electrons pass through a coaxial line, the path through the inner of the line being the drift space. Here the fields are in antiphase, so only modes 1, 5, 9, etc., can be excited. In the ordinary klystron the feed-back coupling can be so arranged that either series of modes appears, or in certain circumstances both series.

It will be appreciated from the form of eqn. (13) that the radio-frequency current, which depends on

$$\cos(\omega t_2 - S) = \cos(\omega t_0 + S - S),$$

or, in other words, it leads the gap voltage by 90 degrees. In the case of the amplifier this current drives the output, so the output voltage is in antiphase. In the case of an oscillator, the radio-frequency current only passes through its maximum at the same instant as the radio-frequency voltage passes its maximum for a particular, discrete voltage. For other neighbouring voltages the current either lags or leads the voltage. Following ordinary circuit theory we can split the current into in-phase and reactive parts, the in-phase part interchanging power and the reactive part altering the frequency of the output circuit. In complex form the fundamental component of the output current is $2I_0 J_1\left(\frac{S\alpha_1\beta_1}{2}\right)\exp[j(\omega t_2 - S)]$.

We can now write down an expression for the electronic admittance, which the beam applies to the terminals of the catcher gap. We have

$$Y_e = \frac{\beta_2 i_2}{V_2} = \frac{2\beta_2 I_0 J_1\left(\dfrac{S\alpha_1\beta_1}{2}\right)\exp[j(\omega t_2 - S)]}{V_2 \exp[j(\omega t_2 + \gamma - \frac{1}{2}\pi)]}$$

$$= 2\beta_2 \frac{I_0}{V_2} J_1\left(\frac{S\alpha_1\beta_1}{2}\right)\exp[j(\tfrac{1}{2}\pi - S - \gamma)]$$

$$= Y_0 \exp[j(\tfrac{1}{2}\pi - S - \gamma)]. \qquad (20)$$

It is useful to evaluate Y_0 for very small signals, such that

$$\frac{S\alpha_1\beta_1}{2} \to 0. \quad \text{Then} \quad J_1\left(\frac{S\alpha_1\beta_1}{2}\right) \to \frac{S\alpha_1\beta_1}{4}.$$

Thus, for small signals and equal gaps,

$$Y_0 = \frac{I_0}{V_2}\frac{S\alpha_1\beta^2}{2}. \qquad (21)$$

These relations are of importance in studying the automatic frequency control properties of reflex klystrons and in the study of frequency stability in ordinary klystrons.

We now calculate the transconductance of the velocity-modulated tube. In an ordinary valve the transconductance appears in the form $i_a = g_m V_g$. For the velocity-modulated tube we obtain the transconductance in the same way,

$$g_m = \frac{2\beta I_0}{V_1} J_1\left(\frac{S\alpha_1\beta}{2}\right). \qquad (22)$$

In the small signal approximation we have

$$g_{ms} = \frac{I_0}{V_1}\frac{S\beta^2 V_1}{2V_0}$$

$$= \frac{\beta^2 I_0 S}{2V_0} \text{ amp./V.} \qquad (23)$$

Eqn. (23) shows that it is necessary to have the maximum possible current for a given voltage and the largest possible value of S† if the largest possible value of g_{ms} is to be obtained.

† In fact, space charge forces which tend to debunch the beam set an upper limit to the useful value of S. We shall say something more about this later on.

In velocity-modulated tubes g_m is much smaller than in conventional valves. As an example, a 40 W. output amplifier for $\lambda = 10$ cm. had $I_0 = 120$ mA., $S = \frac{5}{2}\pi$, $V_0 = 2$ kV., $\beta^2 = \frac{1}{2}$. From these figures $g_{ms} = 118$ μmhos. This compares very badly with modern high-slope pentodes with $g_m = 7000$ μmhos at $I_0 = 10$ mA. It should be noted that the requirement for minimum beam impedance is not independent of the requirements for large S. In electrostatic focusing systems, for example, we have $I_0 \propto V_0^{\frac{3}{2}} s^{-2}$, therefore

$$g_{ms} \propto \frac{\beta^2 V_0^{\frac{3}{2}} \omega s}{2V_0 s^2 u_0},$$

or

$$g_{ms} \propto \frac{\beta^2 \omega}{2s} \sqrt{\left(\frac{m}{2e}\right)}.$$

For this special case, therefore, the drift length should be made short rather than long. In existing tubes the requirement that a reasonably low beam impedance should be obtained limits the drift length to a shorter value than does the effect of space-charge debunching. The use of modified focusing arrangements may alter this situation.

Returning now to eqn. (22) it is interesting to study the way in which the transconductance behaves as the amplitude increases. If we put $\frac{1}{2}S\alpha\beta = \theta$ and insert (23) in (22) we have

$$g_m = g_{ms} \frac{2J_1(\theta)}{\theta}. \tag{24}$$

This function of θ is shown in fig. 12.6; it is a slowly decreasing function which is still over 60 per cent of the initial value at $\theta = 1{\cdot}84$. For $\theta < 0{\cdot}6$ the error involved in putting $2J_1(\theta) = 1$ is less than 5 per cent.

The ideas developed in this section can be used to provide quite a good idea of the operation of most velocity-modulated tubes; particularly the low-power reflex klystrons used as local oscillators, which fulfil the small signal conditions rather well. For higher power tubes the oscillation amplitudes are greater and there are very substantial departures from the small signal theory, and we must now consider some of the complicating factors.

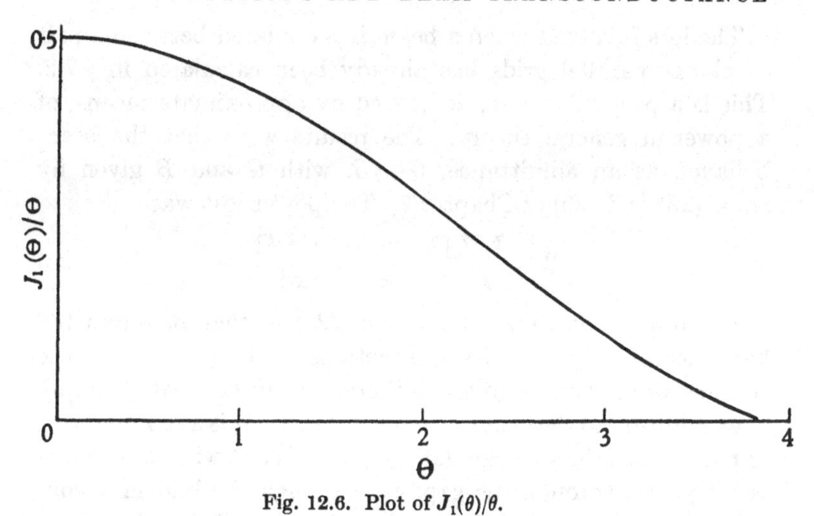

Fig. 12.6. Plot of $J_1(\theta)/\theta$.

12.4. Beam loading

The first departure from the ideal conditions postulated in § 12.2 which we shall study is the introduction of beam loading. When the beam passing through a pair of modulating grids is initially unbunched, the theory of § 12.2 indicates that on the average there is only a very small power interchange between the buncher and the beam. The physical reason for this is that, although the electrons which are accelerated in the gap gain a sizeable amount of energy, an exactly equal number of electrons is decelerated, each electron losing as much energy as an electron of the other group gained. The net energy interchange is therefore zero. Unless the gap is infinitesimally short or the electron velocity infinitely high, the real situation will be that the beam will have a small but finite degree of density modulation when it emerges from the gap, or, in other words, the electrons which are accelerated gain more energy than is lost by those which are decelerated. The energy gained by the beam is then greater than the loss to the resonator in the negative half-cycle. To summarize, the condition for zero loss is that two finite amounts of energy shall be equal and opposite. The process of velocity modulation makes the amounts unequal so that quite important losses can occur.

The loss involved when a beam is modulated between a pair of planar parallel grids has already been calculated in § 7.2. This is a particular case, developed by approximate means, of a powerful general theory. The results were that the beam behaved as an admittance, $G+jB$, with G and B given by eqns. (28) and (29) of Chapter 7. The power lost was

$$P_b = \frac{V_1^2 I_0}{2V_0}\left[\frac{1-\cos\phi}{\phi^2} - \frac{\sin\phi}{2\phi}\right]. \tag{25}$$

The important thing about eqn. (25) is that it shows the beam losses to be quadratic functions of V, just as are the circuit losses. The beam losses therefore can be treated simply as an addition to the resonator conductance. It is also important to realize that the same amount of power is lost when a bunched beam passes through the gap, even though the bunching contributes an amount of power given by eqn. (14). Thus, considering a two-resonator klystron oscillator, the energy balance is as follows:

Energy in load = Energy converted from d.c. – (energy lost in driving input circuit + energy lost in bunching beam + energy lost in driving output circuit + energy lost in debunching beam in output gap).

It is clear that if the valve is not designed so as to minimize all these sources of loss, the energy available for use outside the valve will be much less than the energy actually generated.

We shall not develop the general theory of beam loading here. It was worked out by Petrie, Strachey and Wallis in various unpublished reports. Full details are contained in a paper by Fremlin, Petrie *et al.*,† and other derivations are given by Pierce,‡ Beck,§ etc. The result is as follows:

If P_l = the power lost in bunching,

$$P_l = \frac{I_0}{4V_0}\xi\,|A|\frac{\partial |A|}{\partial \xi}, \tag{26}$$

† Fremlin, Petrie *et al.*, *J. Instn. Elect. Engrs.* **93**, pt. IIIA, no. 5 (1946), 868.
‡ Pierce and Shepherd, *Bell Syst. Tech. J.* **26** (1947), 460.
§ Beck, *Velocity-modulated Thermionic Tubes*, Cambridge University Press, 1948.

where $\xi = \omega/u_0$,

A = Fourier transform of the field in the gap

$$= \int_{-\infty}^{+\infty} E(z)\, e^{j\xi z} dz.$$

For single gaps $|A|$ reduces to the modulation coefficient, β times V^1. The expression $\xi \frac{\partial |A|}{\partial \xi}$ is usually written as γV_1, and is called the second gap coefficient. Since it is basically the rate of change of β with respect to velocity, it expresses the increased gain in energy of those electrons which are speeded up in the gap.

If we inset β and γ and divide eqn. (26) by $I_0 V_0$, the d.c. power in the beam, the diode efficiency of the gap is given by

$$\eta_D = \frac{\alpha^2}{4} \beta\gamma, \tag{27}$$

and in this form we shall make a good deal of use of it. Another useful form is the beam-loading resistance defined by

$$P_l = \frac{V_1^2}{2R_b},$$

therefore

$$R_b = \frac{2V_0}{I_0} \frac{1}{\beta\gamma} = \frac{2R_0}{\beta\gamma}. \tag{28}$$

The calculation of $\beta\gamma$ is considered for various practical arrangements in the references already given. It is simple to verify that if $E_1(z)$ is put equal to V_1/d, eqn. (26) leads to the same result as eqn. (25). Eqn. (26) neglects terms of the order α^3 and higher powers of α, but further investigation of the tractable planar case shows that the coefficients of these terms are in any case small for all ordinary transit angles. We may take eqn. (26) then to be a valid result except for the rather special case of very high modulating voltage combined with a long gap, which is not used in practice, although some theoretical studies have been made†.

† Muller and Rostas, *Helv. Phys. Acta*, **13** (1940), 435. Marcum, *J. Appl. Phys.* **17** (1946), 4.

12.5. Bunching. More accurate theory

In § 12.2 we used the first approximation to the binomial expansion of eqn. (7) to go to eqn. (8). This may or may not be a valid step, and we now proceed to investigate the question. The second approximation to ωt_2† is

$$\omega t_2 = \omega t_0 + \frac{\omega s}{u_0}\left(1 - \frac{\alpha\beta}{2}\sin\omega t_0 + \frac{\alpha^2\beta^2}{4}\sin^2\omega t_0 - \frac{\alpha^3\beta^3}{8}\sin^3\omega t_0,\ \text{etc.}\right). \quad (29)$$

We can use this expression for t_2 to calculate the coefficients of the Fourier expansion of the catcher current. To save writing, we put $\omega t_0 = \theta$, $\frac{1}{2}S\alpha\beta = B$. Limiting ourselves to the fundamental component we get

$$a_1 = \frac{I_0}{\pi}\int_{-\pi}^{+\pi}\cos\left(\theta - B\sin\theta + \frac{B^2}{S}\sin^2\theta - \frac{B^3}{S^2}\sin^3\theta\right)d\theta, \quad (30)$$

$$b_1 = \frac{I_0}{\pi}\int_{-\pi}^{\pi}\sin\left(\theta - B\sin\theta + \frac{B^2}{S}\sin^2\theta - \frac{B^3}{S^2}\sin^3\theta\right)d\theta. \quad (31)$$

We shall reject all terms of order S^{-3} and above. We can now expand $\cos\left[(\theta - B\sin\theta) + \left(\frac{B^2\sin^2\theta}{S} - \frac{B^3\sin^3\theta}{S^2}\right)\right]$ as follows:

$$\cos(\theta - B\sin\theta)\left(1 - \frac{B^4\sin^4\theta}{2S^2}\right) - \sin(\theta - B\sin\theta)\left(\frac{B^2\sin^2\theta}{S} - \frac{B^3\sin^3\theta}{S^2}\right),$$

therefore

$$a_1 = \frac{I_0}{\pi}\int_{-\pi}^{+\pi}\cos(\theta - B\sin\theta)\left(1 - \frac{B^4\sin^4\theta}{2S^2}\right)d\theta - \frac{I_0}{\pi}\int_{-\pi}^{+\pi}\sin(\theta - B\sin\theta)\left(\frac{B^2\sin^2\theta}{S} - \frac{B^3\sin^3\theta}{S^2}\right)d\theta$$

$$= I_0\left(2J_1(B) - \frac{B^4}{2\pi S^2}\int_{-\pi}^{\pi}\sin^4\theta\cos(\theta - B\sin\theta)d\theta - O + \frac{B^3}{\pi S^2}\int_{-\pi}^{+\pi}\sin(\theta - B\sin\theta)\sin^3\theta\,d\theta\right).$$

† A fuller discussion is given by Pierce and Shepherd (loc. cit.). Their results are slightly different from those below because they use an approximation for the input velocity.

Similarly,

$$b_1 = \frac{I_0 B^2}{\pi S}\int_{-\pi}^{+\pi} \cos(\theta - B\sin\theta)\sin^2\theta\, d\theta.$$

Even without carrying out the integrations it is obvious that the condition for the analysis of § 12.2 to be valid is that $\frac{B^2}{S} < 1$, which makes b_1 small and the correction terms in a_1 even smaller.

Then, for the Webster approximation to be valid we have to make $\frac{1}{4}S\alpha^2\beta^2 < 1$. If we take 0·25 as a reasonable value, the condition is just

$$\alpha^2\beta^2 < \frac{1}{S}. \tag{32}$$

For a mode 5 klystron the expression (32) shows that $\alpha\beta = 0{\cdot}36$ is the top limit for the small signal theory. This is encouraging because $\alpha\beta = 0{\cdot}36$ is quite close to the value required† to obtain maximum efficiency from such a valve.

Let us now return to the evaluation of the integrals. We have

$$2J_1(B) = \frac{1}{\pi}\int_{-\pi}^{+\pi} \cos(\theta - B\sin\theta)\, d\theta.$$

Differentiating with respect to B

$$2J_1'(B) = \frac{1}{\pi}\int_{-\pi}^{+\pi} \sin(\theta - B\sin\theta)\sin\theta\, d\theta$$

$$-2J_1''(B) = \frac{1}{\pi}\int_{-\pi}^{+\pi} \cos(\theta - B\sin\theta)\sin^2\theta\, d\theta,$$

and so on. Using these we finally obtain

$$a_1 = I_0\left(2J_1(B) - \frac{B^4}{S^2}J_1'''' - \frac{2B^3}{S^2}J_1'''\right), \tag{33}$$

$$b_1 = -I_0\frac{2B^2}{S}J_1''(B), \tag{34}$$

or

$$i_2 = I_0\left\{2J_1(B) + \frac{j2B^2}{S}J_1''(B) - \frac{1}{S^2}\left(2B^3 J_1'''(B) + B^4 J_1''''(B)\right)\right\}. \tag{35}$$

† Beck, op. cit. p. 55.

The derivatives can be evaluated by the well-known recurrence relations, $J_1' = \frac{1}{2}(J_0 - J_2)$, etc., to be found in books on Bessel functions.

This analysis has been carried further by Pierce and Shepherd, who show that the departures from the Webster theory are unimportant, even for small S, when $B < 1{\cdot}8$. It will be noted that the major effect is the introduction of a small reactive current so that the phase conditions discussed in § 12.3 will be not quite accurate. Since the phase conditions are of relatively little importance—they only serve to determine the centre point of the operating voltage range and a 5–10 per cent variation from the calculated value is tolerable—and the theory of the power production need not be modified, we can use the Webster approximation in all our future work.

The questions studied above have been dealt with in a different manner by Wallis,† who comes to the same conclusion, viz., that the Webster approximation is quite adequate for engineering purposes.

12.6. Space-charge debunching

As yet we have not considered the effect of space-charge forces on the bunching of the beam. Obviously there is no longitudinal space-charge force in a uniform beam, but when local variations of electron density are set up, space-charge forces come into being. They act in such a sense as to diminish the density variations, i.e. in opposition to the bunching. Moreover, since the initial current density in the beam must be a multiplying factor in the expression for the space-charge force, the debunching will increase with the initial current. As we have seen, high transconductance means high beam current, and therefore maximum debunching effect. It is therefore very important to obtain reliable estimates of the degree of debunching which obtains in a given tube. Unfortunately, the difficulties involved in giving a theoretical analysis are very great, and only some rather special cases have been studied.

† Fremlin *et al.*, loc. cit. (pp. 901 *et seq.*).

In his original work, Webster† calculated the effects of space charge in a beam of infinite cross-section. The result is that i_2 is reduced from $2I_0J_1(\frac{1}{2}S\alpha\beta)$ to

$$i_2 = 2I_0J_1\left(\frac{S\alpha\beta}{2} \times \frac{\sin h_0 S}{h_0 S}\right), \tag{36}$$

where

$$h_0^2 = \frac{eI_0}{m\epsilon_0 \sum u_0^3}$$

$$= \frac{3\pi \times 10^4 I_0}{\sum V_0^{3/2}}, \tag{37}$$

$$\sum = \text{beam area (m.}^2\text{)}.$$

This expression neglects the image forces induced by the bunches in the walls of the tunnel. These forces tend to reduce the debunching, so that eqn. (36) considerably underestimates the value of i_2 unless the beam is a thin pencil in a very wide tunnel. This arrangement is never used in practice.

In dealing with more practical cases some special assumptions have to be made. It is always assumed that the electrons travel along parallel rectilinear paths with no convergence or divergence. A first approximation to this state of affairs can be obtained experimentally by using a very strong longitudinal magnetic field, but as we have seen in § 6.3, this will depress the potential at the centre of the beam so that peripheral electrons travel faster than axial ones. This will introduce phase differences between the radio-frequency current carried by different beam segments, and therefore will partially invalidate the requirements of the analysis. In theoretical work this last objection is usually overcome by assuming that a thread of positive ions exists along the axis of the beam and that the charge density of positive ions just cancels the charge density of electrons. It can be demonstrated that such a situation can exist under assumed conditions of gas pressure,‡ etc., but whether such ion beams exist in real valves seems doubtful. The important point is to set up a model which is sufficiently simple for analysis, and this model ought to be realizable in

† Webster, loc. cit.

‡ Spangenberg, Field and Helm, *Electr. Commun.* **24** (1947), 108

practice, even if only in rather special circumstances. Finally, it is necessary to assume that electrons do not overtake in the drift space.

Using these assumptions together with the assumption that the drift tube is much longer than its diameter, and that the wave-length and also that the depth of modulation is small, Warnecke† and his co-workers have worked out the case for a cylindrical beam in a cylindrical tunnel. To a good degree of approximation,

$$i_2 = 2J_1\left(\frac{S\alpha\beta}{2}\frac{\sin h_1 S}{h_1 S}\right), \tag{38}$$

where
$$h_1^2 = h_0^2\left[1 - \frac{2I_1(A)}{AI_0(A)}F_1\right], \tag{39}$$

B = normalized tunnel radius,

A = normalized beam radius $= \dfrac{\omega a}{u_0} = \dfrac{2\pi a}{S_0}$,

S_0 = distance travelled per cycle by an unmodulated electron,

a = beam radius in same units as S_0;

I_0, I_1, K_0, K_1 = modified Bessel functions,

$$F_1 = \left(\frac{K_1(A)\,I_0(B) - K_0(B)\,I_1(A)}{K_1(A)\,I_0(A) - K_0(A)\,I_1(A)}\right)\frac{I_0(A)}{I_0(B)}.$$

Fig. 12.7 shows h_1/h_0 plotted as a function of a for several values of b/a. It is seen that h_1 is always less than h_0, and in many practical cases will be quite a small proportion of h_0. As an example, a 7 cm. valve might have $a = 0{\cdot}125$ cm., $S_0 = 0{\cdot}5$ cm. Thus $A = \frac{1}{2}\pi$ and $h_0/h_1 \doteqdot 0{\cdot}4$ for $b/a = 1{\cdot}0$. In this case the Webster theory would limit the drift length to a value only 40 per cent as great as would the more refined theory.

The case of a rectangular beam has been studied by Feenberg and Feldman‡ under somewhat more general conditions. Their result is

$$h_1^2 = h_0^2[1 + \coth A \coth(B - A)]^{-1}, \tag{40}$$

† Warnecke, Bernier and Guénard, *J. Phys. Radium*, series 8, **4** (1943) 96 and 116. Warnecke, Guénard and Fauve, *Ann. Radioélect.* **2** (1947) 224.

‡ Feenberg and Feldman, *J. Appl. Phys.* **17** (1946), 1052.

where A = normalized beam width, B = normalized tunnel width. For gaps without grids

$$i_2 \doteqdot I_0 \frac{\alpha\beta^2}{2}\left\{\frac{\sinh 2A}{2A} \frac{1}{\cosh^2 B}\right\} \frac{2\pi}{S_0 h_1} \sin(h_1 s). \tag{41}$$

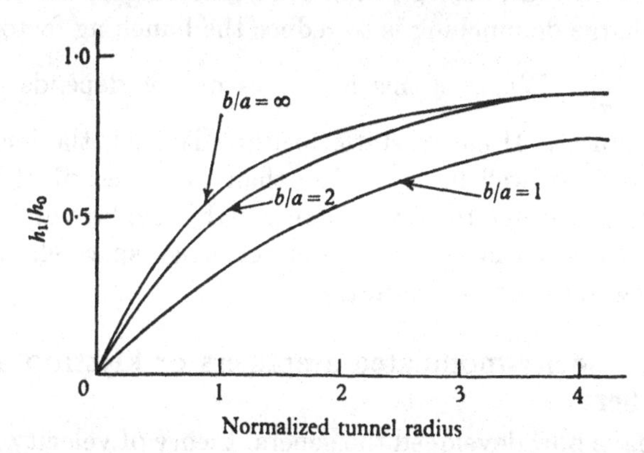

Fig. 12.7. Space-charge debunching factor corrected for finite beam diameter.

For gaps with grids

$$i_2 = I_0 \frac{\alpha\beta}{2}\left[\frac{\tanh A}{A} \frac{2\pi}{s_0} \exp(-jh's) \frac{\sin(h_1 s)}{h_1} + \left(1 - \frac{\tanh A}{A}\right) \frac{2\pi}{s_0} \frac{\sin h_0 s}{h_0} + j\left(A \frac{\partial}{\partial A} \frac{\tanh A}{A}\right)\{\exp(-jh's)\cos(h_1 s) - \cos(h_0 s)\}\right], \tag{42}$$

where

$$h' = h_0^2 a \frac{\partial}{\partial A}\left(\frac{h_1}{h_0}\right)^2. \tag{43}$$

These results also indicate that the effect of debunching is much less marked in practical valves than eqns. (36) and (37) would indicate.

It is now time to examine the question of the maximum length which can be used in an electron beam. Eqns. (36), (38) and (41) all involve terms of the form $\sin hs/h$ which has its first maximum at $hs = \frac{1}{2}\pi$. Thus, it is no use making s longer than $\pi/2h$, where the generalized debunching factor h is properly defined for the geometry under discussion. With this condition inserted,

$$i_2 = 2I_0 J_1\left(\frac{S\alpha\beta}{\pi}\right). \tag{44}$$

We see then that, for optimum drift tube length, the effect of space-charge debunching is to reduce the bunching factor from $\frac{S\alpha\beta}{2}$ to $\frac{S\alpha\beta}{\pi}$. Whether this is serious or not depends on the circumstances. If electrostatic focusing is used, the length of the valve may well be much less than that prescribed by debunching, in order to obtain a reasonable current and power input. In the case of magnetic focusing, space-charge debunching may be a hindrance.

12.7. Velocity-modulated amplifiers or klystron amplifiers

We have now developed the general theory of velocity modulation far enough to allow us to calculate the performance of some complete valves with a considerable degree of confidence that the results will not be seriously in error, in spite of the approximate nature of the analysis. The first specific example we take up is the ordinary klystron amplifier, as shown in fig. 12.1. Before calculating the performance of this device it should be explained that klystron amplifiers have a certain practical importance at the present time. They are not useful as low level (signal) amplifiers because they produce too much noise, so that a system using a klystron radio-frequency amplifier preceding a crystal mixer actually gives a worse noise factor than a crystal alone. For medium power outputs, of the order of a few watts, travelling wave tubes have much greater bandwidth, and some special triodes are also useful at wave-lengths above 6 cm. (e.g. Bell 1553). But for powers higher than this, the klystron has the great advantage that the structure is inherently capable of handling large powers and, as we shall see, the higher the power input the better the performance becomes. Apart from possible uses in communication systems it seems likely that klystrons may be used to provide power for driving microwave linear accelerators and similar particle accelerators because they provide stability of frequency against load changes, a feature noticeably lacking in systems relying on magnetrons. In the analysis that follows we shall be mainly concerned with communication aspects of klystron operation.

We assume that the resonators are identical with parameters $\omega_0, Q_0, R_c, Q_0 = \omega_0 CR_c$. The drift length is s, normalized to S radians. The depths of modulation at the gaps are $\alpha_1 = V_1/V_0$, $\alpha_2 = V_2/V_0$, α_2 being the output gap. Furthermore, we suppose that the input resonator is correctly matched to the input transmission line. When this is the case, the loaded Q of the buncher resonator is $Q_L = \frac{1}{2}Q_1$, Q_1 being the resonator Q measured when the beam is switched on.

The theory has to be based on observable quantities, in this case the input and output powers, since there is no way of measuring voltages and currents at the frequencies considered.

The input power W_1 produces a peak voltage V_1 across the resonator gap and across the loss conductance representing the electron beam. The resonator term is simply $V_1^2/2R_c$, while the second term is given by eqn. (27). Therefore

$$W_1 = \frac{V_1^2}{2R_c} + \frac{I_0 V_0}{4}\alpha_1^2 \beta\gamma. \tag{45}$$

We can arrange this as

$$\left.\begin{aligned} W_1 &= (\alpha_1\beta)^2 \frac{I_0 V_0}{8} P, \\ P &= \frac{4V_0}{I_0 \beta^2 R_c} + \frac{2\gamma}{\beta}. \end{aligned}\right\} \tag{46}$$

The available output power is given using eqns. (14) and (27) as

$$W_2 = I_0 V_0 \left[\alpha_2 \beta J_1\left(\frac{S\alpha_1\beta}{2}\right) - (\alpha_2\beta)^2 \frac{P}{8}\right]. \tag{47}$$

The phase term does not appear because the output phase adjusts itself so as to extract maximum power from the beam.

We now consider the maximum power output which can be obtained (*a*) by increasing the excitation, and (*b*) by correctly adjusting the coupling to the load.

Maximizing eqn. (47) by putting $\partial W_2/\partial\alpha_1 = 0$ yields

$$J_1'\left(\frac{S\alpha_1\beta}{2}\right) = 0 \quad \text{or} \quad \frac{S\alpha_1\beta}{2} = 1{\cdot}84,$$

i.e. the value of the bunching parameter for which the radio-frequency current at the catcher gap is a maximum. Using $\dfrac{\partial W_2}{\partial\alpha_2} = 0$ gives

$$P = 4\frac{J_1(\frac{1}{2}S\alpha_1\beta)}{\alpha_2\beta},$$

therefore

$$W_2 = \frac{I_0V_0}{2}\left[\alpha_2\beta J_1\left(\frac{S\alpha_1\beta}{2}\right)\right]. \tag{48}$$

Putting the optimum bunching parameter into this we get

$$\alpha_2\beta_{\text{opt.}} = \frac{2{\cdot}336}{P}.$$

Inserting this and the other expressions into eqn. (47) gives

$$W_{2\,\text{max.}} = \frac{0{\cdot}68I_0V_0}{P}, \tag{49}$$

or

$$\eta_{0\,\text{max.}} = \frac{0{\cdot}68}{P}. \tag{50}$$

Eqns. (48) and (49) give the power output and efficiency under optimum conditions of circuit and drive adjustment. Now let us consider the gain. First, consider that the output coupling has been correctly adjusted for maximum power output at some arbitrary drive. Substituting for $\alpha_2\beta$ in terms of P in eqn. (48) we get

$$W_2 = 2\frac{I_0V_0}{P}J_1^2\left(\frac{S\alpha_1\beta}{2}\right), \tag{51}$$

or

$$G_p = \frac{W_2}{W_1} = \frac{16J_1^2(\frac{1}{2}S\alpha_1\beta)}{P^2(\alpha_1\beta)^2}. \tag{52}$$

The maximum power gain is obtained for small driving power. Putting in the asymptotic value of J_1 the small signal gain turns out to be

$$G_{p\,\text{max.}} = \left(\frac{S}{P}\right)^2. \tag{53}$$

We see that P must be small for large gain and large power output. Thus the beam impedance must be made as low as possible to obtain a good valve.

The power gain at the drive giving maximum output can be written down using eqns. (46), (48) and $(\frac{1}{2}S\alpha_1\beta) = 1{\cdot}84$. The value is

$$G_{p\,\text{opt.}} = 0{\cdot}402(S/P)^2. \tag{54}$$

The gain at optimum drive is then about 4 db. below the maximum gain.

The next design parameter we need to calculate is the load resistance into which the maximum power can be coupled. By a simple transformation of eqn. (48) we find

$$W_2 = \frac{I_0 V_0}{8}(\alpha_2\beta)^2 P.$$

If R_L = optimum load resistance,

$$W_2 = \frac{V_2^2}{2R_L},$$

therefore

$$R_L = \frac{4V_0}{I_0\beta^2 P}. \tag{55}$$

This can be written in a much more easily recognized way if we write P as follows, using the beam loss defined by eqn. (28):

$$P = \frac{4V_0}{I_0\beta^2}\left(\frac{1}{R_c}+\frac{1}{R_b}\right),$$

$$R_L = \left(\frac{1}{R_c}+\frac{1}{R_b}\right)^{-1},$$

or

$$G_L = G_c + G_b. \tag{56}$$

Eqn. (56) shows that, as one would expect, the maximum power output is obtained when the load is equal to the parallel combination of the beam-loss resistance and the circuit-loss resistance. As in the case of the input resonator, the loaded Q of the catcher is half the Q measured with no load, but the beam switched on.

Finally, we can derive some interesting results involving the band-width, a parameter which is of prime importance in communication problems. By definition

$$Q_1 = \frac{\omega_0 C}{G_c + G_b} \quad \text{and} \quad Q_L = \frac{\omega_0 C}{2(G_c + G_b)}$$

for matching, therefore

$$P = \frac{2V_0 \omega_0 C}{I_0 \beta^2 Q_L}. \tag{57}$$

Now Q_L is equal to $\omega_0/\Delta\omega$, where $\Delta\omega$ is the band-width between the -3 db. points of the input or the output stage alone. As is well known, the band-width of n similar stages tuned to the same frequency is $(2^{1/n}-1)^{\frac{1}{2}}$ times the band-width of a single stage. For the klystron, $n = 2$ and the factor is 0·643. If we now write $\Delta\omega_2$ for the overall band-width of our klystron.

$$P = \frac{3{\cdot}11\, V_0 C \,\Delta\omega_2}{I_0 \beta^2}. \tag{58}$$

Using eqn. (58) the maximum power gain becomes

$$G_p = \left(\frac{S I_0 \beta^2}{3{\cdot}11\, V_0 C\, \Delta\omega_2}\right)^2, \tag{59}$$

or the voltage gain × band-width product, since the amplifier working between equal impedances is

$$a\,\Delta\omega_2 = \frac{S I_0 \beta^2}{3{\cdot}11\, V_0 C} \tag{60 a}$$

$$\propto g_{ms}/C \tag{60 b}$$

Eqn. (59) takes no account of space-charge debunching; to do this all that is necessary is to replace S by $(S \sin hs/hs)$, or, at the optimum drift tube length, by $2S/\pi$. In discussing eqn. (59) it is important to remember that the parameters on the right-hand side are not independent of one another. For space-charge limited guns $I_0 \propto V_0^{\frac{3}{2}}$ so the gain × band-width product is directly proportional to $V_0^{\frac{1}{2}}$. Again β^2 and C are not independent. If we assume that the resonator is designed with nearly all its capacitance at the gap (which is clearly a desirable state of affairs) and the gap is closed by circular grids of radius r, the gap spacing being d,

$$C = \frac{1{\cdot}11r^2}{4d}\,pF \text{ and } \beta = \frac{2\sin \omega d/2u_0}{\omega d/u_0},$$

therefore
$$\frac{\beta^2}{C} \propto \frac{\sin^2 \omega d/2u_0}{d} \tag{61}$$

The optimum spacing is then given by the solution of $\phi = \tan \frac{1}{2}\phi$, where $\phi = \dfrac{\omega d}{u_0}$,

therefore
$$\phi_{\text{opt.}} \doteqdot 2{\cdot}33\, rad. \tag{62}$$

Finally, the perveance (the constant in $I_0 = kV_0^{\frac{3}{2}}$) depends on the gap dimensions and therefore on the gap capacitance. The exact dependence is different for the different types of focusing. With these considerations in mind the discussion of eqn. (59) can be carried considerably farther once a specific structure has been decided on, but we shall not do this here.

12.8. The reflex klystron. (1) Bunching theory

The most important type of velocity-modulated tube at the present time is the reflex klystron. These are used as local oscillators in microwave superheterodynes and as frequency-modulated transmitters in low- and medium-power microwave communication systems. This pre-eminent position is due to several causes, among which simplicity, ease of obtaining fairly large deviation frequency modulation, and the absence of focusing magnets are the most important.

In the reflex klystron the bunching conditions are very slightly different from those which we have discussed so far. If we restrict ourselves, for a moment, to systems in which the negative field between the second grid and the reflector electrode is a linear function of distance, it is easily seen that the electrons which are accelerated travel farther and take longer to return to the gap than do those which were decelerated. Thus, the centre of the bunch in a reflex is the unmodulated electron which passed the gap at the time the field changed from accelerating to decelerating, i.e. there is a phase difference of π with respect to the two-resonator klystron. On the other hand, the direction of electron motion is reversed by the reflector so that on the return journey a field in the positive (initially accelerating) direction extracts energy from the beam and a second phase difference of π is introduced. The relations are shown in fig. 12.8, and it is obvious that oscillations appear for $(4n+3)\pi/2 = S$ as before. If we now write down the velocity after the second grid, for the electron which has the shortest

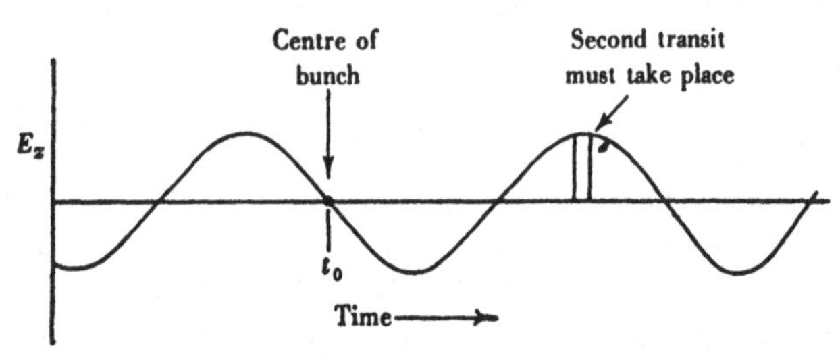

Fig. 12.8. Field relations in the reflex klystron.

transit, as in § 12.2, we have

$$u_1 = u_0[1-(\alpha\beta/2)\sin\omega t_0]. \tag{63}$$

The transit time in the retarding field is

$$\tau = \frac{2u_1 ml}{e(V_0-V_r)},$$

where $\quad l$ = grid-reflector spacing,

$\qquad V_r$ = reflector voltage (usually negative).

Now if $\tau_0 = \dfrac{2u_0 ml}{e(V_0-V_r)}$ = transit time of unmodulated electron,

$$\begin{aligned} t_2 &= t_0+\tau \\ &= t_0+\tau_0[1-(\alpha\beta/2)\sin\omega t_0], \end{aligned} \tag{64}$$

and

$$\frac{dt_2}{dt_0} = 1-\frac{S\alpha\beta}{2}\cos\omega t_0, \tag{65}$$

where S = total transit angle in the reflecting field. Since eqn. (64) is exactly the same as the corresponding equation in § 12.2 the analysis developed in that section applies also to the reflex klystron. It should be mentioned at this point that if the electrons drift through an angle θ_d before entering a retarding field in which the transit angle is θ_r, S in eqn. (65) is $(\theta_r - 2\theta_d)$, but S in the phase equation is $(\theta_r + 2\theta_d)$. This is because the two types of bunching are opposed to one another.

In general, the reflector field is not a linear function of distance from the second grid, so we next extend our analysis to cover this case. The modification has considerable importance, as several characteristics are changed by its introduction. We repeat the analysis given in Beck.† The radio-frequency current is given by $i_2 = I_0 dt_2/dt_0$ as usual.

But $$t_2 = t_0 + \tau,$$

therefore $$i_2 = \frac{I_0}{1 + d\tau/dt_0}. \tag{66}$$

If the amplitude of the h.f. voltage on the gap is V_1 we can write $\tau = f(V_1)$ and expand, using Maclaurin's theorem. Therefore

$$\tau = f(0) + V_1(f'(V_1))_0 + \frac{V_1^2}{2}(f''(V_1))_0 + O(V_1^3)$$

$$\doteqdot f(0) + V_1(f'(V_1))_0$$

if we neglect all terms of $O(V_1^2)$,

therefore $$\tau = \tau_0 + V_1\left(\frac{d\tau}{dV_1}\right)_0,$$

and $$\frac{d\tau}{dt_0} = \frac{dV_1}{dt_0}\left(\frac{d\tau}{dV_1}\right)_0. \tag{67}$$

Since V_1 is the voltage actually modulating the beam, it is β times the voltage on the circuit. Therefore

$$\frac{dV_1}{dt_0} = \omega\beta V_c \cos\omega t_0. \tag{68}$$

Inserting eqns. (68) and (67) into eqn. (66) we get

$$i_2 = \frac{I_0}{1 + \omega\beta V_c \cos\omega t_0\,(d\tau/dV_1)_0}. \tag{69}$$

† Beck, op. cit. pp. 119–22.

We next determine $(d\tau/dV_1)_0$ as follows. If the reflector field is some function of z, say $\phi(z)$, we can write the following equation of motion for an electron which gained energy V volts in the gap,

$$\tau = 2\left(\frac{2e}{m}\right)^{-\frac{1}{2}} \int_{z_1}^{z_2} (\phi(z)+V)^{-\frac{1}{2}}\,dz, \tag{70}$$

where z_1, z_2 = coordinates of gap centre and point of turn around. If this is written in dimensionless quantities using $Q = \phi(z)/V_0$, $q = V/V_0$, $Z = z/l_1$, $\Delta Z = \Delta_z/l_1$, where l_1 = distance from gap centre to the zero equipotential,

$$\tau = 2\left(\frac{2e}{m}\right)^{-\frac{1}{2}} V_0^{-\frac{1}{2}} l_1 \int_{-\Delta Z}^{l_1} (Q+q)^{-\frac{1}{2}}\,dZ,$$

and
$$\tau_0 = 2\left(\frac{2e}{m}\right)^{-\frac{1}{2}} V_0^{-\frac{1}{2}} l_1 \int_{0}^{l_1} Q^{-\frac{1}{2}}\,dZ.$$

Then
$$\left(\frac{d\tau}{dV_1}\right)_0 = 2\left(\frac{2e}{m}\right)^{-\frac{1}{2}} V_0^{-\frac{3}{2}} l_1 \frac{d}{dq}\left[\int_{-\Delta Z}^{l_1} (Q+q)^{-\frac{1}{2}}\,dZ\right]. \tag{71}$$

Using eqn. (71)

$$\left(\frac{d\tau}{dV_1}\right)_0 = \frac{\tau_0}{V_0} \frac{\dfrac{d}{dq}\left[\displaystyle\int_{-\Delta Z}^{l_1} (Q+q)^{-\frac{1}{2}}\,dZ\right]}{\displaystyle\int_{0}^{l_1} Q^{-\frac{1}{2}}\,dZ} = \frac{\tau_0}{V_0} R. \tag{72}$$

R is called the reflector factor and is a numerical constant which can be determined graphically from electrolytic trough measurements. For a linear field, long in comparison with the gap, $R = \frac{1}{2}$. For this value, substituting eqn. (72) into (69) yields the ordinary Webster expression for i_2, but in general the bunching factor is $(RS\alpha\beta)$ instead of $(\frac{1}{2}S\alpha\beta)$. Since $R > \frac{1}{2}$ for many common fields, the radio-frequency current in a reflex reaches its maximum for lower gap voltages than in an ordinary klystron. In other words, the transconductance is bigger by the factor of $2R$.

12.9. The reflex klystron. (2) Power output and efficiency

We now study the power production in the reflex klystron. Since the beam traverses the same gap twice, $V_1 = V_2$ and the term due to diode losses has to be doubled. Using eqns. (14) and (27) the conversion efficiency is

$$\eta_c = -\tfrac{1}{2}\beta\gamma\alpha^2 - \alpha\beta J_1(RS\alpha\beta)\sin S, \tag{73}$$

$$S \doteqdot (n+\tfrac{3}{4})\,2\pi.$$

With this value of S, $\sin S$ is negative, so the second term represents power delivered to the resonator. The output efficiency is

$$\eta_0 = \alpha\beta J_1(RS\alpha\beta) - \frac{\alpha^2}{2}\left(\frac{V_0}{I_0 R_c} + \beta\gamma\right). \tag{74}$$

We define P_R as

$$\frac{2V_0}{I_0 R_c \beta^2} + \frac{2\gamma}{\beta}, \tag{75}$$

therefore
$$\eta_0 = -\alpha\beta J_1(RS\alpha\beta)\sin S - \left(\frac{\alpha\beta}{2}\right)^2 P_R. \tag{76}$$

The next step is to optimize η_0 with respect to S, α and R. This means physically that we choose the best mode and the best adjustment of reflector voltage in the mode, the correct coupling to the load and the best shape for the reflector field. To save writing we put $\alpha\beta = \alpha'$ and $RS\alpha' = \theta$. Then

$$\frac{\partial \eta_0}{\partial S} = -\alpha' J_1(\theta)\cos S - \alpha' \sin S\left[\frac{J_1(\theta) - \theta J_2(\theta)}{\theta}\right] R\alpha'$$

$$= 0, \quad \text{for a maximum,}$$

therefore
$$S \cot S = \frac{J_2(\theta) - J_0(\theta)}{J_2(\theta) + J_0(\theta)}, \tag{77}$$

and
$$\frac{\partial \eta_0}{\partial \alpha} = -\beta J_1(\theta)\sin S - \alpha'[J_0(\theta) - J_2(\theta)]\frac{RS\beta}{2}\sin S - \alpha'\frac{\beta P_R}{2},$$

therefore
$$P_R = -2RS\sin S J_0(\theta),$$

using the identity
$$J_1(\theta) = \tfrac{1}{2}\theta(J_0(\theta) + J_2(\theta)). \tag{78}$$

Finally,
$$\frac{\partial \eta_0}{\partial R} = -\alpha' J_1'(\theta) S\alpha' \sin S,$$

therefore
$$\theta = 1{\cdot}84,$$

and
$$R_{\text{opt.}} = \frac{1{\cdot}84}{S\alpha'}. \tag{79}$$

If we put eqn. (78) back into eqn. (76) we obtain the following expression for the output efficiency,

$$\eta_0 = -\frac{\alpha'}{2}\sin S\theta J_2(\theta)$$

$$= -\frac{\theta^2}{2R}J_2(\theta)\frac{\sin S}{S}. \tag{80}$$

The numerical technique for handling these rather involved equations is fairly straightforward. Eqn. (77) is solved graphically and yields a set of pairs of (S, θ) values, all very close to the central values calculated from the mode number. The pairs of (S, θ) are then inserted into eqn. (78) to give the corresponding value of P_R. Then the same (S, θ) values are put into eqn. (80) to give η_0. This can be done for several fixed R values, so that, finally, we can plot a family of η_0, P_R curves with R as parameter for each mode. P_R only depends on the geometry of the valve; so, once this has been done, the optimum output from any valve can be calculated given the current, voltage, gap dimensions and resonator shunt resistance.

An important parameter in reflex klystron design is the 'starting current'. This is the value of beam current which is just large enough to produce an oscillation of vanishingly small amplitude in the circuit. It can be found by putting $\eta_0 = 0$ and $\sin S = -1$ in eqn. (76). Therefore

$$\alpha'\frac{RS\alpha'}{2} = \left(\frac{\alpha'}{2}\right)^2 P_R,$$

or

$$P_R = 2RS.$$

Neglecting the term $2\gamma/\beta$ in P_R, we get

$$I_{0s} \geqslant \frac{V_0}{\beta^2 R_c SR}. \tag{81}$$

From eqns. (80) and (81) we see that the efficiency and starting current are inversely proportional to R. This means that a valve designed for very low-power output and minimum input should use a reflector with a large R value, while an efficient valve for medium-power output should have a small R reflector.

We shall discuss the design considerations somewhat more fully after the theory of the frequency pulling has been given.

12.10. The reflex klystron. (3) Frequency pulling

It has now proved possible to give a much more detailed theory of the frequency pulling than was written down in our earlier work. This account given below follows an unpublished S.T.C. report.

Near any given resonant frequency a cavity resonator can be considered as equivalent to a high Q parallel resonant circuit. The admittance of such a circuit is (fig. 12.9)

$$Y_c = G_c + j\omega_0 C\left(\frac{\omega}{\omega_0} - \frac{\omega_0}{\omega}\right)$$

$$\doteqdot G_c + 2jC\Delta\omega. \qquad (82)$$

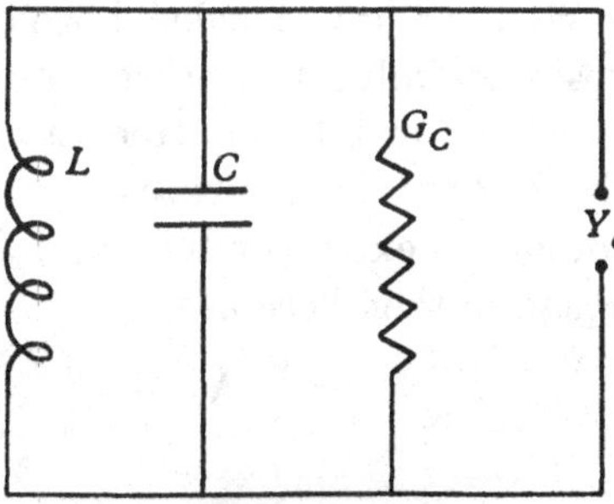

Fig. 12.9. Resonator equivalent circuit.

Eqn. (82) is correct for all derivations encountered in practical cases. We now wish to find out what happens to the oscillation frequency of this circuit when the electronic admittance Y_e is connected to the terminals. Y_e is, in general, complex, and is controlled by the reflector voltage. Rearranging eqns. (20) and (21) for the reflex

$$Y_e = 2\beta^2 RS \frac{I_0}{V_0} \frac{J_1(\theta)}{\theta} \exp\left[-j(S - \tfrac{1}{2}\pi)\right]. \qquad (83)$$

Now, the condition for oscillation is that $Y_e = -Y_c$.

If we let $S = (4n+3)\pi/2 + \delta$,

$$Y_e = Y_{0R} \frac{J_1(\theta)}{\theta} (\cos\delta - j\sin\delta), \qquad (84)$$

where $$Y_{0R} = 2\beta^2 RS \frac{I_0}{V_0},$$

constant for a given valve and mode number. Equating real

and imaginary parts of Y_e and $-Y_c$ from eqns. (84) and (82) we get

$$G_c = Y_{0R}\frac{J_1(\theta)}{\theta}\cos\delta,$$

$$\Delta\omega = -\frac{Y_{0R}}{2C}\frac{J_1(\theta)}{\theta}\sin\delta,$$

or
$$\Delta\omega = -\frac{G_c}{2C}\tan\delta.$$

We have not considered the existence of a load. Let us suppose the valve is worked into either a well-matched broad-band load or into an attenuator of about 10 db. In either case the load will be non-reactive, and we can take account of it by adding an extra conductance, G_L, across the circuit. The last equation then becomes

$$\Delta\omega = -\left(\frac{G_c+G_L}{2C}\right)\tan\delta,$$

or
$$\frac{\Delta\omega}{\omega_0} = \frac{\Delta f}{f_0} = -\frac{1}{2Q_L}\tan\delta. \qquad (85)\dagger$$

Since it may be assumed, at any rate for small δ, that $\delta \propto V_R$, eqn. (85) indicates that the curve of frequency derivation versus reflector voltage is roughly a tangent curve. It also indicates that the loaded Q must be small for a large frequency deviation and reasonable linearity. It does not, however, connect the frequency pulling with the valve parameters, and with a self-oscillator such as the reflex there is clearly a maximum load, and minimum Q, beyond which the valve will fail to oscillate. We next calculate the maximum value of frequency deviation, i.e. the value at which the power output falls to zero. At these points

$$\delta = \pm\delta_1, \quad \frac{J_1(\theta)}{\theta} \to \frac{1}{2},$$

therefore
$$G_L + G_c = \frac{Y_{0R}}{2}\cos\delta_1.$$

† Eqn. (85) shows that increasing the transit time by lowering the reflection voltage decreases the frequency. We should expect this since the time separation between bunches becomes greater, pulling the resonator to a lower frequency.

Using the relation $\tan^2\delta_1 = \sec^2\delta_1 - 1$, we find

$$\Delta\omega_1 = \pm\frac{1}{4C}\sqrt{[Y_{0R}^2 - 4(G_L+G_c)^2]}. \tag{86}$$

Eqn. (86) still requires experimental knowledge of G_L before it can be used.

Now we turn to the question of calculating the frequency pulling from the valve parameters alone. We first ask the question, what will be the frequency deviation if we adjust the valve to two points on either side of the maximum power such that the new power is a fraction n^2 of the original value (n^2 in practice is usually $\frac{1}{2}$, i.e. -3 db.)? Remembering that the voltage amplitude at the gap varies as $\sqrt{\text{power}}$, we have

$$\left.\begin{aligned} G_L+G_c &= Y_{0R}\frac{J_1(n\theta_1)}{n\theta_1}\cos\delta_n, \\ C\,\Delta\omega_n &= Y_{0R}\frac{J_1(n\theta_1)}{n\theta_1}\sin\delta_n. \end{aligned}\right\} \tag{87}$$

where θ_1 = value of θ at the centre of the mode. Squaring, adding and inserting the central value for (G_L+G_c) we get

$$\Delta\omega_n = \frac{Y_{0R}}{C}\sqrt{\left(\frac{J_1^2(n\theta_1)}{(n\theta_1)^2} - \frac{J_1^2(\theta_1)}{\theta_1^2}\right)},$$

or

$$\frac{\Delta\omega_n}{\omega_0} = \frac{\Delta f_n}{f_0} = \frac{Y_0}{\omega_0 C}\sqrt{\left(\frac{J_1^2(n\theta_1)}{(n\theta_1)^2} - \frac{J_1^2(\theta_1)}{\theta_1^2}\right)}. \tag{88}$$

We now insert the unloaded Q of the resonator, $Q_0 = \omega C/G_c$ and a quantity which we may call the working Q, $Q_w = \omega_0/\Delta\omega_n$. Using these, together with P_R but neglecting the beam loading term in P_R we obtain

$$\frac{Q_0}{Q_w} = \frac{4RS}{P_R}\sqrt{\left(\frac{J_1^2(n\theta_1)}{(n\theta_1)^2} - \frac{J_1^2(\theta_1)}{\theta_1^2}\right)}. \tag{89}$$

The right-hand side of eqn. (89) is a numerical factor, usually greater than unity, which gives the ratio of unloaded Q to working Q.

In the last section we deduced a relation (78) between R, S and P. Using this we finally derive

$$\frac{Q_0}{Q_w} = -\frac{2}{\sin S}\frac{1}{J_0(\theta_1)}\sqrt{\left(\frac{J_1^2(n\theta_1)}{(n\theta_1)^2} - \frac{J_1^2(\theta_1)}{\theta_1^2}\right)}. \tag{90}$$

This equation involves only the valve parameters, and not, explicitly, the load. Since this relation only involves S and θ

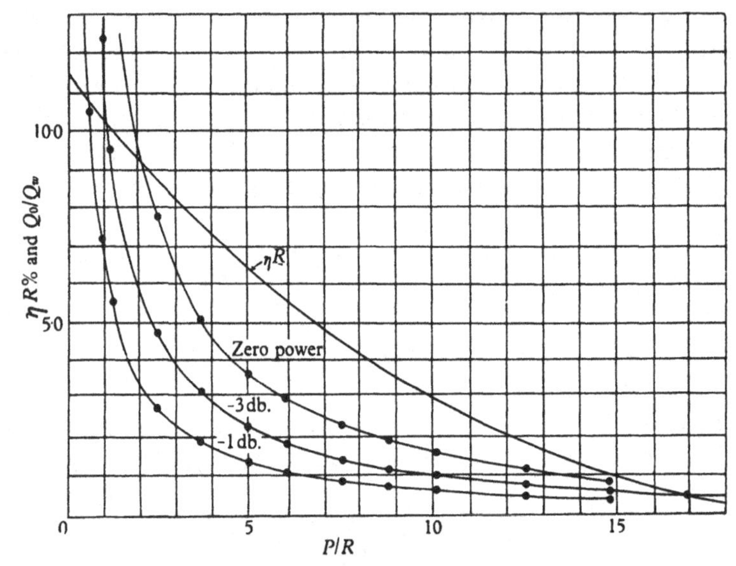

Fig. 12.10. Performance chart for reflex klystron, mode 7.

on the right-hand side, the pairs of (S, θ) already calculated can be inserted to obtain the values of Q_0/Q_w. Each value of Q_0/Q_w thus calculated is that corresponding to the value of P_R already found for the pair (S, θ). We can thus present the theory of power production and frequency pulling in a simple way by plotting $\eta_0 R$, and Q_0/Q_w against P/R for each mode. Figs. 12.10, 12.11 and 12.12 show these plots for modes 7, 11, 15 for the -1 db., -3 db. and zero power points.

In calculating these curves, it has been assumed that the frequency deviation versus reflector voltage curve is symmetrical, which is only true for $S \gg \delta$, since the electronic admittance varies as S. For low modes the curves will therefore

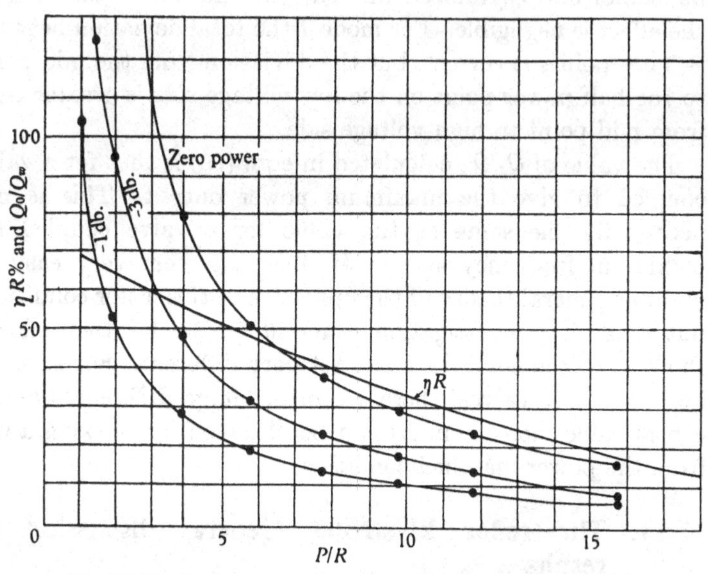

Fig. 12.11. Performance chart for reflex klystron, mode 11.

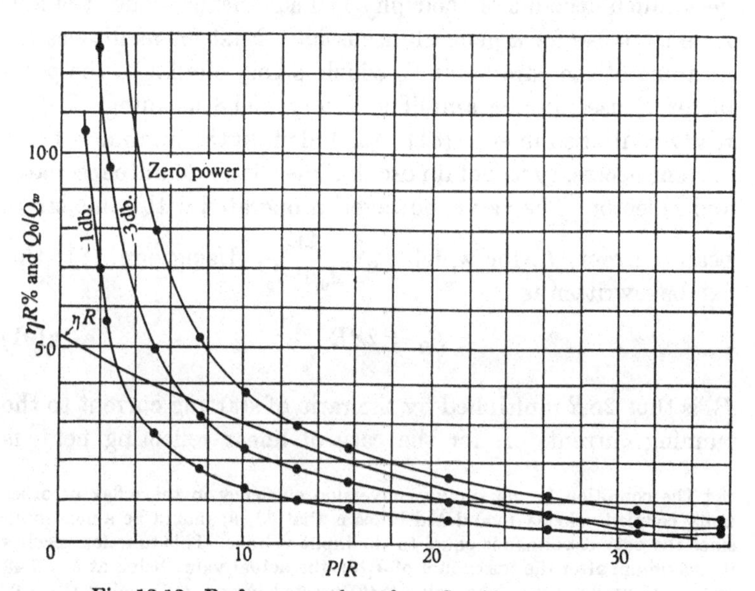

Fig. 12.12. Performance chart for reflex klystron, mode 15.

be somewhat asymmetrical, but for modes 11 and higher the effect is negligible. For mode 7 the total deviation between -3 db. points is correct, but the deviation from the mid-point to the half-power point on the low-voltage side is greater than from mid-point to high-voltage side.

The value of Q_0/Q_w calculated in eqn. (90) is that for a valve coupled to give the maximum power output. This is not necessarily the same as the value for a valve coupled for maximum frequency swing. It does not seem very easy to obtain a general theory of the operation in this latter condition, but numerical investigation round $\theta_{\text{max.}}$ (2·4 for the reflex)† shows that the conditions are not very different. For small θ, i.e. low-power valves, there are rather large differences and a considerable increase in Δf can be obtained by working away from the power-matched condition.

12.11. The reflex klystron. General discussion of results

We have derived several mathematical results without taking very much account of their physical significance, and it is now time to pause for a general discussion. First let us discuss the meaning of the parameter P_R which is very similar to the parameter P used in the amplifier theory. The meaning of P_R is really contained in eqn. (81), which defines the minimum beam current necessary to obtain oscillations with a definite resonator and reflector. The valve, however, is operated with some larger beam current, I_0, for which $P_R \doteqdot \dfrac{2V_0}{I_0\beta^2 R_c}$. Using eqn. (81) this can be rewritten as

$$P_R = 2SR\frac{I_{0S}}{I_0}. \tag{91}$$

P_R is thus $2SR$ multiplied by the ratio of starting current to the running current, or, for the case of linear reflecting field, is

† The condition for maximum conversion efficiency in the reflex or other unity coupled velocity-modulated tubes is that $\theta J_1(\theta)$ should be a maximum since the output voltage is equal to the input voltage. This function reaches its maximum after the maximum of $J_1(\theta)$, the actual value being at $\theta = 2{\cdot}43$ instead of at $\theta = 1{\cdot}84$. The value 2·43 inserted in eqn. (78) makes $P_R = 0$, i.e. $\theta = 2{\cdot}43$ also gives the maximum output efficiency.

simply S times this ratio. Since the graphs show that P_R must be small for an efficient valve, eqn. (91) tells us (*a*) that the starting current must be made as small as possible by correct resonator design, and (*b*) that the running current must be made as large as possible. The latter is usually much the more useful engineering approach.

The frequency pulling also increases with decreasing P_R, so that a valve designed for high efficiency will normally have a large frequency pulling. The physical reason for this is that the susceptance developed by the beam is directly proportional to the current, and so small P_R means large susceptance. In some cases it may be undesirable to have large frequency pulling; for instance, if the load is reactive and has to be tightly coupled to the valve. The frequency pulling can be reduced without any ill effect on the power output by increasing the gap capacitance.

The reflector factor R† also plays an important role in the operation of the valve. When the main requirement is to design a valve which will oscillate for the minimum input power, or at the highest possible frequency where small gap dimensions and low resonator shunt impedance make it difficult to obtain a low P_R value, it is advisable to use a reflector with a fairly large R as this reduces the starting current. On the other hand, high power valves should be worked with optimum R, firstly, because of the resulting improvement in efficiency, and secondly, because variations of input, electrode spacings, etc. from tube to tube introduce smaller variations in radio-frequency characteristics when R is optimized. If it is necessary to increase the frequency swing above the value corresponding with optimum R, a higher R value may be used. In the above we are inferring that the R value may be freely chosen but in fact this is not always the case because the requirement that the reflector should have the correct focal properties may be more important. We have also tacitly assumed that all the electrons follow the same path in the reflector. This may be far from true in practice, particularly when the beam diameter

† The calculation of R values is dealt with by Barford and Manifold, *J. Instn. Elect. Engrs.*, **94**, pt. III (1947), 302.

is a substantial proportion of the reflector diameter, in which case peripheral electrons may travel a longer or shorter time than axial electrons. This is a very undesirable effect because if the peripheral electrons return to the gap out of phase with the axial ones, they will reduce the available power or may even stop the tube from oscillating. If the axial electrons have a transit time τ_0 which is correct for the desired mode, all electrons with transit times $\tau_0 \pm 1/4f_0$ will contribute to the oscillation, or, in terms of transit angle, the angles must be in the range

$$S \pm \tfrac{1}{2}\pi. \tag{92}$$

The allowable percentage path difference therefore decreases with increasing S. Deviations of this type can cause serious difficulties, particularly in low-power oscillators at the maximum frequencies used to-day.

The reader is referred to a long paper by Pierce and Shepherd† for very much fuller information on the design of reflex klystrons, and for many details of the series of valves designed by the Bell Laboratories.

12.12. The reflex klystron. Hysteresis

The reflex klystron is subject to an objectionable hysteresis phenomenon which is sufficiently important to be shortly discussed. The observations are the following. Fig. 12.13 shows a typical power-reflector voltage characteristic with marked hysteresis. As the reflector is made more negative, the power output falls smoothly until point A is reached where the oscillation drops abruptly to zero at B. The reflector is taken to C and then made more positive; instead of oscillation starting at B no power output is obtained until D is reached, where the valve suddenly starts to oscillate with a large amplitude E. The phenomenon may be observed at either end, or both ends, of the characteristic, and the oscillation may drop smoothly to zero in one direction but jump violently in the other. Phenomena of this type are particularly objectionable in radar systems which use the reflector characteristic for search automatic frequency control.

† Pierce and Shepherd, *Bell Syst. Tech. J.* **26** (1947), 460.

Hysteresis can be of two general types, either due to the valve alone or due to the combination of the valve with an unsuitable circuit. Here we shall only concern ourselves with hysteresis which cannot be eliminated even when the valve works into a well-matched resistive load, and is therefore a fundamental property of the valve. Hysteresis of this type is due to unwanted phenomena of various kinds which all produce the same effect, viz. a variation of electronic admittance with

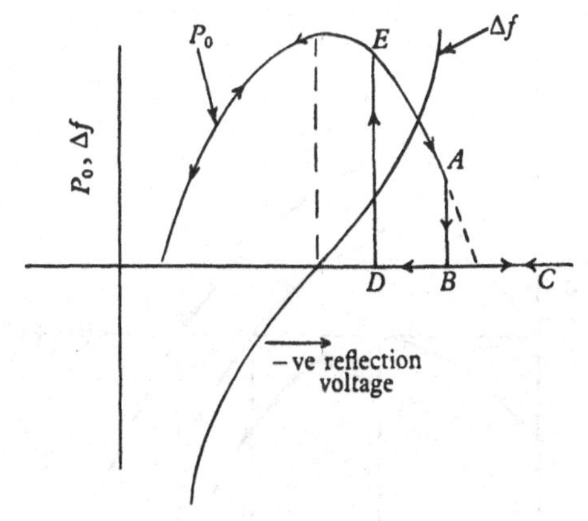

Fig. 12.13. Illustrating output curve for valve with hysteresis.

amplitude which is not a monotonic decreasing function of the amplitude as it is in the ideal theory. Consider that the curve of Y_e/Y_0 has a maximum for some value of V_1 different from zero. Fig. 12.14 shows such a curve contrasted with the dotted curve which represents $J_1(\theta)/\theta$. Suppose also that at the centre voltage of the mode the valve oscillates sufficiently strongly for the operating point to be on the high side of the maximum, as is shown by the intercept with the horizontal line representing the negative of the total circuit conductance. The reflector is made more negative, thereby transposing the whole curve vertically downwards, since we can consider I_R as having changed to $I_R \cos \delta$. Oscillation continues until the maximum of the curve just touches $-(G_L + G_c)$, i.e. to $V_R = V_B$ in fig. 12.14. When the reflector voltage is increased again it is not possible for the oscillation to build up for any voltage below $V_R = V_D$, the value at which the zero amplitude admittance is equal to $-(G_L+G_c)$, because there is a net circuit damping until the latter value is reached. When the valve does start to oscillate, the amplitude immediately builds up to a considerable value, proportional to $\sqrt{P_E}$.

We have now seen that a curve of the type shown must give rise to hysteresis, but we have not said how such a characteristic can be caused. In fact, this type of characteristic can be

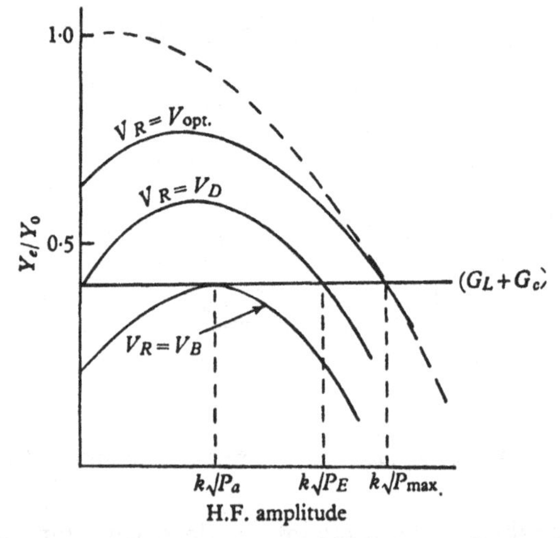

Fig. 12.14. Admittance curves which can cause hysteresis.

caused in several ways. One way, which has been rather fully treated by Pierce and Shepherd,† is the case of valves in which some of the electrons focused back into the cathode-anode space are not collected on cathode or screen and return through the valve to make a second, or even multiple, transit through the gap. Making reasonable assumptions these authors show that the postulated type of curve results. In some valves hysteresis was cured by reducing or eliminating the possibility of multiple transits. It should not be thought, however, that hysteresis is invariably due to multiple transits. Another type of defect which can cause the same trouble is the following.

† Pierce and Shepherd, loc. cit.

If the reflector is so designed that the current focused back is a rapid function of reflector voltage in the working region, the same thing can result. Fig. 12.15 shows a curve of working or running current (the current which makes a double transit through the gap) for such a valve. V_m denotes the mode centre. For increasing V_R and small amplitudes of oscillation the working current falls off according to the curve of fig. 12.15; but when the oscillation amplitude is large, the working current is

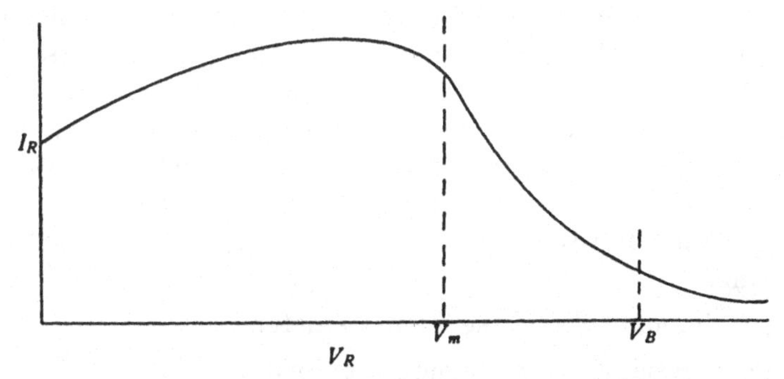

Fig. 12.15. Reflector characteristic causing hysteresis.

given by a dynamic curve which can be prepared from fig. 12.15. It is clear, for instance, that at V_B working current will increase with the amplitude of the oscillation and thus give rise to a curve of the type causing hysteresis. In this case it will be seen that the curve of Y_e/Y_0 against V_1 and V_R changes its shape as well as its scale.

Yet another cause is found in some valves in which the cathode current changes as the valve goes into and out of oscillation. If the electron optics are such that the number of electrons getting back into the cathode space decreases suddenly when oscillation starts, the cathode current will increase and will stay at the higher value until oscillation stops. Oscillation will then not start at the same V_R value but at some lower value giving a higher Y_e for the reduced total current.

There are, no doubt, other causes of hysteresis in reflex klystrons. Those described above have all been observed, and it is usually fairly simple to eliminate the defect once the physics of the particular problem have been elucidated. It is not always easy to find out which phenomenon is the cause of the trouble, and remedies designed to prevent one sort of hysteresis sometimes introduce others.

12.13. Noise in klystron amplifiers

To conclude our brief description of velocity-modulation tubes we shall study the noise produced by a gridded klystron amplifier. Klystrons are 'noisy', and as such are useless as input amplifiers, except for special purposes where noise is immaterial. The fundamental reason for the noisiness of klystrons lies in the expression for mutual conductance derived in § 12.3. The discussion there shows that the current required to obtain a useful transconductance is very much higher than in a triode, and therefore the shot noise is very much more important. The sources of noise in the klystron are the following:

(1) thermal noise in the input resonator,

(2) thermal noise in the output resonator,

(3) shot noise induced into the input resonator by the beam and amplified together with the signal.

(4) shot noise in the output resonator.

Since an amplifier is of little practical value unless it has a gain of 10 db. or more, it is permissible to neglect items (2) and (4) in what follows. Let us suppose that the input resonator is matched to an aerial of effective temperature T_a and that the transformed aerial conductance at the terminals of the gap is G_a. The circuit conductance is G_c and the beam conductance is G_b. Then, for a matched aerial,

$$G_a = G_c + G_b.$$

Then the total available noise power in the first resonator can easily be written down. It is

$$kT_a \Delta f + kT \Delta f + \tfrac{1}{2}\beta^2 \frac{2eI_0 \Gamma^2 \Delta f}{2(G_c + G_b)}. \tag{93}$$

The noise power available from the aerial at room temperature is $kT\Delta f$.

The noise factor of the klystron is then (93) divided by $kT\Delta f$, or

$$\text{N.f.} = 1 + \frac{T_a}{T} + \frac{\beta^2 I_0 \Gamma^2}{(G_c + G_b)} \frac{e}{2kT}. \tag{94}$$

At $T = 290° \text{K.}, \quad \frac{e}{2kT} = 20 \frac{20}{\text{V}}.,$

therefore

$$\text{N.f.} = 1 + \frac{T_a}{T} + \frac{20\beta^2 I_0 \Gamma^2}{(G_c + G_b)}. \tag{95}$$

It is clear that there is very little one can do about eqn. (95). If I_0 and β^2 are made small and $(G_c + G_b)$ large, the noise will be small but the gain will also be minimized. The best possible amplifier klystron from the noise point of view is one with a long low-current beam. Such a valve has a very low output and efficiency, but these quantities are not of much importance in this application. Successful valves of this type have not been made.†

Three resonator klystrons can be made to have much bigger amplifications than two resonator tubes; but the central problem still exists, since the input noise is still the dominating factor. Measurements made by Barlow‡ show that a three-resonator amplifier had a minimum noise figure of 26 db. against a crystal figure of 19 db. at 3000 Mc./sec. (This is a very high figure for the crystal; 7–10 db. would be more normal.) This result indicates the general order of deterioration of noise figure found in practice.

† Appendix 3 discusses recent work which gives information on practical methods of minimizing Γ. If these results are used, it seems likely that klystrons with noise factors less than 10 db. can be built.

‡ Quoted in Hamilton *et al.*, M.I.T. series, vol. 7, p. 259.

Chapter 13

TRIODES AT ULTRA-HIGH FREQUENCIES

We are now in a position to inquire more deeply into the behaviour of triodes at frequencies where the transit angles are no longer small in comparison with a half-cycle of the impressed voltage. Triodes will amplify, and therefore oscillate, in such conditions, but not with the same orders of power output and efficiency observed at lower frequencies. At the present time theory is not sufficiently advanced for us to explain the complete operation of the triode, particularly when conditions nearly corresponding to those of class C operation at low frequencies are reached, but by use of the ideas developed in Chapters 7 and 12 we can give a fairly adequate model.

Before discussing the more theoretical parts of the subject we shall say a few words about the tube structures and circuits used. Continuous-wave tubes for frequencies above 1000 Mc./sec. are, with very few exceptions, made with planar geometry. When the cylindrical geometry is used the ratio of grid diameter to cathode diameter is small enough to allow us to use planar expressions. Several types of construction are in use, but all agree in mounting the grid on a metal disk sealed through the glass wall. Disks of copper, 50 : 50 nickel iron or of kovar are used, and the General Electric Co. use silver-plated iron. The planar cathode is mounted very close to the grid and the anode at a distance several times as great. The anode presents a simple problem; but the cathode, which has to be in good electrical contact with the rest of the cathode circuit but should be thermally insulated, is much more awkward, and much ingenuity has been applied to the design of suitable structures. In addition to the difficulties mentioned, the thermal expansion of the cathode after switching on may materially reduce the cathode-grid spacing, and other variables, such as the shrinking of the cathode coating on activation, must also be considered.

Finally, it is usual to find considerable difficulty in activating cathodes which are placed very close to a close-meshed grid. Other problems arise in connexion with the grid. This must be tensioned enough to ensure that it does not buckle when hot, but not so much that the wires break. Grid emission is likely to be a problem, as the barium contaminated grid inevitably runs hot. In some cases grids grow well-defined insulating or semiconducting layers which finally render the grid almost ineffective. Notwithstanding these technical difficulties, the last ten years have seen enormous advances in the design of triodes, and they are now useful at frequencies up to 4000 Mc./sec. Typical ultra-high-frequency triodes are the American General Electric Co. 'Lighthouse' series, the best known of which are the 2C39 and 2C43, and among English valves the General Electric Co. series ranging from the small CV90 and near derivatives up to the ACT23 capable of delivering 200 W. at 1000 Mc./sec.

Turning to the circuit, it is important to note that triode amplifiers for the frequencies under discussion are of the grounded grid, otherwise known as grid separation, type. As is well known this circuit has the following properties.

(1) It is much easier to build a valve to fit into it, special inverted valves being necessary for the common cathode circuit.

(2) It is much more stable than the earthed cathode circuit because the main stray feed-back is through C_{ak} which is much less than C_{ag}.

(3) The power gain is much lower than for grounded cathode triodes.

(4) The input impedance of the valve is very low, approx. $1/g_m$, and thus the input circuit is heavily damped. This means that the input is nearly aperiodic, and only one tuned circuit is needed per stage instead of two. This is important practically because of the reduction in tuning controls and in communication systems, because of the reduction in the number of tuned circuits in an amplifier chain and consequent improvement in band-width.

13.1. Grounded grid amplifiers

In view of the importance of the grounded grid stage, it is worth while developing a few of its more important properties in an elementary manner. The most powerful method for developing the necessary relations is by means of the matrix theory of active quadripoles which is given in the references.†

The grounded grid triode is shown in fig. 13.1. If the voltage

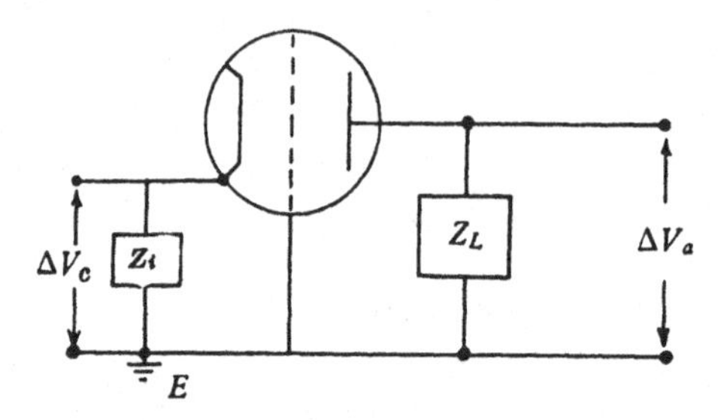

Fig. 13.1. Schematic grounded grid triode stage.

V_c increases, the anode current decreases and the anode voltage increases. Neglecting the feed-back through C_{ak} we have

$$\Delta i_a = \frac{\partial i_a}{\partial V_g}\Delta V_c - \frac{\partial i_a}{\partial V_a}\Delta V_a = g_m.\Delta V_c - \frac{\Delta V_a}{r_a}. \tag{1}$$

But

$$\Delta V_a = \Delta i_a.Z_L - \Delta V_c, \tag{2}$$

therefore

$$\Delta i_a\left(1+\frac{Z_L}{r_a}\right) = \Delta V_c\left(g_m+\frac{1}{r_a}\right), \tag{3}$$

The input admittance of the tube alone is then

$$Y_{iT} = \frac{\Delta i_a}{\Delta V_c} = \frac{g_m + 1/r_a}{1+Z_L/r_a} = \frac{(g_m+1/r_a)Y_L}{Y_L+1/r_a}. \tag{4}$$

The right-hand bracket of eqn. (3) reduces to g_m when $Z_L < r_a$ and $\mu \gg 1$. The admittance given by eqn. (3) appears in shunt across Z_i, so that the total admittance is

$$Y_{in} = Y_i + \frac{(g_m+1/r_a)Y_L}{Y_L+1/r_a}. \tag{5}$$

† Strecker and Feldtkeller, *Elekt. Nachr.-Tech.* **6** (1929), 93. Brown and Bennett, *Proc. Inst. Radio Engrs., N.Y.* **36** (1948), 844. M.I.T. series, vol. **18**, ed. Valley and Waldman. Kleen, *Elekt. Nachr.-Tech.* **20** (1943), 140. Rothe, *Arch. elekt. Übertragung*, **3** (1949), 233.

At very high frequencies there is an additional term in eqn. (5) due to the transit-time admittance of the cathode-grid space. The susceptive part of this will be tuned out by the susceptive part of Y_i, so there is only an additional conductance G_τ to be inserted in eqn. (5) to obtain the final result.

The voltage gain immediately follows from eqn. (3); it is

$$A = \frac{(g_m + 1/r_a)}{Y_L + 1/r_a}. \tag{6}$$

This is slightly bigger than the corresponding expression for the common cathode triode, which is $\frac{g_m}{Y_L + 1/r_a}$. However, the voltage gain is not an observable quantity at microwave frequencies and the proper parameter is the power gain. Let us calculate this for the situation of fig. 13.2. Here, a transmission line of imped-

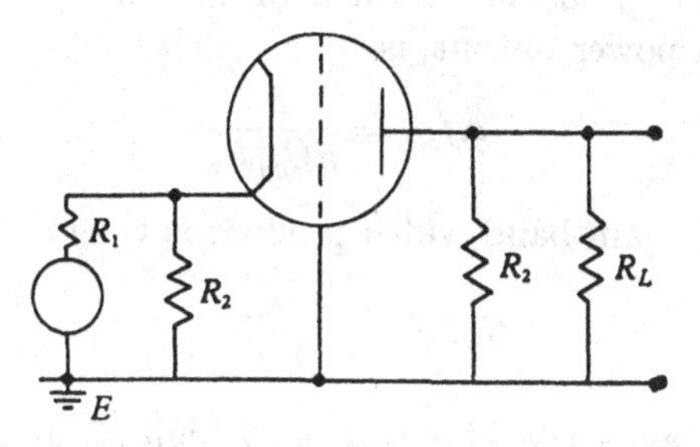

Fig. 13.2. Schematic tuned grounded grid stage at resonance.

ance R_1 is matched to the total input impedance of the valve. The output is tuned to resonance R_2 representing the losses of the output circuit. R_L represents the useful load. The input power P_1 is

$$P_1 = \frac{V_1^2}{2R_1}, \quad V_1 = \text{peak input voltage}.$$

Also
$$P_2 = \frac{V_2^2}{2}\left(\frac{1}{R_2} + \frac{1}{R_L}\right).$$

The output power
$$P_0 = \frac{V_2^2}{2R_L}, \tag{7}$$

and the power gain
$$G_p = A^2 \frac{R_1}{R_L}. \tag{8}$$

Neglecting $1/r_a$ in eqn. (6), which is nearly always permissible,

$$G_p = \frac{g_m^2 R_1 R_2^2 R_L}{(R_L + R_2)^2} \tag{9}$$

Eqn. (9) is a maximum for $R_L = R_2$ when

$$G_p = \frac{g_m^2 R_1 R_2}{4}. \tag{10}$$

A further simplification is possible for the case when the input-circuit conductance and the transit-time conductance are much smaller than g_m. R_1 is then $1/g_m$ and

$$G_p = \frac{g_m R_2}{4}. \tag{11}$$

If the only capacitance in the output circuit is the grid-anode capacitance, C_{ga}, the band-width of the circuit, when loaded for optimum power output, is

$$\Delta f_{\text{opt.}} = \frac{1}{\pi C_{ga} R_2} \tag{12}$$

The optimum gain band-width product is thus

$$G_p \Delta f_{\text{opt.}} = \frac{g_m}{4\pi C_{ga}}. \tag{13}$$

Eqn. (13) stresses the importance of high g_m and low-output capacitance when large values of gain × band-width product are required.

Finally, let us deduce the output impedance of the valve. When the valve anode circuit is open circuited, the voltage gain is $(\mu+1)$ from eqn. (6) and the input voltage is $\frac{I}{Y_i + G_\tau}$, since, according to eqn. (4), the tube input admittance is zero. Thus, the open-circuit output voltage is $\frac{(\mu+1)I}{Y_i + G_\tau}$. When the anode load is short-circuited, the input impedance is decreased to $[Y_i + G_\tau + (g_m + 1/r_a)]$. The voltage across the input is then $\frac{I}{[Y_i + G_\tau + (g_m + 1/r_a)]}$. Of the input current, some flows through the valve and the short circuit, the rest through the input circuit. The current through the valve is $V_{\text{in}}(g_m+1/r_a)$, where V_{in} is the voltage just deduced. The output admittance is the short-circuit current divided by the open-circuit voltage, therefore

$$Y_0 = \frac{I(g_m+1/r_a)}{Y_i+G_\tau+(g_m+1/r_a)} \div \frac{(\mu+1)I}{Y_i+G_\tau}$$

$$= \left(\frac{g_m+1/r_a}{\mu+1}\right)\frac{Y_i+G_\tau}{Y_i+G_\tau+(g_m+1/r_a)}$$

$$= \frac{1/r_a(Y_i+G_\tau)}{Y_i+G_\tau+(g_m+1/r_a)}. \tag{14}$$

The total admittance of the valve plus resonator is thus Y_0+Y_L, Y_L = resonator admittance.

13.2. Triode oscillator circuits

For the sake of completeness we must say a little about oscillator circuits used at microwaves. The possible types of oscillator, excluding those based on mutual inductance, for practical reasons, have been enumerated by Bell *et al.*† and other authors.‡ The general triode circuit is shown in fig. 13.3, where the Y's are the total admittances between the electrodes. By an obvious rewriting of eqn. (14) the total admittance from anode to grid is

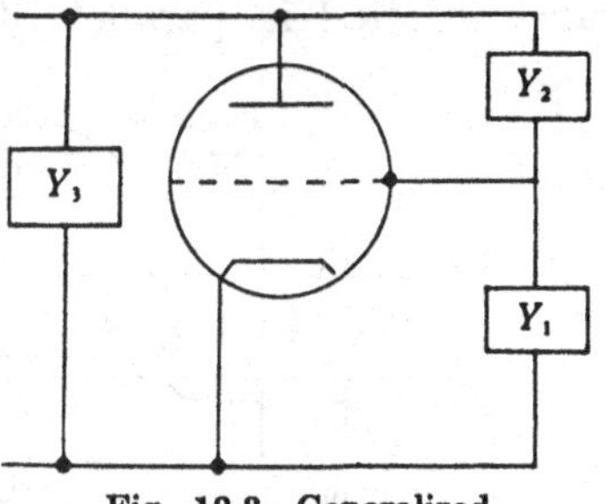

Fig. 13.3. Generalized triode oscillator.

$$Y_2' = Y_2 + \frac{Y_1 Y_3}{Y_1+Y_3+g_m}.$$

Separating the Y's into real and imaginary parts,

$$G_2' = G_2 + \frac{G_1G_3(G_1+G_3+g_m)+G_3B_1^2+G_1B_3^2-g_mB_1B_3}{(G_1+G_3+g_m)^2+(B_1+B_3)^2}. \tag{15}$$

† Bell, Gavin, James and Warren, *J. Instn. Elect. Engrs.* **93**, pt. IIIA, no. 5 (1946), 833.

‡ Gurewitsch and Whinnery, *Proc. Inst. Radio Engrs., N.Y.* **35** (1947), 462. Kuper, M.I.T. series, vol. **7**, chap. 7. Gavin, *Wireless Engr.* **25** (1948), 315.

If the circuit is to maintain oscillations, G_2' must be zero or negative. The last term in the numerator is the only one which can give rise to a negative conductance, and we observe that B_1 and B_3 must be of the same sign for oscillation. G_3' can be evaluated in the same manner and it must be of the opposite sign to the other two B's. Thus the only arrangements which can oscillate are:

	B_1	B_2	B_3
Class 1	C	L	C
Class 2	L	C	L

It is found, both theoretically and experimentally, that a given valve will oscillate at a higher limiting frequency in class (1) circuits than in class (2).

A typical example of class (1) oscillators is the grounded grid or grid-separation oscillator in which feed-back is applied to the amplifier of § 13.1, the feed-back susceptance C being supplied by the anode-cathode capacitance, or by extra capacitance inserted in parallel. This circuit has certain practical

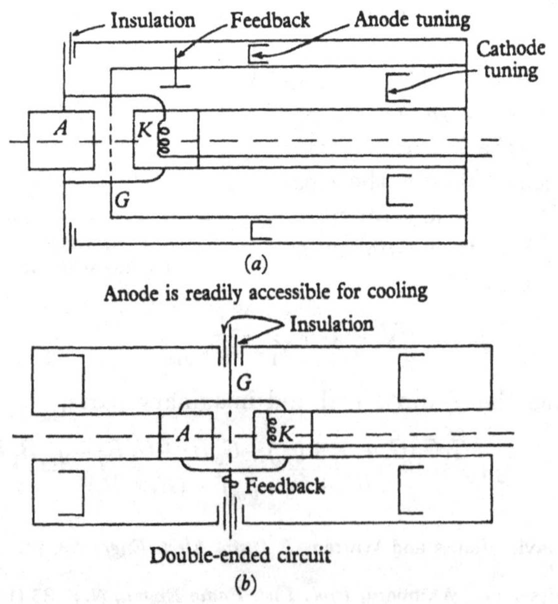

Fig. 13.4. Types of oscillator cavity.

disadvantages, since the anode is built into a tuned circuit and is usually at the end of a deep re-entrant which makes it awkward to cool. A better circuit from this point of view is the common grid-earthed anode circuit, which is usually 'folded back' as shown in fig. 13.4. American tubes often use the grid-separation grounded cathode circuit. For practical details the reader is referred to the works cited.

13.3. Triode electronics at ultra-high frequencies

We can now pass to the discussion of electronic phenomena in ultra-high-frequency triodes. We can divide the valve into two spaces in which interaction between field and beam takes place, regarding the grid as a division between the cathode-grid space and the grid-anode space. In the cathode-grid space the electron beam is density modulated or, more accurately, is density modulated with the addition of a small velocity modulation imposed on it. This process requires that power be taken from the input circuit and given to the beam, whereas in the low-frequency case the grid circuit does not absorb power unless the grid is driven positive. To calculate this power we can either calculate the resistive component of the beam-input impedance or, if we prefer it, the current induced in the grid. This current is not in antiphase with the grid voltage and therefore a loss component appears. It is more convenient to calculate the resistive component of the beam impedance, and we follow this course here. The density-modulated beam passes through the grid and enters into the grid-anode space. Here it is accelerated by a relatively intense d.c. field, and the modulation of density induces a radio-frequency current into the grid-anode circuit, exactly as did the density modulation in the beam built up in the drift space of a velocity-modulated tube. Since there is a transit angle in the grid-anode gap, we expect that the current induced into the grid-anode circuit will be less than the radio-frequency component of the beam current by a factor which has the same physical significance as the coupling factor of velocity-modulation theory. This expectation is justified and the factor reduces to β as defined in Chapter 12, when grid and anode are at the same direct voltage. We can thus look on the operation of an ultra-high-frequency triode as a system which in density modulation is directly applied to the beam and the output is taken by induction from electron bunches in the grid-anode gap.

Following this scheme we have to calculate the following quantities: (*a*) the input-beam admittance, (*b*) the conduction-current density produced at the grid plane by a specified grid voltage, and (*c*) the generalized gap factor for the case of different velocities at the entrance and exit planes. The mathematical apparatus has all been developed in Chapter 7 in the form of the Llewellyn electronic equations. However, we must make one qualifying remark: Llewellyn's equations, as he has emphasized in all his writings, only apply to the case of an electron beam of uniform velocity and in which no overtaking occurs. The question of overtaking is not of much practical importance here; but the fact that a monochromatic electron beam is assumed is of considerable importance, since it means that the theory only applies to the region between potential minimum and anode and not to the region between cathode and potential minimum. The Llewellyn theory cannot be expected, therefore, to give accurate results for the input loading when the distance cathode-potential minimum is comparable with the cathode-grid spacing. We shall have a little to say about experimental work on this subject later on; here we merely note the qualification, since it is sometimes inferred that this failure is a reason for questioning the whole Llewellyn analysis. We have already said that no adequate theory for multi-velocity streams is in existence, and as yet no acceptable theory for the cathode-potential minimum region has been propounded.

It may be asked, however, why it is not possible to achieve our ends in practice by reducing the grid-cathode transit angle so that it remains small even at the highest frequency desired. The grid-cathode transit angle depends on the spacing and on the equivalent grid-plane voltage. Obviously mechanical factors limit the smallness of the spacing, but why is not the equivalent voltage increased? The answer is that there is a relation between the transit angle and the cathode-current density, so that the allowable maximum current density, which for modern oxide cathodes is of the order 1 A./cm., determines the grid-plane voltage. We can easily calculate the relation for the ideal case of zero emission velocity. Eqn. (85) of Chapter 6 gives, for this case,

$$\tau = \frac{3s}{u_a}, \quad u_a = 5{\cdot}95 \times 10^7 \sqrt{V_{eq}} \text{ cm. sec.}$$

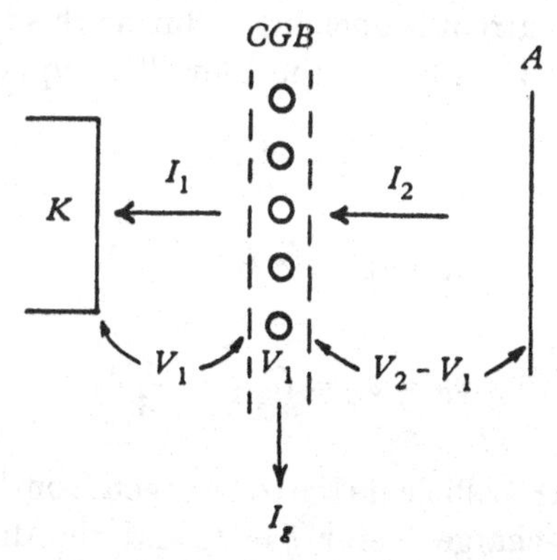

Fig. 13.5. Notation for the Llewellyn electronics equations.

The diode formula is

$$J_0 = \frac{2{\cdot}336 \times 10^{-6}}{d^2} V_{eq}^{\frac{3}{2}},$$

therefore

$$\tau = 6{\cdot}68 \times 10^{-10} \left(\frac{d}{J_0}\right)^{\frac{1}{3}} \text{ sec.}$$

Thus, to halve the transit angle we must increase the cathode-current density by eight or divide the grid-cathode spacing by the same amount. This result is considerably modified by consideration of the potential minimum, but the essential point is the close interaction between emission and transit angle which explains the extreme importance of improving cathode performance.

The notation for the Llewellyn analysis is shown in fig. 13.5. V_1 and V_2 are the a.c. voltages across the gaps, I_1 and I_2 are the total current densities, i.e. the sum of conduction and displacement current densities, in the gaps. The grid is negative, and so there is no conduction current in the grid lead; there is, however, a displacement current I_g in this lead. The cathode is taken as the zero for direct voltages, and the electrons are assumed to leave it with zero initial velocity, the a.c. component of conduction current being zero at the cathode surface. Two imaginary planes, C and B, are located infinitely close to the grid wires in such a way that the perturbation due to the wires is included in the length CB; but the transit angle across CB is so short that the conduction current is unchanged during this transit. Taking the a.c. voltage V_1 as given, the Llewellyn equations give

$$V_1 = A_1 I_1, \tag{16}$$

$$q_C = q_B = D_1 I_1 = \frac{D_1 V_1}{A_1}, \tag{17}$$

$$v_C = v_B = G_1 I_1 = \frac{G_1 V_1}{A_1}. \tag{18}$$

The values of the coefficients are to be taken from Table 2, p. 208, with the space-charge factor $\zeta = 1$, and the direct potential, required for the evaluation of the velocity and transit angle, is taken as the effective potential of the grid plane. The theory yields an accurate expression for this quantity, but an approximation, valid at high frequencies, is $V_{eg} = \dfrac{\mu V_g + V_a}{\mu + 1 + d_2/d_1}$, an expression already used in Chapter 11. The term A, evaluated under these conditions, thus gives the input impedance of the triode. Furthermore, from eqn. (17) the term D_1/A_1 represents the mutual conductance of the triode, for it gives the value of the conduction-current density at entrance to the output gap. Turning now to the output gap we have in general

$$\left.\begin{aligned} V_2 - V_1 &= A_2 I_2 + B_2 q_B + C_2 v_B, \\ q_a &= D_2 I_2 + E_2 q_B + F_2 v_B, \\ v_a &= G_2 I_2 + H_2 q_B + I_2 v_B. \end{aligned}\right\} \tag{19}$$

In the grid-anode space of the triode it is a reasonable first approximation to take $\zeta = 0$. This makes $C_2 = D_2 = F_2 = 0$, so the first two of eqns. (19) become

$$\left.\begin{aligned} V_2 - V_1 &= A_2 I_2 + B_2 q_B, \\ I_2 &= \frac{V_2 - V_1}{A_2} - \frac{B_2 q_B}{A_2}, \\ q_a &= E_2 q_B. \end{aligned}\right\} \tag{20}$$

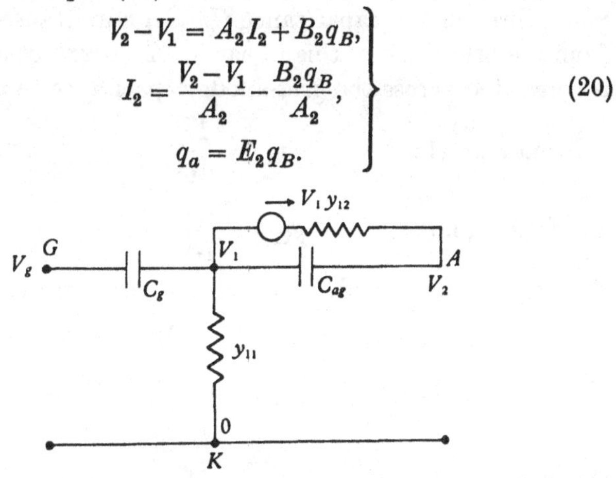

Fig. 13.6. Triode network from Llewellyn's equation.

Since the term q_B in eqns. (20) represents the radio-frequency conduction current driving the output gap (cf. § 7.1), we can recognize B_2/A_2 as the generalized coupling factor, and $1/A_2$ as the gap admittance y_{22}. The final expression for the radio-frequency current induced in the output gap is thus

$$q_0 = \frac{B_2 D_1}{A_2 A_1} V_1. \tag{21}$$

It is clear that we might discuss q_0/V_1, the transadmittance, as a generalization of the mutual conductance, but it seems clearer to separate out the effects due to the two gaps and maintain our distinction between mutual conductance and coupling factor.

As yet we have said nothing about the grid current and the way in which the various quantities measured at the two imaginary planes are related to the grid wires, where measurements are made. Since the total current is continuous through the valve,

$$I_g = I_1 - I_2, \tag{22}$$

and, as we are still discussing negative-grid triodes, I_g is a displacement current and the grid wires are connected to the planes C, B by a condenser.

We can now draw an equivalent circuit for our ultra-high-frequency triode, which is done in fig. 13.6 (identical with fig. 7.7). The grid terminal is connected to the equivalent grid plane through the capacitance C_g. An admittance y_{11} extends from the other side of this to earth. A current generator $V_1 . y_{12}$ is connected across the grid-anode capacitance to the anode.

From eqn. (16)
$$y_{11} = \frac{1}{A_1}, \tag{23}$$

and from eqn. (21)
$$y_{21} = \frac{B_2 D_1}{A_2 A_1}. \tag{24}$$

Writing down the node equation for the currents, we get

$$V_1 y_{11} + (V_1 - V_g) y_g + (V_1 - V_2) y_{22} + V_1 y_{12} = 0,$$

or
$$V_1(y_{11} + y_{22} + y_g + y_{21}) = V_g y_g + V_2 y_{22};$$

therefore
$$V_1 = \frac{V_2 + \left(\frac{y_g}{y_{22}}\right) V_g}{1 + \frac{y_g}{y_{22}} + \frac{y_{11}}{y_{22}}\left(1 + \frac{y_{21}}{y_{11}}\right)}. \tag{25}$$

Comparing eqn. (25) with the expression for V_{eg} obtained at low frequencies but including space charge,

$$V_1 = \frac{V_2 + \mu V_g}{1 + \mu + \frac{4}{3} d_2/d_1} \tag{26}$$

we see that a generalized μ is y_g/y_{22}. Thus

$$\mu = C_g/C_{ag}, \quad \text{or} \quad C_g = \mu C_{ag}. \tag{27}$$

In all ordinary ultra-high-frequency triodes C_{ag} is a fairly large value, of the order of a few pF. and μ is large, usually greater than 100, so that C_g is of the order 100 pF. and C_g has little effect except as a slight phase shift in the grid current.

We can now discuss the values of the admittances y_{11} and y_{21}. The former, y_{11}, has already been discussed in Chapter 7. From eqn. (56) of § 7.5

$$y_{11} = \frac{1}{r_c\left[\frac{2}{\beta_1} + \frac{12S_1}{\beta_1^4}\right]}$$

$$= \frac{g_0}{\left[\frac{2}{\beta_1} + \frac{12S_1}{\beta_1^4}\right]}. \tag{28}$$

By definition g_0 is the conductance of the diode which has a cathode-anode spacing d_1 and the direct voltage V_1, given by eqn. (26) with d.c. voltages substituted for the a.c. voltages, applied to the anode. The d.c. density is

$$I = 2{\cdot}336 \times 10^{-6} V_1^{\frac{3}{2}}/d_1^2,$$

therefore

$$g_0 = 3{\cdot}504 \times 10^{-6} V_1^{\frac{1}{2}}/d_1^2. \tag{29}$$

The denominator expresses the dependence on transit angle. Values can be determined by substitution of the values for β and S given in Table 2. The result is shown in fig. 13.7. An

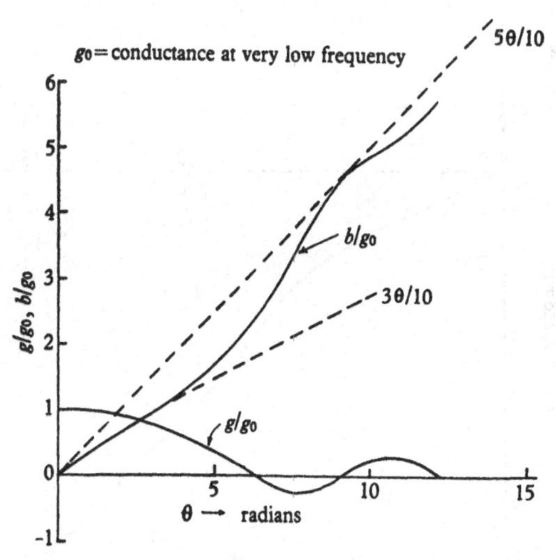

Fig. 13.7. Variation of input conductance and susceptance with frequency.

approximation, valid for angles up to about $\frac{1}{3}\pi$, is

$$y_{11} = g_0\left(1 + j\frac{3\theta}{10}\right).$$

A rough approximation, valid from $\theta = 0$ to $\theta \doteqdot \pi$, is

$$y_{11} = g_0\left\{\cos\frac{\theta}{4} + j\frac{3\theta}{10}\right\}. \tag{30}$$

Now

$$y_{21} = \frac{D_1}{A_1}\frac{B_2}{A_2}.$$

First consider the factor depending only on the first gap. From the table

$$\frac{D_1}{A_1} = g_0\left[\frac{2P_1}{\beta_1^2\left(\dfrac{2}{\beta_1}+\dfrac{12S_1}{\beta_1^4}\right)}\right]$$

$$= g_0\frac{(1-e^{-\beta_1}-\beta_1 e^{-\beta_1})}{\beta_1+(6/\beta_1^2)(2-2e^{-\beta_1}-\beta_1-\beta_1 e^{\beta_1})}.$$

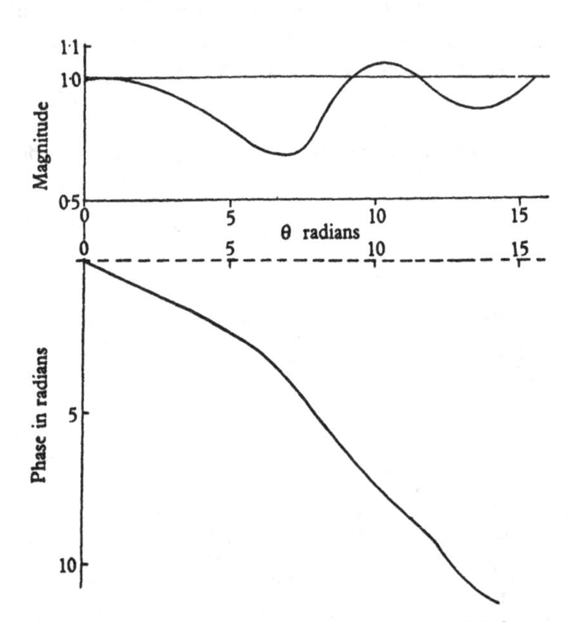

Fig. 13.8. Variation of mutual conductance with frequency.

At very low frequencies $\beta_1 = j\theta_1 \to 0$,

$$\frac{D_1}{A_1} \to g_0. \tag{31}$$

At very high frequencies $\beta_1 \to \infty$,

$$\frac{D_1}{A_1} \to -g_0 e^{-\beta_1} = -g_0(\cos\theta - j\sin\theta). \tag{32}$$

Therefore
$$\left|\frac{D_1}{A_1}\right| = g_0.$$

Fig. 13.8 shows the modulus and phase of g/g_0 plotted as a function of θ.

The factors depending on the second gap can now be discussed.

$$\frac{B_2}{A_2} = \frac{2/\beta_2^2}{u_B + u_a}[u_B(P_2 - \beta_2 Q_2) - u_a P_2], \tag{33}$$

where u_B and u_a represent the velocities at plane B and the anode respectively:

$$\frac{B_2}{A_2} = \frac{2/\beta_2^2}{u_B + u_a}[u_B(1 - e^{-\beta_2} - \beta_2) - u_a(1 - e^{-\beta_2} - \beta_2 e^{-\beta_2})].$$

For small β_2, this reduces to -1. Therefore y_{21} is essentially negative in the notation used. For large β_2 the value is

$$-\frac{2}{\beta_2}\left(\frac{u_B - u_a e^{-\beta_2}}{u_B + u_a}\right). \tag{34}$$

In the special case $u_B = u_a$, eqn. (34) reduces to

$$-e^{-\frac{1}{2}j\theta_2}\frac{\sin\frac{1}{2}\theta_2}{\frac{1}{2}\theta_2}. \tag{35}$$

This will be recognized as the coupling factor used in velocity-modulated tube theory. Owing to the fact that the actual velocities appear in eqn. (33) it is not easy to give a graphical plot which is of general utility. Since the average electron velocity in the output gap is many times that in the input gap, the output-transit angle is less than the input-transit angle, in spite of the greater spacing. It can be shown, without much difficulty, that in these circumstances the magnitude of y_{12} is not reduced, although there is a noticeable phase shift.

Another special case of some utility is the case when the velocity at the grid plane is negligible in comparison with the velocity at the anode plane. In this case

$$\frac{B_2}{A_2} = -\frac{2P_2}{\beta_2^2} = \frac{-2(1 - e^{-\beta_2} - \beta_2 e^{\beta_2})}{\beta_2^2}.$$

Lastly, $y_{22} = 1/A_2$. For small currents $\zeta \to 0$, y_{22} reduces to $j\omega C_{ag}$. This follows immediately from the definition of A_2,

i.e. $$A_2 = \frac{1}{\epsilon}(u_a + u_b)\frac{T^2}{2\beta_2}.$$

For zero space charge

$$T = \frac{2d_2}{(u_a + u_b)}, \quad \beta_2 = j\omega T_2,$$

therefore $$A_2 = \frac{d_2}{j\epsilon\omega}.$$

But the capacitance per unit area $= \dfrac{\epsilon}{d_2}$, so $A_2 = \dfrac{1}{j\omega C_{ag}}$.

When the current density is so high that ζ cannot be taken as zero, it can be shown that A_2 is given by

$$A_2 = \frac{12r_a}{\theta_2^4}\left[2(1-\cos\theta_2) - \theta_2 \sin\theta_2 - j\left\{\frac{\theta_2^3}{6} + \theta_2(1+\cos\theta_2) - 2\sin\theta_2\right\}\right]$$

$$-\frac{j}{\omega C_{ag}} \frac{1-\zeta}{(1-\zeta/3)},$$

where $$r_a = \frac{2}{3}\zeta^2\left(\frac{\sqrt{V_1}+\sqrt{V_a}}{I_a}\right).$$

These results are generalizations of eqns. (30) and (54) of Chapter 7, so that the curves of fig. 13.7 can be used to determine the frequency dependence of y_{22}. The extra term due to the capacitance of the gap must not be forgotten, as it ordinarily forms the major element of the network.

We have now determined all the quantities necessary for the theoretical discussion of triode circuits at ultra-high-frequency subject to the assumptions made. We see that the interesting fact emerges that the high-frequency mutual conductance does not vary very much from the low-frequency value.

In the next section we discuss the overall combination of valve and circuit in a very general way which gives a clear insight into the manner in which the operation of the valve depends on both electronic and circuit parameters.

13.4. The combination of valve and circuit

In this section we treat the combination of valve and circuit as an active four-pole with the notation for current and voltage directions shown in fig. 13.9. We are not here concerned with the general theory of such networks, which is discussed in the references cited earlier, but the formalization which results from the application of the theory allows the results to be expressed in the simplest and most general forms. Peterson†

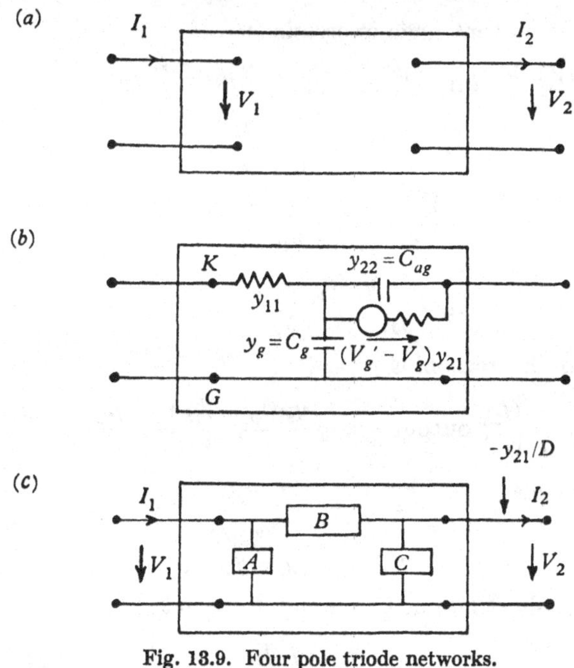

Fig. 13.9. Four pole triode networks.

has given a very full discussion of the three possible triode networks, grounded cathode, grounded grid, and grounded anode. Here we confine ourselves to the grounded-grid circuit. From fig. 13.9 we obtain

$$\left.\begin{aligned} I_1 &= c_{11}V_1 + c_{12}V_2, \\ I_2 &= c_{21}V_1 + c_{22}V_2, \end{aligned}\right\} \tag{36}$$

where c_{11} = input admittance with output short-circuited,
$-c_{22}$ = output admittance with input short-circuited,
$-c_{12}$ = feedback admittance with input short-circuited,
c_{21} = transfer admittance with output short-circuited.

† Peterson, *Bell Syst. Tech. J.* **27** (1948), 593.

We require to reduce the actual network, which for grounded-grid connexions is given in fig. 13.9 *b* to the π network of fig. 13.9 *c*. First let us calculate c_{11}:

$$c_{11} = \left(\frac{I_1}{V_1}\right) \text{output s.c.} = \frac{(V_1 - V_g')\,y_{11}}{V_1}.$$

But, from the current node expressions,

$$(V_1 - V_g')\,y_{11} = V_g'\,y_g + (V_g' - V_1)\,y_{21} + V_g'\,y_{22},$$

therefore
$$(V_1 - V_g') = \frac{V_g'\,(y_g + y_{22})}{y_{11} + y_{21}},$$

and
$$V_1 = \frac{V_g'\,(y_g + y_{11} + y_{22} + y_{21})}{y_{11} + y_{21}},$$

therefore
$$c_{11} = \frac{y_{11}(y_g + y_{22})}{y_{11} + y_{22} + y_g + y_{21}}. \tag{37}$$

Following similar reasoning

$$c_{21} = \left(\frac{I_2}{V_1}\right) \text{output s.c.} = \frac{y_{21}(V_g' - V_1) + V_g'\,y_{22}}{V_1}$$

Reduction yields
$$c_{21} = \frac{-y_{21}\left(y_g - \dfrac{y_{11}y_{22}}{y_{21}}\right)}{y_{11} + y_{22} + y_g + y_{21}}. \tag{38}$$

Applying the short circuit to the input, we get

$$c_{22} = -\frac{y_{22} + y_{11}/\mu}{D}, \tag{39}$$

$$c_{12} = -\frac{y_{11}}{\mu D}, \tag{40}$$

where
$$D = \frac{y_g + y_{11} + y_{21} + y_{22}}{y_g}. \tag{41}$$

The four relations for the *c*'s can now be used to determine the elements of the equivalent π of fig. 13.9 *c*.

Clearly
$$c_{11} = Y_A + Y_B, \quad -c_{12} = Y_B, \quad c_{21} = Y_C + Y_B.$$

From these we find

$$Y_A = y_{11}/D, \quad Y_B = y_{11}/\mu D, \quad Y_C = y_{22}/D. \tag{42}$$

For some purposes an equivalent circuit based on a constant-voltage generator is more convenient. This is shown in fig. 13.10. Finally, it is worth noting that for the planar grid valves used at microwave frequencies μ is very large, between 100 and 400, and therefore terms containing $1/\mu$ may be neglected. The equivalent network reduces to that shown in Table 7, and the input admittance does not depend on the load.

In his paper, Peterson gives expressions for the various parameters which are valid at moderate frequencies. These are not

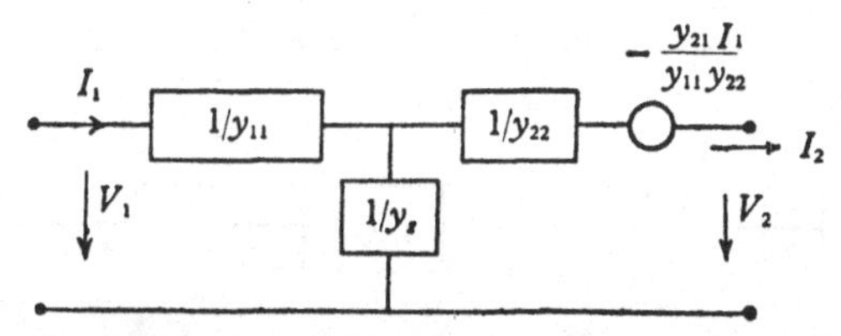

Fig. 13.10. Equivalent network for grounded grid triode.

reproduced here as they do not apply to the regions which are of most interest. At high frequencies it is not possible to deduce approximate expressions having any general validity, and so the best technique is to use the curves and data of Table 2 to calculate the y's for the frequency and geometry considered, and to use these values to obtain the required results from this section. The results for grounded-grid and grounded-cathode circuits are compared in Table 7.

13.5. The limits of validity of the Llewellyn analysis

It has been found experimentally that the Llewellyn analysis does not give the correct value for the conductance of very close-spaced diodes at very high frequencies. Some of the first measurements on this question were made by Smyth,† who suggested that the reason for the much higher admittance observed was damping introduced by those electrons which have only sufficient energy on leaving the cathode surface to reach a point in the neighbourhood of the potential minimum but not to pass it. A simple calculation shows that such

† Smyth, *Nature, Lond.*, **157** (1946), 841.

TABLE 7. NETWORK REPRESENTATION FOR HIGH μ TRIODES

The networks are valid for all frequencies but the y's are functions of frequency.

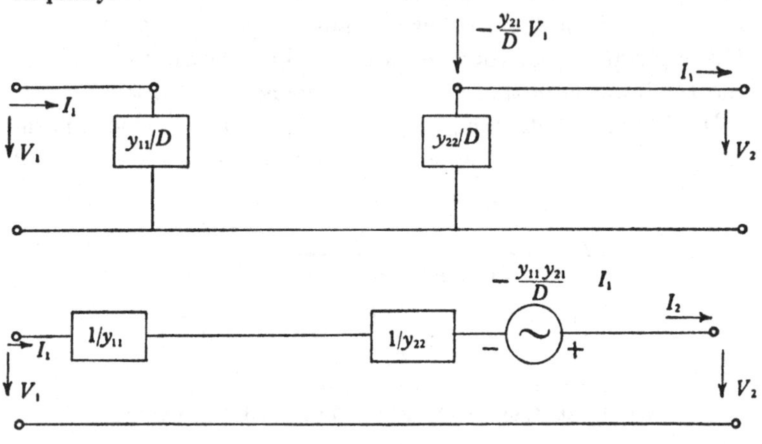

1. Grounded Grid.

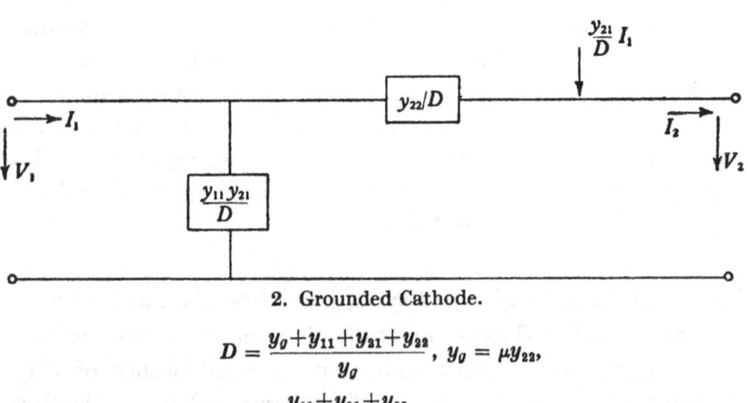

2. Grounded Cathode.

$$D = \frac{y_g + y_{11} + y_{21} + y_{22}}{y_g}, \quad y_g = \mu y_{22},$$

therefore

$$D = 1 + \frac{y_{11} + y_{21} + y_{22}}{\mu y_{22}} \to 1 \text{ as } \mu \to \infty.$$

electrons can have considerable transit angles and can absorb measurable amounts of power. Similar results on triodes have been reported by Lavoo,† and for extremely close-spaced diodes by Diemer and Knoll.‡ In all cases the observed

† Lavoo, *Proc. Inst. Radio Engrs., N.Y.* **35** (1947), 1248.
‡ Diemer and Knoll, *Physica*, **15** (1949), 459.

conductances are greater than theoretical values, the disagreement being most marked when the conditions are such that the potential minimum is very close to the anode. In view of this finding, it is very satisfactory that a careful investigation of the performance of the Bell Laboratories 1553 triode by Robertson† shows that the low-frequency value of the transadmittance compares very well with the transadmittance measured at 4060 Mc./sec. although the input conductance is again considerably higher than the theoretical value. These experiments are particularly important because they show that the triode-input conductance behaves in essentially the same way as a special test diode with a spacing equal to the 0·00065 in. cathode-grid spacing of the triode. The effects due to deflexion by the grid wires, irregularity of field, etc., are therefore not very important in this particular case. The input susceptance of the triode behaves differently from that of the diode, since instead of falling to 60 per cent of the value at current cut-off, it falls to a much lower value. A correlation between the depth of the minimum and the cathode activity was observed, and it is suggested that variations in the position and the depth of the potential minimum cause the effect. It was also observed that the value of y_{21} beyond current cut-off was 9000 mhos, instead of about 180 μmhos estimated from C_{ag}. A grid with a mesh instead of parallel wires showed a much smaller transadmittance at and beyond cut-off. An investigation showed that the effect was due to electromagnetic coupling through the grid apertures and not to inductance appearing in series with the grid wires which might have modified y_g.

Experimentally, then, it may be taken as proved that Llewellyn's analysis applies well to the input circuits of valves, if the spacing is large in comparison to the distance to the minimum, and that it gives substantially correct values for the transadmittance. In the case of close-spaced valves, the input conductance may be much higher than the calculated value.

Theoretical interpretations of the increased input damping have been given, but they do not agree with the observations. The main difference is that the increased damping is not

§ Robertson, *Bell Syst. Tech. J.* **28** (1949), 619, 647.

directly proportional to the saturated emission as it theoretically should be. Papers by Begovich† and by Knipp‡ contain the most relevant results.§ It should also be remembered that some of the damping may be due to series resistance in the core-semiconductor interface of the cathode and in the coating itself. Finally, because of the practical difficulty of providing a good electrical connexion to the cathode and at the same time retaining sufficient thermal insulation, there is the possibility of series resistance and inductance in the cathode lead. For all these reasons, departures from the theoretical values are much more likely to be found in the grid-cathode circuit than in the grid-anode circuit.

13.6. Comparison between grounded grid and grounded cathode stages

Having described experiments which give a confirmation of a considerable part of the Llewellyn analysis, we are now able to make a meaningful comparison between the behaviour of grounded grid and grounded cathode stages at ultra-high-frequency and to determine the frequency limitations on the performance of both types of stage.

First, consider some approximate expressions valid for small-transit angles and large values of anode-load impedance. For the grounded cathode stage the input admittance is $\frac{y_{11}+y_{21}}{D}$. Since we are restricting ourselves to large values of μ, $D \doteqdot 1$. The input admittance at fairly low frequencies is then

$$g_0\left(\cos\frac{\theta}{4}+j\frac{3\theta}{10}\right)-g_0 = g_0\left(\cos\frac{\theta}{4}-1+j\frac{3\theta}{10}\right).$$

For small θ, this is $\frac{-g_0\theta^2}{32}+j\frac{3\theta}{10}$. But the ordinary mutual conductance $g_m = -g_0$, so the real part of the input admittance is given by $\frac{g_m\omega^2\tau^2}{32}$. This is in very fair agreement with the well-

† Begovich, *J. Appl. Phys.* **20** (1949), 457.
‡ Knipp, *J. Appl. Phys.* **20** (1949), 425.
§ See also, Gray, *Bell Syst. Tech. J.* **30** (1951), 830.

known result of Ferris using the small θ expansion of all the exponentials. Ferris gives $R_e(y_{11}) = \frac{g_m \omega^2 \tau^2}{20}$ which is easily verified by using the correct values of y_{11} and y_{21} for $\theta_1 \to 0$. When this is done, the leading term in the imaginary part turns out to be $\frac{4}{3}C_0$, C_0 = cold input capacitance, instead of 0·6 C_0 which is the diode value. For the grounded grid, however, the input admittance

$$\div y_{11} = g_0\left(\cos\frac{\theta}{4} + j\frac{3\theta}{10}\right).$$

The real part of the admittance therefore diminishes as the frequency increases, according to the Llewellyn theory. We have seen that in practice this may not be true, but even so the increased admittance must be at least 0·1 g_0 to have an appreciable effect; whereas in the case of the grounded-cathode circuit, the damping becomes objectionable directly it becomes comparable with the loss resistance of the input-tuned circuit, i.e. a value of several kΩ.

For large values of θ the comparison is even more in favour of the grounded-grid triode, since the input admittance of the grounded-cathode stage tends to

$$j\frac{\theta}{2} - g_0 = g_m + j\frac{\theta}{2}.$$

For the grounded grid $y_{\text{in}} \to j\frac{\theta}{2}$. In a sense, the stages have changed roles, since the conventional stage is now very heavily damped and the grounded-grid stage lightly damped.

The transfer admittance is identical for both stages, except for the change of sign.

The output admittances can also be shown to be nearly identical. For matched input, the grounded-cathode stage has an output admittance y_{out}, given by

$$y_{\text{out}} = \frac{2y_{22}(y_{11} + y_{21})}{y_{22} + 2(y_{11} + y_{21})}.$$

Now $y_{22} = j\omega C_{ag}$, and in all normal valves $C_{ag} \ll C_{gk}$. Since the capacitative part of y_{11} is at least 0·6 C_{gk}, even for small θ,

$y_{22} \ll 2(y_{11}+y_{21})$. Thus $y_{\text{out}} \doteqdot y_{22}$. The same value holds for the grounded-grid stage in the present approximation.

The last important quantity is the power gain. To calculate this we consider a constant-current generator matched to the input (fig. 13.11). The tuned circuits include the susceptive

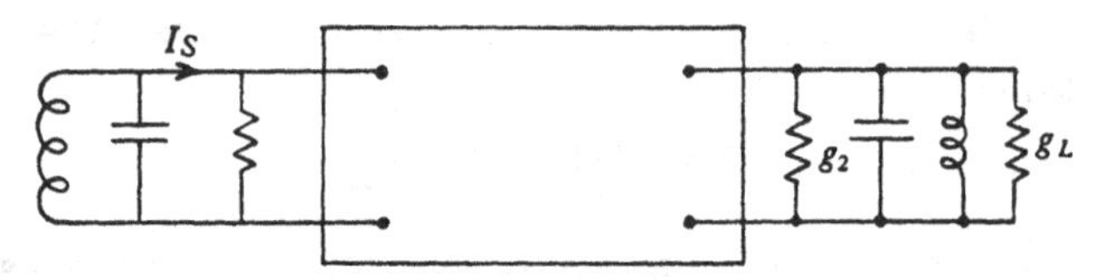

Fig. 13.11. Loaded four pole network.

parts of the valve input and output impedances. The loss conductances of the tuned circuits are g_1 and g_2, and the final load conductance, in which the useful power is dissipated, is g_L. Consider first the grounded-cathode circuit. The available power input from the constant-current generator I_s is

$$P_s = \frac{I_s^2}{4g_1},$$

therefore

$$V_1 = \frac{I_s}{2g_1}.$$

The output current is then $-\frac{y_{21} I_s}{2g_1}$, and to the approximation we are using the output current divides between g_2 and g_L, since y_{22} is a pure susceptance. The power gain is found to be

$$G_p = \frac{y_{21}^2 g_L}{g_1(g_2+g_L)^2}.$$

For $g_2 = g_L$, i.e. for maximum power output,

$$G_p = \frac{y_{21}^2}{4g_1 g_L}.$$

The expression for the grounded-grid stage is formally the same, as is shown by eqn. (10). The difference is the value of g_1, which in the grounded-cathode case is determined by the required band-width. In the grounded-cathode stage we can assume equal input and output Q's as a reasonable basis for comparison. The overall band-width is then 0·643 times the band-width of the individual circuits. Then we have

$$g_1 = \frac{\pi \Delta f C_{gk}}{0{\cdot}643}, \quad g_L = \frac{\pi \Delta f C_{ag}}{0{\cdot}643},$$

Then $$G_p \Delta f^2 = \frac{0{\cdot}1 g_m^2}{\pi^2 C_{gm} C_{ag}}.$$

For the grounded-grid stage the input band-width is almost always much bigger than the overall band-width and

$$G_p \Delta f = \frac{g_m}{4\pi C_{ag}}.$$

These expressions are equal if $\Delta f = \dfrac{0{\cdot}4 g_m}{C_{gk}}$. This expression ranges in value from several tens to a few hundreds of megacycles for modern valves. The gain band-width product is therefore greater for the grounded-cathode stage than for the grounded-grid stage. In practice, however, this only holds at frequencies sufficiently low to allow pentodes or tetrodes to be used, since, at the frequencies with which we are concerned in this chapter, it is not possible to neutralize a triode stage of the grounded-cathode type, and the constructional difficulties have so far prevented the use of multigrid valves at frequencies above about 500–1000 Mc./sec. In spite of the theoretically somewhat lower gain × band-width product, the grounded-grid stage is seen to have many advantages over the grounded-cathode stage, particularly when very wide band-widths, which are just beginning to be used in microwave systems, are in question.

To conclude this section we must inquire whether there are any other factors which cause the gain of the grounded-grid stage to fall as the frequency increases. We have seen that the input conductance is greater than the calculated value by an amount depending on variables such as the cathode temperature and d_{gk}, but whose frequency dependence is not known. Apart from this, the main reason why the performance of grounded-grid valves falls off is that the circuit losses increase at least as fast as $f^{\frac{1}{2}}$. Owing to the necessity of insulating grid from anode, it is difficult to design the output circuit so as to have a sufficiently high Q when unloaded, and the difficulty becomes greater as the frequency is raised. It is often found that relatively minor changes in anode-circuit design cause large variations in efficiency and, if the stated performance is not obtained from a valve, the anode circuit is often the cause of the trouble. The reader should consult the papers of Foster,† Bell *et al.* and van der Ziel,‡ and Volume 7 of the M.I.T. series for further practical information on valves and circuits.

13.7. High efficiency operation

The work has so far dealt with low efficiency or class A amplification. It is an experimental fact that it has not proved generally possible to operate microwave triodes in the conditions of high efficiency, known as class B and class C operation at lower frequencies. It seems to be common experience, however, that some individual valves give particularly good results, for a very short time.

The reasons for the failure are not too well understood, but some exploratory work has been done by Wang§ and by Gundlach,|| while the differential analyser has been used to study the operation of the 2C39 by Whinnery and Jamieson.¶ In this last study a set of conditions was found which should give a power gain of about 9 and an efficiency of 50 per cent at 3000 Mc./sec., whereas the 2C39 does not yield anything like these figures at 3000 Mc./sec., the best valves having efficiencies of under 20 per cent. The differential analyser study showed one possible reason for the discrepancy, namely, that many electrons acquire a large velocity component parallel to the cathode surface, since the grid-cathode field is decidedly non-uniform. If conditions are unfavourable, many of these electrons are turned back to the cathode after having reached

† Foster, *J. Instn. Elect. Engrs.* **93**, pt. IIIA, no. 5 (1946), 868.
‡ Van der Ziel, *Philips Res. Rep.* **1** (1946), 381.
§ Wang, *Proc. Inst. Radio Engrs., N.Y.* **28** (1941), 200.
|| Gundlach, *Funk u. Ton*, nos. 8, 9 and 10 (1948), 407, 454, 516.
¶ Whinnery and Jamieson, *Proc. Inst. Radio Engrs., N.Y.* **36** (1948), 76.

a point in the vicinity of the grid. These electrons constitute both a loss of energy, which appears as excessive cathode heating, and a diminution in the working current.

Another point of difficulty is that the output coupling factor would be very low if the output voltage swing reached class C conditions. For higher efficiency the electron 'bunch' must reach the anode when the instantaneous anode voltage is not much above the instantaneous grid voltage. This means that the average electron velocity might be five to ten times smaller than in the class A triode. This, in general, would mean a serious reduction in the value of the output coupling factor.

The most fundamental difficulty is that the angle over which the electron flow is allowed to take place cannot be reduced enough to attain really high-efficiency operation. The angle of flow should be of the order 60–90° for class C conditions, but if this angle is made smaller than the cathode-grid transit angle, the field will change direction before the electron 'bunch' reaches the grid plane and the bunch will be retarded, and may even be returned to the cathode. The best operating conditions are therefore those for which the flow angle is somewhat greater than the cathode-grid transit angle. If the latter is much more than 60°, class C operation will not be attainable, and we must therefore conclude that the highest frequency for which a particular valve will operate in class C is considerably lower than the frequency at which the class A performance of the same valve shows a marked deterioration. Taking the 2C39 as example, the tube will give 70 per cent efficiency at frequencies up to about 500 Mc./sec., 50 per cent at 1000 Mc./sec., and only 10–15 per cent at 3000 Mc./sec. At higher frequencies the performance deteriorates very rapidly.

Lastly, as the frequency is raised, it becomes more and more difficult to make the circuit losses sufficiently small to allow high-output amplitudes to build up. This is partly because the circuit becomes worse as the frequency increases and partly because of the above-mentioned decrease in the working current which means that a higher impedance is needed to match the valve.

13.8. Noise in ultra-high-frequency triodes

To conclude this chapter we give a very brief outline of the noise behaviour of ultra-high-frequency triodes. Strictly speaking there are two distinct aspects of the behaviour which we should study, first, the behaviour of common cathode triodes (or, in practical circuits, pentodes) at frequencies sufficiently high to cause a real component in the tube-input impedance, and secondly, noise behaviour of grounded-grid triodes at microwave frequencies. To save space we refer the reader to the excellent treatment of the first subject given by Schremp, Twiss and Beers,† and merely quote the results for the noise figure, which are:

$$\text{Common cathode: n.f.} = 1+\frac{g_1}{g_a}+\frac{5g_\tau}{g_a}+\frac{R_{eq}|Y_a+g_\tau|^2}{g_a}, \quad (43)$$

$$\text{Grounded grid: n.f.} = 1+\frac{g_1}{g_a}+\frac{5g_\tau}{g_a}+\frac{R_{eq}\mu^2|Y_a+g_\tau|^2}{g_a(\mu+1)^2}, \quad (44)$$

where g_a = aerial conductance,
g_τ = transit time conductance,
R_{eq} = noise equivalent resistance of valve.

These are practically identical, since $\mu \gg 1$, for all useful valves, but it must be remembered that the value of g_τ is not the same for the same valve in different circuits and, in fact, g_τ is negative, in the Llewellyn approximation, for the grounded-grid stage. In practice, the grounded-grid tube may be worse than the grounded-cathode stage because the power gain may not be sufficient to make the second stage noise negligible, particularly for not too large a band-width. Beers and Twiss give the following figures for two 6J4 stages:

Neutralized triode-grounded grid, $f = 180$ Mc./sec.,
$\Delta f = 2{\cdot}5$, N.F. = 5·5 db.

Grounded grid-grounded grid, $f = 180$ Mc./sec.,
$\Delta f = 2{\cdot}5$, N.F. = 7·0 db.

In this case the band-width is rather small in comparison with the mid-band frequency, and a larger band-width would reduce the difference. To conclude this part of the discussion we point

† Vol. 18, M.I.T. series, *Vacuum Tube Amplifiers*, chaps. 12 and 13.

out the fact, originally noted by Strutt, that once the input impedance departs from a pure reactance it is possible to improve the noise figure at a slight loss of signal by slightly detuning the input circuit so that there is a small net capacitative susceptance, i.e. the circuit is tuned to too low a frequency. By this means the phase angle between the noise voltage induced in the grid circuit and the noise-conduction current in the anode circuit can be adjusted so as to minimize the noise figure.

Now consider much higher frequencies. In spite of its inadequacies the Llewellyn-Rack analysis is the only useful tool for this work, and we apply it to the triode of fig. 13.12, considering the short-circuit conditions, i.e. we assume that there are large capacitances connected from grid to cathode and from grid to anode. The analysis of § 8.4 can then be applied to give the initial velocity variation at the plane of the minimum, and the Llewellyn equations then yield

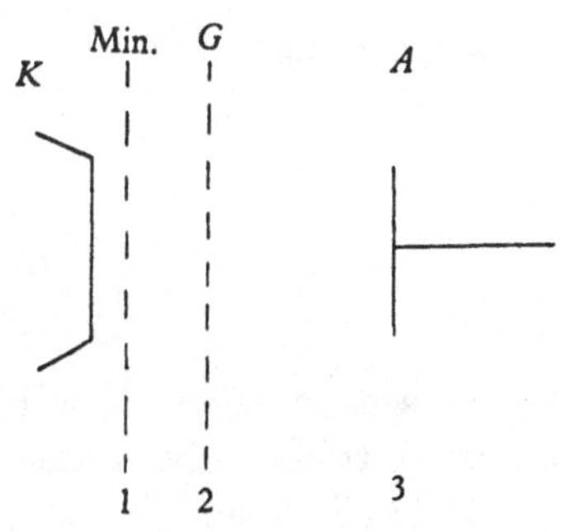

Fig. 13.12. Notation for noise analysis.

$$I_{1n} = -\tilde{v}_1 \frac{B_1}{A_1},$$

$$q_{2n} = \tilde{v}_1\left(F_1 - \frac{D_1 C_1}{A_1}\right),$$

$$v_{2n} = \tilde{v}_1\left(K_1 - \frac{G_1 C_1}{A_1}\right).$$

From the grid-anode region we have

$$I_{2n} = \tilde{v}_1\left[-\frac{B_2}{A_2}\left(F_1 - \frac{D_1 C_1}{A_1}\right) - \frac{C_2}{A_2}\left(K_1 - \frac{G_1 C_1}{A_1}\right)\right]. \qquad (45)$$

This expression allows us to evaluate the noise current induced in the grid-anode circuit when no external noise sources are connected to the valve. The general expressions are very complicated to handle, and we can only deal with the special cases of very small or very large transit angles here.

For small-transit angles (always assuming $\zeta = 1$ for the cathode region),

$$F_1 - \frac{D_1 C_1}{A_1} = \frac{I}{u_1} e^{-j\theta_1}(3 + j\,\theta_1),$$

$$K_1 - \frac{G_1 C_1}{A_1} = j\frac{2\theta_1}{15} e^{-j\theta_1}.$$

For large-transit angles,

$$F_1 - \frac{D_1 C_1}{A_1} = j\frac{I\theta_1}{u_1} e^{-j\theta_1}, \tag{46}$$

$$K_1 - \frac{G_1 C_1}{A_1} = -e^{-j\theta_1}. \tag{47}$$

In the second region it is usual to take $\zeta = 0$, and for most practical triodes the transit angle from grid to cathode is rather small. Then

$$-\frac{B_2}{A_2} \doteqdot \left(1 + j\frac{2\theta_2}{3}\right) e^{-j\theta_2} \tag{48}$$

and $C_2/A_2 = 0$. For ζ small but not zero

$$-\frac{C_2}{A_2} = \frac{I}{u_2}\left(1 + j\frac{2\theta_2}{3}\right) e^{-j\theta_2}. \tag{49}$$

For microwaves the value of transit angle in the grid-cathode space will assume a value which is intermediate between the cases considered, since it will be of the order π radians. For this case it is better to go back to the original expressions and plot curves giving the variation of the required quantity. However, as an example we evaluate the mean square noise current when θ_1 is large and θ_2 small. For this case

$$\begin{aligned} I_n^2 &= \frac{4kTI}{V_{eq}}\left(1 - \frac{\pi}{4}\right)\Delta f\left[\theta_1 + \left(\frac{u_1}{u_2}\right)^2\right] \\ &= 2eI\Gamma^2\Delta f, \end{aligned} \tag{50}$$

where

$$\Gamma^2 = \Gamma_0^2\left[\left(\frac{\theta_1}{3}\right)^2 + \left(\frac{u_1}{3u_2}\right)^2\right]. \tag{51}$$

Here we have used the results for $\tilde{v}_1^2$ and Γ_0^2 derived in § 8.4. It can be seen that the space-charge smoothing factor decreases as the transit angle from cathode grid is made larger, as we might expect. Many other interesting relations can be derived from the Llewellyn noise analysis, but the experimental data on the noise performance of triodes at microwave frequencies are rather meagre, and it would be unwise to proceed much further until it has been established whether the theory is in reasonable agreement with careful experiments.

Chapter 14

TRAVELLING-WAVE TUBES AND BEAM INTERACTION TUBES

We now proceed to discuss two of the more recent developments in ultra-high-frequency amplifiers. The travelling-wave tube was invented by Kompfner at Birmingham and Oxford Universities during the 1939–45 war, and was then brought to a useful state of development by Pierce at the Bell Laboratories in 1945–6. The second type of tube, in which there is no circuit, but simply two interacting electron beams, seems to have been invented simultaneously by Haeff at the U.S. Naval Research Laboratory and by groups at the R.C.A. and Bell Laboratories. A large amount of theoretical and experimental development has by now been applied to the travelling-wave tube, and it is possible to give a fairly well-informed appreciation of its practical advantages and disadvantages. The beam interaction tube has not been so deeply studied, and its future uses are a matter for speculation, but the principles used are of great interest and may have considerable bearing on the subjects of noise, oscillation build-up in magnetrons and galactic radio noise. A brief account is thus worthy of inclusion here.

14.1. Travelling-wave tube. General description

In the chapter on velocity-modulation devices we saw that there are rather stringent limits to the useful length between the two gaps of an amplifier, caused by space-charge repulsion between the bunched electrons. Some of the deleterious effects of space charge can be overcome by using more than two gaps with their associated resonators. Since the resonators have to be tuned, rather accurately in most cases, such a scheme rapidly becomes too complicated, and in practice not more than three or four gaps can be used. Kompfner started with the idea of increasing the number of gaps without limit, so that the electron beam moved in a continuous h.f. field, and of eliminating the sharply tuned resonator. This was done by enclosing the beam in a wire helix, on which the h.f. field propagated in the direction of the beam. Without any elaborate mathematics it can be seen that, for a helix of mean diameter d and t turns per cm., the wave must move πdt cm. on the wire to advance 1 cm. axially. The z velocity component is therefore reduced to approx. $c/\pi dt$ cm./sec. and if $\pi dt \doteqdot 10$, the beam velocity will be equal to the z component of the wave velocity for a voltage of about 2·5 kV. It is thus possible to obtain a condition in which the electrons travel at very nearly the z directed wave velocity, even for moderate h.t. voltages. When this is the case the electrons travel in a h.f. field which varies sinusoidally at the impressed frequency but which is of constant amplitude. The electrons which enter the helix, when the h.f. field at the entrance to the helix is accelerating, tend to catch up the electrons which entered earlier. Bunches therefore tend to form about the electrons which enter as the h.f. field changes from negative to positive. If the beam velocity is not exactly equal to the wave velocity the bunches, when formed, will travel slowly through the h.f. field, and it may be expected that if they go forward through the field, the field will gain energy from the beam, thus amplifying the helix wave. Conversely, if the beam falls back with respect to the field, the field will give more energy to the beam, and the h.f. wave on the helix will be attenuated.

It is not too easy to form a mental picture of this process, since it would appear that the bunching and energy conversion processes should occur simultaneously and the device should therefore fail to function. Analysis shows that the energy given to the beam near the input is sufficient to attenuate the exciting wave fairly rapidly. Bunching then takes place in a substantially field-free space, and at the end of this space the bunches create a new field, in phase with their centres, which grows exponentially in the direction of the beam and at the same time tends to restore the beam to the original unbunched state. This account of the process is made more realistic by the fact that some modern tubes actually have a large, lumped attenuation, one-third to one-half of the way along the helix. If such an attenuation is greater than 60 db. the circuit behaves much as though it were in two separate parts, with matched terminations at the attenuator and input and output transformers to the drive and load. Actually, it will be shown that there is not one wave on the helix, as we have assumed above, but four. Three of these propagate in the direction of the beam, and the fourth in the opposite direction. Only one of the forward waves is eventually amplified, the others merely help to satisfy the boundary conditions. The existence of the backward wave means that the travelling-wave tube can oscillate if the loop gain is high enough, even if there is no mismatch at the output. It will be shown that attenuation in the helix is necessary if the tube is not to self-oscillate at very moderate values of forward gain. At first sight it seems wrong that introducing attenuation should increase the gain. In fact, it does not do so, but it allows the beam current to be so much increased that the loss due to the attenuation is more than compensated and an improvement in forward gain results. It is easy to see that this can be the case if a lumped attenuator is considered, for it is clearly possible to locate this at a position where the input driving wave has been attenuated to a negligible amplitude by interaction with the beam. The lumped attenuator then has practically no effect on the forward wave but a marked effect on the backward wave, and the gain can be increased by increasing the current or the length of the helix.

The original tubes made by Kompfner were without attenuation and used very low beam currents. These tubes gave some gain and an indication that the noise figure was not too high, but they were of no value from an engineering viewpoint. Pierce designed the first useful tubes using much higher beam currents and attenuating helices and soon realized that the most important practical feature of the travelling wave tube was the enormous band-width over which amplification can be obtained. Improved designs for the devices used for coupling the input and output to the helix have been developed, and to-day travelling-wave tubes with band-widths of 1400 Mc./sec. at a centre frequency of 4000 Mc./sec. are available. It should be stressed that these band-widths are a property of the helix and not of the travelling-wave tube in general. Other circuits which have been suggested for travelling-wave tubes do not exhibit this feature, but some have other valuable properties such as much increased capability of dissipating power.

Fig. 14.1 shows a typical helix circuit travelling-wave tube using wave-guide input and output. An electron gun produces a beam which is prevented from diverging by an axial magnetic

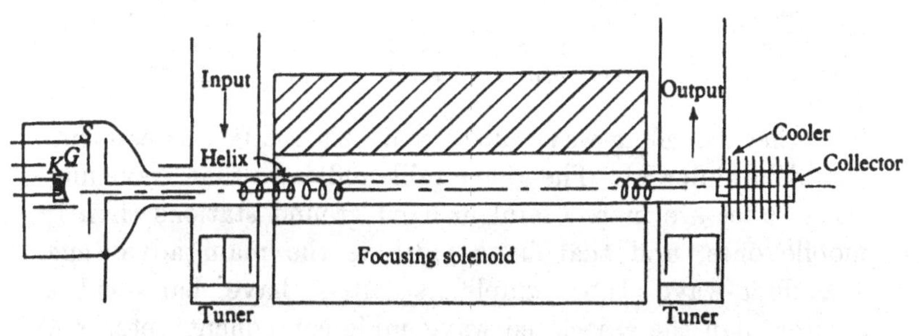

Fig. 14.1. Schematic travelling-wave amplifier.

field produced by a long solenoid surrounding the travelling-wave tube. The helix is coupled to the guide by means of some kind of transformer, many examples having been described. A similar transformer couples the other end of the helix to the output guide. Beyond this point the electrons are removed by some form of collector, which may be air- or water-cooled. Tubes of this type are capable of outputs up to several watts. The efficiency, expressed as the ratio output-power/beam d.c. power, is 2–5 per cent; but the power required to operate the solenoid is large, and the overall efficiency may be about half these figures. The gain is a function of the output power, being lower for high powers. Typical figures for the 4000 Mc./sec. band are 25–30 db. for a 100 mW. valve, 15–20 db. for a 1 W. tube, and less for still higher powers.

These figures are for valves essentially designed as output amplifiers. An input amplifier would be designed differently, as the gain per unit current is then the main noise determining factor.

We may summarize the properties of the helix type travelling-wave tube and compare them with other types of ultra-high-frequency amplifier as follows:

Property	Travelling-wave tube	Triode	Klystron
Gain	Fair	Fair	Good
Band-width	Excellent	Fair	Poor
Voltage	Medium	Low	Medium–low
Output	Medium	Medium–low	Good
Noise	Fair	Fair	Very poor
Efficiency	Poor	Good	Fair–good
Bulk	Very poor	Good	Fair
Weight	Very poor	Good	Fair

For other travelling-wave tube circuits the relative assessments would be different. The above table indicates that travelling-wave tubes are more useful in fixed ground stations than in mobile ones, and that band-width is the main advantage. Travelling-wave tube amplifiers often have band-widths greater than the associated wave-guide components, etc., and it is therefore difficult to use them to the best advantage.

14.2. Theory of travelling-wave tubes

As in the case of velocity-modulated tubes, two types of theory have been applied to the travelling-wave tube. The first consists in the application of Maxwell's equations† to the tube, which is divided into two regions, one of which contains all the charges, the other being charge free. Matching conditions at the boundary between the charge-containing region and the second region determine the observed values of the fields. Theories of this type have the advantage that they are, in principle, capable of including all the relevant phenomena, space-charge effects, transverse fields, magnetic effects and so on. Against this, the amount of work in obtaining useful solutions is very large, and it has to be repeated each time a new physical phenomenon is included in the analysis. Dynamical theories, essentially similar to the theory we developed in Chapter 12, are less general, but they give results for much

† Hahn, *Gen. Elect. Rev.* **42** (1939), 258, 497. Rydbeck, *Ericsson Technics*, no. 46 (1948). Chu and Jackson, *Proc. Inst. Radio Engrs., N.Y.* **36** (1948), 853.

less work, and new physical situations can be included in the initial solutions by perturbation methods or as fairly simple extensions of the theory. From an engineering viewpoint the dynamical theories seem to offer very many advantages and will therefore be adopted here. The dynamical theory divides naturally into two parts, the calculation of the effect of the field on the beam, and the calculation of the reverse effect, the fields induced by a given system of charges. The calculation of the field system produced by a given circuit appears as a third distinct problem.

14.3. Electron beam in an extended field

Let us now introduce the necessary notation. Direct-current quantities are denoted by capital letters and the suffix 0, i.e. I_0, V_0. The corresponding a.c. quantities are given the suffix 1. The propagation constant is $\Gamma = \alpha + j\beta$; α = attenuation constant, $\beta = \omega/u$ = phase constant, $\beta_e = \omega/u_0$ = phase constant for the electrons. Γ_0 is used for the propagation constant of the helix with no beam. We use ρ_0 to mean the initial charge per unit beam length, instead of the charge density which it normally denotes. Thus $\rho_0 = -I_0/u_0$. We now consider a one-dimensional system. At any point along the beam we have

$$u = u_0 + u_1\,e^{-\Gamma z + j\omega t}, \quad \rho = \rho_0 + \rho_1\,e^{-\Gamma z + j\omega t},$$

and

$$i = \rho u = I_0 + i_1\,e^{-\Gamma z + j\omega t}.$$

We now make the assumption which limits us to small signals but also gives us the great advantage of handling only linear equations. This assumption is the neglect of cross-products of a.c. quantities so that we put

$$i_1 = (u_0 \rho_1 + u_1 \rho_0)\,e^{-\Gamma z + j\omega t}.$$

Next we calculate the a.c. quantities using the equation of motion and the equation of continuity:

$$\frac{du}{dt} = -\frac{e}{m} E(z)\,e^{j\omega t}, \tag{1}$$

$$\frac{\partial(\rho u)}{\partial z} + \frac{\partial \rho}{\partial t} = 0. \tag{2}$$

Further, the derivative in eqn. (1) must be taken along the trajectory of the electron, so we have

$$\frac{d}{dt} = u_0\frac{\partial}{\partial z}+\frac{\partial}{\partial t}. \tag{3}$$

Then
$$u_0\frac{\partial u_1}{\partial z}+j\omega u_1 = -\frac{e}{m}E(z). \tag{4}$$

Using our approximate i_1 in (2), we get

$$\frac{\partial i_1}{\partial z}+j\omega\rho_1=0. \tag{5}$$

Carrying out the indicated differentiations with respect to z, and solving for u_1 and ρ_1, the following results are obtained:

$$\left.\begin{aligned} u_1 &= \frac{-\frac{e}{m}E(z)}{(j\omega-\Gamma u_0)}, \\ \rho_1 &= -j\frac{\Gamma i_1}{\omega}. \end{aligned}\right\} \tag{6}$$

When these are inserted into the original expression for i_1, there results

$$i_1\left(1+j\frac{\Gamma u_0}{\omega}\right) = \frac{I_0(e/m)E(z)}{u_0^2[j(\omega/u_0)-\Gamma]},$$

or, putting $u_0^2 = 2(e/m)V_0$ and $\omega/u_0 = \beta_e$ and rearranging

$$i_1 = \frac{j\beta_e}{(j\beta_e-\Gamma)^2}\frac{I_0E(z)}{2V_0}. \tag{7}$$

We may, if we like, insert $E(z) = -\partial V/\partial z = \Gamma V$ into eqn. (7) to yield

$$i_1 = \frac{jI_0\beta_e\Gamma V}{2V_0(j\beta_e-\Gamma)^2}. \tag{8}$$

Eqn. (8) is of some interest because the parameter $\partial i_1/\partial V$ is of the nature of a generalized mutual conductance.

Eqns. (7) and (8) are the first major results of the theory since they give the radio-frequency current as a function of the impressed field. We next have to obtain a relation giving the field at any point along the circuit due to an arbitrary distribution of current in the beam. This can be done without reference to the actual circuit geometry by assuming a coupling impedance relating the axial field component to the current in the circuit. The determination of the coupling efficiency depends, of course, on the actual geometry of the circuit.

14.4. The reaction of the beam on the circuit

In fig. 14.2 we show a line of length l, terminated at either end in the characteristic impedance. A generator is located at

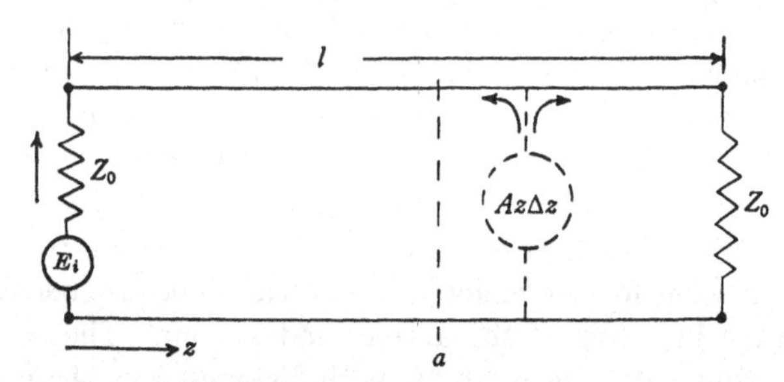

Fig. 14.2. A line excited by a distribution of current sources.

the origin, and in addition the line is excited by an arbitrary distribution of sources of intensity $A(z)\Delta z$. Consider the field at a plane a. There are three field components, the field at a due to the generator E_i, the part of the field propagating to the right from all the sources between O and a, and last, the part of the field propagating to the left from the sources between a and l. Therefore

$$E(a) = E_i\,\mathrm{e}^{-\Gamma_0 a} + \frac{1}{2}\int_0^a A(z)\,\mathrm{e}^{-\Gamma_0(a-z)}dz + \frac{1}{2}\int_a^l A(z)\,\mathrm{e}^{\Gamma_0(a-z)}\,dz.$$

Differentiating twice, we obtain

$$\frac{d^2E(z)}{dz^2} - \Gamma_0^2 E(z) = -\Gamma_0 A(z), \tag{9}$$

together with the appropriate boundary conditions. Now when the beam is present $d^2E(z)/dz^2 = \Gamma^2 E(z)$ so eqn. (9) becomes

$$(\Gamma^2 - \Gamma_0^2)E(z) = -\Gamma_0 A(z). \tag{10}$$

We next introduce our coupling impedance by putting

$$A(z) = -\Gamma^2 Z i_1,$$

or

$$(\Gamma^2 - \Gamma_0^2)E(z) = \Gamma_0 \Gamma^2 Z i_1. \tag{11}$$

Eqn. (11) is the complement of eqn. (7), so the problem is reduced to finding the values of Γ which satisfy these two simultaneous equations, therefore

$$\frac{I_0 Z}{2V_0} = -\frac{(\Gamma^2 - \Gamma_0^2)(j\beta_e - \Gamma)^2}{\Gamma_0 \Gamma^2 j\beta_e}. \tag{12}$$

Before discussing the solution of eqn. (12) we must digress for a moment to discuss the validity of the reasoning leading to eqn. (11). The derivation has tacitly assumed that, even when the beam is on, all the modes of propagation are 'active' ones, that is, that they extend throughout the length of the helix. This is approximately true for small-beam currents and small gains, but it is not true in general. When the beam is dense and the bunching pronounced, it is not possible to expand the expression for the exact shape of the radio-frequency current wave in terms of the active modes alone. The 'passive' modes, which attenuate rapidly with distance from the point of excitation, must also be included. The situation is analogous to that met in wave-guide theory when it is desired to find the effect of a sudden discontinuity, such as an iris or a stub. An expansion in terms of active modes alone will not represent the field correctly, or, more strictly, the boundary conditions cannot be satisfied at the discontinuity by such an expansion. The passive modes have to be included and can be represented by a lumped susceptance at the discontinuity. In the same way the passive modes contribute a lumped susceptance term to eqn. (11) in the travelling-wave tube. The magnitude of the term depends on the beam current, and for small currents we can neglect it for the moment. The effect of the correction will be considered later.

14.5. The solution of the propagation equation

Since eqn. (12) is of the fourth degree there are four waves, as we said in the introduction. In the solution of eqn. (12) we are guided by the fact that we wish to find a wave which shows amplification when the electronic velocity is about equal to the unperturbed wave velocity, i.e. $\Gamma_0 \doteqdot j\beta_e$. Let us start by putting

$\Gamma_0 = j\beta_e$, $\Gamma = \Gamma_0 - \zeta$. Putting $I_0 Z/4V_0 = C^3$, following Pierce,† we get

$$2C^3 = \frac{-(-\zeta)(2j\beta_e - \zeta)(\zeta)^2}{-\beta_e^2(-\beta_e^2 - 2j\beta_e\zeta + \zeta^2)}.$$

But $$\zeta^2 \ll 2j\beta_e\zeta \quad \text{and} \quad \zeta \ll 2j\beta_e,$$

therefore $$C^3 = \frac{j\zeta^3}{\beta_e^3} \quad \text{or} \quad \zeta = (-j)^{\frac{1}{3}}\beta_e C. \tag{13}$$

Inserting the three roots of $(-j)$ we obtain

$$\left.\begin{aligned}
\zeta_1 &= \beta_e C\left(\frac{\sqrt{3}}{2} - \frac{j}{2}\right), \\
\text{therefore} \quad \Gamma_1 &= \beta_e\left[j\left(1 + \frac{C}{2}\right) - \frac{\sqrt{(3)}C}{2}\right], \\
\zeta_2 &= \beta_e C\left(-\frac{\sqrt{3}}{2} - \frac{j}{2}\right), \\
\text{therefore} \quad \Gamma_2 &= \beta_e\left[j\left(1 + \frac{C}{2}\right) + \frac{\sqrt{(3)}C}{2}\right], \\
\zeta_3 = \beta_e C(j), \quad &\text{therefore} \quad \Gamma_3 = \beta_e[j(1 + C)],
\end{aligned}\right\} \tag{14}$$

and, if these are put back into the original equation, we find

$$\zeta_4 = \frac{-j\beta_e C^3}{4}, \quad \text{therefore} \quad \Gamma_4 = -\beta_e\left[j\left(1 - \frac{C^3}{4}\right)\right]. \tag{15}$$

It must be remembered that C^3 is usually quite small, so that $C \gg C^3$. Eqn. (14) shows that Γ_1 is an amplified wave since $R_e(\Gamma_1)$ is negative. The phase velocity is very little different from that of the unperturbed wave. Γ_2 has the same velocity as Γ_1 but is attenuated at the same rate as Γ_1 is amplified. Γ_3 travels slightly faster, but its amplitude does not vary along the circuit. The fourth wave Γ_4 is in the backward direction and its velocity is very nearly that of the unperturbed backward wave.

These solutions relate to a very special case, and we now seek to extend them. Accordingly put $\Gamma_0 = j\beta_e + jb + a$. This means that the circuit has an attenuation of a nepers per unit

† Pierce, *Proc. Inst. Radio Engrs., N.Y.* **35** (1947), 111.

length, and that the electron velocity is no longer equal to the velocity of the free wave but departs from it by an amount proportional to b. Both a and b are considered small compared with β_e but not necessarily smaller than $\beta_e C$. Putting these expressions into eqn. (12) we obtain

$$2C^3 = \frac{-(-\zeta - jb - a)(2j\beta_e + jb + a - \zeta)\zeta^2}{j\beta_e(j\beta_e + jb + a)(j\beta_e - \zeta)^2},$$

or, approximating,

$$\beta_e^3 C^3 = \zeta^2(j\zeta - b + ja). \tag{16}$$

Rewriting, we obtain

$$\zeta^3 + \zeta^2(a + jb) + (j\beta_e^3 C^3) = 0. \tag{17}$$

If we put† $y = \zeta + \dfrac{a + jb}{3}$ in eqn. (17) we find

$$y^3 - \frac{(a + jb)^2 y}{3} + (j\beta_e^3 C^3) + \frac{2(a + jb)^3}{27} = 0. \tag{18}$$

When $(a + jb)^2 y < j\beta_e^3 C^3$, the solution of eqn. (18) is very closely

$$y = (-j)^{\frac{1}{3}} \beta_e C,$$

or

$$\zeta = \zeta_0 - \left(\frac{a + jb}{3}\right).$$

Then

$$\Gamma_1 = j\beta_e - \zeta_1 = j\beta_e + \frac{a + jb}{3} - \zeta_{01}$$

$$= -\beta_e C\left[\left(\frac{\sqrt{3}}{2} - \frac{a}{3\beta_e C}\right) - j\left(\frac{1}{2} + \frac{1}{C} + \frac{b}{3\beta_e C}\right)\right], \tag{19}$$

with corresponding expressions for the other waves. Eqn. (19) shows an interesting fact, namely, that only one-third of the total attenuation present in the circuit appears in the expression for the negative attenuation of the amplified wave. While the numerical factor is not, in general, one-third, only a fraction of the attenuation is found in less restricted expressions for Γ_1,

† This substitution is simply the first step of Tartaglia's method for extracting the roots of cubic equations. See any good text on higher algebra.

which we now proceed to derive. Returning to eqn. (17), put

$$\zeta = u + jv,$$

then

$$u^3 + 3ju^2v - 3uv^2 - jv^3 + (u^2 + 2juv - v^2)(a + jb) + j\beta_e^3 C^3 = 0.$$

Equating real and imaginary parts

$$\left.\begin{aligned} (u^2 - v^2)(u+a) - 2uv(v+b) &= 0, \\ (u^2 - v^2)(v+b) + 2uv(u+a) &= \beta_e^3 C^3. \end{aligned}\right\} \qquad (20)$$

Since all the terms on the left-hand side are of the third order, we can plot these relations conveniently in units of $\beta_e C$, starting from the approximate values for a and b small.

Pierce† has done this, and his curves (Fig. 14.3) show a number of interesting points. First, the form of eqn. (19) is not generally correct, since the variations of b affects the real part of Γ_1 as well as the imaginary part. This is particularly noticeable for $a = 0$. In this case $R_e(\Gamma_1)$ falls sharply to 0 at $b \doteqdot 1{\cdot}9\ \beta_e C$. When $a \neq 0$, the tube continues to amplify, though with very small gain, for much greater values of b, up to approx. 5 $\beta_e C$. As one would expect, eqn. (19) gives much better values for $I_m(\Gamma_1)$ than for $R_e(\Gamma_1)$, and the general behaviour of $I_m(\Gamma_1)$ is that predicted by eqn. (19). Actually, for the special case $a = 0$, a better solution to eqn. (18) is $y = (-j)^{\frac{1}{3}}\left(\beta_e^3 C^3 - \frac{2b^3}{27}\right)^{\frac{1}{3}}$. This makes the forward gain zero for $b \doteqdot 2{\cdot}4\ \beta_e C$ which is not very different from the more correct value given above. We may suppose that the general form of Γ_1 is

$$\Gamma_1 = -\beta_e C\left[\left(\frac{\sqrt{3}}{2} - k_1 b^3 - \frac{a}{3\beta_e C}\right) - j\left(\frac{1}{2} + \frac{1}{C} + \frac{b}{3\beta_e C} + k_2 a^3\right)\right], \quad (21)$$

where k_1 and k_2 are constants. It is unnecessary to discuss the precise values of Γ_1 more closely since the whole derivation is approximate, and it is not worth while solving the equations more correctly until more of the physical phenomena taking part in the operation of the tube have been included in the theory. To summarize the results of this analysis, the interaction

† Pierce, loc. cit.

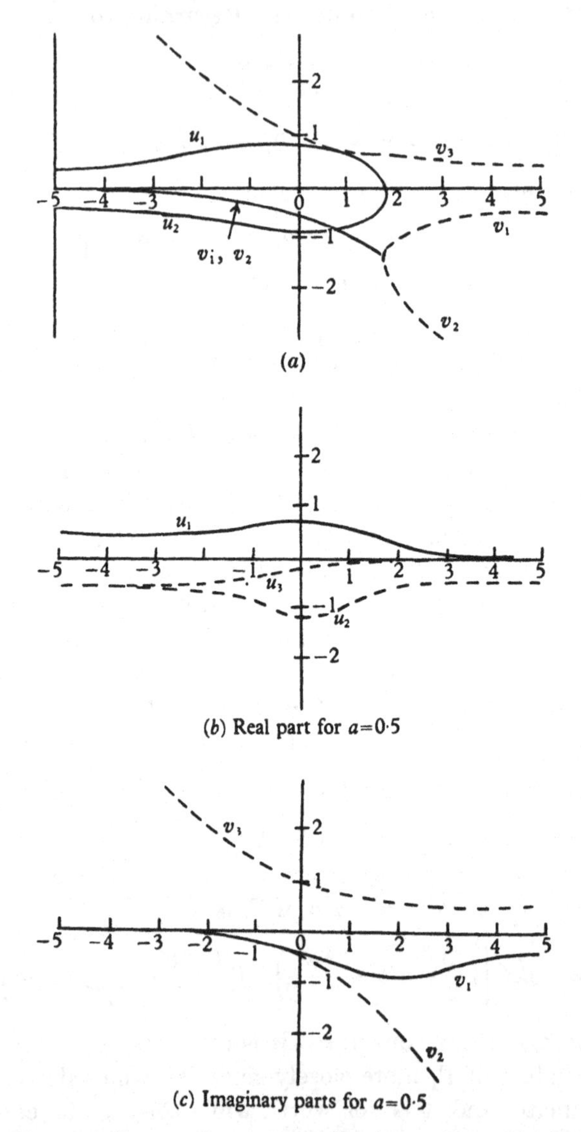

(b) Real part for $a=0{\cdot}5$

(c) Imaginary parts for $a=0{\cdot}5$

Fig. 14.3. Propagation constants for driven waves. (*a*) Real and imaginary parts of propagation constants when the circuit attenuation is zero. (*b*) Real part for $a = 0{\cdot}5$. (*c*) Imaginary parts for $a = 0{\cdot}5$. Data from J. R. Pierce, *Proc. Inst. Radio Engrs., N.Y.* 35 (1947), 111.

of the beam with the field causes the original unperturbed wave system to break up into a system of four waves, three forward in the direction of motion of the beam, and the fourth in the opposite direction. It must be emphasized that this wave is quite independent of matching; a mismatch at the output will modify the amplitude and phase of the backward wave, but in no circumstances can the backward wave be eliminated over a wide band of frequencies, although cancellation is possible over a narrow band.

14.6. The initial amplitude and the gain

We must now discuss the amplitude of the waves set up at the input. As yet we have no knowledge of how the input power divides between the forward waves. The answer to this problem is contained in the initial conditions for the differential eqn. (9) and in eqns. (7) or (8). We assume that the backward wave is very small at the input. The first condition is that

$$E_1+E_2+E_3 = E_i, \tag{22}$$

where E_i = amplitude of the impressed wave. From eqn. (7), as the conduction current is zero at the input,

$$\frac{E_1}{\zeta_1^2}+\frac{E_2}{\zeta_2^2}+\frac{E_3}{\zeta_3^2} = 0. \tag{23}$$

Also the radio-frequency velocity is zero at the input, so from the former of eqns. (6),

$$\frac{E_1}{\zeta_1}+\frac{E_2}{\zeta_2}+\frac{E_3}{\zeta_3} = 0. \tag{24}$$

Some lengthy, but straightforward, algebra leads to the solution for E_1,

$$E_1 = \frac{E_i}{(1-\zeta_2/\zeta_1)(1-\zeta_3/\zeta_1)}. \tag{25}$$

The values of E_2 and E_3 are obtained by permuting the ζ's. Using the first approximations to the ζ's, given by eqns. (14),

$$E_1 = \frac{E_i}{\left(1+\frac{\sqrt{(3)}+j}{\sqrt{(3)}-j}\right)\left(1-\frac{2j}{\sqrt{(3)}-j}\right)} = \frac{E_i}{3}. \tag{26}$$

E_2 and E_3 are also equal to $\frac{1}{3}E_i$. Thus at the input, the amplitudes of the three waves are equal. As the waves and beam move along the helix, E_1 increases in amplitude, E_2 decreases and E_3 remains the same. After a distance z, still using the first approximation,

$$\left.\begin{aligned} E_{1z} &= \frac{E_i}{3}\exp\left\{\beta_e Cz\left[\frac{\sqrt{3}}{2}-j\left(\frac{1}{2}+\frac{1}{C}\right)\right]\right\}, \\ E_{2z} &= \frac{E_i}{3}\exp\left\{\beta_e Cz\left[-\frac{\sqrt{3}}{2}-j\left(\frac{1}{2}+\frac{1}{C}\right)\right]\right\}, \\ E_{3z} &= \frac{E_i}{3}\exp\left\{\beta_e Cz\left[-j\left(1+\frac{1}{C}\right)\right]\right\}, \end{aligned}\right\} \tag{27}$$

so that if z is large, the first wave will be very much larger than either of the others. Since the circuit is continuous, the input and output impedances are equal, and the power gain is simply

$$G = 20\log_{10}\left(\frac{E_0}{E_i}\right), \quad E_0 = \text{field at output.}$$

For large z, $E_0 \doteqdot E_1$, therefore

$$G = \left[20\log_{10}\frac{1}{3} + 47{\cdot}3 \times \frac{\sqrt{3}}{2}\beta_e Cl\right]\text{db.}$$

$$= [41CL - 9{\cdot}54]\text{db.} \tag{28}$$

where $L = \omega l/u_0$ = normalized helix length (see Chapter 12). Eqn. (28) is accurate for a long, loss-free helix when the electron velocity is near to the free-wave velocity, i.e. it is the maximum gain for a long loss-free helix.

When the helix is not free from loss, and the beam velocity is not adjusted for a maximum, eqn. (25) still holds but eqns. (26) and (27) are, naturally, no longer correct. Using eqns. (20) we easily find the gain for a long tube,

$$G = 47{\cdot}3uCL - 20\log_{10}\left|1-\frac{\zeta_2}{\zeta_1}\right|\left|1-\frac{\zeta_3}{\zeta_1}\right|\text{db.} \tag{29}$$

Using the approximations leading to eqn. (19) we can obtain

the following result for the value of $|E_1|$:

$$|E_1| = \frac{E_i}{3}\left(1+\frac{a^2+b^2}{9(\beta_e C)^2}+\frac{b-\sqrt{(3)}a}{3\beta_e C}\right), \tag{30}$$

$$G = 47{\cdot}3CL-\frac{8{\cdot}66al}{3}-20\log_{10}\left(\frac{1}{3}+\frac{b-\sqrt{(3)}a}{9(\beta_e C)}+\frac{a^2+b^2}{27(\beta_e C)^2}\right). \tag{31}$$

For $a = 0$, eqn. (30) makes the division term $-9{\cdot}54$ db. for $b = 0$, $-6{\cdot}36$ db. for $b = 1$ and $-11{\cdot}74$ db. for $b = -1$. These figures agree well with those obtained from the more accurate expression (29). Eqn. (30) can therefore be considered as valid for a and b between $\pm\beta_e C$. The term $8{\cdot}66al/3$ is one-third of the forward attenuation when there are no electrons flowing.

We have now completed the calculations giving the gain as a function of the helix parameters a and b and the beam parameter C, in which everything is known except the coupling impedance Z. As yet there appears to be no limit to the gain which can be obtained as we increase C. This is not the case, because the backward wave sets a limit to the forward gain, since if the loop gain is sufficiently great, the valve will burst into oscillation.

14.7. The limit of gain and oscillation

The conditions for oscillation in travelling-wave tubes are those found in all feed-back amplifiers, viz. that the loop gain must be unity, and the phase change between input and feed-back must be $2n\pi$. The total power feed-back is the vector sum of the component due to mismatching of the output circuit and that due to the backward wave. The two components can be made to cancel at a definite frequency, but it is not, in general, possible to achieve cancellation over a broad band. The question of oscillation is rather intimately tied up with that of band-width, since the power feed-back from the mismatch increases as the frequency departs from the mid-band. If the reflexion coefficient at the mid-band frequency is 0·02 (V.S.W.R. = 1·02) the reflexion coefficient at the half-power points will be 0·18, or about 4 per cent of the power reaching the output will be reflected back to the input. In a tube without added attenuation this would mean that the gain would have to be restricted to about 13 db. if the whole band-width of the output circuit were to be used. If attenuation were inserted, the gain could be increased, at the cost of using more current, to approximately 13 db. plus the inserted attenuation. Actually matters are not quite so happy, but it is a fact that the addition of attenuation does permit higher useful gains. Introducing attenuation is not the only method of increasing the stability of travelling-wave tubes; another expedient which is often useful is to choose the circuit dimensions so that it is dispersive, i.e. the phase velocity becomes a function of frequency; by this device the gain at a fixed voltage can be reduced for frequencies beyond a desired range. In some types of circuit the dispersion is so great that the band-width is reduced to values which are too small to be useful. An example of this type of failure is the disk-loaded circular wave-guide.

Let us now take up the discussion of the backward wave. The simplest method of obtaining the amplitude of the wave sent back to the input is to return to the integral equation for the amplitudes of the waves, the equation before eqn. (9). The amplitude, observed at the input, of the backward wave is the last term with $a = 0$, that is,

$$E_{iB} = \frac{1}{2}\int_0^l A(z)\,e^{-\Gamma_0 z}\,dz.$$

Putting in the values of $A(z)$ and i from eqn. (7) we obtain

$$E_{iB} = \frac{1}{2}\int_0^l \frac{j\beta_e\Gamma^2}{(j\beta_e-\Gamma)^2}\cdot\frac{I_0 Z}{2V_0}\,E(z)\,e^{-\Gamma_0 z}\,dz. \tag{32}$$

We now use eqn. (12) for $I_0Z/2V_0 = 2C^3$, to obtain

$$E_{iB} = \frac{1}{2}\int_0^l \frac{\Gamma^2-\Gamma_0^2}{\Gamma_0}E(z)e^{\Gamma_0 z}dz = \frac{1}{2}\int_0^l \frac{(\Gamma-\Gamma_0)(\Gamma+\Gamma_0)}{\Gamma_0}E(z)\,e^{\Gamma_0 z}dz.$$

But
$$E(z) = \frac{E_i}{3}\sum_1^3 e^{-\Gamma_n z}, \tag{33}$$

therefore $$E_{iB} = \frac{E_i}{6}\int_0^l \sum_1^3 \frac{(\Gamma_n - \Gamma_0)(\Gamma_n + \Gamma_0)}{\Gamma_0} e^{-(\Gamma_n+\Gamma_0)z} dz$$

$$= \frac{E_i}{6}\sum_1^3 \frac{\Gamma_0 - \Gamma_n}{\Gamma_0}(e^{-l(\Gamma_n+\Gamma_0)} - 1)$$

$$= \frac{E_i}{6}\left[\sum_1^3 \frac{\Gamma_0 - \Gamma_n}{\Gamma_0} e^{-l(\Gamma_n+\Gamma_0)} + \sum_1^3 \frac{\Gamma_n - \Gamma_0}{\Gamma_0}\right]. \quad (34)$$

For the special case of beam velocity equal to the unperturbed wave velocity and zero attenuation, the second sum is identically zero. Also, since only the current at the output end makes a contribution, it will usually be sufficient to consider only the amplified wave. For this case then

$$E_{iB} = \frac{E_i}{6}\frac{\Gamma_0 - \Gamma_1}{\Gamma_0} e^{-l(\Gamma_0+\Gamma_1)}.$$

But $\frac{1}{3}E_i e^{-l\Gamma_1}$ is the output wave E_0, therefore

$$\left|\frac{E_{iB}}{E_i}\right| = \left|\frac{E_0}{E_i}\right|\left|\frac{\Gamma_0 - \Gamma_1}{2\Gamma_0}\right|, \quad (35)$$

$$20\log_{10}\left|\frac{E_{iB}}{E_i}\right| = G + 20\log_{10}\left(\frac{C}{2}\right).$$

Since the tube will just oscillate for $|E_{iB}/E_i| = 1$, the condition on C for oscillation is that

$$\log_{10}\left(\frac{2}{C}\right) = \frac{G}{20}. \quad (36)$$

For a gain of 13 db., eqn. (36) gives $C = 0{\cdot}447$ (which is much larger than is used in the small-power tubes made as input amplifiers). If such C value is obtained, eqn. (36) shows that l cannot be increased to give more than 13 db. gain without oscillation. A comparison with the figures obtained at the beginning of this section for oscillation due to mismatch indicates that oscillation will usually be due to the latter and not to the fourth wave. When attenuation is added, it is easy to see that the equivalent of eqn. (36) will then permit much bigger values of C or that there will be much less feed-back for a given C. However, the mismatch power is also attenuated, and the ratio of mismatch power to fourth wave power will be more or less constant. In this case also oscillation is likely to be due to mismatching. In some cases radio-frequency power may be transferred from the output to the input by modes of the helix and, possibly, any surrounding metal tube, other than the slow waves which are desired.

In the above we have assumed that oscillation is an undesirable feature of the travelling-wave tube. In some cases it may be useful to make a travelling-wave tube oscillate to act as a source of power. For instance, it has been suggested that the travelling-wave tube may be a better solution to the local oscillator problem at millimetre wave-lengths than are the reflex klystrons now used. The travelling-wave tube may be made to oscillate by mismatching or by providing a feed-back path external to the tube. The latter solution is obviously more desirable. The difficulty is that the feed-back path must be so arranged that the rate of phase change with wave-length is equal and opposite to that in the travelling-wave tube, so that the phase condition for oscillation can be satisfied over a band of frequencies. Since the travelling-wave tube circuit is many wave-lengths long, this is not easy to do. If the travelling-wave tube circuit length is l and the phase velocity V, with feed-back path $L'\lambda$ we must have

$$\frac{\omega l}{v} + 2\pi L' + \phi = 2\pi N, \quad N = \text{an integer},$$

where ϕ is a correction term due to the couplings. If ω is changed to $\omega + \Delta\omega$,

$$\Delta\omega\frac{l}{v} + 2\pi\Delta L' = 0 \quad \text{or} \quad \Delta f\frac{l}{v} + \Delta L' = 0,$$

therefore

$$\frac{dL'}{df} = -\frac{l}{v}. \tag{37}$$

If $l = 30$ cm., $v = c/20$, eqn. (37) shows that the external feed-back path must change in phase by 360° for a 50 Mc./sec. frequency change. Such a figure is rather large to be convenient for a wave-guide feed-back path and rather small for a resonant feed-back path. It was assumed in deriving eqn. (37) that the circuit was not dispersive over the frequency band considered, i.e. that the phase velocity is independent of the frequency.

For more details and some estimates of performance for travelling-wave tube oscillators, the reader is referred to a paper by Döhler, Kleen and Palluel.† Another treatment, using a very different approach, has been given by Van Iperen.‡

14.8. The efficiency of travelling-wave tubes

The efficiency of the travelling-wave tube is very easily evaluated if we assume that the maximum permissible power input and the circuit impedance are given. The gain is calculable from the theory already discussed, and the efficiency follows. To calculate the maximum efficiency a more refined analysis is necessary in which large signal effects are included. Before discussing such a theory we can derive a very simple result due to Pierce. Assuming that the maximum alternating conduction current just equals the d.c. beam current, an assumption which is made plausible by considering that if this limit is exceeded the electron motion becomes considerably more complicated, we obtain from eqn. (7)

$$|E_{\text{max.}}| = \frac{2V_0|\zeta_1^2|}{\beta_e}. \tag{38}$$

From the definition of the coupling impedance the power in the circuit is

$$|i|^2(-\Gamma^2)\,2Z = \frac{|E|^2}{-\Gamma^2 2Z}.$$

For $\Gamma = j\beta_e$, therefore

$$P = \frac{2V_0^2|\zeta_1^2|^2}{\beta_e^4 Z}.$$

But

$$|\zeta_1^2| = \beta_e^2 C^2 \quad \text{and} \quad Z = \frac{4V_0C^3}{I_0},$$

therefore

$$P = V_0 I_0 \frac{C}{2}$$

or

$$\hat{\eta} = \frac{C}{2}. \tag{39}$$

† Döhler, Kleen and Palluel, *Ann. Radioélect.* **4** (1949), 68.
‡ Van Iperen, *Philips Res. Rep.* **4** (1949), 20.

In this approximation the maximum efficiency is about $\frac{1}{2}C$, which might be as large as 10 per cent but is unlikely to exceed this figure. In fact, however, bigger efficiencies up to 25 per cent have been claimed, and it is clear that eqn. (39) is not more than a very rough indication of what is to be expected.

Some discussion of the large signal theory has been given by Brillouin.† He shows that the solution dealt with in earlier articles varies into another solution with lower gain per unit current as the current increases. A final state in which the electron beam is broken up into a series of discrete bunches, the electrons in each bunch having substantially the same velocity, may be reached.

The efficiency has also been discussed by Döhler and Kleen.‡ They include the effects of circuit attenuation and of beam velocities different from the wave velocity. It turns out that the maximum electronic efficiency occurs when the beam is travelling slightly more slowly than the wave, but that the circuit efficiency is maximum for the beam moving faster than the wave. This disadvantage can be overcome by using localized attenuation rather than distributed attenuation.

14.9. The helical circuit

The performance of helical circuits has been investigated by many authors dating back to the earliest days of wireless.§ The exact theory for helices of arbitrary pitch wound with circular wire has not been given, but an exact theory is available for a cylinder on whose surface flow is supposed to be possible only at a definite angle to the axis. A tightly wound helix of flat tape is a physical approximation to such a cylinder and is also a reasonable approximation to the round wire often used in real tubes. Before outlining the results of this analysis as given by Schelkunoff and Pierce, we note the major results

† Brillouin, *Phys. Rev.* **74** (1948), 90; *J. Appl. Phys.* **20** (1949), 1196.

‡ Döhler and Kleen, *Ann. Radioélect.* **4** (1949), 216.

§ Hertz, 'Electrical Waves' gives experimental results. An important early theoretical work is Pocklington, *Proc. Camb. Phil. Soc.* 9, (1897), 324. Other early analyses are Lenz, *Ann. Phys., Lpz.* **43** (1914), 749; Drude, *Ann. Phys., Lpz.* **9** (1902), 590. The analysis due to Schelkunoff and Pierce is usually quoted at present. This is given in Pierce, *Proc. Inst. Radio Engrs., N.Y.* **35** (1947), 111. Roubine, *Onde élect.* **27** (May 1947), 203, gives another treatment.

obtained by Pocklington. He showed that (*a*) if $2\pi/\lambda_0$ is not small compared with the mean radius of the helix, the phase velocity on the wire is c, the velocity of light; (*b*) if the wavelength is long enough for λr to be much greater than a^2, where r = radius of wire and a = mean radius of helix, the phase velocity along the *axis* of the helix equals c (which is physically obvious as the helix is, in this case, little different from a straight wire); (*c*) if $\lambda r \doteqdot a^2$ the phase velocity is intermediate between these two values. In practical travelling wave tube circuits we are concerned with cases (*a*) and (*c*), since we desire

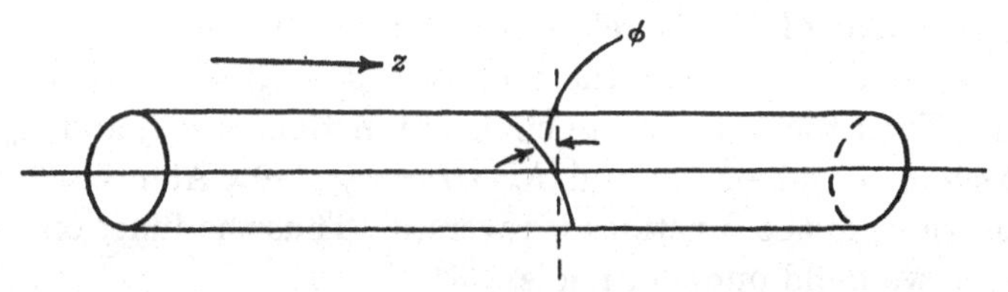

Fig. 14.4. Notation for the helix.

to obtain a slow wave so that an interaction can be obtained with electrons moving with velocities obtained by acceleration through a few kV. (2·5 kV. corresponds to about $c/10$). It turns out that a helix for $\lambda = 2\pi$ should have a diameter of 2–3 mm., so that case (*c*) is often encountered and is important because of the new effects produced by dispersion.

Let us now consider the helically conducting cylinder of fig. 14.4, where the current is constrained to flow at an angle ϕ with a plane normal to the z axis. We assume a lossless helix throughout the section. This problem in electromagnetic theory is usually solved by introducing a Hertzian vector $\boldsymbol{\pi}$ which completely determines the fields through the equations,

$$\mathbf{E} = \nabla \times \nabla \times \boldsymbol{\pi}, \quad \mathbf{H} = \left(\epsilon \frac{\partial}{\partial t} + \sigma\right) \nabla \times \boldsymbol{\pi}. \tag{40}$$

For our case $\epsilon = \epsilon_0$, and σ, the conductivity of the medium, = 0. By requiring that $\boldsymbol{\pi}$ shall only possess a component in the z direction, the restriction to transverse magnetic waves can be introduced. A second Hertzian vector $\boldsymbol{\pi}_1$ defined by eqns. (40) with **E** and **H** interchanged, gives the transverse electric modes.

It can then be proved that the field is a superposition of waves of the form†

$$\psi_n = e^{jn\theta} Z_n[r\sqrt{(\beta_0^2-\beta^2)}]\, e^{\pm j\beta z + j\omega t}. \tag{41}$$

Here θ = angular coordinate,

r = radial coordinate,

$\beta_0 = \omega/c$ = phase constant for a wave with velocity of light,

$\beta = \omega/v$ = phase constant for the wave component,

Z_n = generalized Bessel function.

For our case we require that the wave function ψ_n must be independent of θ. Therefore $n = 0$ and $e^{jn\theta} = \cos 0 = 1$. Also $\beta^2 > \beta_0^2$ as $c > v$, so that the root is $j\sqrt{(\beta^2-\beta_0^2)} = j\gamma$, where γ is real. Then the appropriate circular functions are I_0 and K_0, of which $I_0 \to 1$ as $\gamma r \to 0$ and $K_0 \to \infty$ as $\gamma r \to 0$. At infinity the behaviour of the functions is reversed. Thus the function from which we build our solution is

$$\psi_0 = [AI_0(\gamma r) + BK_0(\gamma r)]\, e^{\pm j\beta z + j\omega t}. \tag{42}$$

The field is divided into two regions: that inside the helix, for which $B = 0$ since the region contains $r = 0$ and the solution must remain finite at the origin; and the region outside the helix, $r > a$, where $A = 0$ since the function must remain finite as $r \to \infty$. The helical conductor acts as a boundary between the regions. On the helix the following boundary conditions must be satisfied. The current flow is in the ϕ direction. The tangential electric field must be perpendicular to the direction of current flow, since there is no loss in the helix, and it must be continuous. The field components for the total field found by superposing transverse magnetic and transverse electric waves, are

$$\left.\begin{aligned} E_r &= -j\beta \sum^n a_n \frac{\partial \psi_n}{\partial r} - \frac{\mu\omega}{r} \sum^n n b_n \psi_n, \\ E_\theta &= -\frac{\beta}{r} \sum^n n a_n \psi_n + j\mu\omega \sum^n b_n \frac{\partial \psi_n}{\partial r}, \\ E_z &= -\gamma^2 \sum^n a_n \psi_n; \end{aligned}\right\} \tag{43}$$

† Stratton, *Electromagnetic Theory*, McGraw Hill, chap. I, §§ 6·1–6·4 1941.

$$\left.\begin{aligned} H_r &= \frac{\beta_0^2}{\mu\omega}\frac{1}{r}\sum^n n a_n \psi_n + j\beta \sum^n b_n \frac{\partial \psi_n}{\partial r}, \\ H_\theta &= -\frac{j\beta_0^2}{\mu\omega}\sum^n a_n \frac{\partial \psi_n}{\partial r} - \frac{\beta}{r}\sum^n n b_n \psi_n, \\ H_z &= -\gamma^2 \sum^n b_n \psi_n, \end{aligned}\right\} \tag{44}$$

with $n = 0$ for our special case. At the wire, the boundary conditions give the following results:

$$r = a - \epsilon,\ \epsilon \to 0; \quad E_{z_1}\sin\phi + E_{\theta_1}\cos\phi = 0; \tag{45 a}$$

$$r = a + \epsilon,\ \epsilon \to 0; \quad E_{z_2}\sin\phi + E_{\theta_2}\cos\phi = 0; \tag{45 b}$$

also $$E_{z_1} = E_{z_2}, \quad E_{\theta_1} = E_{\theta_2}.$$

Finally,

$$H_{z_1}\sin\phi + H_{\theta_1}\cos\phi = H_{z_2}\sin\phi + H_{\theta_2}\cos\phi. \tag{46}$$

Let a_{01}, a_{02} be the coefficients of E inside and outside the helix respectively, and b_{01}, b_{02} the coefficients of E_θ (or H_z). The continuity relations then show that

$$a_{02} = a_{01}\frac{I_0(\gamma_a)}{K_0(\gamma_a)}, \quad b_{02} = -b_{01}\frac{I_1(\gamma_a)}{K_1(\gamma_a)}. \tag{47}$$

Eqn. (45 *a*) gives $\dfrac{b_{01}}{a_{01}} = \dfrac{-j\gamma}{\mu\omega}\tan\phi \,.\, \dfrac{I_0(\gamma a)}{I_1(\gamma a)}$, which can be put into eqn. (46) to yield

$$\gamma^2 = \beta_0^2 \cot^2\phi \left[\frac{I_1(\gamma a) K_1(\gamma a)}{I_0(\gamma a) \,.\, K_0(\gamma a)}\right]. \tag{48}$$

Thus γ is determined by the helix dimensions, and the four adjustable constants can all be expressed in terms of a single one, say a_{01}. We put $-\gamma^2 . a_{01} = E_z(0)$, the field along the axis. Explicit relations for the field components are given in Table 8. By putting the asymptotic expressions for the cylinder functions into eqn. (48) it can be shown that for γa large (i.e. $\gamma a > 3$ about) $I_0 \to I_1$ and $K_0 \to K_1$. The square bracket is thus unity, and for $\gamma a > 3$ we have $\gamma = \beta_0 \cot\phi$, which is equivalent to $v = c\tan\phi$. This proves that slow waves propagate with a velocity, in the direction of the helix, equal to that of light,

TABLE 8. EXPLICIT EXPRESSIONS FOR THE FIELD COMPONENTS OF THE HELIX WAVE

The components are derived from eqns. (43) and (44) of Chapter 14, using the value of the field on the axis where $\gamma r \to 0$, $I_0(\gamma r) \to 1$,

therefore $$E_z(O) = -\gamma^2 a_{01}.$$

(1) Inside the helix :

$$E_r = j\beta \frac{E_z(O)}{\gamma} I_1(\gamma r) \exp[j(\omega t - \beta z)]$$

$$E_\theta = -E_z(O) \tan\phi \frac{I_0(\gamma a)}{I_1(\gamma a)} I_1(\gamma r) \exp[j(\omega t - \beta z)]$$

$$E_z = E_z(O) I_0(\gamma r) \exp[j(\omega t - \beta z)]$$

$$H_r = E_z(O) \sqrt{\frac{\epsilon}{\mu}} \frac{\beta}{\beta_0} \tan\phi \frac{I_0(\gamma a)}{I_1(\gamma a)} I_1(\gamma r) \exp[j(\omega t - \beta z)]$$

$$H_\theta = -jE_z(O) \sqrt{\frac{\epsilon}{\mu}} \frac{\beta_0}{\gamma} I_1(\gamma r) \exp[j(\omega t - \beta z)]$$

$$H_z = -jE_z(O) \sqrt{\frac{\epsilon}{\mu}} \frac{\gamma}{\beta_0} \frac{I_0(\gamma a)}{I_1(\gamma a)} I_1(\gamma r) \exp[j(\omega t - \beta_z)]$$

(2) Outside the helix :

$$E_r = -jE_z(O) \frac{\beta}{\gamma} \frac{I_0(\gamma a)}{K_0(\gamma a)} K_0(\gamma a) \exp[j(\omega t - \beta_z)]$$

$$E_\theta = -E_z(O) \tan\phi \frac{I_0(\gamma a)}{K_1(\gamma a)} K_1(\gamma r) \exp[j(\omega t - \beta_z)]$$

$$E_z = E_z(O) \frac{I_0(\gamma a)}{K_0(\gamma a)} K_0(\gamma r) \exp[j(\omega t - \beta_z)]$$

$$H_r = E_z(O) \sqrt{\frac{\epsilon}{\mu}} \tan\phi \frac{I_0(\gamma a)}{K_1(\gamma a)} K_1(\gamma r) \exp[j(\omega t - \beta_z)]$$

$$H_\theta = -jE_z(O) \sqrt{\frac{\epsilon}{\mu}} \frac{\beta_0}{\gamma} \frac{I_0(\gamma a)}{K_0(\gamma a)} K_1(\gamma r) \exp[j(\omega t - \beta_z)]$$

$$H_z = jE_z(O) \sqrt{\frac{\epsilon}{\mu}} \tan\phi \frac{I_0(\gamma a)}{K_1(\gamma a)} K_0(\gamma r) \exp[j(\omega t - \beta_z)]$$

Numerically, $$\sqrt{\frac{\epsilon}{\mu}} = \frac{1}{120\pi} \text{ ohms.}$$

---

when γa is large enough. For smaller values of γa we rewrite eqn. (48) in a form which can be plotted, that is,

$$(\gamma a)^2 F(\gamma a) = (\beta_0 a \cot\phi)^2, \tag{49}$$

where $$F(\gamma a) = \frac{I_0(\gamma a) K_0(\gamma a)}{I_1(\gamma a) K_1(\gamma a)}. \tag{50}$$

This relationship is plotted in fig. 14.5. The figure shows that a given helix is non-dispersive so long as $\gamma a > 0{\cdot}8$ about. In non-dispersive helices the band-width is normally determined by the coupling devices.

The adjustable constant a_{01} has to be determined from the power input to the helix, assumed to be known. The power flow is related to the electromagnetic fields by Poynting's

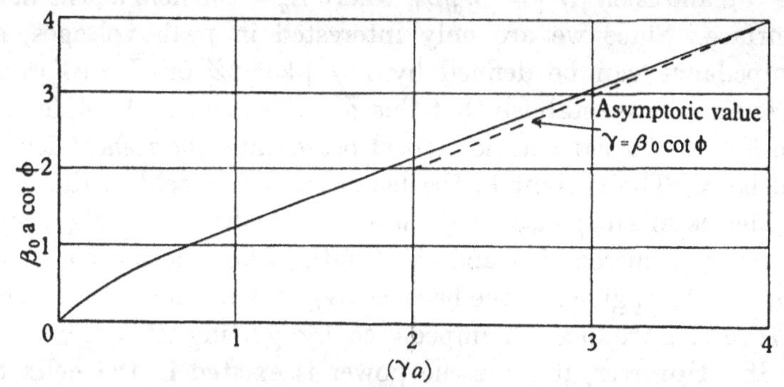

Fig. 14.5. Radial propagation constant for helical circuits.

theorem which states that the power is given by

$$P = \tfrac{1}{2} Re \int \mathbf{E} \times \mathbf{H}\, ds,$$

taken over the plane normal to the axis. Both the interior and exterior regions contribute, so

$$P = \frac{2\pi}{2} Re \left[\int_0^a (E_r H_\theta^* - E_\theta H_r^*) r dr + \int_a^\infty (E_r H_\theta^* - E_\theta H_r^*) r dr \right] \quad (51)$$

$$= E_z^2(0) \frac{\pi \beta \beta_0^2}{\gamma^2 \mu \omega} \left[\left(1 + \frac{I_1 K_0}{I_0 K_1} (\gamma a) \right) \int_0^a I_1^2(\gamma r) r dr \right.$$

$$\left. + \left(\frac{I_0}{K_0} (\gamma a) \right)^2 \left(1 + \frac{I_1 K_0}{I_0 K_1} (\gamma a) \right) \int_a^\infty K_1^2(\gamma r) r dr. \right] \quad (52)$$

In eqn. (52) (γa) after a group of symbols representing cylinder functions means that they have the common argument γa. Evaluating the definite integrals we get

$$P = E_z^2(0) \frac{\pi \beta \beta_0^2 a^2}{2\gamma^2 \mu \omega} \left\{ \left[1 + \frac{I_0 K_1}{I_1 K_0} (\gamma a) \right] [I_1^2(\gamma a) - I_0 I_2(\gamma a)] \right.$$

$$\left. + \left[\frac{I_0}{K_0} (\gamma a) \right]^2 \left[1 + \frac{I_1 K_0}{I_0 K_1} (\gamma a) \right] [K_0 K_2(\gamma a) - K_1^2(\gamma a)] \right\}. \quad (53)$$

Since $\beta_0 = \omega/c \quad \frac{\beta_0}{\omega} = \sqrt{(\mu\epsilon_0)}$ or $\frac{\beta_0}{\mu\omega} = \sqrt{\left(\frac{\epsilon_0}{\mu}\right)} = \frac{1}{377}$ ohms.

Eqn. (53) gives the relation between the power flowing in the circuit and the axial field. For a low-loss circuit we can put $\Gamma = \beta$ and then $|V| = |E_a/\beta|$, where E_a = the field at the helix surface. Since we are only interested in peak voltages, an impedance can be defined by $P = |V|^2/2Z$ or $Z = |V|^2/2P$. We must now establish that this Z is the same as the Z defined in § 14.4. Consider a modulated beam entering a short length of helix. The current in the beam induces a field in the helix which is in antiphase with the existing current. If the radio-frequency current is i and the length of helix is short enough, the power P_1 given to the helix is $\frac{1}{2}iV_L$, V_L longitudinal 'voltage'. We can introduce an impedance by putting $P_1 = V_L^2/2Z$, or $\frac{1}{2}i^2Z$. However, if the same power is excited in the helix by another method the voltage must be the same and the effects are reciprocal. The impedance defined above is thus the same as that of § 14.4. Then

$$Z = \left|\frac{E^2_{(a)}}{2\beta^2 P}\right| = \frac{120\gamma^4}{\beta^2\beta_0}\frac{I_0^2(\gamma a)}{(\gamma a)^2}[F_1(\gamma a)]^{-1}, \tag{54}$$

where $F_1(\gamma a)$ is the quantity in the bracket in eqn. (53).

$I_0^2(\gamma a)[F_1(\gamma a)]^{-1}$ is plotted in fig. 14.6. It will be noted that this function $\to \frac{1}{4}\gamma a$ for $\gamma a > 3$. We have already shown that, in the same approximation, $\gamma \to \beta_0 \cot\phi$, and for the values of ϕ used to obtain slow waves this makes γ very nearly equal to β. Using this fact we find that a useful approximation for Z is $Z = \dfrac{30\cot\phi}{\gamma a}$.

Experimental studies on the fields in helices and on the helix impedance have been made by Cutler† and by Jessel and Wallauschek.‡ The agreement with theory is fair, the theoretical impedance being somewhat larger than those observed.

In eqn. (54) the impedance is evaluated at the helix surface and thus is correct only for a thin hollow cylindrical beam of

† Cutler, *Proc. Inst. Radio Engrs., N.Y.* **36** (1948), 230.
‡ Jessel and Wallauschek, *Ann. Télécommun.* **3** (1948), 291.

the same radius as that of the helix. If a hollow beam of smaller mean radius b is used, eqn. (54) must be multiplied by the factor $\dfrac{I_0^2(\gamma b)}{I_0^2(\gamma a)}$. If the beam is solid and of radius c, we

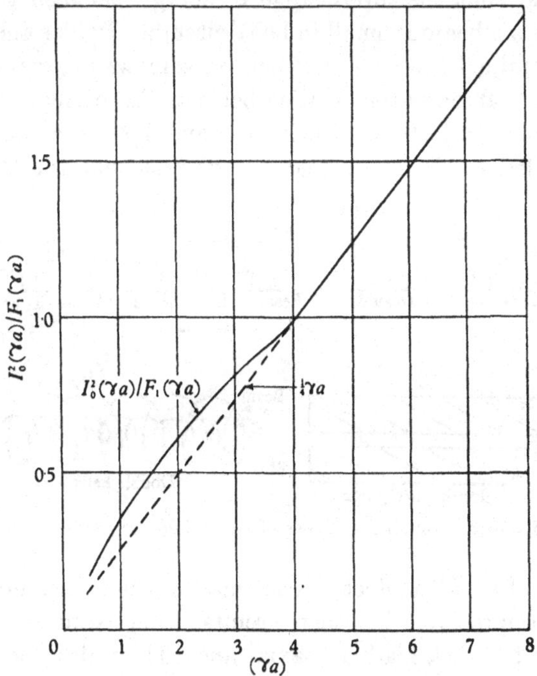

Fig. 14.6. Impedance parameter for helix.

can evaluate the mean-square field acting on the beam by integration:

$$E_{\text{eff}}^2 = E_z^2(0)\frac{2\pi}{\pi c^2}\int_0^c rI_0^2(\gamma r)\,dr = E_z^2(0)[I_0^2(\gamma c)-I_1^2(\gamma c)], \tag{55}$$

so the multiplying factor for this case is $\dfrac{I_0^2(\gamma c)-I_1^2(\gamma c)}{I_0^2(\gamma a)}$. The resulting expressions for Z can be used together with the known beam current and voltage to calculate C. For a given helix, Z is a function of ω, through γ and β, so that a variation of gain with frequency is introduced independently of dispersion. It is a simple matter, however, to choose the helix dimensions so that Z is independent of ω in the desired operating range.

The electric field also has a radial component which is zero on the axis and a maximum at a. The beam can interact with the radial field as well as with the axial field. Pierce† has worked out the consequences of this type of interaction and has found that the interchange of energy between the radial field and the beam is small in helix circuits. Döhler and Kleen‡ have examined the same problem in somewhat greater detail, and find that cross-terms arise because the oscillations about the original trajectory, which are caused by the radial field, are coupled to the axial field by the fact that the latter is a

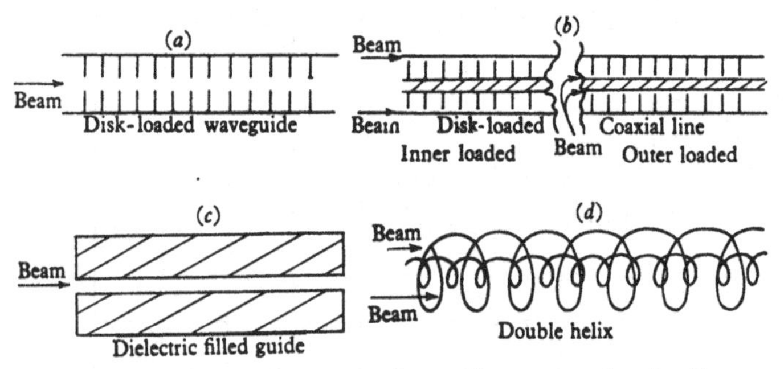

Fig. 14.7. Some other types of travelling-wave tube circuit.

function of r. This effect is also small in the helical circuit but may be appreciable in other circuits. One example of such a circuit is the disk-loaded coaxial line. They also consider the effect of radial field in causing electron current to the circuit, which is thus lost from the beam.

Various circuits other than the helix have been studied to determine their usefulness in travelling-wave tubes. Some examples are shown in fig. 14.7. The disk-loaded wave-guide has been closely studied because of its use in linear accelerators. As a travelling-wave tube circuit it has the insuperable defect that the dispersion is so great that the band-width is reduced to the order of 0·1 per cent. The disk-loaded coaxial with disks on the inner axis has been tried§ and seems to have had a

† Pierce, *Bell Syst. Tech. J.* **27** (1948), 732.

‡ Döhler and Kleen, *Ann. Radioélect.* **4** (1949), 76.

§ Field, *Proc. Inst. Radio Engrs., N.Y.* **37** (1949), 34. Dewey, Parzen and Marchese, *Proc. Inst. Radio Engrs., N.Y.* **39** (1951), 153.

certain degree of success. By correct choice of dimensions, useful band-widths can be obtained, but only about 1 per cent of those obtained from helical tubes. The major advantages of such structures are the ease with which they can be made to dissipate large amounts of power and their mechanical rigidity, which overcomes the manufacturing difficulty of supporting the flimsy helix. The gain per unit length also tends to be higher than that of the helix. Pierce† has given valuable general criteria for assessing the performance of circuits. If the group velocity is u_g and the energy stored $= W$/unit length, the power is $P = Wu_g$, therefore

$$\left(\frac{E^2}{\beta^2 P}\right) = \frac{E^2}{\beta^2 W u_g} \quad \text{and} \quad C = \left(\frac{E^2 I_0}{8\beta^2 W u_g V_0}\right)^{\frac{1}{3}}. \tag{56}$$

In this relation W is the energy stored when the field is E. Eqn. (56) shows that u_g must be small for large C. On the other hand

$$u_g = \frac{\partial \omega}{\partial \beta} = u_p\left(1 - \frac{\omega}{u_p} \cdot \frac{\partial u_p}{\partial \omega}\right),$$

where u_p = phase velocity. For a fixed u_p, i.e. fixed beam voltage, u_g can only be made small by using a dispersive circuit. But we have already seen that gain is only possible for $du_p/u_p \doteq \pm 0{\cdot}02$, so that the band-width gets smaller as the velocity difference allowed gets larger. By finding relations giving the stored energy, Pierce has discussed various circuits in very general terms.

14.10. Some factors omitted in the simple theory

We must now say a little about the factors omitted from the simple travelling-wave tube theory given above. The most important of these is the effect of space charge. Space charge can make its presence felt in several ways. First, there are the gross effects such as defocusing the beam, with consequent electron loss to the helix. With properly designed guns and solenoids this can be overcome, but only at the cost of fairly careful adjustment. Secondly, there are the effects of space charge debunching mentioned in Chapter 12. These are less

† Pierce, *Proc. Inst. Radio Engrs., N.Y.* **37** (1949), 510.

important in the travelling-wave tube because the a.c. velocities are fairly small and the bunching probably never becomes sufficiently pronounced to produce large space-charge deviations. Lastly, there are the effects of passive modes already noted in § 14.4. To discuss this question we must consider the system of field and beam from the viewpoint of Maxwell's equations. The electric field is then formed by an expansion in terms of the normal modes which now relate to the circuit plus electrons. This gives rise to an equation analogous to eqn. (11) but with extra terms. Since we know the approximate values of Γ for which gain is to be expected, this new equation can be solved for Z, which turns out complex instead of real. The imaginary part increases as the current increases, since it depends on $\Gamma - \Gamma_0$, i.e. on the gain. The details of this calculation are given by Bernier.† Pierce‡ has also studied the effects of varying $I_m(Z)$ which can be summed up as reducing the gain, increasing the beam velocity for which the gain is a maximum, and reducing the gain to zero for beam velocities below the maximum. (In the simple theory there is always some gain in this case.) These papers do not deal with the actual calculation of the passive mode parameters. This has been covered recently by Fletcher,§ who matches the radial impedances, derived from field theory by the use of boundary conditions $E_{z_i} = E_{z_o}$ and $H_{z_o} - H_{z_i} = i/2\pi b$ (o = outside the beam, i = inside), with those of Pierce's theory to obtain the following equation:

$$-\frac{\Gamma^2\Gamma_0 Z}{\Gamma^2-\Gamma_0^2}-\frac{2jQZ\Gamma^2}{\beta_e}=\frac{1}{2\pi b Y_c}, \tag{57}$$

where Y_c = an impedance given by the field theory, Q = passive mode parameter. Eqn. (57) can be solved for Q. The result is somewhat elaborate, but is fully graphed in the reference given.

The question of space charge has also been considered by Döhler and Kleen‖ by a direct kinetic method. In their treatment it appears that the space charge acts to increase or

† Bernier, *Ann. Radioélect.* **2** (1947), 87, Appendix.
‡ Pierce, *Proc. Inst. Radio Engrs., N.Y.* **36** (1948), 993.
§ Fletcher, *Proc. Inst. Radio Engrs., N.Y.* **38** (1950), 413.
‖ Döhler and Kleen, *Ann. Radioélect.* **3** (1948), 184.

decrease the value of the circuit attenuation depending on the relative velocity of the field and the beam.

Perhaps the major outstanding problem in travelling wave tube theory is to determine how the small signal theory goes over into the intermediate and large signal cases.

14.11. Noise in travelling-wave tubes

Kompfner† and Pierce‡ in their earliest work both derived the same expression for the noise factor of the travelling-wave tube. The derivation of this expression is simple, it is assumed that the noise current in the beam at entry to the helix is given by

$$\overline{i^2} = 2eI_0\gamma^2\Delta f,$$

where γ^2 = space charge smoothing factor (using a small letter to prevent confusion with the propagation constant). This can be put into eqn. (23) which is solved with eqns. (22) and (24) for $E_i = 0$, $u_1 = 0$. The noise from the input circuit gives rise to an input of $kT\Delta f$ watts and, knowing the gain, the output due to this noise is easily calculated. The result is

$$F = \frac{|E_{IA}|^2 + |E_{IN}|^2}{|E_{IA}|^2}. \qquad (58)$$

where E_{IA} = output amplitude due to thermal noise, E_{IN} = output amplitude due to shot noise. For the case of beam velocity = unperturbed wave velocity,

$$F = 2\gamma^2 \frac{2eV_0}{kT} C$$

$$\doteqdot 80\gamma^2 V_0 C, \qquad (59)$$

if $T = 290°$ K.

It was soon found that experimental valves constructed to give low noise figures, i.e. with low voltage and small C (low current) gave values of γ^2 in eqn. (59) of 0·02 or smaller. These figures are lower than low-frequency values of γ^2, which is not to be expected. Also the behaviour of the noise figure with voltage and current does not agree with the forecast of eqn. (59).

† Kompfner, *Wireless Engr.*, **24** (1947), 255.
‡ Pierce, *Proc. Inst. Radio Engrs., N.Y.* 35 (1947) 111.

Smullin,† following suggestions of Pierce, has applied the Llewellyn analysis to the travelling-wave tube. The electrons pass through the space-charge limited space between the cathode and first anode and then through a constant potential region before they reach the commencement of the helix. Applying the analysis in the way that we have done in the last section of Chapter 13, Smullin finds a very different answer in which the tube parameters appear differently and which depends on the ratio of the transit angles cathode-anode, anode entry-helix commencement. However, Smullin's analysis is also open to objection because he considers that the space charge in the second region is zero. The space-charge parameter depends on the current density which need not be small even for the small beam currents used in low noise travelling-wave tubes (a few hundred μA.). The neglect of space charge, which should reduce the noise, means that Smullin's result is an upper limit to the noise factor. The Llewellyn results are, of course, derived for purely electrostatic fields, and it is not immediately obvious how they should be modified to allow for intense magnetic focusing fields. However, the indication that the transit angles in the gun-start of helix region are important seems to be a step in the right direction and some improvement in noise performance, which is already within a few decibels of 10 cm. crystals, is to be looked for.‡

14.12. Space-charge wave tubes

Soon after the general theory of the travelling-wave tube was widely appreciated, a new type of amplifier was simultaneously invented in several laboratories.§ In this amplifier there is no circuit joining the input to the output, the waves propagate solely on the electron beam. The electron beam consists of dual, or a plurality of, elementary beams travelling

† Smullin, M.I.T., *Research Lab. of Electronics Progress Report*, Jan. 1950. *Research Lab. of Electronics Technical Report*, No. 142, 1949 (this report contains some errors which are corrected in the *Progress Report*).

‡ An account of recent work is given below, pp. 559–65.

§ Haeff, *Proc. Inst. Radio Engrs., N.Y.* **37** (1949), 4. Nergaard, *R.C.A. Rev.* **9** (1948), 585. Pierce and Hebenstreit, *Bell Syst. Tech. J.* **28** (1949), 33. Pierce, *Proc. Inst. Radio Engrs., N.Y.* **37** (1949), 980.

with slightly differing velocities. Velocity modulation is impressed on the beams. The beams moving parallel to each other build up a density modulation, and for certain values of the velocity difference growing waves are found to propagate. Fig. 14.8 shows a schematic form of a two-beam tube

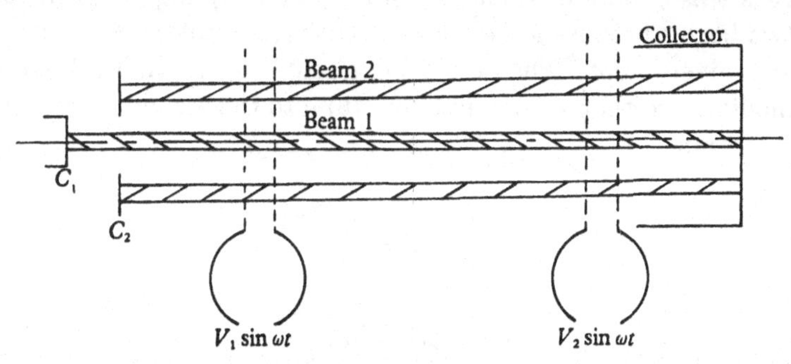

Fig. 14.8. Space-charge wave tube with two beams.

in which a cylindrical beam from C_1 moves inside an annular beam from C_2, and the potential difference between C_1 and C_2 regulates the velocity difference between the beams. Fig. 14.9

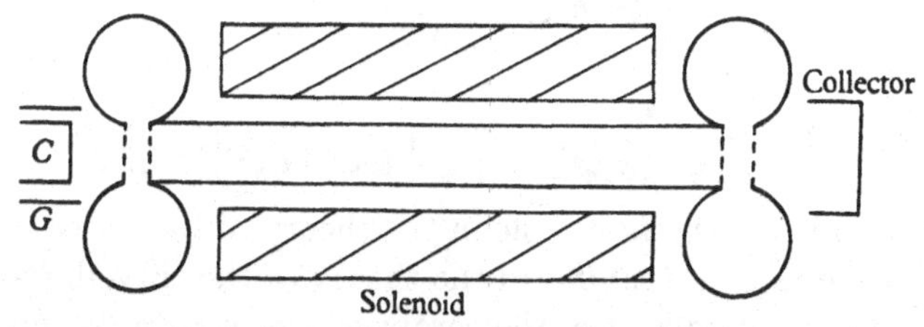

Fig. 14.9. Space-charge wave tube with magnetic field.

shows another type in which the axial magnetic field is used to focus an intense electron beam through the long tunnel. As we saw in Chapter 6, the potential varies across the beam in such a case, being a minimum on the axis. This continuous distribution of velocity is simply the limiting case of a large number of elementary beams, each with a predetermined velocity.

For the sake of concreteness let us confine ourselves to the case of two beams. The suffices a and b refer to the two beams.

The equations required for the analysis are (1) and (2) of this chapter and Poisson's equation which in rectangular coordinates is

$$\frac{\partial^2 V}{\partial z^2} = -\frac{\rho}{\epsilon_0}. \tag{60}$$

It is this equation which introduces the coupling between the two beams because ρ is a scalar quantity, the total space charge at a given point. The derivation of eqn. (6), (7) and (8) is not modified, so that we can put eqn. (8) into the second of eqn. (6) to obtain

$$\begin{aligned} \rho_{1a} &= \text{alternating component of } \rho_a \\ &= \frac{\beta_{ea}}{\omega(j\beta_{ea}-\Gamma)^2}\frac{I_{0a}}{2V_{0a}}\Gamma^2 V \\ &= \frac{-u_{0a}}{(\omega+ju_{0a}\Gamma)^2}\frac{I_{0a}}{2V_{0a}}\Gamma^2 V. \end{aligned} \tag{61}$$

Similarly,

$$\rho_{1b} = \frac{-u_{0b}}{(\omega+ju_{0b}\Gamma)^2}\frac{I_{0b}}{2V_{0b}}\Gamma^2 V. \tag{62}$$

Taking these with eqn. (60) gives

$$\Gamma^2 V = -\frac{1}{\epsilon_0}(\rho_{1a}+\rho_{1b}),$$

$$\text{therefore } 1 = \left[\frac{u_{0a}}{(\omega+ju_{0a}\Gamma)^2}\frac{I_{0a}}{2\epsilon_0 V_{0a}} + \frac{u_{0b}}{(\omega+ju_{0b}\Gamma)^2}\frac{I_{0b}}{2\epsilon_0 V_{0b}}\right]. \tag{63}$$

Eqn. (63) plays the same role in the theory of the space-charge wave tube as eqn. (12) did in the theory of the travelling-wave tube, for it determines the propagation constants for the allowed waves. Haeff's† method of procedure is to put

$$\frac{u_{0a} I_{0a}}{2\epsilon_0 V_{0a}} = \frac{e}{m}\frac{\rho_{1a}}{\epsilon_0} = \omega_a^2$$

with a similar expression for ω_b^2. These expressions are recognizable as the electron plasma frequencies. Then, if there is only one homogeneous beam,

$$\Gamma_1 = j\left(\frac{\omega}{u_{0a}} \pm \frac{\omega_a}{u_{0a}}\right).$$

† Haeff, loc. cit.

Thus two waves propagate along the beam with velocities just above and below the beam velocity. To solve eqn. (63) he puts $u_{0a} = u_0 + \delta$, $u_{0b} = u_0 - \delta$, and $\Gamma = j\dfrac{\omega}{u_0} + j\gamma$. Eqn. (63) then becomes

$$1 = \frac{\omega_1^2}{(\delta\omega/u_0 + \gamma u_0)^2} + \frac{\omega_2^2}{(\delta\omega/u_0 - \gamma u_0)^2}. \tag{64}$$

Eqn. (64) is easily solved if $\omega_1 = \omega_2$, for in this case

$$\gamma\frac{u_0}{\omega_1} = \pm\sqrt{\left\{\left(\frac{\delta\omega}{u_0\omega_1}\right)^2 + 1 \pm \sqrt{\left[4\left(\frac{\delta\omega}{u_0\omega_1}\right)^2 + 1\right]}\right\}}. \tag{65}$$

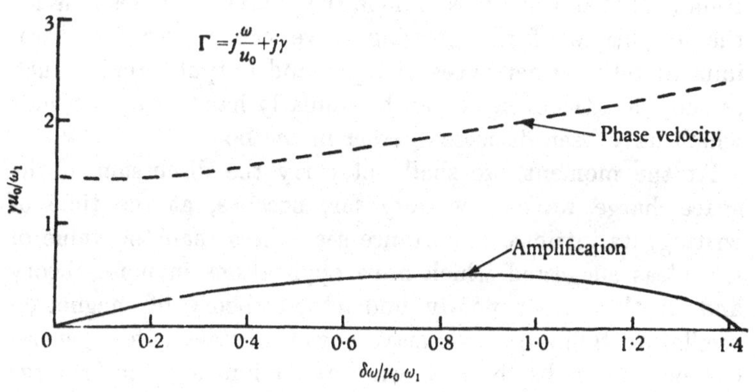

Fig. 14.10. Propagation constant for two-beam space-charge wave tube.

The left-hand side of eqn. (65) is proportional to the perturbation introduced into the propagation constant and the right-hand side is a function of δ, which for a fixed mean velocity u_0 is half the difference in velocity between the two beams. Eqn. (65) is plotted in fig. 14.10. For some values of δ, Γ has a negative real part, and therefore an amplified wave can be propagated down the tube. The negative attenuation takes its maximum value of 0·5 when

$$\frac{\delta\omega}{u_0\omega_1} = \frac{\sqrt{3}}{2}.$$

More general solutions of eqn. (63) in which $\omega_1 \neq \omega_2$ have been discussed by Nergaard.†

† Nergaard, loc. cit.

To deduce the performance of the tube we must introduce a set of boundary conditions, which of course depend on the particular geometry used. Taking the structure of fig. 14.8 as an example, we could take the zero plane at the centre of the first h.f. circuit. Velocity modulations are imposed on the beam at this point and their values can be written down in terms of the impressed voltage and the gap dimensions, using the expressions given in Chapter 12. Then at $a = 0$ we have prescribed values of u_1a and u_1b and the additional fact that $\rho_{1a} = \rho_{1b} = 0$. These relations determine all the arbitrary constants, so that when this is done the complete expressions for the amplitude of the growing wave can be written down immediately. Other types of input and output circuits, such as short lengths of helix, can be similarly handled by methods which have been discussed earlier in the book.

At the moment, we shall not carry the discussion of the space-charge wave tube very far, because, at the time of writing, its intrinsic importance seems less than the value of the ideas suggested which have applications in noise theory and in the rather poorly understood theory of magnetron oscillation build-up. We may, however, state a few general conclusions. It has been shown that the limiting gain for large current densities is 27·3 db./wave-length. The usefulness of the tube is somewhat limited by the fact that it tends to give maximum output at rather low voltages which make it difficult to obtain enough beam current to work at reasonable power levels. It is therefore to be regarded as a signal amplifier rather than as a power amplifier. It also becomes clear from fig. 14.10 that only a limited degree of velocity variation is permitted if the space charge waves are to be amplified. Saturation with increasing input is therefore to be expected and is, in fact, marked in the tube described by Hollenberg.† The electronic band-width is very large, so that the overall band-width is determined by the input and output circuits. Finally, the tube is noisy, at any rate according to the experimental results of Haeff,‡ for no theory of the noise has yet been given.

† Hollenberg, *Bell Syst. Tech. J.* **28** (1949), 52.
‡ Haeff, loc. cit.

14.13. Additional bibliography for this chapter

This chapter has been devoted to one particular type of theory. The field theory has not been discussed for reasons given in the introductory sections. However, the theory of travelling-wave tubes and space-charge wave tubes is not finally established and we give an extended bibliography in order that readers can study other approaches to the problem.

14.13.1. *Field theory*

Hahn, *Gen. Elect. Rev.* **42** (1939), 258, 497.
Ramo, *Proc. Inst. Radio Engrs., N.Y.* **27** (1939), 757.
Ramo, *Phys. Rev.* **56** (1939), 276.
Rydbeck, *Ericsson Technics*, no. 46 (1948) (reprinted as Rep. No. 1 from Research Lab. of Electronics, Chalmers University, Gothenburg).
Chu and Jackson, *Proc. Inst. Radio Engrs., N.Y.* **36** (1948), 853.
Laplume, *Onde élect.* **29** (1949), 66.
Lapostolle, *Ann. Télécommun.* **3** (1948), 57.
Lapostolle, *Ann. Télécommun.* **3** (1948), 257.

14.13.2. *Kinematic theory*

Döhler and Kleen, *Ann. Radioélect.* **2** (1947), 232.
Döhler and Kleen, *Ann. Radioélect.* **3** (1948), 124.

14.13.3. *Experimental results on tubes and circuits*

Cutler, *Proc. Inst. Radio Engrs., N.Y.* **36** (1948), 230.
Kompfner, *Proc. Inst. Radio Engrs., N.Y.* **35** (1947), 124.
Tomner, *Trans. Chalmers Tech. Gothenburg* no. 67, (1948).
Bruck, *Ann. Radioélect.* **4** (1949), 222.
Rogers, *Elect. Commun.* **26** (1949), 144.
Blanc-Lapierre and Kuhner, *Ann. Telécommun.* **3** (1948), 259.
Bryant, *Electr. Commun.* **27** (1950), 277.

14.13.4. *Theory of wave guide circuits*

Chu and Hansen, *J. Appl. Phys.* **18** (1947), 999.
Brillouin, *J. Appl. Phys.* **19** (1948), 1023.
Pierce, *Bell Syst. Tech. J.* **29** (1950), 189.
Walkinshaw, *Proc. Phys. Soc.* **61** (1948), 246.

14.13.5. *Space-charge wave tubes*

Guénard, Berterottière and Döhler, *Bull. Soc. Fr. Élect.* **9** (1949), 543; *Ann Radioélect.* **4** (1949), 171.
Macfarlane and Hay, *Proc. Phys. Soc.* **63** (1950), 409.

14.13.6. *Comprehensive Studies*

J. R. Pierce, *Traveling-Wave Tubes.* Van Nostrand, N.Y., 1950.
R. Kompfner, 'Travelling Wave Tubes', *Rep. Progress Physics*, **15** (1952), 275.

Chapter 15

MAGNETRONS

15.1. Introductory

The magnetron has been known as a generator of microwaves since its invention in 1921 by Hull†, but it was only in the late thirties that magnetrons were transformed from laboratory devices to practical oscillators for high powers, by the addition of cavity circuits built integrally with the anode. Early types of magnetron, using either solid anodes or split anodes, functioned in several ways of which the most important were the dynatron régime and space-charge modes. An early example of the use of magnetrons at about 1 cm. wave-length is given by Cleeton and Williams.‡

The original magnetron was simply a cylindrical diode with an axial magnetic field. If the magnetic field is increased, at a constant V_a, the anode current remains constant until it drops rather suddenly to zero. This critical point is called 'cut-off'. Fig. 15.1 shows a single-anode magnetron with a family of characteristics taken at various values of V_a. It can be shown that the paths of the electrons are cycloids when the field is adjusted for cut-off. Oscillations are detected if a resonant circuit is connected between the filament and anode, and the dimensions of the device are made such that the transit time round the cycloidal orbit is approximately equal to the period of the circuit.

Dynatron oscillations are normally encountered in split-anode magnetrons working with magnetic fields considerably greater than that at cut-off. Fig. 15.1 *d* shows a two-gap magnetron, with an indication of the field between the two segments at a particular instant of time. A d.c. bias is applied to the segments and the valve yields static characteristics of the form shown. Since more current goes to the plate at a

† Hull, *Phys. Rev.* **18** (1921), 31.
‡ Cleeton and Williams, *Phys. Rev.* **50** (1936), 1091.

lower potential, the device has a negative resistance characteristic and will sustain oscillations in a resonant circuit, connected between the two anode segments. Often more than one pair of segments is used.

Other types of oscillation, which in the light of present-day knowledge can be ascribed to travelling wave interactions, were observed. Since these are the types of oscillation used in

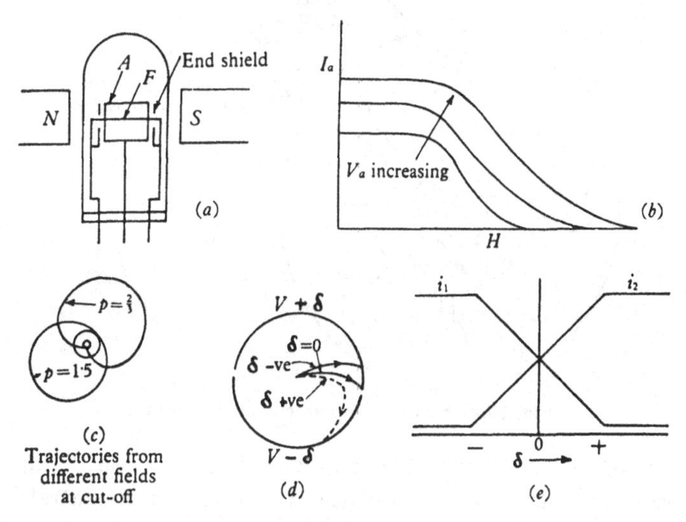

Fig. 15.1. Magnetron with cut-off characteristic and orbits for assumed field variation.

cavity magnetrons we shall describe them in some detail later on.

A good description of the state of knowledge on magnetrons immediately before the war will be found in Groos.† The great step forward was made by Randall and Boot at the University of Birmingham, who tried a magnetron of the form shown in fig. 15.2.‡ Much the same scheme had been described in various patents,§ but apparently without any practical realization.

† Groos, *Theorie und Technik der Dezimeterwellen*, S. Hirzel, Leipzig, 1937.

‡ Randall and Boot, *J. Instn. Elect. Engrs.* **93**, pt. IIIA, no. 5 (1946), 928. Megaw, *J. Instn. Elect. Engrs., N.Y.* **93**, pt. IIIA, no. 5 (1946), 977.).

§ Samuel, U.S.P. 2,063,341 (1947). Clavier, B.P. 540, 232 (1941

Russian investigators† had also worked with successful continuous-wave magnetrons of the same form, obtaining results which were extremely good at that date. However, these disclosures were not known to the Birmingham investigators whose interests were, in any case, mainly in the direction of pulsed generators. This fact had an important result in that

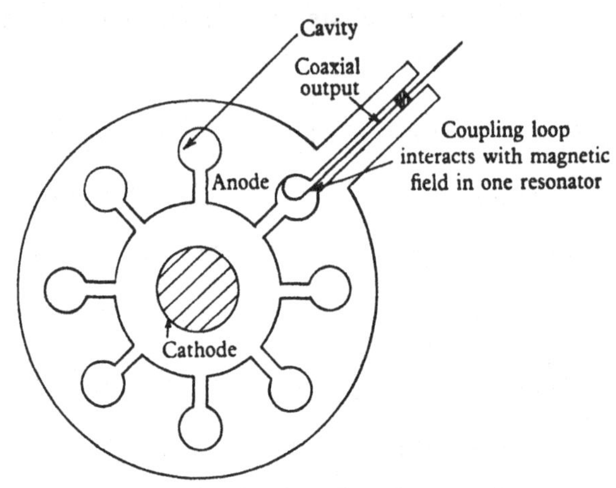

Fig. 15.2. Section through cavity magnetron.

it led to the trial of oxide-coated cathodes whose diameters were quite a large fraction of the diameter of the central hole in the anode block. Large efficiencies were obtained in this condition, for reasons now understood, but not then obvious. Since 1940 most magnetrons have been of the cavity type and enormous advances have been made. At a wave-length of 10 cm., pulse powers of at least 3 megawatts can be obtained. Magnetrons working at wave-lengths well below 1 cm. have been made. Continuous-wave results are not so spectacular, but outputs of 250 W. at approx. 3 cm. have been obtained. The efficiency of pulsed magnetrons is high, values of up to 70 or 80 per cent having been obtained. The major defects of magnetrons are that they are noisy, which limits their use as local oscillators, and that they are not easy to modulate either for amplitude modulation or broad-band frequency modulation.

† Alekserev and Maliarov, *J. Tech. Phys. U.S.S.R.* **10** (1940), 1292; translated in *Proc. Inst. Radio Engrs., N.Y.* **32** (1944), 136.

Small values of frequency deviation are not difficult to obtain. Magnetrons are therefore used as the transmitting oscillators in radar sets, as sources of pulsed energy for linear accelerators, etc., for pulsed communication systems in a few instances and as continuous-wave sources where high power and efficiency is important and noise is no object. The magnetron principle has often been studied with a view to its use in amplifiers. Some success seems at last to attend these efforts, but it is too early to assess the value of such devices.

From the theoretical point of view, the magnetron is by far the most difficult device used in electronics. A useful theory must be a large-signal theory for obvious reasons. The large values of magnetic field and space charge used mean that nothing can be neglected, so that the theory is necessarily involved. The main problem is to obtain a theory which shows how the oscillations build up from noise, since it is fairly easy to obtain enough experimental data to explain the operation in the stable state. In spite of all this, progress made since the war in understanding travelling-wave tubes and space-charge wave tubes has introduced beneficial new ideas, and it may be hoped that the production of adequate theories is now only a question of time. However, the present state of magnetron theory is by no means adequate for the full discussion of even simple cases. For instance, the d.c. conditions in the magnetron operated beyond cut-off are still a matter for argument. In this chapter we can therefore do little more than indicate the main ideas of magnetron operation without much detail.

15.2. Magnetrons without space charge

As a preliminary step we shall treat electron motions in magnetrons, assuming that the saturated emission is so low and the anode voltage so high that space charge can be neglected. These assumptions are very far from the truth in modern magnetrons, and we must be wary of carrying over results into the discussion of real magnetrons. The relative simplicity of this analysis makes it possible to obtain a qualitative understanding without too much effort, and also allows us to introduce some of the mathematics useful in more realistic analyses.

Consider first the planar magnetron (fig. 15.3) with a positive voltage V_a applied to the anode and a magnetic field in the direction indicated by the vector **B**. We briefly discussed this

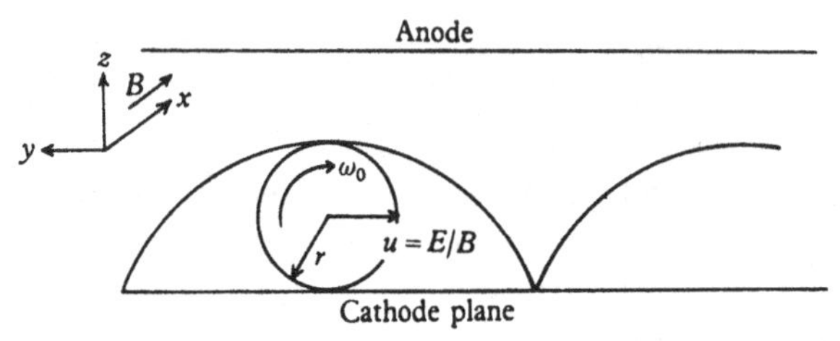

Fig. 15.3. Electron motion in planar magnetron.

situation in Chapter 5, and we there found that for zero initial velocity

$$\left.\begin{aligned} y &= \frac{at}{\omega_0} - \frac{a}{\omega_0^2}\sin\omega_0 t, \\ z &= \frac{a}{\omega_0^2}(1-\cos\omega_0 t), \\ x &= 0, \end{aligned}\right\} \tag{1}$$

where

$$a = \frac{e}{m}E_z, \quad \omega_0 = \frac{e}{m}B_x. \tag{2}$$

These expressions are the equations of a cycloid generated by a point on the circumference of a circle radius a/ω_0^2 rolling on the cathode plane. The angular velocity of a point on the circumference of the circle is ω_0, and the axis of the circle is translated in the direction parallel to the cathode plane with a linear velocity $u = \dfrac{a}{\omega_0} = \dfrac{E_z}{B_x}$. The frequency f_0 corresponding to ω_0 is termed the cyclotron frequency, as it is the frequency of rotation of an electron of arbitrary velocity injected into a uniform magnetic field when the electric field is zero.

When the initial velocities are not zero, the trajectories can easily be calculated using the initial conditions $\dot{x} = u_{0x}$, $\dot{y} = u_{0y}$, $\dot{z} = u_{0z}$ at $t = 0$ in the analysis given in Chapter 5. The results show that the paths are then trochoidal, i.e. they are generated by a point on an arm extending from the axis of the generating circle. Owing to the large electric fields used in practical magnetrons, the initial velocities are negligible and we need discuss this case no further, although these paths have a certain interest in the treatment of magnetic focusing.

Next consider a cylindrical magnetron whose cathode is a thin cylinder radius r_c extending along the z axis. The magnetic field is along the same axis and the electric field is radial. We use cylinder coordinates r, θ, z. As always the basic equation is the Lorentz force law

$$F = -e(\mathbf{E}+\mathbf{u}\times\mathbf{B}) = e\,\mathrm{grad}\,V - e\mathbf{u}\times\mathbf{B}. \qquad (3)$$

Writing this for our coordinate system and remembering that

$$\frac{\partial V}{\partial \theta} = \frac{\partial V}{\partial z} = B_r = B_\theta = 0,$$

we obtain†

$$\ddot{r} - r\dot{\theta}^2 = \frac{e}{m}\left(\frac{\partial V}{\partial r} - r\dot{\theta}B_z\right), \qquad (4)$$

$$\frac{1}{r}\frac{d}{dt}\left(r^2\dot{\theta}\right) = \frac{e}{m}\dot{r}B_z. \qquad (5)$$

Putting $\omega_L = eB_z/2m$ (the Larmor pulsatance) to save writing, we integrate eqn. (5) to give $r^2\dot{\theta} = \omega_L r^2 + C$. Or, if we assume that the electrons have no initial velocities,

$$\dot{\theta} = \omega_L\left(1 - \frac{r_c^2}{r^2}\right). \qquad (6)$$

Thus the angular velocity is always less than ω_L and is smaller the larger r_c becomes. Eqn. (6) is then put into eqn. (4) giving

$$\ddot{r} + r\omega_L^2\left(1 - \frac{r_c^4}{r^4}\right) = \frac{e}{m}\frac{dV}{dr}. \qquad (7)$$

Now $\quad V(r) = V_a\dfrac{\ln(r/r_c)}{\ln(r_a/r_c)}, \quad$ so $\quad \dfrac{dV}{dr} = \dfrac{V_a}{\ln(r_a/r_c)}\dfrac{1}{r} = \dfrac{k}{r},$

or

$$\ddot{r} + r\omega_L^2\left(1 - \frac{r_c^4}{r^4}\right) = \frac{e}{m}\frac{k}{r}. \qquad (8)$$

This equation can easily be integrated once by writing $p = \dot{r}$.

† In cylindrical coordinates $a_r = a_x\cos\theta + a_y\sin\theta$, $a_\theta = a_x\sin\theta + a_y\cos\theta$. The results quoted follow from differentiating $x = r\cos\theta$, $y = r\sin\theta$ twice with respect to t.

Then $$\ddot{r} = \frac{dp}{dt} = \frac{dp}{dr}\frac{dr}{dt} = p\frac{dp}{dr}.$$

Thus $$\int p\,dp = \int\left[\frac{e}{m}\frac{k}{r} - \omega_L^2 r\left(1 - \frac{r_c^4}{r^4}\right)\right]dr,$$

or $$p^2 = 2\frac{e}{m}k\ln r - \omega_L^2 r^2 - \omega_L^2\frac{r_c^4}{r^2} + C.$$

Again we assume that the initial velocities are zero, so that $p = 0$ at $r = r_c$. Solving for C and substituting we find

$$p^2 = (\dot{r})^2 = 2\frac{e}{m}k\ln\left(\frac{r}{r_c}\right) - \omega_L^2 r^2\left(1 - \frac{r_c^2}{r^2}\right)^2. \tag{9}$$

Eqn. (9) can be used directly to determine the cut-off condition, for by definition cut-off occurs when $\dot{r} = 0$ on the anode surface, i.e. the electrons just graze the anode. For this condition eqn. (9) gives

$$2\frac{e}{m}V_a = \omega_L^2 r_a^2\left(1 - \frac{r_c^2}{r_a^2}\right)^2. \tag{10}$$

For a given magnetron the values of V_a and B_z which cause the anode current to drop to zero are related by the parabolic expression (10). This curve is often referred to as the Hull cut-off parabola, as it was first investigated by Hull.† Numerically, for radii in cm., induction in gauss,

$$V_a = \frac{B_{zc}^2 r_a^2}{45{\cdot}5}\left(1 - \frac{r_c^2}{r_a^2}\right). \tag{11}$$

Eqn. (10) is important in the study of magnetrons in which the space charge is not negligible. This is because the necessary and sufficient condition for eqn. (10) to hold is that the potential V in the anode-cathode space should be a function of r alone, and eqn. (10) does not depend on any particular assumption as to the functional form of $V(r)$.

Returning to the integration of eqn. (9) we find

$$t = \int\frac{dr}{\sqrt{\{2(e/m)k\ln(r/r_c) - \omega_L^2 r^2[1 - (r_c^2/r^2)]^2\}}},$$
$$k = V_a\ln(r_c/r_a). \tag{12}$$

† Hull, *Phys. Rev.* **18** (1921), 31.

To proceed further we must approximate. This is done by noting that if the cathode is much smaller than the anode, r_c^2/r^2 is much less than unity over most of the range of integration. The second term under the radical is thus put equal to $\omega_L^2 r^2$. This argument is not strictly correct as the electron is travelling slowly in the region where r is small. Secondly, we may without error omit the bracket in eqn. (10) as $(r_c/r_a)^2 \ll 1$ certainly. Finally, instead of putting $V(r) = V_a \ln (r/r_0)$, let us put $V(r) = V_a(r/r_a)^p$ $(0 < p < 2)$, and study the integral as a function of p. Making these substitutions eqn. (12) becomes

$$\omega_L t = \int_0^{R_a} \frac{dR_a}{\sqrt{(R_a{}^p - R_a^2)}}, \quad R_a = r/r_a, \tag{13}$$

or

$$R_a = \left(\sin \frac{2-p}{2} \omega_L t\right)(2/2-p). \tag{14}$$

Putting $p = \frac{2}{3}$ we get $R_a = (\sin \frac{2}{3}\omega_L t)^{\frac{3}{2}}$, a result originally given by Hull. This is the equation of a cardioid. Values of p near $\frac{2}{3}$ give similar curves, up to $p = 1$ for which $R_a = \frac{1}{2}(1 - \cos \omega_L t)$.

For $p > 1$ the cardioids become convoluted as shown in fig. 15.1 *c*.

For $p = \frac{3}{2}$, $R_a = \frac{1}{4}(1 - \cos \omega_L t/2)^2$ which is sketched.

These results on orbits are of little direct interest in real magnetrons, except that they indicate that the trajectories depend very much on the way in which the radial field varies. In view of this indication, and the fact that the assumptions made are open to objection, it is hardly surprising that transit-time calculations made on the basis of these orbits are seriously in error.

To conclude these remarks on orbits let us calculate the angular velocity for an electron in a stable circular orbit. The equation of motion is (4) with $\ddot{r} = 0$, i.e.

$$\dot{\theta}^2 - 2\omega_L \dot{\theta} + \frac{2\omega_L}{rB_z} E_r = 0, \tag{15}$$

or

$$\dot{\theta} = \omega_L + \sqrt{\left(\omega_L^2 + \frac{2\omega_L}{rB_z} E_r\right)}$$

$$\doteqdot 2\omega_L + \frac{E_r}{rB_z}. \tag{16}$$

The first term represents the contribution made by circular motion round the line of translation, while the second term is due to the translation. The linear translation velocity is thus $u_r = r\,\dot{\theta}(t) = E_r/B_z$, very similar to the planar result.

15.3. Magnetrons with space charge

The problem of the magnetron with space charge is much more complicated than the case dealt with above, not only because of the increased mathematical difficulties in handling the expressions, but also because the question of the stability of the solutions derived is a complicated matter. This question arises in the following way. A possible method of procedure in the space-charge limited case is to use the Langmuir solution for the cylinder diode (Chapter 6) to obtain $V(r)$. This is physically correct if we consider that the anode voltage is switched on and the anode current reaches its equilibrium value before the magnetic field is increased from zero. For magnetic fields well below cut-off, the anode current remains constant at the zero field value, and all that happens is that the electron paths are curved instead of being radial. At cut-off the anode current suddenly drops to zero, and all the electrons which originally reached the anode now return to the cathode. After the period of the transient, which occurs as B_z goes through cut-off, conditions must be as follows. As many electrons leave the cathode as return to it, because the net current is zero and each of these contributions must be half the zero field anode current since the total space charge is the same as it was originally. The state in which there is no emission and no space charge is not allowed because there is no magnetic force acting at the cathode surface where the velocities are zero and, however large the magnetic field, the electrons must leave the cathode surface before they can be influenced by it. In a real valve the initial velocities are not zero, and what happens is that the number of electrons with velocities in a certain range fluctuates so that when there is an excess of high initial velocity electrons, more low-velocity ones are returned to the cathode and vice versa.

As far as the orbits are concerned there are two possibilities, first orbits which are broadly similar to those in the absence of space charge, and secondly, circular orbits round the cathode. The first case is a double-stream flow, and the second one a single-stream flow, the effect of the fluctuation being taken up as a slight expansion or contraction of the space-charge ring as a whole. The case of circular orbits has been fully investigated by Brillouin† and is usually described as the Brillouin steady state. A very convincing case for the existence of this state can be built up for the system we are considering by plotting the trajectories with gradually increasing B_z. It is not so obvious that this is the correct state when the more practical case of fixed B_z and suddenly applied V_a is considered. On the whole, the Brillouin steady state is accepted as representing the space-charge conditions beyond cut-off, but it cannot be regarded as proven without doubt that this is the case.

We now take up the discussion of the Brillouin steady state. This is done by introducing an apparent spherical potential function $P(r)$ such that

$$m\ddot{r} = e\frac{\partial P}{\partial r}.$$

Comparing this with eqn. (7) or eqn. (9) we see that

$$eP(r) = eV(r) - \tfrac{1}{2}m\omega_L^2 r^2[1-(r_c^2/r^2)]^2. \tag{17}$$

$\dot{\theta}$ is still given by eqn. (6). The study of the motion is thus reduced to the study of the function $P(r)$. We now consider the potential distribution in a space-charge limited system. In our coordinates Poisson's equation reduces to

$$\frac{1}{r}\frac{\partial}{\partial r}\left(r\frac{\partial V}{\partial r}\right) = -\frac{\rho}{\epsilon_0},$$

where ρ is defined by $I = 2\pi r\rho\dot{r}$, I = current per unit cathode length. Now $\dot{r}$ is given by eqn. (9) which we have already seen

† Brillouin, *Electr. Commun.* **20** (1941), 112; *Phys. Rev.* **60** (1941), 385 (steady state); **62** (1942), 166 (split-anode magnetron); **63** (1943), 127; *Proc. Instn. Radio Engrs.* **32** (1944), 216 (review of results).

to be independent of the existence of space charge. Therefore

$$\frac{\partial}{\partial r}\left(r\frac{\partial V}{\partial r}\right) = -\frac{1}{\epsilon_0}\frac{I}{2\pi\dot{r}} = -\frac{I}{2\pi\epsilon_0}\frac{1}{\left[2\frac{e}{m}V(r) - \omega_L^2 r^2\left(1-\frac{r_c^2}{r^2}\right)^2\right]^{\frac{1}{2}}}. \quad (18)$$

This has to be solved subject to the conditions that

$$V(r_c) = 0, \quad \left.\frac{\partial V}{\partial r}\right|_{r=r_c} = 0,$$

which are those used in the simple theory of space-charge limitation, when the initial velocities are neglected. Brillouin starts the discussion by taking the current $I = 0$, i.e. the magnetron is cut-off. There are two possibilities, (a) $\dot{r} = 0$, $\rho \neq 0$, (b) $\rho = 0$, $\dot{r} \neq 0$; the second is of no interest, as it simply relates to the cylinder condenser and gives the solution $V = A\ln r + C$. The first case gives $P(r) = 0$ or

$$V_0(r) = \frac{m}{2e}\omega_L^2 r^2\left(1-\frac{r_c^2}{r^2}\right)^2. \quad (19)$$

$V_0(r)$ is the maximum potential for which a given magnetron can be cut-off as has already been said. The space-charge density is easily evaluated by putting this expression into Poisson's equation and carrying out the indicated differentiations. The result is

$$\rho_0(r) = -\frac{2\epsilon_0 m\omega_L^2}{e}\left(1+\frac{r_c^4}{r^4}\right). \quad (20)$$

These equations give the full results for the actual cut-off voltage; if the anode voltage is below the cut-off voltage the space charge does not extend out to the anode but only to some smaller radius a. The potential from a to r must be logarithmic and the constants can be evaluated by making the potential and field continuous at the boundary $r = a$. Thus

$$\left.\begin{aligned} \frac{m}{2e}\omega_L^2 a^2\left(1-\frac{r_c^2}{a^2}\right)^2 &= A\ln a + C, \\ \frac{m}{e}\omega_L^2 a\left(1-\frac{r_c^4}{a^4}\right) &= \frac{A}{a}. \end{aligned}\right\} \quad (21)$$

This completes the discussion for the magnetron in which the magnetic field is greater than that required for cut-off. For fields below cut-off, a rigorous solution is rather difficult. Brillouin divides the anode-cathode region into two zones, one near to the cathode and the other far distant from it. In the first zone the magnetic forces are small because the velocities are small, the solution is then only a perturbation of the solution for the Langmuir diode. Far from the cathode the magnetic forces are much bigger than the electrostatic ones. A solution neglecting the latter then holds for large radii. The intermediate region naturally presents difficulties, but it is easy enough to get a general idea of the behaviour in this zone. We first observe that $eP(r)+eV_0(r) = eV(r)$ from eqns. (17) and (19). Therefore

$$\frac{\partial}{\partial r}\left\{r\frac{\partial}{\partial r}[P(r)+V_0(r)]\right\} = -\frac{I}{2\pi\epsilon_0}\sqrt{\left(\frac{2m}{e}\right)}\frac{1}{\sqrt{P}}. \tag{22}$$

But $V_0(r)$ is given by eqn. (19). Inserting this we have

$$\sqrt{(P)}\left[\frac{2m\omega_L^2}{e}\left(r+\frac{r_c^4}{r^3}\right)+\frac{\partial}{\partial r}\left(r\frac{\partial P}{\partial r}\right)\right] = -\frac{I}{2\pi\epsilon_0}\sqrt{\left(\frac{2m}{e}\right)}. \tag{23}$$

This equation has to be solved for P subject to the conditions $P(r_c) = 0$, $\left.\frac{\partial P}{\partial r}\right|_{r=r_c} = 0$. The solution is carried out by approximation. Near the cathode we can assume a power expansion. Put $P(r) = A(r-r_c)^n$. Then, since $r \doteqdot r_c$,

$$\sqrt{[A(r-r_c)^n]}\left[4\frac{m\omega_L^2}{e}r_c+nA(r-r_c)^{n-1}+nA(n-1)r_c(r-r_c)^{n-2}\right]$$
$$= \frac{-I}{2\pi\epsilon_0}\sqrt{\left(\frac{2m}{e}\right)}.$$

If we put $n = 4/3$, suggested by the space-charge limited planar diode, to which the solution must reduce for $r-r_c$ small, we find that the first two terms are zero and P is given by

$$P(r) = \frac{1}{2}\left(\frac{-9}{r_c}\frac{I}{2\pi\epsilon_0}\right)^{\frac{2}{3}}(r-r_c)^{\frac{4}{3}}\left(\frac{m}{e}\right)^{\frac{1}{3}}. \tag{24}$$

This approximation is only true for $r/r_c < 1{\cdot}5$. A better approximation is obtained by using Langmuir's solution for the cylindrical diode

$$P(r) = \frac{1}{2}\left(\frac{-9I}{2\pi\epsilon_0}\right)^{\frac{2}{3}}\left(\frac{m}{e}\right)^{\frac{1}{3}}(\beta^2 r)^{\frac{2}{3}}. \tag{25}$$

The numeric β^2 has been defined in Chapter 6.

The second region is that in which the magnetic forces are the only important ones. Then $\frac{\partial}{\partial r}\left(r\frac{\partial P}{\partial r}\right)$ can be neglected and P_∞ is given by

$$P_\infty = \frac{I^2}{8\pi^2\epsilon_0^2\omega_L^4}\frac{e}{m}\left(r+\frac{r_c^4}{r^3}\right)^{-2}. \tag{26}$$

Now $\frac{\partial}{\partial r}\left(r\frac{\partial P_\infty}{\partial r}\right) = \frac{4P}{r}$, since in the derivation we have assumed that r is large and thus $\frac{r_c^4}{r^3}$ is negligible. The condition for eqn. (26) to be valid is then that

$$\frac{I^2}{2\pi^2\epsilon_0^2\omega_L^4}\frac{e}{m}\frac{1}{r^3} \ll \frac{2m\omega_L^2 r}{e},$$

or

$$\left(\frac{I}{2\pi\epsilon_0\omega_L^3}\frac{e}{m}\right) \ll r^2. \tag{27}$$

Brillouin uses the symbol L^2 for the characteristic radius $\frac{I}{2\pi\epsilon_0\omega_L^3}\frac{e}{m}$. We use the more appropriate R^2. Numerically $R^2 = \frac{2{\cdot}3\times10^6 I}{H^3}$ cm.^{2}, I in amp./cm., length H in oersteds. Inserting values encountered in modern magnetrons, R is a few millimetres. As the magnetron dimensions are scaled down to shorter wave-lengths, R tends to remain a constant proportion of the cathode-anode space. In fig. 15.4 we have indicated Langmuir's distribution and the form of P for three values of I, showing how the Langmuir distribution must vary into the asymptotic distribution with increasing r.† Clearly the larger I is the greater is the radius at which the potential distribution departs from the Langmuir distribution. If $R < r_c$

† This is discussed in detail by Brillouin, *Electr. Commun.* **20** (1941), 112.

we find, by equating $P(r)$ with $P(\infty)$, that $R^2 \doteq 3\beta(2r_c)^{\frac{3}{2}} r^{\frac{1}{2}}$. This gives the value of r beyond which the Langmuir distribution is seriously in error.

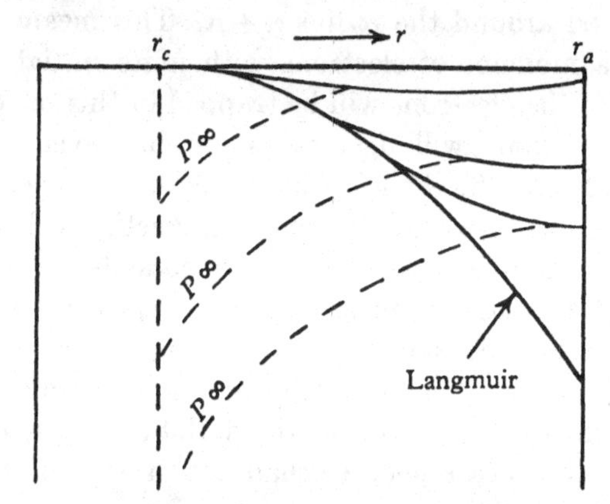

Fig. 15.4. Brillouin's potential distribution.

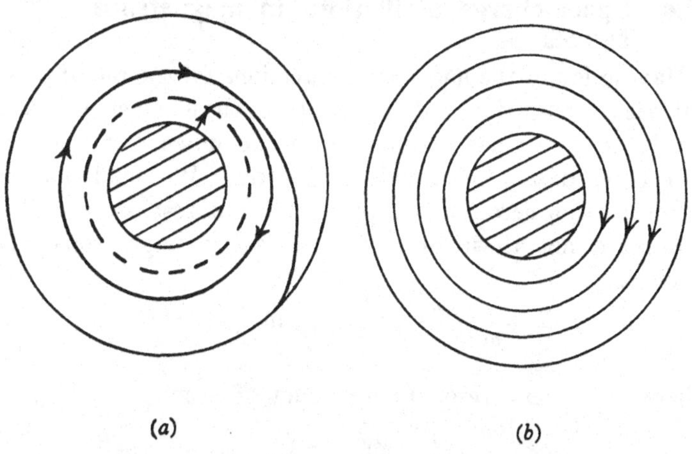

Fig. 15.5. Orbit formation according to Brillouin.

The physical meaning of $P(r)$ is that it is the amount by which the potential at any point in the magnetron is depressed below the potential $V_0(r)$, valid in the absence of space charge. Since $eP(r)$ is the radial energy the electrons are accelerated from r_c to R and then retarded. Fig. 15.5 a shows an orbit for an electron just reaching the anode; fig. 15.5 b shows orbits in a cut-off magnetron, each electron moving with the linear velocity $r.B$ Another point which is worthy of mention is that $P(r)$ exhibits a minimum when R is small, the minimum being located around the radius r_c+R. This means, when we think for a moment of electrons with finite initial velocities, that some of the electrons will be trapped in this region so that the anode current will decrease somewhat even before the magnetron is cut off. This is what is observed experimentally.

We have not answered the question whether other types of motion, e.g. double stream flow, are possible. We refer the reader to the discussion in *Microwave Magnetrons*.† The conclusion is that such states can only exist in magnetrons for which $r_a/r_c > 2{\cdot}2$ and in these only for limited ranges of anode voltage. Since the question of the stability of these types of flow is a very delicate one, we shall not pursue the topic here.

15.4. Space-charge oscillations in magnetrons. Small signals

Having found the necessary expressions for the radial potential it is a straightforward application of the small signal theory developed in earlier chapters to write down the electronic impedance of a single-anode magnetron. We shall not give the derivation here;‡ it is of only minor interest, since the processes are not those operating in cavity magnetrons. The result is

$$R_a \doteqdot \frac{4\omega}{\omega^2(r)-\omega^2}\left(\frac{1}{r_c^2}-\frac{1}{r^2}\right)\sin\frac{\omega\tau}{2}\,e^{-j(\omega\tau/2)}, \tag{28}$$

where ω = resonant frequency of circuit,

$$\omega^2(r) = 2\omega_L^2\left(1+\frac{r_c^4}{r^4}\right),$$

τ = period of the oscillation.

$R_a \to \infty$ as $\omega^2(r) \to \omega^2$. A more exact theory would not show this property, but R_a will still have a large maximum in

† *Microwave Magnetrons*, M.I.T. series, vol. **6** chap. 6, by L. R. Walker.
‡ See Brillouin, *Phys. Rev.* **60** (1941), 385; **63** (1943), 127.

this region, i.e. where $\omega \doteqdot \sqrt{(2)}\omega_L$.

$$R_e(R_a) = \frac{2\omega}{\omega^2(r)-\omega^2}\left(\frac{1}{r_c^2}-\frac{1}{r^2}\right)\sin\omega\tau. \tag{29}$$

The behaviour is sketched in fig. 15.6, where the function crosses the axis at $\omega = \frac{n\pi}{\tau}$. Oscillations are possible for all the regions in which $R_e(R_a)$ is negative, but the most powerful ones

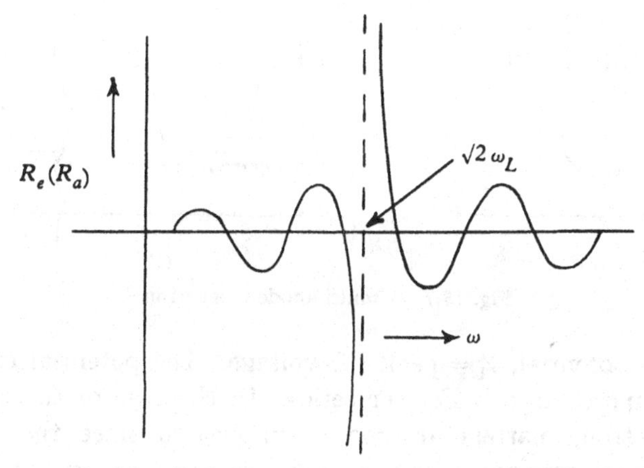

Fig. 15.6. Impedance for space-charge oscillations.

are those for which ω is a little less than $\sqrt{(2)}\omega_L$. This type of oscillation can be shown to correspond to a radial expansion and contraction of the space charge, i.e. an oscillation with cylindrical symmetry. The other oscillations represent more complicated motions of the space charge cloud.

It is known experimentally that electronic oscillations of this type are not efficient mechanisms for the transfer of d.c. energy into r.f. energy. This is physically rather easy to understand, since there is no coupling with the tangential motion of the electrons which in the case of circular orbits accounts for practically the whole k.e. We expect, therefore, that to generate r.f. power with efficiency it will be necessary to operate on the tangential velocity. Before discussing this type of magnetron, which is that used to-day, we must digress somewhat to discuss the circuit problem.

15.5. Circuits for cavity magnetrons

One very obvious difference between the simple cylindrical magnetron which we have hitherto discussed and the multi-anode magnetron or the modern cavity magnetron is the fact that the r.f. field distribution varies round the anode. Consider the six-anode magnetron of fig. 15.7. Since the connexion constrains neighbouring segments to be 180° out of phase, the potential round the anode is as shown, where V_0 = d.c.

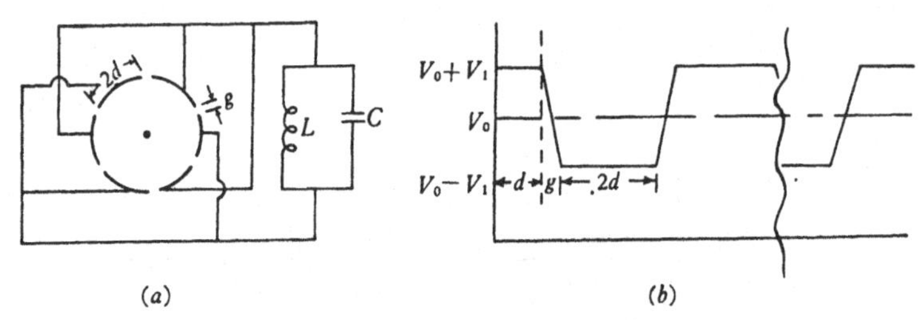

Fig. 15.7. A multi anode magnetron.

anode potential, V_1 = peak r.f. voltage. The potential can be decomposed into a Fourier series. In the case of the cavity magnetron, matters are more complicated since the phase difference between adjacent lands is not necessarily 180° or π radians, although nearly all modern magnetrons are designed to work on the π mode and the other modes only appear as unwanted perturbations. Since each cavity may be represented by a mesh in a chain of resonant circuits, there are very many resonances permitted to the whole block and only one of these gives the desired π mode. A change of mode is accompanied therefore by a change of frequency as well as a drop in output due to the less favourable electronic interaction. In most practical applications the frequency change is more objectionable than the change in power. The purposes of magnetron circuit theory are therefore to determine, at any rate approximately, the few lowest frequencies of the block; to determine the frequency of the π mode; to suggest, if necessary, methods by which the π mode frequency can be shifted away from the frequencies of interfering modes in order to make the excitation of these less likely at the specified working point (voltage, current and field); and lastly to give information about the r.f. field in the interaction space.

The behaviour of magnetron blocks has been extensively studied by equivalent circuit methods and by direct use of Maxwell's equations. The latter approach is much more useful, but it is very lengthy; so, since it is not difficult, we refer the

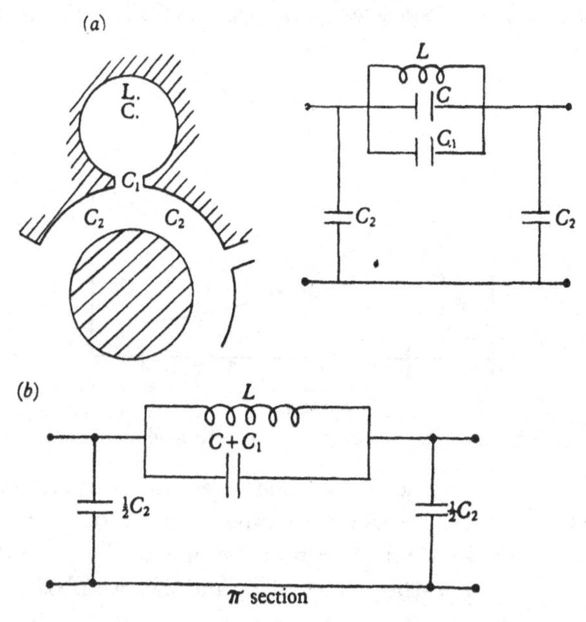

Fig. 15.8. Equivalent circuit of resonant cavity.

reader to the excellent treatments by Kroll and Walker† and by Kroll and Lamb.‡ Here we only deal with the simplest possible equivalent circuit method (fig. 15.8) in which each cavity is represented by a parallel L.C. circuit with an extra capacitance C_1 across the lips and equal capacitances C_2 to cathode. Such a treatment clearly neglects the infinite number of resonances actually occurring in a distributed circuit. These resonances are at much higher frequencies than those of interest in the present work. The block is represented by a ring of

† Kroll and Walker, *Microwave Magnetrons*, M.I.T. series, vol. **6**.
‡ Kroll and Lamb, *J. Appl. Phys.* **19** (1948), 166.

identical π sections (fig. 15.8 *b*), one for each cavity. The phase constant of each section is given by

$$\cos\phi = 1 + Z_1/2Z_2 \tag{30}$$

Z_1 = impedance of the parallel L, $(C+C_1)$ circuit,

Z_2 = impedance of the capacitance $C_2/2$.

This result can be found in any text-book on filter theory (e.g. Guillemin. *Communication Networks*, vol. 2, p. 179), if results

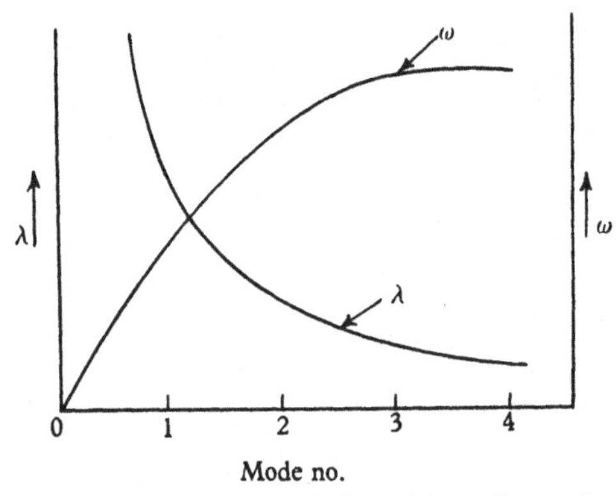

Fig. 15.9. Frequency variation with mode number.

for a non-dissipative network are inserted in the equations there deduced. Since the network is closed on itself only the phases $\phi = 2\pi n/N$ are allowed, N = number of cavities, $n = 0, 1, 2, \ldots, N/2$. Putting in the values of Z_1 and $2Z_2$ we obtain

$$\cos\phi = \frac{1 - \dfrac{\omega^2}{\omega_R^2} - \dfrac{1}{2}\dfrac{\omega^2}{\omega_R^2}\dfrac{C_2}{C_1+C}}{1 - \dfrac{\omega^2}{\omega_R^2}}, \quad \omega_R^2 = \frac{1}{\sqrt{L(C+C_1)}},$$

or, using the condition on ϕ,

$$\left(\frac{\omega}{\omega_R}\right)^2 = \frac{1}{1 + \dfrac{C_2}{2[1-\cos(2\pi n/N)](C+C_1)}},$$

i.e.

$$\omega = \omega_R \sqrt{\frac{1}{1 + \dfrac{\rho}{2[1-\cos(2\pi n/N)]}}}, \quad \rho = \frac{C_2}{C+C_1}. \tag{31}$$

Fig. 15.9 shows the behaviour of eqn. (31) for an 8-cavity

magnetron, which indicates that there is very little difference in frequency between the lowest wave-length modes, i.e. that the π mode, $n = N/2$, is very close in frequency to the $n = 3$ mode. By adjustment of the parameters eqn. (31) agrees reasonably well with experiment. This fact makes the simple N-cavity magnetron block very susceptible to mode changes and the early NT 98 and CV 41 magnetrons were very variable in efficiency which ranged from 10 to over 40 per cent for this reason.

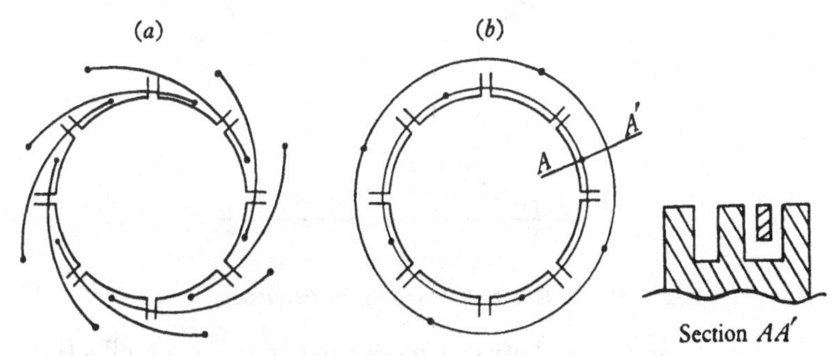

Fig. 15.10. Types of strapping. Dots represent a join between strap and segment.

In 1941 Sayers at Birmingham University introduced a technique for increasing the separation between the desired π mode and the adjacent modes. This expedient is known as strapping, and the essential point is that alternate anode segments are joined together by metallic bars located in the end space of the magnetron. For modes other than the π mode, current flows in the straps and constitutes a damping load on the resonator so that, in addition to the frequency shift caused by the mutual reactance elements introduced into alternate ring sections, the circuit losses for the undesired modes are increased. In this way the operation of the magnetron can be stabilized on the π mode. The use of strapping in NT 98 valves immediately brought the efficiency up to 50–60 per cent, since the valve operated invariably on the π mode and the input could also be increased without mode changes occurring. Various types of strapping can be used, some of which are shown in fig. 15.10. Fig. 15.10 *a* is the so-called echelon strapping and fig. 15.10 *b* the double-ring strapping which is commonly used to-day. The strapping system is broken on the bottom end of the resonator in such a way that the wave patterns due to the unwanted modes are correctly loaded. This is important because the unwanted modes turn out to be doublet modes and, if one component is not loaded, moding may still occur. The

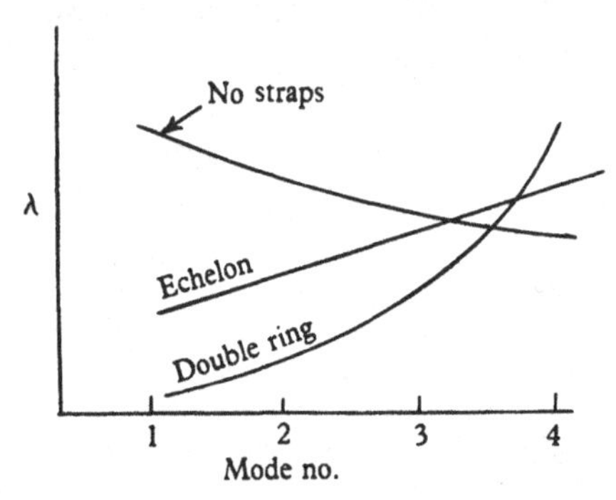

Fig. 15.11. Effect of strapping on frequency curve.

degree of separation between modes depends on the effectiveness of the strapping, the double-ring strapping being the most effective. Fig. 15.11 indicates the way in which the resonant wave-lengths of a magnetron block change as the strapping is applied. It can be seen that a very considerable increase in spacing between the π mode and the $n = 3$ mode is obtained.

The behaviour of the strapping can be understood qualitatively as follows. For the π mode the straps merely increase the parallel capacitance of each resonator slightly since no r.f. current flows in them. In the other modes current flows in the straps, which is equivalent to decreasing the inductance of the individual resonators. Thus the decrease in inductance offsets the increased capacitance and experiment shows that there is a net increase in frequency for the non-π modes.

For wave-lengths below 3 cm., e.g. for K band (1·25 cm.), the block and the straps become so small that constructional difficulties are very formidable. The Columbia Radiation Laboratory† group overcame this difficulty by making a block

† Millman and Nordsieck, *J. Appl. Phys.* **19** (1948), 156. Hollenberg, Kroll and Millman, *J. Appl. Phys.* **19** (1948), 624.

with alternate large and small resonators sketched in fig. 15.12. This block is known as the 'Rising Sun' and it introduces a very different type of behaviour. The resonances occur in two

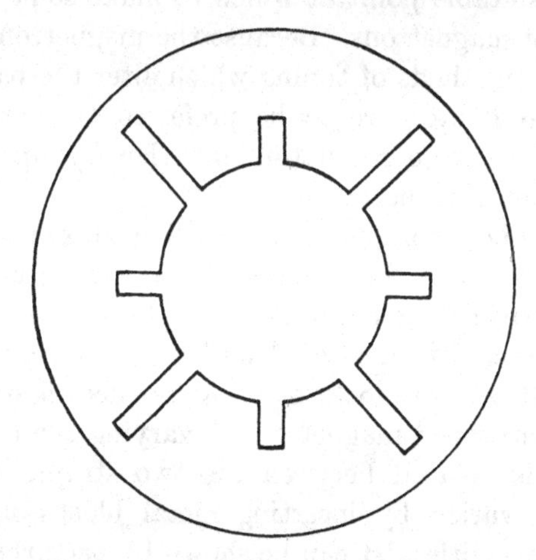

Fig. 15.12. Rising Sun magnetron block.

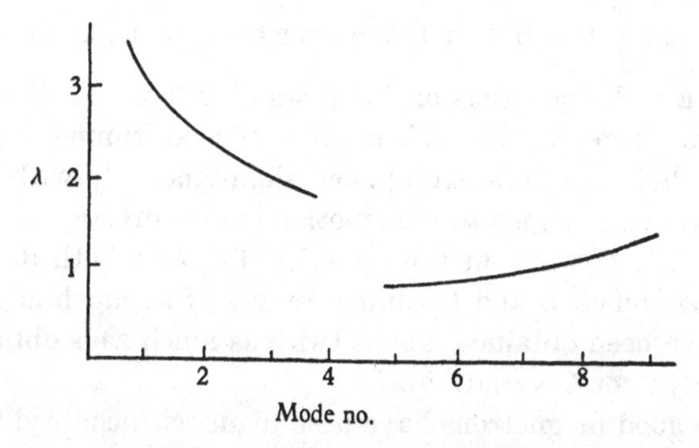

Fig. 15.13. Frequency variation for a Rising Sun block.

groups, one set belonging to the large resonators and the other to the small, as shown in fig. 15.13, where λ_0 is the resonant wave-length of the smaller resonator. Fig. 15.13 is drawn for an 18 resonator block, i.e. $n = 9$ is the π mode which is seen to be well separated from the $n = 8$ and $n = 4$ modes. Because of its simplicity, the Rising Sun block is now used for high power 3 cm. magnetrons, as well as for the very shortest wave-length magnetrons. The full theory of the Rising Sun magnetron is given in the references cited, and we shall not discuss it here.

This is a suitable point at which to make some remarks on the tuning of magnetrons. Because the magnetron circuit is a coupled ring, methods of tuning which alter the parameters of every unit resonator are to be preferred to methods which simply alter a single resonator, and thereby upset the electronic behaviour of the valve.

Tuning can be accomplished by varying either the inductance or the capacitance of the cavities. Metal rods inserted into the cavities decrease their inductance and lower the wave-length. The capacitance can be altered by bringing a ring up towards the open ends of the block; or a bigger effect is obtained in a double-ring strapped magnetron by varying the insertion of a metal cylinder placed between the two straps. Capacitance can also be varied by inserting metal blocks into the gap region of the cavities. It can be shown by perturbation theory that the change in reactance produced by a metal insert depends on $\int(E^2-H^2).dv$ taken over the volume of the insert. Where $E>H$, the insertion increases C and where $H>E$ the insertion decreases L. When an extended tuning range is desired, both methods are applied simultaneously with mechanical motions ganged so that motion in one direction increases both L and C while in the opposite direction both decrease. Using combined L and C tuning, ranges of as much as 30 per cent have been obtained, about twice as much as is obtainable with only L or C variation.

Some good magnetrons have been made with coupled tuning elements but this expedient is only workable over ranges of the order ± 5 per cent. This is, however, sufficient for radar purposes where the problem is usually one of keeping within a fairly wide band and the receiver is tuned to a particular transmitter. More extensive band-widths are necessary for communication and other purposes, and the tuning problem is of more importance in c.w. magnetron design than it is in pulsed magnetrons.

It should be clear from what has been said above that a small variation in strap spacing will change the wave-length. This is found experimentally to be the case, and deformation of the straps is a well-known method of pre-setting the oscillation frequency of fixed tuned magnetrons so as to reduce the importance of mechanical tolerances in the resonator dimensions.

15.6. The radio-frequency field in the interaction space

Having established the main features of the behaviour of the circuits we must discuss qualitatively the behaviour of the

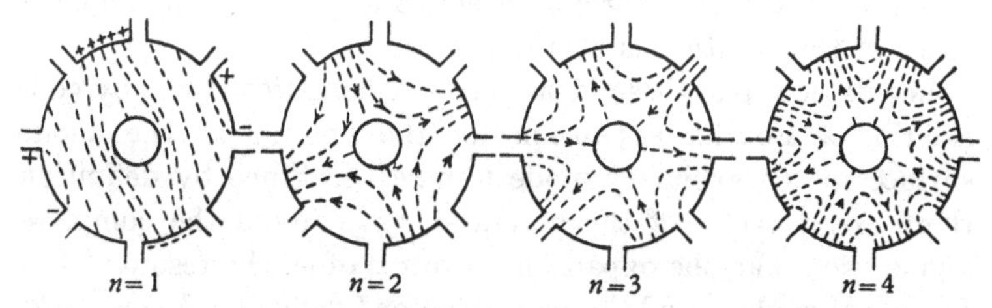

Fig. 15.14. Electric field distribution for several modes.

r.f. fields in the interaction space. For the π mode, the spatial variation of the field at the anode surface is as shown in fig. 15.7 *b*, since there is very little potential drop across the anode segments. The field on the anode surface for oscillation in other modes is determined by the condition that the variation of potential after a complete rotation round the cathode must return to the initial value, i.e. the potential wave has n complete cycles. Clearly n can only take on the values 0, 1, 2, ..., $N/2$ where N = number of resonators. These values of n give the field periodicities at the corresponding frequencies given by eqn. (31). Fig. 15.14 shows the approximate configuration of the interaction field for several of the modes. It should be noted at this point that the r.f. field of the π mode does not penetrate so far into the interaction space as do the lower modes. Thus the interaction between the electrons and the π mode is only strong when the cathode approaches the anode fairly closely. For a thin cathode the low modes have strong components near the cathode surface. This is one of the reasons why pre-war magnetrons with many anode segments and thin cathodes operated erratically and with low efficiencies.

The field configurations sketched in fig. 15.14 are, of course, rotating so that, taking the π mode, for example, during one half-cycle of the oscillation frequency the field configuration will have rotated by one anode segment. It is important to understand the peculiarities of the π mode: a rotation through an anode segment simply reverses the sense of the fields and does not alter the shape of the field; moreover, it is immaterial in which direction we consider the rotation to have taken place. For the other modes the configuration is not simply changed in sense by rotation through a single anode segment and the sense of the rotation is also important.

Let us now put these ideas into mathematical form by considering a Fourier expansion for the voltage at the anode surface for an arbitrary mode number n. Since by definition there are n cycles of the potential wave round the complete anode, we make the expansion in terms of ω, the resonant frequency of mode n, and the spatial factor k defined in $k = n+pN$, $p = 0, \pm 1, \pm 2$, etc. Each value of k therefore defines a spatial harmonic of the fundamental mode n. These harmonics are termed Hartree harmonics because they were first extensively investigated by Professor Hartree's group at Manchester University. The Fourier series for V_1 is then

$$V_1 = \sum_k A_k \, e^{j(\omega t - k\theta + \phi)} + \sum_k e^{j(\omega t + k\theta + \delta)}, \tag{32}$$

where θ = angular coordinate, ϕ and δ are phase factors. The amplitudes A_k and B_k are, in general, not identical. A little manipulation yields

$$V_1 = \sum (A_k - B_k) \cos(\omega t - k\theta + \phi) + 2\sum B_k \cos\left(\frac{\phi-\delta}{2}\right) \times \cos\left(\omega t + \frac{\phi+\delta}{2}\right) \cos k\theta + 2\sum B_k \sin\left(\frac{\phi-\delta}{2}\right) \times \cos\left(\omega t + \frac{\phi+\delta}{2}\right) \sin k\theta. \tag{33}$$

Eqn. (32) represents the anode a.c. potential as two travelling waves moving in opposite directions, while eqn. (33) exhibits V

as a travelling wave superposed on two standing waves. It is fairly easy to demonstrate that the amplitudes of the Hartree harmonics fall off rapidly as one approaches the cathode surface, so that only the fundamental is of importance in this region. Eqns. (32) and (33) contain four arbitrary constants A_k, B_k, ϕ and δ which can be fixed by comparison with the electromagnetic field equations. On the other hand, the solutions of the mesh equations of the equivalent circuit theory only give rise to two such constants unless each solution is assumed to be a degenerate doublet. (A solution of a wave equation is said to be degenerate if there are two possible oscillations at the same frequency.) On further investigation, however, it is easily seen that only the modes other than $n = 0$ or $n = N/2$ are degenerate, the 0 and $N/2$ modes representing simple standing waves. The $n = 0$ mode is trivial because there is no space variation in field, but the pure standing wave nature of the $N/2$ mode is very important. It means that when the magnetron is operated on the π mode there is little chance of interaction with harmonic fields in the operating voltage range, since the velocities favouring interaction with the harmonics are very different from those favouring the fundamental.

15.7. The interaction of the electrons with the field

We now ask the question, what are the favourable conditions for interaction between the electrons and the field? First consider a standing wave mode. On the basis of the considerations discussed in Chapter 12, we should expect that a favourable condition for interaction would exist when the electrons reach the centres of consecutive gaps at instants so related that the field is always acting on the electrons in the same sense. Now in the magnetron we note that if the r.f. field at the cathode accelerates an electron, the magnetic forces are stronger than they would be in the absence of r.f. field, so that the trajectory is at all points more curved than it would be with zero field. The increase in curvature is also greater the larger the field. Thus the electrons, which leave the cathode when the r.f. field is accelerating, follow paths more curved than the neutral reference electrons and are turned back to the cathode if the field is strong enough. Therefore in the magnetron there is a fundamental mechanism which operates so that electrons which gain energy from the field are very soon returned to the cathode, the excess energy being given up as back heating. On the other hand, those electrons which enter the r.f. field when it is in the decelerating direction are moving more slowly than they would be if there were no field. Their trajectories are less curved and they move out towards the anode. If the circumferential velocity is adjusted so that they reach the centre of a second gap when the field is also decelerating, some more energy is lost to the field and the electrons drift further towards the anode. Thus, an electron which enters the field when the latter is decelerating, drifts out towards the anode giving up part of its energy to the field at each interaction. If there are many such interactions, the electron will reach the anode with only a small proportion of the energy corresponding to the d.c. anode voltage and since the unfavourable electrons have been rejected, the process of energy conversion should be very efficient; as it is experimentally known to be.

The interaction with a travelling wave can be explained in exactly the same way, for we can describe the interaction in a coordinate system rotating with the same angular velocity as the wave. Then, at the optimum circumferential voltage, the wave and electron velocities are equal and the electrons give up maximum energy to the beam. The electronic behaviour is sketched in fig. 15.15. It will be obvious that the field will gain energy from the beam if the electron is travelling circumferentially with $\frac{1}{2}$, $\frac{1}{3}$, etc. of the wave velocity, for in this case it will be decelerated correspondingly by each second, third, resonator, and so on. Naturally this process would be less efficient than the fundamental interaction. This point is made clear by fig. 15.16, which shows the standing and travelling wave components for the $n = 2$ mode in an eight-resonator magnetron, chosen as being the simplest travelling wave case. Here the lines going through the dots represent favourable electrons in a space time plot. For the standing wave case,

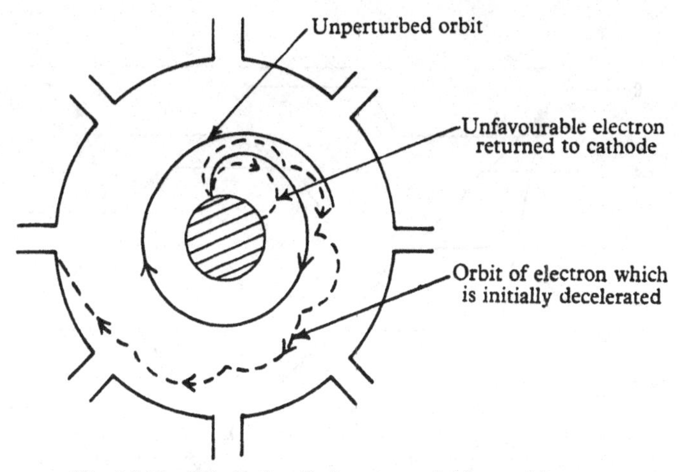

Fig. 15.15. Effect of radio frequency field on orbits.

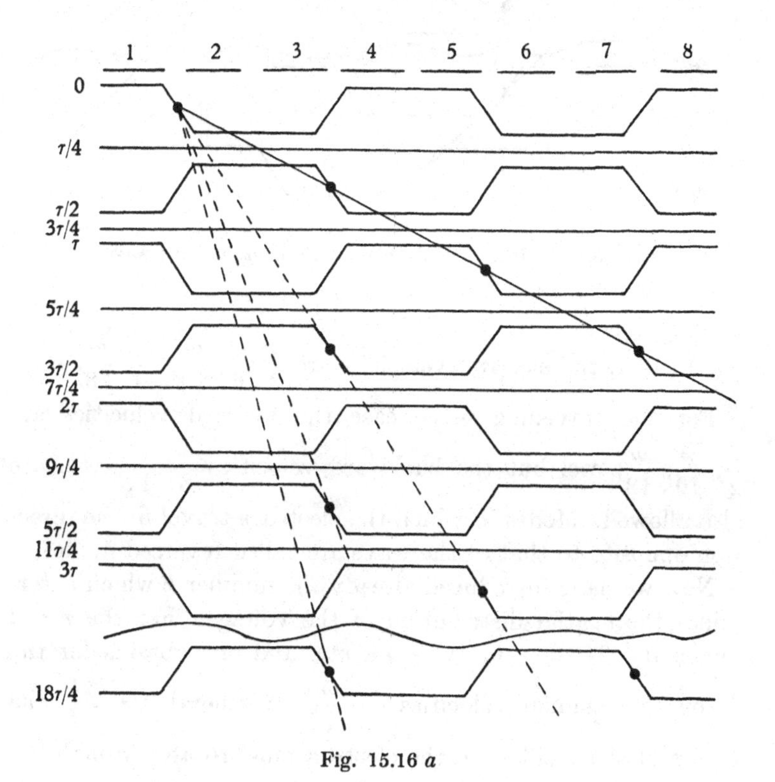

Fig. 15.16 *a*

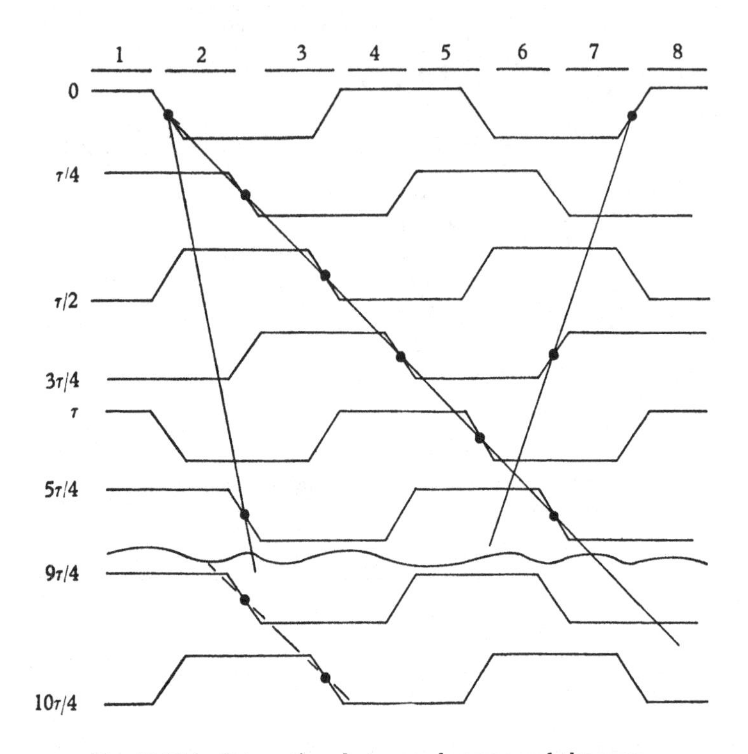

Fig. 15.16 *b*. Interactions between electrons and the wave.

fig. 15.16 *a*, the electron velocities are $\frac{\omega}{2}, \frac{\omega}{6}, \frac{\omega}{10}, \frac{\omega}{14}, \frac{\omega}{18}$, etc.

For the travelling wave case the forward velocities are $\frac{\omega}{2}, \frac{\omega}{10}, \frac{\omega}{18}$, etc. but the backward velocities $\frac{\omega}{6}, \frac{\omega}{14}$, etc. are also allowed. Modes in which the electrons travel in the direction opposite to that of the wave are called reversed modes.

Now we have introduced already the number k which determines the spatial distribution of the voltage. For the $n = 2$ mode, $k = 2, 10, -6, 18, -14$, etc. and one obtains for the favourable angular velocities $\dot{\theta} = \frac{\omega}{k}$. In general if $\tau = \frac{1}{f}$, the period of the oscillation, the electron must rotate through the arc subtending π/k radians in $\tau/2$ seconds for a correct phasing, i.e.

$$\dot{\theta} = \frac{2\pi}{k\tau} = \frac{\omega}{k}, \tag{34}$$

proving that the above result is general. We now wish to use eqn. (34) to obtain a relation between the observable parameters for the tube. An approximate expression is easily obtained. Using the translation term in eqn. (16), evaluated half-way across the interaction space, we find that the angular velocity of the electron round the cathode is given by

$$\frac{2E_r}{B(r_a+r_c)},$$

but we assume that E_r is constant and is given by the linear value $\dfrac{V}{r_a-r_c}$. Thus

$$\dot{\theta} = \frac{2V}{B(r_a^2-r_c^2)} = \frac{\omega}{k}, \tag{35}$$

or

$$V = \frac{\omega}{2\,|\,k\,|}\,Br_a^2\left[1-\frac{r_c^2}{r_a^2}\right]. \tag{36}$$

Eqn. (36) was originally derived by Posthumus† in a paper which contains the first treatment of travelling wave magnetrons. A more accurate expression for the translational velocity has been used by Slater‡ to show that eqn. (36) somewhat overestimates V. This is of little consequence, however, since magnetrons work over such wide ranges of voltages and current that it is unnecessary to locate $V_{\text{opt.}}$ accurately. What is more important is the question of the minimum voltage at which oscillations start. We can begin to answer this question with the information already at our disposal by plotting the Hull cut-off parabola and eqn. (36) on a $V \sim B$ basis. Since we know that the operating point must be to the right of the cut-off parabola, it is reasonable to take the point of intersection of the two curves as a first approximation to the operating point. The starting condition has been calculated by Hartree's group but is essentially contained in the paper by

† Posthumus, *Wireless Engr.* **12** (1935), 126.

‡ Slater, *Microwave Electronics*, Van Nostrand, 1950.

Posthumus. Hartree's relation, which Slater derives by simple reasoning, is

$$V_{\text{min.}} = \frac{B\omega}{2n}(r_a^2 - r_c^2) - \frac{m}{2e}\frac{\omega^2}{n^2}r_a^2. \tag{37}$$

In practical units

$$V_{\text{min.}} = 10{,}100\left(\frac{r_a}{\lambda}\right)^2\left[\frac{H\lambda[1-(r_c^2/r_a^2)]}{10{,}700n} - \frac{1}{n^2}\right], \tag{38}$$

H is measured in oersteds.

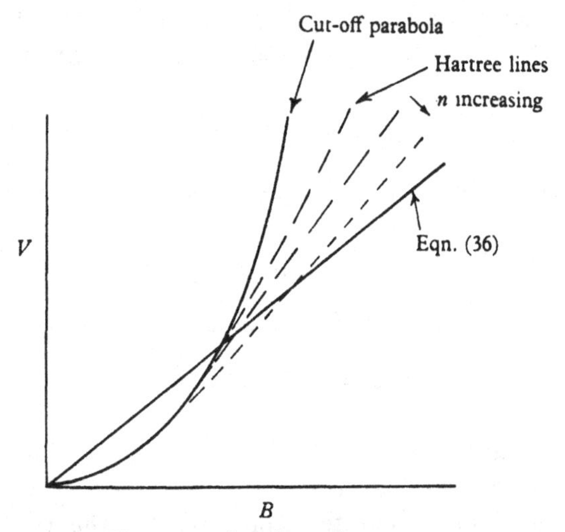

Fig. 15.17. Starting voltage curves.

The first term in eqn. (37) is just the resonance condition (36). The second term is proportional to the additional k.e. gained by an electron acted on by an infinitesimal r.f. field. Eqn. (38) is easily shown to be tangent to the cut-off parabola at the point

$$V_0 = \frac{1}{2}\frac{m}{e}\left(\frac{\omega}{n}r_a\right)^2, \quad B_0 = 2\frac{m}{e}\frac{\omega}{n}\frac{1}{[1-(r_c/r_a)]^2};$$

and the Hartree curve (37) reduces to

$$\frac{V}{V_0} = \frac{2B}{B_0} - 1. \tag{39}$$

Some Hartree lines are indicated in fig. 15.17 showing clearly the way in which increased n reduces the starting voltage at any specified magnetic field. Experimental values of starting voltage usually lie on a tangent line somewhat displaced above or below the theoretical line.

It must be emphasized that it is not the radial component of r.f. electric field which is effective in causing the electrons to reach the anode, but the fact that the circumferential component, by decelerating the electrons in their orbits, deflects them slightly outwards from the d.c. orbit, as has already been discussed. In the case of the infinitesimal r.f. voltage used for

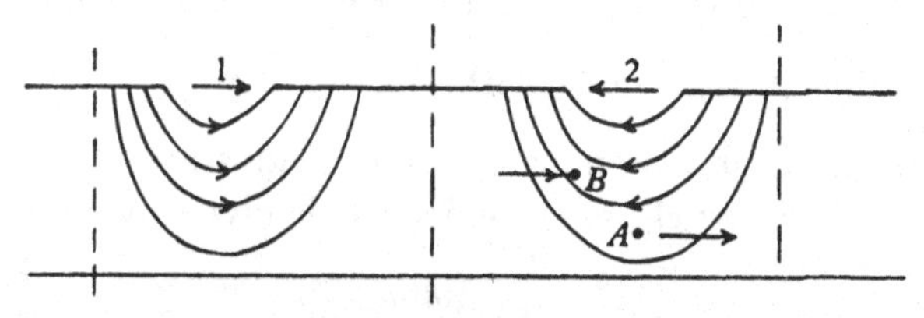

Fig. 15.18. Radio frequency field in anode-cathode space and its phasing effect.

the starting calculation, the radial r.f. field is quite negligible by comparison with the d.c. field.

We have now to discuss another feature of magnetron operation which also has a considerable effect in producing a high efficiency, this is the phasing of the electrons. Fig. 15.18 shows an enlarged and developed sketch of the r.f. field for the π mode in the vicinity of two cavity gaps. An electron at point A is decelerated by the field of gap 2. To the right of the centre line of gap 2 the radial field is directed towards the cathode, and to the left of the centre line it is directed towards the anode so that an electron reaching the gap centre at the instant of maximum r.f. deceleration experiences little radial deflexion (we are now talking of finite r.f. field strengths). However, an electron at B, which will reach the gap centre after the maximum, is shifted radially towards the cathode and, since the angular velocity is $\frac{E}{rB}$, the angular velocity is increased and this electron tends to catch up with electron A. Similar considerations applied to an electron to the right of the centre line show that it is slowed down and therefore a bunch of electrons forms about electron A which is the electron giving maximum energy to the field. This bunching or phase-focusing process contributes materially to the high electronic efficiency of magnetrons.

15.8. The maximum efficiency in a cavity magnetron

We have now described all the more important processes in the cavity magnetron, and on this basis we can make a very elementary and approximate estimate of the electronic efficiency. This is done by assuming that each electron arrives tangentially at the anode with the velocity $\dfrac{E}{B} = \dfrac{\omega r_a}{n}$. If the electrons were acted on by the d.c. field alone, they would arrive with energy eV_a so the efficiency is given by

$$\eta_{\text{max.}} = 1 - \frac{\frac{1}{2}m(\omega r_a/n)^2}{eV_a}. \tag{40}$$

If we write V_s for the starting voltage at a specified B, and V_c for the Hull voltage, eqn. (40) can be transformed† to

$$\eta_{\text{max.}} = 1 - \left[\left(\frac{V_c}{V_s}\right)^{\frac{1}{2}} - \left(\frac{V_c}{V_s} - 1\right)^{\frac{1}{2}}\right]^2. \tag{41}$$

Putting $$Y = \frac{2X}{n} - \frac{1}{n^2} \quad \text{for} \quad V_s,$$

$$Y_c = X^2 \quad \text{for} \quad V_c,$$

$$\eta_{\text{max.}} = 1 - \frac{1}{2nX - 1}, \tag{42}$$

$$X = \frac{H\lambda\left(1 - \frac{r_c^2}{r_a^2}\right)}{21{\cdot}4 \times 10^3}. \tag{43}$$

The value of η is plotted against nX in fig. 15.19, from which it is seen that the efficiency tends to unity as $H \to \infty$ for a fixed set of parameters. Slater‡ prefers to consider each electron as

† Willshaw, Rushforth, Stainsby, Latham, Balls and King, *J. Instn. Elect. Engrs.* **93**, pt. IIIA, no. 5 (1946), 985.

‡ Slater, op. cit.

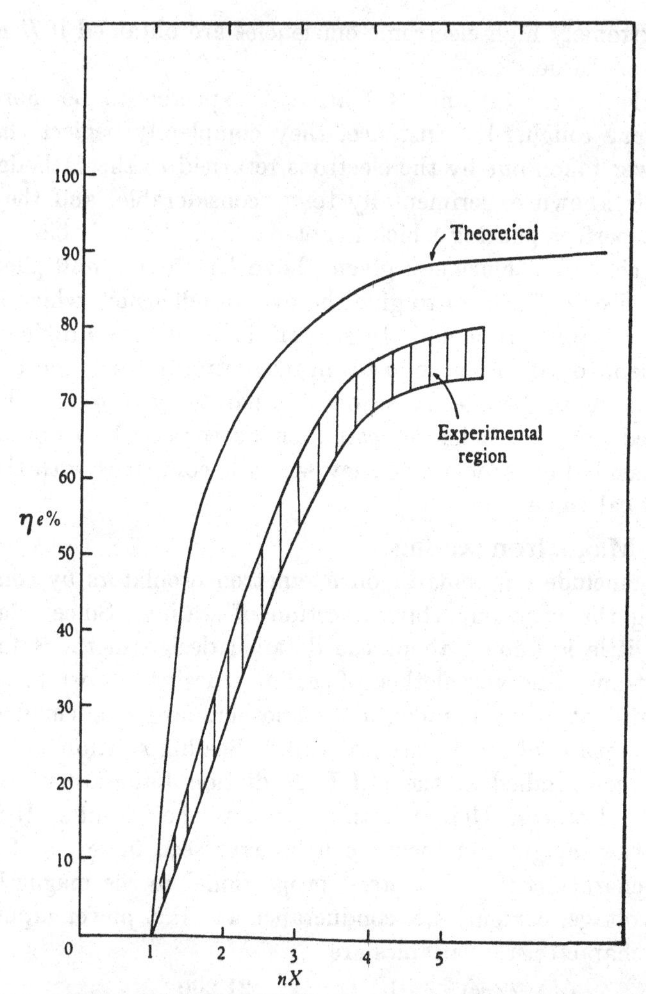

Fig. 15.19. Theoretical and experimental efficiencies. Copied from Willshaw, Rashforth, Stainsby, Latham, Balls and King, *J. Instn. Elect. Engrs.* **93**, pt. IIIA (1946), 992.

arriving at the anode with four times the energy here allowed. His final result is

$$\eta_{\text{max.}} = 1 - \frac{B_c^2 V_a}{B^2 V_c}, \tag{44}$$

which shows the same sort of behaviour but naturally gives lower values. Eqn. (44) is in better agreement with experiment than is eqn. (43). It is interesting that experiment shows that extremely high electronic efficiencies are obtained if H is made very large.

Clearly eqns. (43) and (44) are not expected to be more than very rough. For instance, they completely neglect the r.f. power taken out by the electrons returned to the cathode, which is known experimentally to be considerable, and they assume perfect phasing which is not particularly probable.

The electronic efficiency given above has to be multiplied by the circuit efficiency to give the overall efficiency, which is the measurable parameter. In general, it is much too difficult a problem to calculate the losses in magnetron blocks, and the losses have to be measured on a separate cold test. The observed output efficiencies can then be corrected to obtain the measured electronic efficiency which is compared with the theoretical value.

15.9. Magnetron scaling

We conclude our remarks on magnetron oscillators by considering the very important question of scaling. Since relatively little is known about the detailed design methods for magnetrons, a proven method of scaling a design, known to be successful at one wave-length, to another, larger or smaller, wave-length is of great practical value. Scaling relations have been much studied at the M.I.T. Radiation Laboratory† by Slater and others. Here we can only quote their results. It is found that magnetron theory can be expressed in terms of a set of characteristic parameters, proportional to the magnetic field, voltage, current, d.c. conductance and d.c. power input. These characteristic variables are

$$\mathbf{B} = 2\left(\frac{m}{e}\right)\cdot\left(\frac{2\pi c}{\lambda}\right)\frac{1}{[1-(r_c^2/r_a^2)]} = \frac{21{,}200}{n\lambda[1-(r_c^2/r_a^2)]}\text{ gauss,} \tag{45}$$

$$\mathbf{V} = \frac{1}{2}\left(\frac{m}{e}\right)\cdot\left(\frac{2\pi c}{n\lambda}\right)^2 r_a^2 = 253{,}000\left(\frac{2\pi r_a}{n\lambda}\right)^2\text{ volts,} \tag{46}$$

$$\mathbf{I} = \frac{2\pi a_1}{[1-(r_c^2/r_a^2)]^2[(r_a/r_c)+1]}\frac{m}{e}\left(\frac{2\pi c}{n\lambda}\right)^2 r_a\epsilon_0 h,$$

$$= \frac{8440a_1}{[1-(r_c^2/r_a^2)]^2[(r_a/r_c)+1]}\left(\frac{2\pi r_a}{n\lambda}\right)^2\frac{h}{r_a}\text{amps.,} \tag{47}$$

† *Microwave Magnetrons*, M.I.T. series, vol. 6, pp. 414 *et seq.*

Here a_1 is a function of r_a/r_c which varies little from unity, for usual values of this ratio, and h = height of anode block.

$$\mathbf{G} = \mathbf{I}/\mathbf{V}, \tag{48}$$

$$\mathbf{P} = \mathbf{IV}. \tag{49}$$

The performance of the experimentally good design is measured and plotted in terms of these variables by dividing the observed experimental quantity by the characteristic quantity to obtain the 'normalized' quantity. The observed parameters for the desired new magnetron are then obtained by multiplying the normalized quantities by the characteristic values for the new design. Practically the problem is more often to adjust the design to fit predetermined values, e.g. of **B** and **I**, both of which are limited by present-day materials.

In addition to the factors included above, N, r_c/r_a and the slot width/segment width ratio must be specified. The measured performance chart naturally only relates to a specified load impedance.

As we have said earlier, there is an optimum ratio of r_c/r_a depending on the number of resonators and on the mode. A useful criterion is that

$$\frac{r_c}{r_a} \geqslant \frac{N-4}{N+4}. \tag{50}$$

15.10. Unsolved theoretical problems

We have now very briefly surveyed the existing knowledge of magnetrons and it is worth while listing some of the remaining theoretical problems. We have already mentioned the unsatisfactory state of the d.c. theory. As for the r.f. theory, there is no method of calculating the electronic conductance from purely theoretical considerations. A beginning has been made with a starting current theory, the Bunemann† theory, but this work is of a tentative nature. Another problem of considerable interest, about which little is known, is that of

† O. Bunemann, *Manchester C.V.D. Rep.* no. 37 (1944); Summarized in M.I.T. series, vol. 6, chap. 6.

noise. Magnetrons are known to be extremely noisy, but the basic reasons for this are not yet known. It seems possible that there is a starting mechanism by which the shot noise in the electron current from the cathode is amplified by Haeff's multiple velocity effect (Chapter 14) so that the noise voltage induced in the resonators is much greater than would be expected *a priori*. A similar explanation could partially, at least, account for the anode current observed in cut-off magnetrons in which there is no coherent r.f. oscillation.

15.11. The magnetron amplifier

There have been many attempts to make a magnetron amplifier, none of which has been successful, until very recently when the Société Française Radioélectrique workers† produced such a device by substituting a zigzag wire or flattened helix for the cavities in the magnetron oscillator. By correctly matching the helix a slow travelling wave is set up which interacts with the beam as described in this chapter. Published results indicate that the device is characterized by low gain, good power output, moderate band-width and good efficiency.

Two embodiments have been described which are sketched in figs. 15.20 and 15.21. In the former, an electron beam is injected through a screen electrode into the space between a flattened helical circuit bent round into a cylinder, and a cylindrical cathode inside the helix. The helix is at HT and the field is therefore identical with that in a magnetron except for the short input region. The beam is arranged to move close to the helix so that it is in the strongest possible r.f. field. In the latter type, the actual emitter is a short sector of the inner cylinder and the rest of the system is as before. There is an axial magnetic field in each case.

The operation is almost exactly the same as that of the cavity magnetron in a travelling wave mode described in § 15.7. The linear electron velocity E/B is made nearly equal to that of the slow wave on the circuit, and electron bunches build up in the planes of maximum retarding circumferential

† Brossart and Döhler, *Ann. Radioélect.* **3** (1948), 328. Warnecke, Kleen, Lerbs, Döhler and Huber, *Proc. Inst. Radio Engrs., N.Y.* **38** (1950), 486.

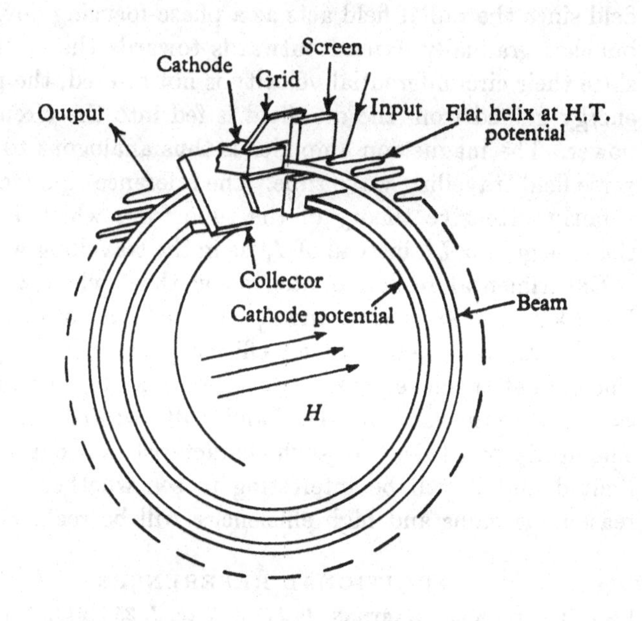

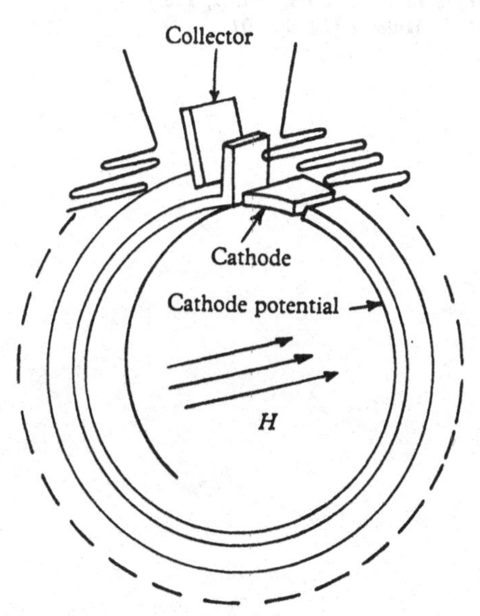

Figs. 15.20, 15.21. Travelling wave magnetron amplifiers.

field since the radial field acts as a phase-focusing device. The bunches gradually extend outwards towards the circuit, and, since their circumferential velocity is not altered, the potential energy gained from the d.c. field is fed into the circuit as r.f. power. The magnetron amplifier is thus analogous to a transverse field travelling wave tube. The references quoted contain a fairly extensive theory of the tube from which it appears that the gain $\propto I_0^{\frac{1}{2}}$, instead of $I_0^{\frac{1}{3}}$ as in the travelling wave tube.

Experimental results quoted show that, when $\lambda \doteqdot 23$ cm., $V_a \doteqdot 2$ kV., $B \doteqdot 880$ gauss, and $I_a \doteqdot 110$ mA., a power output of 55 W. was obtained with an efficiency of over 24 per cent. The gain at this level was, however, only 2 db. For very small signals the gain was 8 db. The band-width was 100–150 Mc./sec. The utility of a tube with such characteristics is naturally very limited and it will be interesting to see whether tubes with reasonable gains and high efficiencies will be realized.

ADDITIONAL REFERENCES

FISK, HAGSTRUM and HARTMAN, *Bell Syst. Tech. J.* **25** (1946), 167.
HOK, G., *Advances in Electronics*, vol. II, 1950.
PIDDUCK, *Quart. J. Math.* **7** (1936), 201.

Chapter 16

PICTURE CONVERTORS AND STORAGE TUBES

It was intended to include a treatment of cathode ray tubes in this chapter, but this intention has been modified for the reason that these devices are well described in many easily available books, and it therefore seemed preferable to devote the space thus gained to the description of some types of storage tube. These latter instruments are at present increasing very rapidly in importance in view of the extensive interest in television, electronic computers and certain specialized types of radar. We may define a storage tube in general terms as a device in which a pattern of events, either spatial or temporal, can be stored in such a way that the information can be subsequently obtained as a time sequence of electrical signals. Examples are the television picture convertors, in which the light from the object falls on a screen consisting of very large numbers of minute photo-sensitive elements. Elements in highly illuminated areas acquire a strong positive charge, those in dark areas a small charge. The screen is subsequently scanned by an electron beam which returns the elements to a state of equilibrium and, in so doing, produces a time sequence of electric signals which can be reconstituted into the picture.

Another type of storage can be exemplified by counter systems. A chain of pulses fed into a conventional counter operates each pair or decade until the chain has been counted. Normally the counter is used with many more pulses entering than there are stable circuit conditions and the information proceeds straight through the device, but if only a few pulses are fed in, there will be no output from the end of the chain. However, the information is not lost, for by examining the condition of the elements of the counter electrically, e.g. by monitoring the several cathode loads, etc., one can find out what information was initially fed in. A sequence of temporal events has thus been stored until it is convenient to utilize the information.

Until the last few years special electronic devices for storage have been confined to television picture convertors, other storage devices being synthesized from large numbers of ordinary receiving valves. Recently, however, the use of special devices has been much extended and we shall here be concerned with some of these special tubes.

16.1. Storage by photo-electric emission

One of the most important and widely used methods of storage is by photo-electric emission. This principle is used in

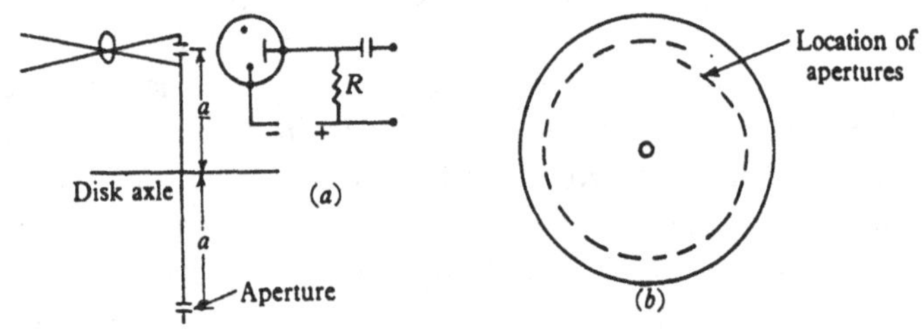

Fig. 16.1. Schematic diagram of Nipkow disk scanner.

modern television picture convertors and in some storage devices for computing. In this section we give a general account of this type of storage, the basic principles of which have already received some attention in Chapter 2.

Before we start this work it is advisable to clarify the great advantages gained by the introduction of storage. To do this we use television as an example, contrasting the performance of an early non-storage system with that of a modern system. A simple and practical mechanical television system is the Nipkow disk shown in fig. 16.1. A disk is perforated with a series of holes which are located at constant angular spacings. The radial distance of the holes from the axis of the disk is arranged so that the difference in radial dimension between successive holes is equal to the diameter of the holes. The optical image is formed on the disk and each hole scans a strip of the image. This strip illuminates a photocell during the time it takes for the aperture to scan the image. The photo-electric cell thus produces a signal which varies with the illumination along the path of the particular hole under the image at the instant under consideration. If the width of the picture is a cm. and the height b, and there are n holes of diameter d, we have $nd = b$ and the mean radius of the spiral of holes is $r = an/2\pi$. Clearly n also equals the number of lines in the picture. If the total luminous flux is B, the flux entering the cell at any instant is $B \times \frac{\pi}{4}\frac{b}{an^2}$. Since n is at least 400 in modern systems, nearly all the original light is wasted in this type of transmitter.

In a modern system the light falls continuously on a mosaic of tiny photo-elements which are insulated from a single metallic plate which acts as signal electrode. The elementary condensers are discharged sequentially by scanning with an electron beam which only remains on the element for an extremely short time. In this case the useful flux is roughly $B\left(1 - \frac{\pi}{4}\frac{b}{an^2}\right)$, if we take d as the spot diameter in this case. Thus the storage system is more efficient than the non-storage system by a factor of the order of n^2, i.e. of the order 10^5. This enormous degree of improvement cannot be obtained in practice, but it can be seen that even if storage systems fall very far short of their theoretical performance, they will still be much better than non-storage systems.

The first practical storage picture convertors were described by Zworykin and his colleagues of the R.C.A. in 1933. Very similar devices were developed in this country by E.M.I. at about the same time and were used in the B.B.C. London television station which was opened at the end of 1936. The R.C.A. tubes are called iconoscopes and the E.M.I. tubes emitrons. Later and improved versions were called the image iconoscope and the super-emitron respectively. All these tubes have features in common, viz. the use of a high voltage (1–1·5 kV.) electron beam to discharge the mosaic of photo-elements. Theory showed that advantages could be obtained by using very low voltage beams to discharge an element at about cathode potential, and this work has resulted in the R.C.A. orthicon and the E.M.I. cathode potential stabilized emitron (C.P.S. emitron). Tubes of the latter class are the most sensitive available to-day and their sensitivities are, in fact, approximately equal to that of the human eye, though the electronic devices have a much more restricted range than that of the eye. When high intensity illumination is available, as in the studio, the earlier tubes still have advantages in definition and freedom from distortion.

16.2. Iconoscopes and emitrons

We next proceed to consider in more detail the operation of the iconoscope or emitron.† The device is shown schematically in fig. 16.2. The mosaic is formed on an insulating layer backed

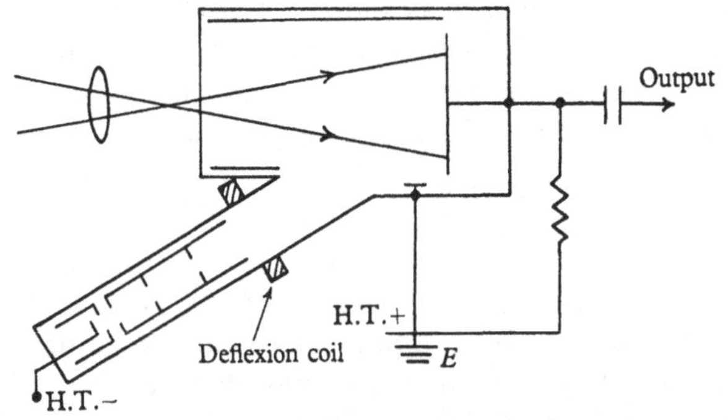

Fig. 16.2. Schematic diagram of an emitron or iconoscope.

by a metal signal plate. Light is focused through the plane end of the tube on to the mosaic, which is also scanned by a low current electron beam from a gun located in a side arm of the main bulb. Appropriate focusing potentials are applied to the gun to produce a very well defined spot with a current of about 0·1 μA. at a voltage of about 1·5 kV. The walls of the main bulb in the vicinity of the mosaic are conductively coated so that photo- and secondary emission can be taken from the mosaic.

† Zworykin, *J. Instn. Elect. Engrs.* **73** (1933), 437. Zworykin, Morton and Flory, *Proc. Inst. Radio Engrs., N.Y.* **25** (1937), 1071. McGee and Lubszynski, *J. Instn. Elect. Engrs.* **84** (1939), 468. Zworykin and Morton, *Television*, John Wiley, 1940. McGee, *Electronics*, Pilot Press, 1947, chap. IV.

The mosaic is made by forming a semi-transparent film of silver on a clean mica sheet and causing it to break up into minute globules of silver. This can be done by baking in air at approx. 700° C. or it can be done by means of an oxygen discharge run in the tube while on the pump. Later in the pumping the mosaic is activated as a photo-electric and secondary emitter by the admission of caesium and baking at low temperatures in the usual way. It may be noted at this point that the optimum photosensitivity cannot be attained as the amount of caesium needed to reach the peak is sufficient to lower the transverse resistance of the mosaic and thereby causes loss of signal and definition. A correctly activated mosaic, however, can be considered as consisting of a very large number (10^6–10^7) of minute condensers having one common plate. The other plate, exposed to the electron beam, is photosensitive and has a fairly high secondary emission coefficient.

To understand the operation of the device, consider first that the mosaic is scanned in the dark. The electrons arrive with a velocity well above the lower value for which $\delta = 1$, so that, as discussed in Chapter 2, the mosaic charges up to approximately the potential of the conducting coating in the bulb. As we know, the potential of the mosaic surface will be slightly below this potential so that on the average the primary and secondary currents are equal. The actual value of this small potential difference depends on the detailed velocity distribution of the secondaries and on the geometry of the tube which naturally determines the field strength at the mosaic surface. McGee gives the experimental figure as 1–2 volts negative to ground. The simple explanation of the signal formation is now as follows. When the mosaic is exposed to light, the mosaic emits photo-electrons some of which are drawn to the coating, leaving the bright areas of the picture less negatively charged than the dark areas. The beam scans the mosaic and discharges each condenser in turn returning it to the -1 volt mean level and the discharge current in the condenser forms the signal current. This process would not be expected to be very efficient since the field needed to saturate the photo-emission from the mosaic is much greater than the existing value, but even with such a low efficiency as 1 per cent the storage tube would be much better than a mechanical system.

In fact, the theory of the iconoscope is a little more complicated, for the reason that the beam does not simply discharge each element but carries it several volts positive during the time it is bombarded. Thus, the succeeding mosaic elements are located in a region of strong positive field (the distances are

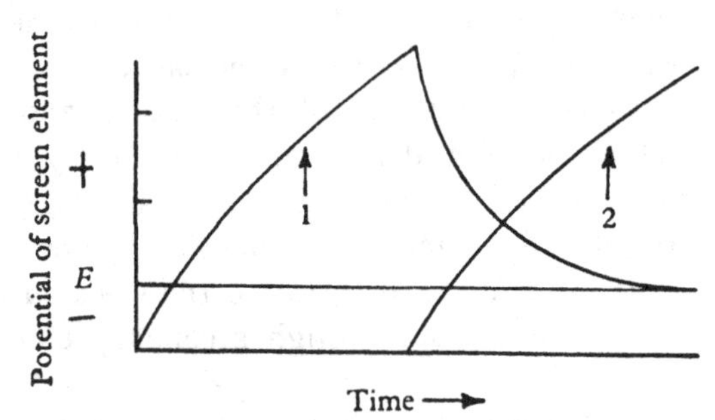

Fig. 16.3. Potential distribution in emitron screen.

so small that very small potentials produce large fields) which tends to produce local saturation of the photo-emission and therefore increased sensitivity in the direction of the scan. The beam carries each element to such a positive potential that the primary and secondary currents to the single element are equal, in the presence of the other elements which tend to repel secondaries towards the element under the beam. Fig. 16.3 demonstrates the charging of two successive elements in the mosaic showing that the field in the direction of motion of the spot tends to saturate the photo-electric emission while behind the spot it tends to suppress the photo-emission entirely. McGee quotes the following figures: the average duration of the bombardment of each element is 0·2 μsec., the average capacitance of each element is 0·05 pF. and the storage takes place in the period of about 0·004 sec. immediately before the element under consideration is scanned by the beam. Thus the photo-electric efficiency is much increased at the expense of a considerably diminution in the duration of the integrating effect.

The operation of the iconoscope is subject to several disturbing influences which we shall not discuss here as they are fully dealt with in the references cited. The main defects of the device are its only moderate sensitivity, poor depth of focus and much background signal. The consideration of depth of focus arises from the consideration that the spot size or picture element must not be made too small or else the lateral capacitance between elements will be comparable with the capacitance to the signal plate, an obviously undesirable state of affairs. The minimum spot size for a reasonable insulator thickness is about 0·010 in., and for a 400 line picture this leads to a 5×4 in. mosaic. A lens of focal length $\sqrt{(5^2+4^2)} = 6{\cdot}5$ in. is required for adequate field coverage, which in turn means that fairly small apertures, $f/3$ or less, must be used.

16.3. The image iconoscope and super-emitron

In these devices a considerable improvement is brought about by separating the functions of photo-emission and secondary emission. They are shown schematically in fig. 16.4. Here the

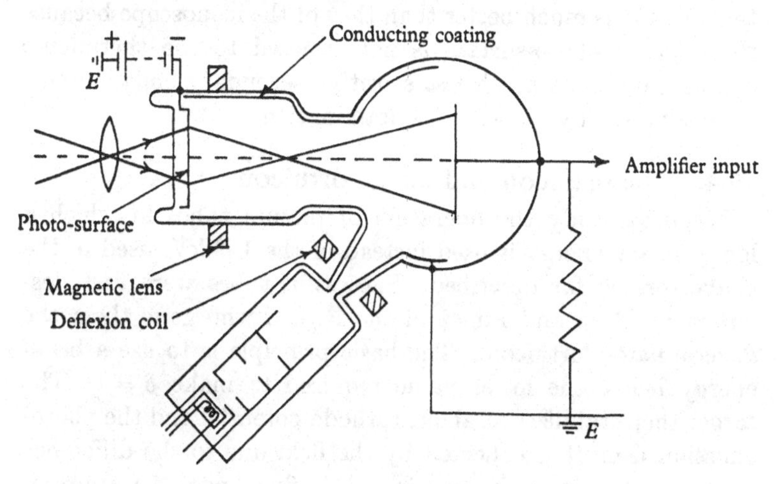

Fig. 16.4. The super-emitron.

picture is optically focused on a transparent photo-surface close to the optically flat window. The photo-electrons are then magnetically focused on the mosaic which is now prepared for maximum secondary emission. The rest of the tube is nearly identical with the normal iconoscope. The photo-surface is held well below the potential of the conducting coating (ground) by a separate bias so that the photo-emission is saturated and the magnetic focusing has a small magnification so that the glass lenses can have shorter focal lengths. Thus the improvements are:

(1) The photo-surface can be activated for optimum sensitivity as leakage is not now a problem. This gives 30–40 μA./lumen as against 10 μA./lumen in the earlier tube.

(2) The photo-electrons reach the secondary emitting mosaic with about 500 volts energy, i.e. the energy for maximum δ, and the mosaic has a δ of about 5.

(3) The first two effects give about a twenty-fold increase in current per unit illumination. The use of magnifying magnetic lenses reduces the current density but allows the optical system to be made cheaper and better.

The overall gain in sensitivity is usually about five times.

An important incidental advantage of this type of tube is that the life is much better than that of the iconoscope because the delicate photo-surface is not exposed to the thermionic cathode, only the much less sensitive secondary emission surface being in any danger of contamination.

16.4. The orthicon and image orthicon

We next study the operation of picture tubes in which a low primary energy is used instead of the 1–2 kV. used in the convertors so far described. Successful tubes were first described by Rose and Iams† of the R.C.A. who gave them the generic name 'orthicon'. The basic principle is to use a beam energy below the lower value required to make $\delta = 1$. The target then stabilizes at about cathode potential and the photo-emission from it is collected by the field due to the difference in potential between the mosaic and the final anode, i.e. approximately the accelerating potential. Thus this scheme has the advantage over the emitron that the photo-emission can be

† Rose and Iams, *Proc. Inst. Radio Engrs., N.Y.* **27** (1939), 547. *R.C.A. Rev.* **4** (1939), 186.

saturated and also the secondary emission is prevented from scattering over the mosaic and partially removing the storage pattern. To offset these advantages one must consider the fact that there are severe practical difficulties in forming and maintaining such low voltage beams. These difficulties have been overcome by the use of magnetic focusing in conjunction with a highly ingenious deflexion system.

An orthicon tube is shown in fig. 16.5. An electron beam is formed by a low voltage gun. The gun is provided with limiting

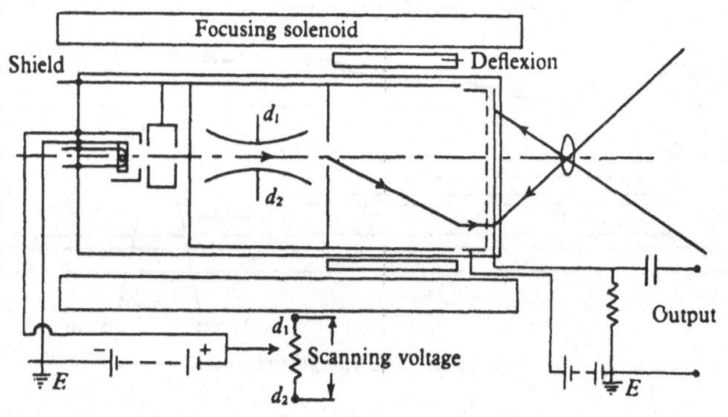

Fig. 16.5. Schematic diagram of an orthicon.

apertures so that the emergent beam consists only of the pencil of electrons whose trajectories are very nearly axial. The beam passes through a hole into the region containing two curved deflexion plates which are electrostatically shielded from the rest of the tube. The beam is thus deflected along one coordinate axis and emerges through a long slot into a region where it is magnetically deflected by a system of coils very similar to the ordinary yoke used in television reception. The object of this rather elaborate deflexion system is to ensure that the beam falls normally on to the surface of the mosaic. The long solenoid surrounding the whole device serves to maintain the initial beam diameter.

The mosaic is transparent; it might be, for instance, formed by coating a sheet of mica with a semi-transparent layer of some refractory metal to act as the signal plate on one side and forming the ordinary photo-emissive mosaic, as used in the emitron, on the other. A loss of light obviously results in this process but it can be made up for by improved efficiency elsewhere. The photo and secondary electrons emitted by the mosaic are collected on a deep ring electrode biased slightly negative to earth so that the primary beam is not intercepted.

In the image orthicon† the arrangement at the right-hand end of the tube is modified by replacing the mosaic by a trans-

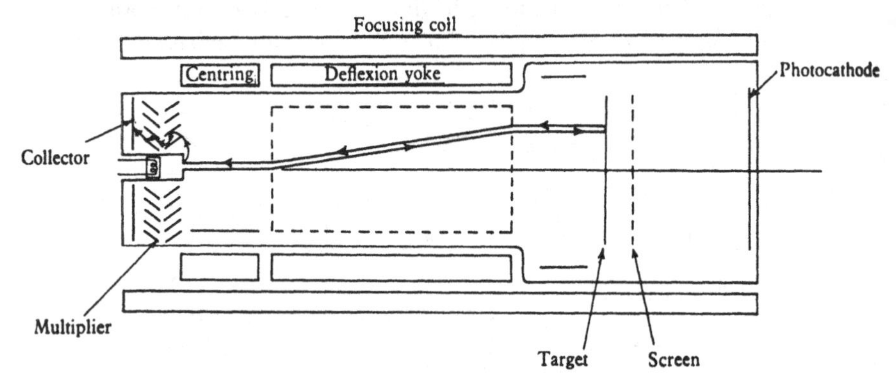

Fig. 16.6. The image orthicon.

parent photo-cathode, the electrons from which are accelerated by a few hundred volts and made to fall on a two-sided mosaic. The secondary electrons from the side of the mosaic bombarded by the photo-electrons are suppressed by means of a fine-meshed grid. The other side of the mosaic is scanned by the low velocity electron beam used in the orthicon. The signal is, however, taken out quite differently; the secondary or reflected electrons from the mosaic return along paths nearly identical with those along which they reached the mosaic and they return to the gun close to the final defining apertures. A seconary emission multiplier of the slat type is built round the gun so that several stages of secondary emission multiplication are built into the tube. Gains of several hundred are obtained in this way, the useful limit being defined by the signal-to-noise ratio. The general arrangements of the image orthicon is shown in fig. 16.6. In modern tubes the electrostatic plates are dispensed with in favour of magnetic deflexion in two directions

† Rose, Weimer and Law, *Proc. Inst. Radio Engrs., N.Y.* **34** (1946), 424.

at right angles. This deflexion field is superposed on the focusing field of 50–100 oersteds. The photo-cathode is held at a potential a few hundred volts negative to ground, while the mosaic and screen are grounded. The beam is accelerated to about 200 volts velocity and the final anode of the emission multiplier is made 1–2 kV. positive so that each stage of secondary emission multiplication may be adjusted to give the maximum δ.

The functioning of the magnetic deflexion system is worthy of a brief description. Fig. 16.7 shows a pair of pancake coils

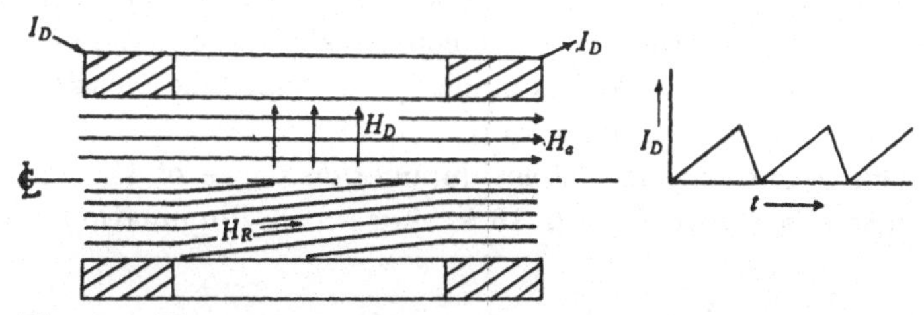

Fig. 16.7. Illustrating the action of the deflexion system superimposed on the focusing field.

producing a vertical magnetic field H_D. The coils are immersed in an axial magnetic field of strength H_a. Above the axis fig. 16.7 indicates the directions of the two component fields while below the axis we have sketched the lines of force of the resultant field drawn, naturally, for a specific value of H_D, i.e. at a specific instant in the scanning cycle. It is easily proved that the trajectory of an electron in such a field is along the field lines of the resultant field. If the coils shown produce the x deflexion, another pair placed at right angles will produce the y deflexion by adding a y component to the resultant field H_R.

Another question of great technical interest in the image orthicon is the target structure. Earlier approaches to the two-sided target problem had been based on various methods of inserting metallic plugs into a lattice-like insulator so that, although the conductivity between opposite faces of the target was good, the lateral conductivity was zero. Early techniques are described by Zworykin and Morton.† In the image orthicon

† Zworykin and Morton. Op cit.

the double-side mosaic consists of a mosaic on a very thin plate of conducting glass only $2\text{–}4 \times 10^{-3}$ mm. thick, the thickness being chosen by the requirement that the charge shall not spread laterally during the storage period ($\frac{1}{25} - \frac{1}{30}$ sec.) by an amount sufficient seriously to impair the resolution.† Since $C \propto t^{-1}$ (t = thickness) and $R \propto t$, the time constant of each elementary condenser depends only on ϵ and ρ for the conducting glass. The mosaic is a disk about $1\frac{1}{2}$ in. in diameter which has to be held flat to within a few thousandths of an inch. It is spaced about 0·002 in. from a fine-meshed grid which must be as flat. A rigid structure which conforms to these requirements and which can be baked is made by attaching the mosaic and screen to metal tensioning rings.

The target output versus illumination curve of the image orthicon is shown in fig. 16.8. For small illuminations the

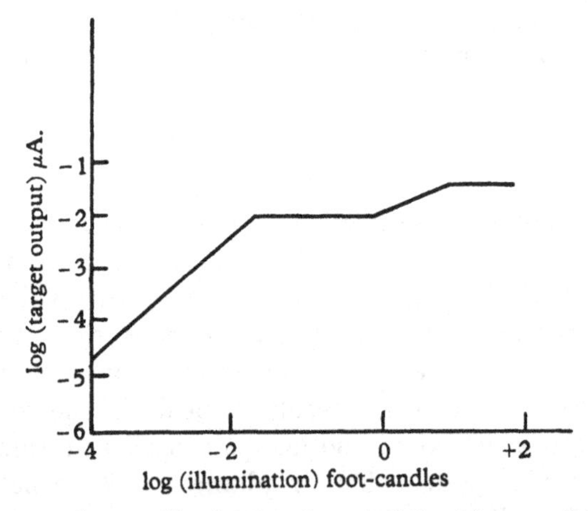

Fig. 16.8. Current-illumination characteristic of image orthicon.

signal is directly proportional to the illumination. The lowest point on the curve is the value of illumination for which the video signal just equals the shot noise in the scanning beam, when the latter is adjusted so that it is just sufficient to discharge the screen, i.e. so that increasing the beam current does

† A study of limiting resolution is given by De Vore, *Proc. Inst. Radio Engrs., N.Y.* **36** (1948), 335.

not make any appreciable difference to the signal amplitude. At the end of the proportional part of the curve the illumination is sufficient to charge the mosaic to saturation, i.e. to the screen potential, in the frame time. Once the curve has flattened out, one would expect that further increase in illumination would merely make the tube produce a uniform white image. This is found not to be the case because the low velocity secondaries from a very intensely illuminated spot tend to discharge the area round the spot, which therefore appears as a white area with a dark halo. Thus the redistribution of secondary emission tends to maintain the contrast of the picture, and the range of illumination over which the device functions is much increased thereby.

16.5. The cathode potential stabilized emitron

The C.P.S. emitron, although very similar to the image orthicon, shows some differences which make it worthy of description. The major point of difference lies in the target,† which consists of a photosensitive mosaic deposited on a 0·002–0·004 in. glass or mica sheet backed by a signal plate which is about 70 per cent transparent. The photosensitive mosaic is formed in a novel way. The technical problem of producing an extremely fine metal mesh has been solved, and a mesh of 1000 lines per inch or 10^6 elements per inch has been produced. Antimony is evaporated on to the target through this mesh, which is then removed. The activation is carried out by evaporating caesium on top of the antimony and the final result is a surface with a sensitivity of 12–15 μA./lumen which has a spectral curve very nearly the same as that of the eye. The standard mosaic is 35×44 mm. and therefore there are about $2\text{–}5 \times 10^6$ elements. Such a mosaic is capable of resolutions better than 1000 lines. The final aperture in the gun is only 0·002 in. in diameter, and this is focused magnetically on the mosaic by a field of about 50 oersteds.

Another difference between the C.P.S. emitron and the orthicon tubes is that the beam is passed through a grid spaced

† McGee, *Proc. Inst. Radio Engrs., N.Y.* **38** (1950), 596; *J. Instn. Elect. Engrs.* **97**, pt. III (1950), 377.

an inch or so away from the mosaic. This grid is at full accelerating potential, about 2·0 kV. when switching on, the beam is cut-off and the mosaic is darkened. The electrode voltages are gradually increased so that the mosaic stays at cathode potential and the beam is then switched on. In the space between the grid and the mosaic it is decelerated and the electrons therefore have little energy when they impinge on the mosaic and $\delta < 1$. The beam thus biases the mosaic back to substantially cathode potential.

The mask is now removed and the picture causes photo-emission from the mosaic. This emission is saturated by the field due to the grid and so the full charge pattern is built up instead of only a small percentage as is the case in the emitron. The beam discharges those areas of the mosaic which are positively charged and thus causes capacitative signals to the signal plate.

The main operating features of the C.P.S. emitron are its exceptionally high resolution, high sensitivity without the complication of electron multiplication, and the smallness of the mosaic which makes it possible to use lenses of short focal length and with large depths of focus. A defect is that there is some blurring of moving objects because the decay time of the mosaic is considerable.

16.6. Storage by secondary emission

We now pass on to a second generic type of storage device in which secondary emission is used as the storage agency, instead of photo-emission. If we consider a sheet of insulator or semi-conductor of low conductivity, backed by a metal signal plate, exposed to a beam of high voltage electrons, with a velocity great enough to give a secondary emission of velocity greater than unity but below the 'sticking' velocity, we see that we can write a message on the plate either by deflecting the beam in a desired manner so that some parts of the screen are charged while others are at the equilibrium potential or by deflecting the beam in a raster and applying the intelligence to be stored in the form of pulse signals which cause 100 per cent amplitude modulation in the beam i.e. in the form of on-off pulses. An ordinary cathode ray tube can be used as a storage device of this type if the outer face of the bulb is metallized, although better results can be obtained with special tubes in which the insulating layer is thinner and more uniform.

This simple type of secondary emission storage tube is somewhat limited in application since the functions of storage and pick-off are performed by a single beam so that a complete raster of intelligence has to be stored before any part can be picked off, which is done by one of the subsequent scans. A more versatile tube is one containing two electron guns, one for use as the storing or writing beam, the other as the pick-off beam. The electron velocities need not be the same and therefore they can be separately adjusted for optimum performance.

Several such tubes have been described in the literature,† and have found application in electronic computers and in radar for moving target indication, etc. One of the most interesting, which embodies a new physical principle, is the 'graphecon' developed by Pensak at the R.C.A. laboratories.‡

16.7. The graphecon

The graphecon is a two-beam secondary emission storage tube developed for use in the 'teleran' system of air navigation in which a ground radar P.P.I. picture of the surroundings of an air-field is projected over a map of the same area. The combination is then transmitted from the ground to the aircraft so that the pilot can follow his own approach to the air-field. The system is somewhat more complicated in fact, e.g. the radar picture is limited to a predetermined height interval and the radar pulses trigger transponder beacons on the planes in its particular height range, the ground transmitter working on a time-sharing multiplex schedule. However, the details of the system are not material to the understanding of the graphecon except in so far as they pose the problem of storing a relatively infrequent group of signals (the radar scan takes about 10 seconds to complete) which has subsequently

† McConnell, *Proc. Inst. Radio Engrs., N.Y.* **35** (1947), 1258. Haeff, *Electronics*, **20** (Sept. 1947), 80.

‡ Pensak, *R.C.A. Rev.* **10** (1949), 59.

to be retransmitted many times before a new set of signals is written on the storage screen.

A double-ended version of the graphecon is shown in fig. 16.9. The writing gun is a standard cathode ray tube gun for 6–10 kV. operation and therefore produces a current of several tens of μA. The pick-off gun is a low current iconoscope gun working at about 1 kV. The storage screen consists of a fine-meshed grid on which a thin continuous film of organic material is

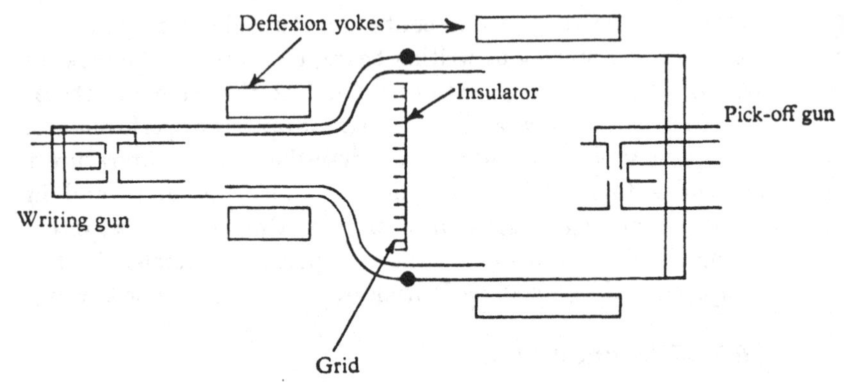

Fig. 16.9. The graphecon.

deposited. A layer of aluminium is then evaporated over the organic layer. The secondary emitter, a thin insulator, is evaporated on to the aluminium. Suitable materials are silica or magnesium fluoride.

The method of operation of the device is as follows. The voltage of the writing beam is made large enough to reduce the secondary emission coefficient to less than one, so that it drives the target slightly negative to the conductive coating on the bulb. The low-voltage pick-off beam on the other hand causes δ to be greater than 1, so that secondaries are set free which discharge those regions which were negatively charged by the writing beam. Thus when the writing potential is just equal to the sticking potential, the target surface is in equilibrium and there is no signal from the pick-off beam. When the writing potential is above the sticking potential, the target is negative to the collector by the difference between the two potentials.

A difference of 20–50 volts is sufficient to saturate the secondary emission from the pick-off beam.

A new phenomenon is used in the double-sided target tube here described. Pensak† has found that if a very thin ($\frac{1}{2}\mu$) insulating film is bombarded with electrons, currents flow in the film which are greater than the bombarding current. It is as though the insulation breaks down with electron multiplication as is, for instance, observed in the case of a gas discharge. The insulation, however, reforms directly the bombardment ceases. The currents flow in the direction of a potential gradient. The high energy beam thus penetrates the insulating target to the side facing the low voltage beam, which is driven positive by the secondary emission except in those regions where it is discharged by the high energy beam. By adjusting the equilibrium potential difference between the target and the collector coating, the discharge process can be controlled so that in saturation conditions it takes place in a single scanning period; or, if desired, only partial discharge takes place and storage over many scanning periods is possible and times of about five minutes have been observed.

It is interesting to note that this tube has been developed in competition with a special storage orthicon‡ which is a standard orthicon (not an image orthicon) in which the target consists of a photosensitive mosaic on the beam side, which is formed on a thin highly insulating glass plate backed with a transparent metal signal electrode. Storage for periods longer than a single frame time is accomplished by using a very thin insulator, thereby increasing the capacitance and by using a smaller beam current than usual.

16.8. The barrier grid storage tube

Another novel type of secondary emission storage tube has been developed by the R.C.A.,§ primarily for signal comparison purposes. This problem might arise, for instance, in the following way. A ground radar station used for early warning of

† Pensak, *Phys. Rev.* **75** (1949), 472.
‡ Forgue, *R.C.A. Rev.* **8** (1947), 633.
§ Jensen, Smith, Mesner and Flory, *R.C.A. Rev.* **9** (1948), 112.

aircraft presents, in addition to the wanted data on aircraft movement, a large number of more or less permanent signals from the terrain in the vicinity of the transmitter. It is highly desirable to remove these permanent signals, called 'clutter', and leave only the signals from moving objects. One way in which this can, in principle, be done is to store the radar picture for a predetermined time and then subtract it from a subsequent picture. The signals due to the fixed objects cancel out, but a moving object will produce signals at different positions in the two scanning cycles which will not cancel, and thus the system discriminates in favour of the desired information. The barrier grid storage tube is fed with a voltage waveform representing the intelligence to be stored. This storage is accomplished by scanning the target with a beam which impresses on it a charge pattern directly proportional to the instantaneous potential. When the scanning cycle has been completed, the target is isolated from the input signal and a second identical scanning cycle reads off the stored information by discharging each element of the target in turn.

To understand the operation in more detail, consider fig. 16.10 *a*. The signal plate is mounted on an insulator which is mounted very close to a fine-meshed grid through which the beam passes to charge or discharge the insulator. The bombarding velocity is above the first critical voltage so that the equilibrium condition of the insulator surface is a few volts negative to the grid, which now determines the surface field, since it is (*a*) close and (*b*) fine-meshed. At equilibrium the number of secondaries which have enough energy to overcome the retarding field of the grid is just equal to the number of electrons in the beam (fig. 16.10 *b*). Those secondaries which cannot penetrate through the grid are returned to the target nearly at their points of origin, since the retarding field is planar and thus charge is hindered from spreading. In fig. 16.10 *c* we show the velocity distribution of the secondaries and the current flowing to or from the grid when the target is not at equilibrium. This latter curve is simply the integral of the former with the scales displaced so that zero current flows for $V = V_e$. Because of the broad flat maximum of the velocity

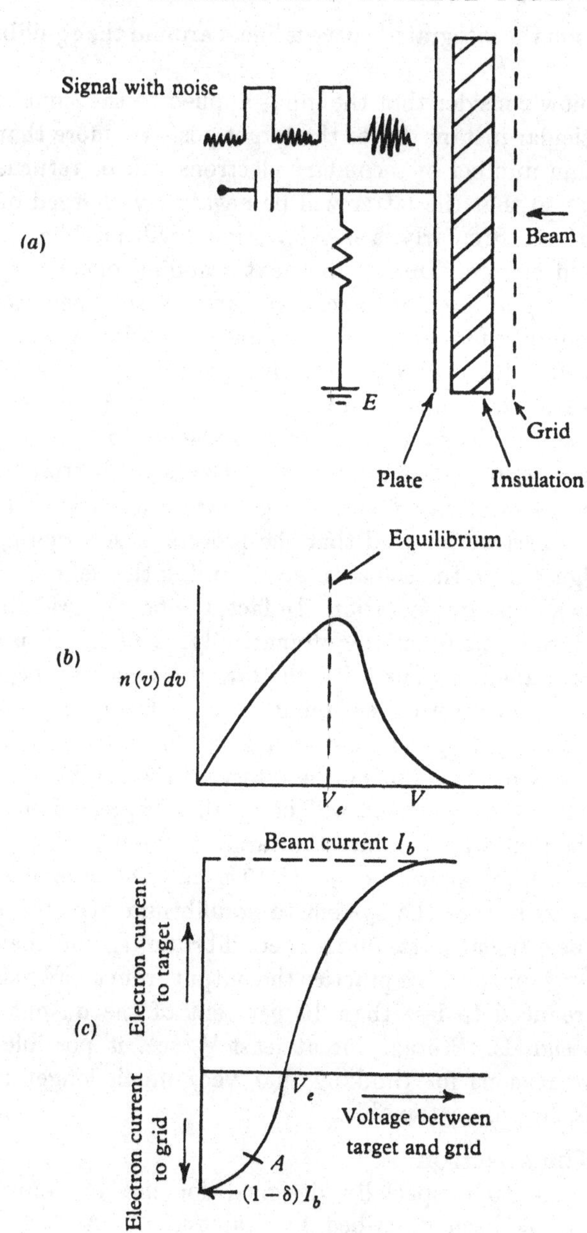

Fig. 16.10. A barrier grid storage tube.

distribution the integrated curve is linear around the equilibrium position.

If we now consider that the input applied to the signal plate at a particular instant drives the target positive, more than the equilibrium number of secondary electrons will be returned to the target so that the latter will be negatively charged on the inner surface. Similarly, a negative signal will produce a positive stored charge. During the next scanning operation with no applied signal each of the charged areas is returned more or less to equilibrium and the secondary emission penetrating back through the grid will be an inverse replica of the stored signals. The edges of the input signals however, are rounded off because the distribution current in the spot is not uniform over it but follows an approximately Gaussian distribution.

In our description of the system for cancellation of permanent echoes we inferred that the process of subtracting the stored signal from the signal received during the next scanning cycle was a separate operation. In fact, the barrier grid storage tube performs this function automatically. This is because the strong permanent signals drive the target a long way negative during the first storing scanning cycle. During the second scanning cycle the permanent signal will drive the negatively charged area positive during the interval in which the beam is on it, and the net current will be small. Expressed in other terms, the first scan will bias the target to some such point as A on the characteristic of fig. 16.10 c, and the second signal will tend to restore the system to equilibrium. On the other hand, a new signal starts off from equilibrium, V_e, and therefore gives a large output. In practice the output from steady signals can be reduced to less than 10 per cent of the output from variable signals. Storage for at least $\frac{1}{50}$ sec. is possible, but there are reasons for thinking that very much longer times, up to 100 hr., are attainable.

16.9. The selectron

A storage tube specially designed for use in computing machines has been described by Rajchman.† Although the

† Rajchman, Harvard Symposium on Calculating Machines (Jan. 8, 1947). *Mathematical Tables and Aids to Computation*, **2** (1947), 359.

storage principle is the same as that used in the tubes already described, the conception of the rest of the device is radically different. The structure of the tube is indicated by the section of fig. 16.11. A concentric geometry is used. The cathode is surrounded by an accelerating grid which draws off the electrons. The electrons move through this accelerator towards

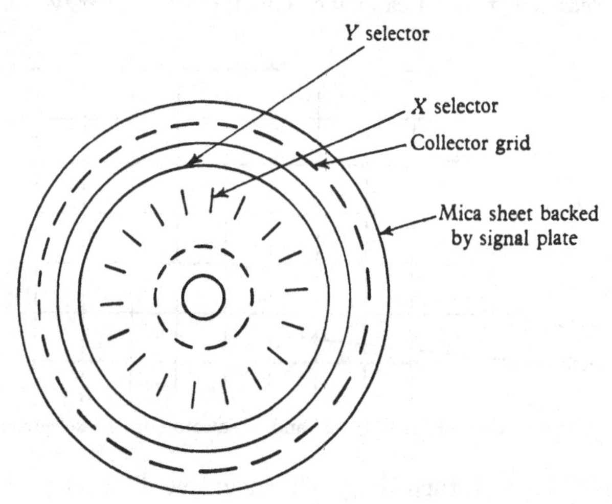

Fig. 16.11. Section through a selectron.

two selecting structures formed of vertical tapes for the *X* selection and horizontal tapes for the *Y* selection. Beyond the selector section there is a mica target backed with a metal signal plate and a collector grid located just in front of the mica. The selector functions as follows (fig. 16.12). In the quiescent state all the tapes are biased negatively so that no electrons can pass through to the screen. If, however, four tapes, two adjacent vertical ones and two adjacent horizontal ones, are pulsed positively, electrons can pass through the shaded area surrounded by positive tapes, all the other gates, including those having one, two or even three positive walls, remaining closed. Thus a charge is stored on the mica sheet behind the single open gate and nowhere else. The sequence of operations in storing a signal is to open all the gates simultaneously, allowing the mica sheet to be bombarded and thus brought to the collector potential. All the gates are then closed except the desired one and simultaneously a pulse is put on the signal plate. A negative going pulse prevents the stream of electrons through the open gate from reaching the screen so that the potential returns to the collector level when the pulse is removed. If, however, the pulse is positive, the electron beam reaches the mica with considerable energy but the

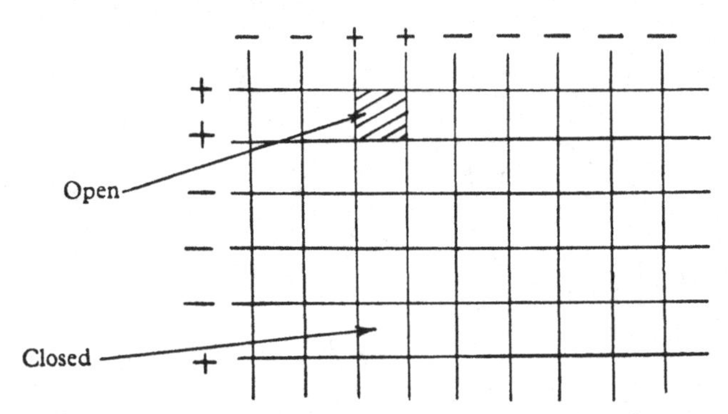

Fig. 16.12. Connexion of the X and Y selectors in the selectron.

secondaries are returned by the negative field between the mica and the collector so that a zero charge is stored under the gate. The signal can be stored on the mica sheet, one element at a time, either as a patch at cathode potential or as a patch at collector potential. Reading the stored signal is performed by closing all the gates except the desired one and determining whether the equilibrium potential is that of the cathode or collector. This can be done either by using the current pulse in the signal plate or by the collector current, or it may be done visually by using a semi-transparent signal plate and a thin coating of phosphor on the mica.

The selector wires are grouped in a special way so that the number of leads which have to be sealed through the bulb is much reduced. Since four positive sides are required to open the gate, this is not difficult and it can be shown that N leads can control $\left(\frac{N}{4}\right)^4$ gates. Experimental tubes using 32 leads to control 4096 gates have been made.

TABLE 9. LANGMUIR'S β^2 FOR CYLINDRICAL DIODES

β^2 applies where $r > r_c$, $-\beta^2$ applies where $r < r_c$.

r/r_c or r_c/r	β^2	$-\beta^2$	r/r_c or r_c/r	β^2	$-\beta^2$
1·00	0·0000	0·0000	3·8	0·6420	5·3795
1·01	0·00010	0·00010	4·0	0·6671	6·0601
1·02	0·00039	0·00040	4·2	0·6902	6·7705
1·04	0·00149	0·00159	4·4	0·7115	7·5096
1·06	0·00324	0·00356	4·6	0·7313	8·2763
1·08	0·00557	0·00630	4·8	0·7496	9·0696
1·10	0·00842	0·00980	5·0	0·7666	9·887
1·15	0·01747	0·02186	5·2	0·7825	10·733
1·20	0·02815	0·03849	5·4	0·7973	11·601
1·30	0·05589	0·08504	5·6	0·8111	12·493
1·40	0·08672	0·14856	5·8	0·8241	13·407
1·50	0·11934	0·2282	6·0	0·8362	14·343
1·60	0·1525	0·3233	6·5	0·8635	16·777
1·70	0·1854	0·4332	7·0	0·8870	19·337
1·80	0·2177	0·5572	7·5	0·9074	22·015
1·90	0·2491	0·6947	8·0	0·9253	24·805
2·0	0·2793	0·8454	8·5	0·9410	27·701
2·1	0·3083	1·0086	9·0	0·9548	30·698
2·2	0·3361	1·1840	9·5	0·9672	33·791
2·3	0·3626	1·3712	10·0	0·9782	36·976
2·4	0·3879	1·5697	12·0	1·0122	50·559
2·5	0·4121	1·7792	16·0	1·0513	81·203
2·6	0·4351	1·9995	20·0	1·0715	115·64
2·7	0·4571	2·2301	40·0	1·0946	327·01
2·8	0·4780	2·4708	80·0	1·0845	867·11
2·9	0·4980	2·7214	100·0	1·0782	1174·9
3·0	0·5170	2·9814	200·0	1·0562	2946·1
3·2	0·5526	3·5293	500·0	1·0307	9502·2
3·4	0·5851	4·1126	∞	1·000	∞
3·6	0·6148	4·7298			

TABLE 10. LANGMUIR'S α^2 FOR SPHERICAL DIODES

(r_c = radius of emitter ; r = radius at any point P ; α^2 applies to case where P is outside emitter, $r > r_c$; $-\alpha^2$ applies to case where P is inside emitter, $r_c > r$.)

r/r_c or r_c/r	α^2	$-\alpha^2$	r/r_c or r_c/r	α^2	$-\alpha^2$
1·0	0·0000	0·0000	6·5	1·385	13·35
1·05	0·0023	0·0024	7·0	1·453	15·35
1·1	0·0086	0·0096	7·5	1·516	17·44
1·15	0·0180	0·0213	8·0	1·575	19·62
1·2	0·0299	0·0372	8·5	1·630	21·89
1·25	0·0437	0·0571	9·0	1·682	24·25
1·3	0·0591	0·0809	9·5	1·731	26·68
1·35	0·0756	0·1084	10	1·777	29·19
1·4	0·0931	0·1396	12	1·938	39·98
1·45	0·1114	0·1740	14	2·073	51·86
1·5	0·1302	0·2118	16	2·189	64·74
1·6	0·1688	0·2968	18	2·289	78·56
1·7	0·208	0·394	20	2·378	93·24
1·8	0·248	0·502	30	2·713	178·2
1·9	0·287	0·621	40	2·944	279·6
2·0	0·326	0·750	50	3·120	395·3
2·1	0·364	0·888	60	3·261	523·6
2·2	0·402	1·036	70	3·380	663·3
2·3	0·438	1·193	80	3·482	813·7
2·4	0·474	1·358	90	3·572	974·1
2·5	0·509	1·531	100	3·652	1144
2·6	0·543	1·712	120	3·788	1509
2·7	0·576	1·901	140	3·903	1907
2·8	0·608	2·098	160	4·002	2333
2·9	0·639	2·302	180	4·089	2790
3·0	0·669	2·512	200	4·166	3270
3·2	0·727	2·954	250	4·329	4582
3·4	0·783	3·421	300	4·462	6031
3·6	0·836	3·913	350	4·573	7610
3·8	0·886	4·429	400	4·669	9303
4·0	0·934	4·968	500	4·829	13,015
4·2	0·979	5·528	600	4·960	
4·4	1·022	6·109	800	5·165	
4·6	1·063	6·712	1000	5·324	
4·8	1·103	7·334	1500	5·610	
5·0	1·141	7·976	2000	5·812	
5·2	1·178	8·636	5000	6·453	
5·4	1·213	9·315	10,000	6·933	
5·6	1·247	10·01	30,000	7·693	
5·8	1·280	10·73	100,000	8·523	
6·0	1·311	11·46			

TABLE 11 (PART 1). TABLE OF η AGAINST $-\xi$

η	$-\xi$	η	$-\xi$
0·0	0·0	1·1	1·656
0·001	0·0629	1·2	1·708
0·002	0·0887	1·3	1·756
0·003	0·1084	1·4	1·801
0·004	0·1250		
0·005	0·1395	1·5	1·842
0·006	0·1527	1·6	1·881
0·007	·01647	1·7	1·917
0·008	0·1759	1·8	1·952
0·009	0·1863	2·0	2·013
0·010	0·1926	2·2	2·068
0·011	0·2056	2·4	2·117
0·012	0·2146	2·6	2·160
0·013	0·2231	2·8	2·199
0·014	0·2314	3	2·234
0·015	0·2393	3·2	2·265
0·02	0·2753	3·4	2·293
0·03	0·3351	3·6	2·318
0·04	0·3848	3·8	2·341
0·05	0·4281	4	2·362
0·06	0·4668	4·5	2·404
0·07	0·5019	5	2·438
0·08	0·5346	5·5	2·463
0·09	0·5655	6	2·483
0·1	0·5941	6·5	2·499
0·2	0·817	7	2·511
0·3	0·979	8	2·528
0·4	1·108	9	2·538
0·5	1·217	10	2·544
0·6	1·312	11	2·548
0·7	1·396	12	2·550
0·8	1·470	15	2·553
0·9	1·538	20	2·554
1·0	1·600	∞	2·554

TABLE 11 (PART 2). TABLE OF ξ AGAINST η

ξ	η	ξ	η
0	0	3·7	2·182
0·01	0·000025	3·8	2·282
0·1	0·00245	3·9	2·384
0·2	0·00964	4·0	2·488
0·25	0·01493	4·2	2·700
0·3	0·02131	4·4	2·915
0·32	0·02413	4·6	3·140
0·34	0·02719	4·8	3·370
0·36	0·03047	5·0	3·604
0·38	0·03388	5·2	3·844
0·4	0·03737	5·4	4·092
0·42	0·04100	5·6	4·343
0·44	0·04482	6·0	4·856
0·46	0·04880	6·5	5·526
0·48	0·05299	7·0	6·224
0·50	0·05726	7·5	6·948
0·55	0·0689	8·0	7·697
0·6	0·0813	8·5	8·471
0·65	0·0946	9·0	9·272
0·7	0·1089	9·5	10·09
0·75	0·1239	10	10·93
0·8	0·1401	11	12·67
0·85	0·1571	12	14·49
0·9	0·1748	13	16·38
0·95	0·1934	14	18·32
1·0	0·2127	15	20·35
1·1	0·2538	16	22·43
1·2	0·2980	17	24·60
1·3	0·3451	18	26·76
1·4	0·3954	19	29·03
1·5	0·4481	20	31·35
1·6	0·5036	21	33·71
1·7	0·5617	22	36·09
1·8	0·6223	23	38·53
1·9	0·6854	24	41·04
2·0	0·7510	25	43·56
2·1	0·8190	26	46·16
2·2	0·8891	27	48·77
2·3	0·9612	28	51·41
2·4	1·036	29	54·14
2·5	1·112	30	56·86
2·6	1·191	32	62·42
2·7	1·272	34	68·09
2·8	1·354	36	74·02
2·9	1·439	40	86·11
3·0	1·526	50	118·6
3·1	1·614	60	153·5
3·2	1·704	80	230·2
3·3	1·796	100	313·8
3·4	1·890	120	403·8
3·5	1·986	140	499·7
3·6	2·083	160	600·9
3·7	2·182		

Appendix 1

RECENT WORK ON OXIDE CATHODES

Chapter I was written in 1949. In the intervening years the oxide cathode has been the subject of much interest and many workers have attempted to elucidate the phenomena concerned. These efforts have not been successful and there is no general agreement on the theory of the conduction process. De Boer's theory is now of only historical interest but the semi-conductor theory now has to share the field with a theory, due to Loosjes and Vink,† which ascribes the emission, at the working temperature, to gaseous conduction through the pores left in the loose mass of crystallites after activation. At low temperatures, below 750° K., Loosjes and Vink believe that the cathode behaves as a semi-conductor. Whether their theory is correct or not, it is clear that the oxide cathode is a much more complicated system than the simple excess semi-conductor which was at first used as a model for its behaviour. It is highly probable that acceptor levels are located only slightly above the filled band. The existence of such levels with consequent 'P' type conductivity considerably complicated the behaviour of a semi-conductor especially when second order effects such as the Hall effect and thermo-electric effect are under discussion. In this Appendix we shall only describe the main experimental evidence for the semi-conductor theory and for the Loosjes-Vink theory and indicate some of the important work on more complicated semi-conductor models.

The major experimental support for the semi-conductor theory is to be found in a paper by Hannay, MacNair and White.‡ In this work the conductivity and thermionic emission of (Ba, Sr)0 cathodes were measured as a function of temperature. The cathodes were very carefully prepared on magnesia sleeves, to eliminate any activation or contamination due to impurities in a metal sleeve. Activation was carried out by

† Loosjes and Vink, *Philips Res. Rep.* **4** (1949), 449.

‡ Hannay, MacNair and White, *J. Appl. Phys.* **20** (1949), 669.

heating the cathode to approximately 1320° K. in methane at a pressure of about 10^{-2} mm. Hg. The methane was then pumped away and the cathode heated in vacuum until a steady state was attained. Platinum wires were embedded in the coating to allow the conductivity to be measured and current was drawn to an anode to measure the thermionic emission. The major conclusion drawn from the measurements are the following.

1. The log (conductivity) $\sim$ $1/T$ curve is made up from two straight lines, one with a slope of 1·1 eV for temperatures below about 1200° K. and the other with slope 0·3–0·45 eV for higher temperatures.

2. Activation with methane increases the conductivity by a factor of up to 3000 and the change of conductivity took place in the first two minutes of exposure to the gas.

3. A plot of log (thermionic emission) $\sim$ log (conductivity) for a given cathode is a straight line with a slope of 45°. It is immaterial whether the cathode is active or inactive, whether it is aged or poisoned, the points still lie on this line.

The possibility of conduction through the pores was considered and the conductivity was measured in tubes containing He at 1 atmosphere pressure. This reduced the mean free path to about 10^{-4} mm. which was smaller than the observed cracks in the coating. Then, if pore conduction were present, the electrons would make collisions with gas molecules and the conductivity would be changed. The changes observed were small and random in sense.

The experimenters conclude that (Ba, Sr)0 is a reduction semi-conductor, probably an electronic semi-conductor with stoichiometric excess of (Ba, Sr) atoms in solid solution. The fact that the conductivity and thermionic emission are directly proportional over three orders of magnitude verifies the theory of Section 1.3.4.

In the work of Loosjes and Vink, the experimental apparatus consisted of two disk cathodes, similar to those used in cathode ray tubes, which were both sprayed and then assembled with the two oxide layers pressed together by springs, so that contraction of the coating during processing could not alter the pressure. The conductivity was measured on an A.C. bridge connected between the cathode supports. The I, V characteristic of the same tubes was studied under pulse conditions. In this work the temperature of the cathode supports, which was measured by means of two thermocouples, was taken as low as 500° K. This figure is about 300° lower than the lowest results reported by Hannay et al. The curves of $\log \sigma \sim 1/T$, for all but the state of poorest activation can be approximated by two straight lines which join at temperatures between 750° K. and 850° K. The slope of the low temperature line is much less than that of the high temperature line. At temperatures in excess of 1050° K. the line sometimes showed another departure from the intermediate temperature line. In Fig. 1 we show two linear functions plotted together with a dotted line representing their sum. The dotted curve gives a very good representation of the variation actually observed by Loosjes and Vink. These authors therefore advanced the hypothesis that the conductivity of oxide cathodes is due to two conduction mechanisms, of different activation energies, operating in parallel. For the low temperature range, the activation energy decreases as the activity of the cathode increases, the limits being 0·6 and 0·09 eV. The activation energy for the intermediate temperature range decreases from 2·2 to 0·8 eV.

Loosjes and Vink then suggest that the conductivity observed in the lower temperature range is that to be expected from a normal excess semi-conductor while the intermediate temperature conductivity, which includes the temperature range in which cathodes are used in practice, is due to conduction through an electron gas filling the pores of the cathode. It should, perhaps, be stressed at this point that the supposition that two conductivities are operating in parallel may well be correct even if the models assumed for the conductivities are completely wrong. Consideration of the mean free path of a thermal electron in the electron gas filling the pores shows that

$$I = \rho \sqrt{\frac{eE}{2m}}\, l \qquad \text{(A.1.1)}$$

where E = electric field strength,

l = mean free path.

The other symbols have their usual meanings.

The measured I, V characteristics were of this form at high enough temperatures. A further experiment was performed to support the hypotheses. Tubes were arranged so that the cathodes could be separated after conductivity measurements had been made, an anode inserted between them and the work function measured. This work function was found to equal, within the experimental error, the activation energy deduced from the slope of the intermediate temperature characteristic (*b*) in fig. A.1.1.

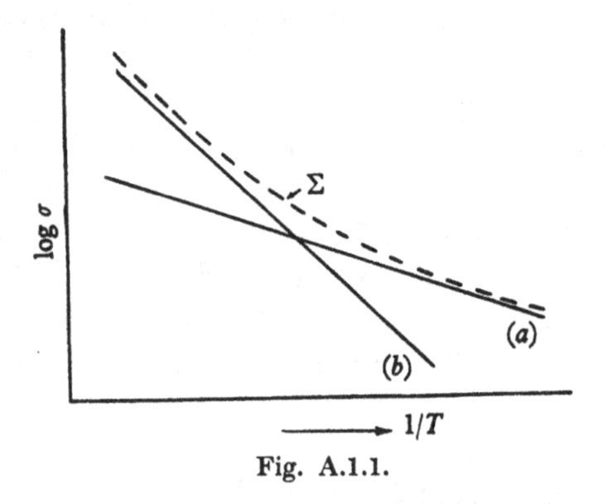

Fig. A.1.1.

The paper concludes by pointing out that Hannay's experiments with helium do not disprove the possibility of pore conduction as the mean free path was still too long to alter the conductivity greatly. It has recently been reported† that measurements of the conductivity in helium at pressures much in excess of one atmosphere are in progress and these should give direct evidence as to the role played by pore conduction.

In the meantime other properties of the oxide cathode have been measured which may throw light on the energetic structure. Examples are the Hall constant, the thermo-electric power, the spectral distribution of the photo-emission, secondary emission and luminescence.

The Hall effect, whose measurement was first reported by Wright,‡ is perhaps the most important of these properties.

† *M.I.T. Research Lab. of Electronics Quarterly Progress Report*, April 1952.
‡ Wright, *Nature*, **164** (1949), 714; *Brit. J. Appl. Phys.* **1** (1950), 150.

These measurements, like those of Hannay *et al.*, were made using sprayed coating on a magnesia base. Gas activation was used when high activities were desired. Simultaneous measurements of conductivity and Hall e.m.f. permit one to calculate the density of free electrons if it is assumed that the coating behaves as a semi-conductor. Wright finds that the cathode exhibits at least two typical states of activity. The first of these which he terms the 'reproducible' state, is at once established after outgassing at 1300° K. For this state

$$\sigma \doteqdot 10^{-3}\ \Omega^{-1}\ \text{cm.}^{-1} \text{ at } 1100^\circ\ \text{K.},\ Js \doteqdot 400\ \text{mA./cm.}^2$$

and the activation energy of the conductivity $\doteqdot 1{\cdot}1$ e.V. The Hall e.m.f. gives $\eta \doteqdot 5 \times 10^{12}$ electrons/cm. at the above temperature. The cathode tends always to return to this state either if current is drawn after poisoning or if current is drawn after activation. The second state is one of high activity and for it

$$\sigma \doteqdot 10^{-2}\ \Omega^{-1}\ \text{cm.}^{-1},\ Js \doteqdot 5{\cdot}0\ \text{A./cm.}^2,\ n \doteqdot 10^{14}.$$

The activation energy of σ is 0·5–0·7 e.V. at about 1000° K. but below 800° K. it drops to 0·2 e.V. Wright states that Hannay *et al.* did not observe this state because they always flashed their cathodes at 1300° K. after gas activation, which returns the cathode to the stable state. He also observes 'P' type conductivity at low temperatures in some circumstances. Wright's results may be summarized as indicating a stable state which may be directly explained by a homogeneous semi-conductor model. The active state, however, requires a more complicated model which may be that of Loosjes and Vink.

In his latest paper Wright† considers the effect of an accelerating field acting at the cathode surface. Since the surface has a relatively low conductivity, the field penetrates into the surface and lowers the level of the conduction band, causing a decrease in net work function. The voltage drop across the coating is also assumed to have a finite value. The energy level diagram is shown in fig. A.1.2. A similar model but without a voltage drop in the coating, has been studied by Morgulis.‡ Wright obtains

† Wright, *Proc. Phys. Soc. B.* **65** (1952), 1934.
‡ Morgulis, *J. Phys. U.S.S.R.* **11** (1947), 67.

the following expression which may be solved for δx.

$$\frac{E}{\kappa} - \frac{J}{\sigma} = -4\left(\frac{2\pi n_0 kT}{\kappa}\right) \sinh \frac{\delta x}{2kT}, \tag{A.1.2}$$

where $E \doteqdot$ field strength, volts/cm.

$\kappa =$ permittivity of coating ($\doteqdot 10$).

δx is defined in fig. A.1.2 and the other symbols have their earlier meanings.

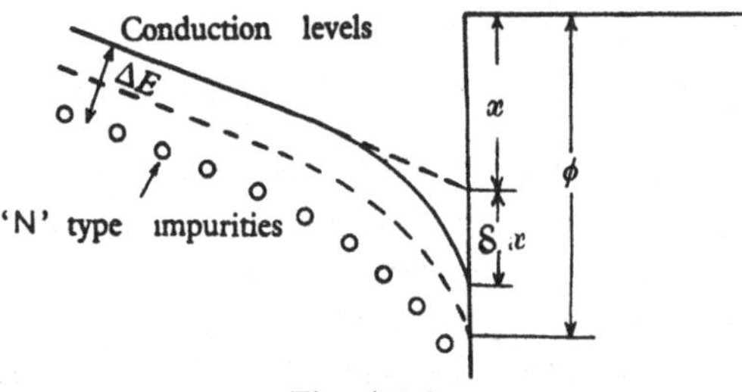

Fig. A.1.2.

This lowering of work function is then added to that due to the Schottky effect (Chapter 1) and the field at the cathode can be determined from the space charge conditions, following Ivey.† Diode I_a—V_a characteristics can then be plotted for assumed values of σ and n_0. A comparison with experimental characteristics then discloses the probable values of these quantities. Thus, an independent method for the estimation of σ and n_0 has been devised.

Wright concludes that

$$n_0 \doteqdot 10^{14} \text{ electrons/cm.}^3 \text{ and } \sigma \doteqdot 3 \times 10^{-2}\ \Omega^{-1}\ \text{cm.}^{-1}$$

are probable values and that n_0 is certainly less than 10^{15}. These values confirm those from Hall effect measurements and indicate that pore conductivity, if present, did not invalidate the earlier measurements.

Experimental work on luminescence‡ provides some direct evidence for locating the energy levels, but the results are conflicting and it is clear that the ordinary oxide cathode is too complicated a system for profitable study as yet. However, a systematic study of the behaviour of large single crystals of

† Ivey, *Phys. Rep.* **76** (1949), 554.
‡ Aitchison, *Nature*, **164** (1949), 1088.

BaO is being undertaken by Sproull† and his co-workers at Cornell University and this work will facilitate the understanding of the more involved problem.

To conclude this review of recent progress we wish to draw attention to the work of the Japanese investigators in this field. These workers have published many experimental studies on the oxide cathode, but their major contribution is the calculation of the thermionic emission from a semi-conductor model modified to include so-called 'frozen holes' some of which are assumed to contain trapped electrons. The concept of hole defects frozen into the crystal structure was introduced into semi-conductor theory by Nijboer.‡ The full consequences for the thermionic emission were calculated by Watanabe, Tokagi and Katsura.§ Their paper shows that if N is the density of frozen holes and n_0 the density of trapped electrons at 0° K., the emission law may be of three types, the Richardson ($T^{\frac{1}{2}}$), Wilson ($T^{\frac{5}{4}}$), or Nijboer (T^2) types, depending on the ratio n_0/N and the temperature. Critical temperature values should be observed at which the ordinary Richardson lines (plots of $ln\ I/T^2 \sim 1/T$) show marked changes of slope. These temperatures represent a change of the dominant emission mechanism. In a well-activated diode a change from the $T^{\frac{1}{2}}$ law to the $T^{\frac{5}{4}}$ law was observed at a temperature of about 850° K. The electron density was about $3 \times 10^{12}/\text{cm.}^3$, in moderate agreement with Wright's work. It will be clear that this work illustrates an alternative hypothesis capable of explaining the results of Loosjes and Vink on the electric conductivity. While the experimental evidence, as has already been said, is not yet complete, we might risk the guess that suitably modified semi-conductor models will explain the facts without making it necessary to include pore conduction.

1.2. Interface resistance effects

Since Chapter 1 was written the effects of the interface resistance between the metal core and the semi-conducting

† Tyler and Sproull, *Phys. Rev.* **83** (1951), 548 ; Bever and Sproull, *Phys. Rev.* **83** (1951), 801 ; DeVore and Dewdney, *Phys. Rev.* **83** (1951), 805.

‡ Nijboer, *Proc. Phys. Soc.* **51** (1939), 575.

§ Watanabe, Tokagi and Katsura, *Techn. Rep. Tohoku Univ.* **14** (1949), 26.

oxide, which were briefly discussed there, have assumed practical importance. It has been found that receiver valves used in pulse circuits where the valve is non-conducting for most of the time may have a very short life due to the growth of the interface resistance.† Moreover, since the interface resistance is shunted by a capacitance the apparent g_m of the valve becomes frequency sensitive, as first pointed out by Raudorf.‡

The effects are explicable if we consider that the valve is represented by the circuit of fig. A.1.3, where R and C are the interface resistance and capacitance respectively. R varies according to the duration of life and operating conditions from zero to a few hundred ohms while C is of the order 0·001–1·0μF. Consideration of these values shows that for frequencies of a few kilocycles, the valve behaves as though a current feedback resistor R were connected in the cathode lead while at frequencies of a few megacycles this resistor is shorted out. The current feedback reduces the stage gain at low frequencies in the usual way.

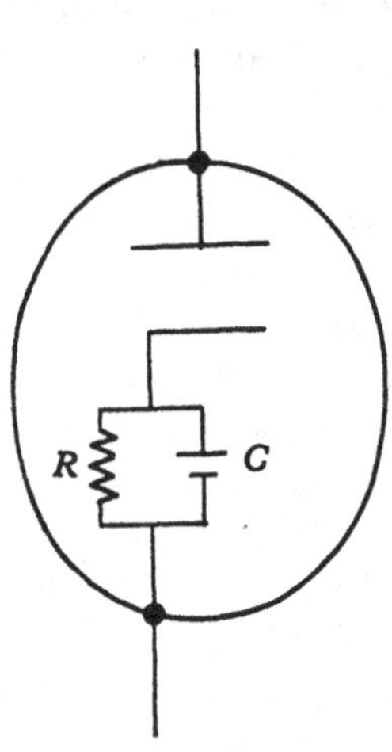

Fig. A.1.3.

Raudorf explained the phenomenon as due to the loosening of the mechanical bond between the sleeve and cathode with consequent appearance of resistance. There may be effects of this sort, but interface resistance growth is observed even in valves for which there is no such mechanical defect. The fundamental mechanism has been given by Eisenstein§ who ascribes the growth of interface resistance to reaction between the activating additions to the core metal and the free barium. In the silicon-containing nickel, used by Eisenstein, the interface compound was Ba_2SiO_4. The interface layer grows thicker with

† Waymouth, *J. Appl. Phys.* **22** (1951), 80; Eisenstein, *J. Appl. Phys.* **22** (1951), 138.

‡ Raudorf, *Wireless Engr.* **26** (1949), 331.

§ Eisenstein, letter in *Wireless Engr.* **27** (1950), 100.

life so that R increases and C decreases with life, as observed. The phenomenon is more pronounced in valves which are operated without a standing current because there is then no electrolytic decomposition of BaO yielding a flow of Ba^{++} ions to the interface layer where they provide an impurity agent for the Ba_2SiO_4 which increases its conductivity.

This picture is probably over-simplified but basically correct. It has an important practical consequence in that it means that cathodes should not be operated at too low a current density. Unpublished work by the author shows that 30 mA./cm.^{2} is about the lower limit.

Work has also been directed towards the elimination of the effect. It would appear that this could be done either by using a very pure nickel core or by using activating agents which produce high-conductivity interfaces. The first possibility has the disadvantage that such cathodes may be difficult to activate under manufacturing conditions. There is some controversy as to whether the use of pure nickel is completely efficacious. The American authors claim that it is; in the author's experience it is not.

The reader is referred to papers by Eaglesfield,† Metson‡ *et al.* and Child§ for further details of the behaviour of various types of valve which have been tested for long periods of time.

1.3. The Lemmens or L cathode

A recent development which is of considerable practical interest is the use of a new type of dispenser cathode. The form most widely known is that using a porous tungsten plug the underside of which is coated with $(BaSr)CO_3$ spray. This cathode was first described by Lemmens, Jansen and Loosjes.|| The tungsten plug is attached to the front of a molybdenum cylinder which is divided into a large and a small compartment. The large compartment contains the heater, and the small one,

† Eaglesfield, *Electr. Commun.* **28** (1951), 95.
‡ Metson, Wagener, Holmes and Child, *J.I.E.E.* **99**, pt. III (1952), 69.
§ Child, *P.O. Elect. Engrs. J.* **44** (1952), 176.
|| Lemmens, Jansen and Loosjes, *Philips Tech. Rev.* **11** (1950), 305. Further particulars are given by Espersen, *Proc. Inst. Radio Engrs., N.Y.* **40** (1952), 248.

which is closed by the W plug, contains the $(BaSr)CO_3$. The cathode temperature is gradually raised, in a good vacuum, to about 1100° C. It is then flashed at 1270° C. for 1 minute. The anode is then outgassed and current drawn from the cathode, still at 1270° C., until the emission reaches 3·0 A./cm.^{2}.

It is supposed that this cathode functions by means of a mono-molecular layer of barium, which diffuses through the pores of the tungsten plug and which is replenished from the relatively large mass of available barium as soon as it has been evaporated from the outside surface. This picture is borne out by the thermionic constants which are

$$\phi = 1{\cdot}6 - 2{\cdot}0 \text{ e.V.}, \quad A = 1 - 15 \text{ A./cm.}^2/\text{deg.}^2.$$

These figures might easily be obtained from a rough surface partially covered with barium. The emission is considerably less than that from an oxide cathode at the same temperature. For instance, at 1000° K. a good oxide cathode gives 1–10 A./cm^2 while the L cathode only gives a few tens of milliamps. However, the L cathode has a useful life at much higher temperatures than the maximum at which the oxide cathode can operate and the life is, in fact, determined by the heater, which, if it is of the normal alumina-coated tungsten wire, cannot be used to heat the cathode above about 1500° K. for more than a few hours. At 1300° K. the emission is 2·0 A./cm.^{2} and the life is determined by the amount of barium available, Espersen quoting a figure of 20,000 hours for a particular cathode. Thus, the cathode can be used to provide C.W. emissions over 10 A./cm.^{2} while the oxide cathode can only be rated at about 1 A./cm.^{2}.

The L cathode is less efficient thermally than the oxide cathode and requires about seven times the power per unit area to reach a given emission density. About 7 watts/cm.^{2} are required to obtain a density of 1 A./cm.^{2}.

Another advantage of the L cathode, which is particularly important in short wave magnetrons and in planar triodes, is that the surface can be machined very flat so that small clearances can be set and maintained in use.

Cathodes, similar to the L cathode, but with emitting agents other than $(BaSr)CO_3$ have been described by Katz and Rau.† These authors compared $(BaSr)CO_3$ with thoria and with a filling of a barium metal alloy. The thoria filled cathode gave $\phi = 2{\cdot}80$ e.V., $A = 100$ A./cm.²/deg.². This corresponds with an emission density of 1A./cm.² at 1640° K. The barium metal alloy cathodes were superior to the L cathode and for them $\phi = 1{\cdot}48$ e.V., $A = 2{\cdot}3$ A./cm.²/deg.². This leads to an emission of about 2·0 A./cm.² at 1200° K., i.e. about 100° cooler than the L cathode for the same density. In these cathodes precautions have to be taken to prevent an excessively rapid evolution of barium and this is done by enclosing the barium alloy in a subsidiary container some distance behind the porous tungsten plug.

It is not yet possible to evaluate fully the practical importance of these cathodes, but their most important use will be in valves for extremely short wavelengths where the cathode dimensions must be small and the current density must be correspondingly high. In these circumstances the large amount of heater power will have to be accepted. Another field of use is in projection cathode ray tubes where the use of a very high current density leads to an overall improvement in the electron optical system.

Nergaard‡ has recently proposed a theory of the oxide cathode in which the donors are mobile and can electrolyse and diffuse. While speculative in nature, this theory appears to throw light on an extremely wide range of phenomena.

† Katz and Rau, *Frequenz*, **5** (1951), 192.
‡ Nergaard, *R.C.A. Rev.* **13** (1952), 464.

Appendix 2

SPACE-CHARGE WAVES

In this section we develop some general results on the propagation of space-charge waves on electron beams. These results are due to Hahn† and to Ramo‡. They provide an alternative and more general approach to the theory of velocity modulation described in Chapter 12. The ballistic approach there used has been preferred by most writers on this subject, mainly because the initial conditions at the gaps can be inserted in a simpler and more natural way and because it is easier to introduce successive refinements in the ballistic theory. However, recent work on noise in travelling wave tubes and other devices with long beams and the development of space-charge wave tubes has shown that the wave approach is of considerable utility. To facilitate the understanding of this work the main results are derived below.

2.1. Derivation of the wave equation

We consider a region in which a beam of electrons moves to right (positive z direction) with a velocity of u_0. The beam is prevented from spreading transversely by a strong magnetic field. This not only prevents d.c. space-charge forces from spreading the beam but also stops the electrons from moving transversely under the action of h.f. fields. In addition, the average negative space-charge density is compensated by an equal positive ion density. This ensures that the potential is uniform throughout the region occupied by the beam and that electrons near the axis have the same velocity as peripheral electrons. Owing to their mass, the positive ions can be assumed stationary in the h.f. fields.

† Hahn, *Gen. Elect. Rev.* **42** (1939), 258 and 497.
‡ Ramo, *Phys. Rev.* **56** (1939), 276.

APPENDIX 2

The following notation is used:

$$\text{d.c. beam current} = I_0,$$
$$\text{d.c. potential} = V_0,$$
$$\text{beam area} = \Sigma,$$
$$\text{d.c. charge density} = \rho_0,$$
$$\text{a.c. charge density} = \rho_1,$$
$$\text{a.c. velocity} = u_1,$$
$$\text{a.c. beam current density} = j_1,$$
$$\text{wave number} = k = \omega/C.$$

we then have

$$\rho_0 = \frac{I_0}{\Sigma u_0} \tag{A.2.1}$$

$$u_0^2 = 2\frac{e}{m}V_0. \tag{A.2.2}$$

We next seek wave functions, of the form $\phi(x, y) \exp(-\Gamma z + j\omega t)$ which are solutions of Maxwell's Equations. If we replace the operators $\partial/\partial t$ and $\partial/\partial z$ by jw and $-\Gamma$ respectively, we can write Maxwell's equations† as

$$\nabla . \mathbf{E} = -\frac{\rho_1}{\epsilon_0} \tag{A.2.3}$$

$$\nabla . \mathbf{B} = 0 \tag{A.2.4}$$

$$\nabla \times \mathbf{E} = -j\omega \mathbf{B} \tag{A.2.5}$$

$$\nabla \times \mathbf{B} = j\frac{\omega}{c^2}\mathbf{E} - \mu_0 \mathbf{j}_1. \tag{A.2.6}$$

The wave equation is derived by taking the curl of eqn. (A.2.5), using the vector relation curl. curl = grad. div $-\nabla^2$ and substituting $\nabla \times \mathbf{B}$ from eqn. (A.2.6). The result is

$$\nabla^2\mathbf{E} + k^2\mathbf{E} = -\frac{\nabla \rho_1}{\epsilon_0} - j\omega\,\mu_0 \mathbf{j}_1. \tag{A.2.7}$$

† See e.g. Stratton, *Electromagnetic Theory*, McGraw Hill, 1941, p. 268.

Now $$\nabla^2\mathbf{E} = \frac{\partial^2\mathbf{E}}{\partial x^2} + \frac{\partial^2\mathbf{E}}{\partial y^2} + \Gamma^2\mathbf{E}.$$

Therefore $$\frac{\partial^2\vec{\mathbf{E}}}{\partial x^2} + \frac{\partial^2\vec{\mathbf{E}}}{\partial y^2} + (\Gamma^2 + k^2)\mathbf{E} = -\frac{\nabla\rho_1}{\epsilon_0} - j\mu_0\omega\mathbf{j}_1 \tag{A.2.8}$$

or for the z component

$$\frac{\partial^2 E_z}{\partial x^2} + \frac{\partial^2 E_z}{\partial y^2} + (\Gamma^2 + k^2)E_z = \frac{\Gamma\rho_1}{\epsilon_0} - j\omega\mu_0 j_1 \tag{A.2.9}$$

To carry the solution further we must introduce the continuity equation and the force equation, as was done in § 14.3. Furthermore, to linearize the expressions we use the small signal assumption, i.e. we neglect the cross product of a.c. quantities in the expression for j_1, putting $j_1 = \rho_0 u_1 + \rho_1 u_0$. This was also done in the analysis of § 14.3. Eqns. (6) of § 14.3 give

$$u_1 = \frac{-(e/m)E_z}{(j\omega - \Gamma u_0)} \tag{A.2.10}$$

$$\rho_1 = -j\frac{\Gamma j_1}{\omega} \tag{A.2.11}$$

and putting these into the expression for j_1, we obtain

$$j_1 = -j\frac{\omega\rho_0(e/m)E_z}{(j\omega - \Gamma u_0)^2} \tag{A.2.12}$$

When eqns. (A.2.11) and (A.2.12) are put into eqn. (A.2.9), they yield

$$\frac{\partial^2 E_z}{\partial x^2} + \frac{\partial^2 E_z}{\partial y^2} + (\Gamma^2 + k^2)\left[\frac{\omega p^2}{(j\omega - \Gamma u_0)^2} + 1\right]E_z = 0 \tag{A.2.13}$$

where $$\omega_p = \frac{I_0(e/m)}{\epsilon_0 u_0 \Sigma}. \tag{A.2.14}$$

The quantity ω_p is often referred to as the plasma frequency for a reason which will become obvious in the next section.

2.2. Solution for an unbounded beam of very large cross section

Since we are assuming that the steady axial magnetic field is very strong, the electrons cannot move in the x or y directions and the wave will propagate so that all electrons in a plane of constant z have the same instantaneous velocity, or, in other words, $\frac{\partial^2 E_z}{\partial x^2} = \frac{\partial^2 E_z}{\partial y^2} = 0$. Eqn. (A.2.13) is then

$$(\Gamma^2 + k^2)\left[\frac{\omega_p^2}{(j\omega - \Gamma u_0)^2} + 1\right] E_z = 0. \tag{A.2.15}$$

The four roots of this quartic are, by inspection,

$$\Gamma_{1,2} = \pm jk \tag{A.2.16}$$

$$\Gamma_{3,4} = j\left(\frac{\omega \pm \omega_p}{u_0}\right). \tag{A.2.17}$$

The first two waves propagated with the velocity of light and are essentially independent of the beam. The remaining pair travel at velocities somewhat above and below the beam velocity. We now suppose that the electrons are given an initial velocity modulation u_{10} at a plane whose coordinates are $z = t = 0$. The two travelling waves Γ_3, Γ_4 can then be combined to give a standing wave

$$u_1 = u_{10} \cos \frac{\omega_p z}{u_0} \exp j\left(\omega t - \frac{\omega z}{u_0}\right). \tag{A.2.18}$$

From eqns. (A.2.10) and (A.2.12)

$$j_1 = j \frac{\omega \rho_0 u_1}{(j\omega - \Gamma u_0)}. \tag{A.2.19}$$

Inserting the values of Γ_3 and Γ_4

$$j_1 = j \frac{I_0 u_{10}}{\Sigma u_0} \cdot \frac{\omega}{\omega_p} \sin \frac{\omega_p z}{u_0} \exp j\left(\omega t - \frac{\omega z}{u_0}\right). \tag{A.2.20}$$

The wavelength of the space charge waves is thus $\frac{u_0}{\omega_p}$, explaining the name given to ω_p.

As a check on these equations, let us compare the results obtained for a short drift length s with those obtained by the ballistic method in Chapter 12. For s small enough to make

$$\sin \frac{\omega_p s}{u_0} \to \frac{\omega_p s}{u_0}$$

$$\Sigma j_1 \to \frac{I_0 u_{10}}{u_0} \cdot \frac{\omega s}{u_0},$$

eqn. (5) of Chapter 12 gives $u_{10} = \dfrac{e\beta_1 V_1}{mu_0}$. Also $\dfrac{\omega s}{u_0} = S$ and $u_0 = 2\dfrac{e}{m}V_0$. Therefore $I_1 = \Sigma j_1 = \dfrac{S\alpha\beta_1}{2}.I_0$.

But this is exactly the result given by the ballistic theory since $2I_0 J_1\left(\dfrac{S\alpha\beta_1}{2}\right) \to I_0\dfrac{S\alpha\beta_1}{2}$ for small values of the argument. The two theories thus agree for short drift lengths, i.e. those for which $\dfrac{\omega_p s}{u_0} < 0{\cdot}2$ radian.

2.3. A cylindrical beam in a cylindrical tunnel

We next study a case of far more practical importance than the infinite beam. This is a cylindrical beam radius b in a tunnel radius a.

We now have to find a solution to the wave equation which makes E_z finite on the axis, zero on the wall of the tunnel, and which obeys the standard boundary conditions at the periphery of the beam $r = b$. The analysis is similar to that of the helix considered in § 14.9.

For the region inside the beam, the wave equation now becomes

$$\frac{1}{r}\frac{\partial}{\partial r}\left(r\frac{\partial E_z}{\partial r}\right) + \gamma_b^2 E_z = 0. \tag{A.2.21}$$

where

$$\gamma_b^2 = (\Gamma^2 + k^2)\left[1 + \frac{\omega_p^2}{(j\omega - \Gamma u_0)^2}\right]. \tag{A.2.22}$$

The solutions to eqn. (A.2.21) are of the form

$$E_z = AJ_0(\gamma_b r) + BY_0(\gamma_b r)$$

and since $Y_0(0) \to -\infty$ we must have $B = 0$, since E_z clearly cannot tend to $\to -\infty$ on the axis.

Therefore, inside the beam,

$$E_z = AJ_0(\gamma_b r). \tag{A.2.23}$$

Outside the beam, ω_p is zero, and we must replace γ_b^2 by γ_0^2, where

$$\gamma_0^2 = (\Gamma^2 + k^2), \tag{A.2.24}$$

the solution becoming

$$E_z = CI_0(\gamma_0 r) + DK_0(\gamma_0 r). \tag{A.2.25}$$

Since $E_z = 0$ at $r = a$,

$$\frac{D}{C} = -\frac{I_0(\gamma_0 a)}{K_0(\gamma_0 a)}. \tag{A.2.26}$$

Two further equations connecting the arbitrary constants are obtained by requiring that the tangential and radial components of the electric field are continuous at $r = b$.† These are

$$\frac{A}{C} = \frac{1}{J_0(\gamma_b b)}\left[I_0(\gamma_0 b) - \frac{I_0(\gamma_0 a)}{K_0(\gamma_0 a)}.K_0(\gamma_0 b)\right] \tag{A.2.27}$$

and

$$\frac{A}{C} = \frac{\gamma_0}{\gamma_b . J_1(\gamma_b b)}\left[I_1(\gamma_0 b) + \frac{I_0(\gamma_0 a)}{K_0(\gamma_0 a)}.K_1(\gamma_0 b)\right]. \tag{A.2.28}$$

Thus, finally,

$$-\gamma_b \frac{J_1(\gamma_b b)}{J_0(\gamma_b b)} = \gamma_0 \left[\frac{I_1(\gamma_0 b)K_0(\gamma_0 a) + I_0(\gamma_0 a)K_1(\gamma_0 b)}{I_0(\gamma_0 b)K_0(\gamma_0 a) - I_0(\gamma_0 a)K_0(\gamma_0 b)}\right]. \tag{A.2.29}$$

As Ramo remarks, the simplest form of solution is that for which $a = b$. We then have merely to satisfy $J_0(\gamma_b b) = 0$, i.e.

$$\gamma_b = \frac{2{\cdot}405}{b}, \quad \frac{5{\cdot}520}{b}, \quad \frac{8{\cdot}654}{b}, \quad \text{etc.}$$

The smallest value of γ_b naturally corresponds with the physical situation in which there are no nodes in the electric field. Having derived a value of γ_b it can be inserted into eqn. (A.2.22) and, all the other quantities being known, a solution for Γ obtained. For practical electron velocities k^2 is much less than Γ^2, which we already know to be approximately ω/u_0 whereas $k = \omega/c$. We can therefore neglect k^2 in the first term of eqn. (A.2.22).

Guided by the value of Γ found in the preceding section, we try the values $\Gamma_{3,4} = j\left(\frac{\omega \pm \omega_c}{u_0}\right)$ where ω_c is a corrected value of ω_p, appropriate to the cylindrical geometry, which we have to determine. Substituting, for Γ_4, in eqn. (A.2.22) we get

$$\gamma_b^2 = -\left(\frac{\omega - \omega_c}{u_0}\right)^2\left[1 - \frac{\omega_p^2}{\omega_c^2}\right]$$

† These equations are misprinted in Ramo's paper, loc. cit. p. 278

If we now assume that $\omega_c \ll \omega$ we obtain

$$\left(\frac{\omega_c}{\omega_p}\right)^2 = \frac{1}{1+(\gamma_b/\beta_e)^2} \tag{A.2.30}$$

where $\beta_e = \omega/u_0$, as in Chapter 14.
For the important practical case of the beam filling the tunnel,

$$\left(\frac{\omega_c}{\omega_p}\right)^2 = \frac{1}{1+\dfrac{5\cdot8}{(\beta_e b)^2}} \tag{A.2.31}$$

The ratio ω_c/ω_p is shown in fig. A.2.1. It is always less than 1

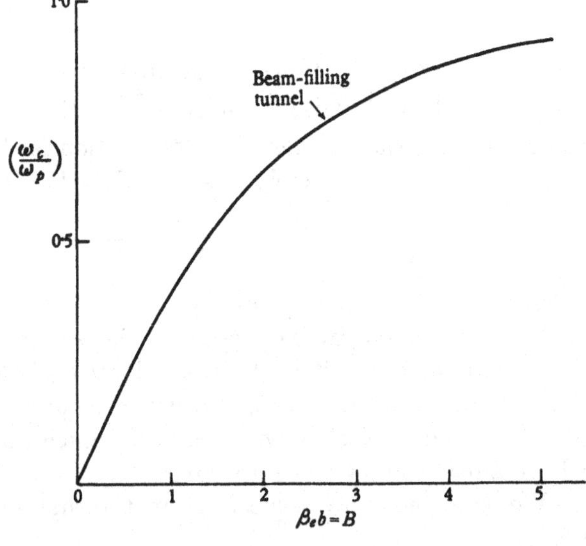

Fig. A.2.1.

and only approaches unity for very large values of b.

We can now discuss the approximation $\omega_c \ll \omega$. If we insert values into the expression for u_0, assuming $J_0 = 10^5$ A./m. and $V_0 = 10$ kV., we find $\omega_p \doteqdot 6\times10^9$. Practical valves work with $0\cdot1 < \omega_c/\omega_p < 0\cdot7$ so that $\omega_c = 0\cdot1\omega$ for frequencies between 1000 and 7000 Mc./sec. The assumed current density is greater than would be encountered in existing valves and the voltage has been assumed at a correspondingly high figure. We may conclude, therefore, that the approximation should be valid for a wide range of practical cases, but that special investigation may be needed at frequencies of a few hundred megacycles.

Having derived an acceptable approximation for ω_c, we can write the expressions for the space-charge standing waves by replacing ω_p by ω_c in eqns. (A.2.18) and (A.2.20). The results are, for a thin beam,

$$u_1 = u_{10} \cos \frac{\omega_c z}{u_0} \exp j\left(\omega t - \frac{\omega z}{u_0}\right), \qquad \text{(A.2.32)}$$

$$j_1 = j \frac{I_0 u_{10}}{\Sigma u_0} \frac{\omega}{\omega_c} \sin \frac{\omega_c z}{u_0} \exp j\left(\omega t - \frac{\omega z}{u_0}\right). \qquad \text{(A.2.33)}$$

It is interesting to compare the results of this section with those given in eqns. (38) and (39) of Chapter 12.

Perhaps the most interesting point about the results is that the space charge waves persist after the first maximum in the current density. In the simple theory subsequent maxima are just as great as the first. In the ballistic theory it is not obvious that this is the case and, in fact, it has usually been assumed necessary to work at the first current maximum. The present results show that this is not necessary and although there may be no gain in transconductance, practical considerations may make it desirable to use the second or a later maximum.

In Appendix 3 we shall see how the space-charge wave theory is used in noise analysis.

Appendix 3

NOISE IN TRAVELLING-WAVE TUBES

In Chapter 14 we briefly discussed the breakdown of the elementary theory of noise in travelling-wave tubes. We also said that a beginning had been made in the application of the Llewellyn-Rack-Peterson noise theory to the travelling-wave tube. This approach has been extended and has resulted in a noise theory which is capable of explaining the most important features of the noise behaviour of long electron beams. It has also enabled low noise travelling-wave tubes to be designed which have lower noise figures than crystal mixers in the 10 cm. band. For comparison, the following figures relate to low noise tubes developed before the theory was elucidated. Harrison† describes a tube for 3000 Mc./sec. with 25 db. gain and 14 db. noise figure for a beam current of 140 μA. In this type the current is drawn from the gun by a focusing electrode at 1250 V. and the helix potential is 650 V. Agdur and Åsdal‡ describe experiments on a 10,000 Mc./sec. valve with 12 db. gain and 17 db. noise figure. The new noise theory is not without difficulties, the most important of which is that it replaces the real electron beam having a Maxwellian velocity distribution by a single velocity stream with the correct mean square current fluctuations. Such a replacement can hardly be justified in devices with long transit angles, but the theory does agree with experimental results to within the limits due to approximations in the treatment and errors in the experiments. Multivelocity streams are now exciting some interest and it is to be hoped that the reasons for the agreement will soon be made clear.

The basic features of the Llewellyn-Rack-Peterson approach, applied by Pierce§ and by other investigators is to replace the complicated u.h.f. valve by a series of diodes. The choice of

† Harrison, *Sylv. Tech.* **3**, 12–16 (April, 1950).

‡ Agdur and Åsdal, *Report No.* 19 *Research Lab. of Electronics*, Chalmers University, Gothenburg, 1951.

§ Pierce, *Traveling-Wave Tubes*, Van Nostrand, 1950, chap. 10.

the diode planes is a matter of commonsense and an obvious choice for a common type of travelling-wave tube is shown in fig. A.3.1. Here the electrons are accelerated from the cathode by a voltage V_0 applied to the anode. They are focused through a tunnel, which forms part of the waveguide coupling system, and eventually reach the helix which is presumed to start at plane (*c*). Clearly, there is some uncertainty as to the precise location of (*c*). The Llewellyn equations, dealt with in Chapter 7, allow us to calculate the electronic behaviour of the

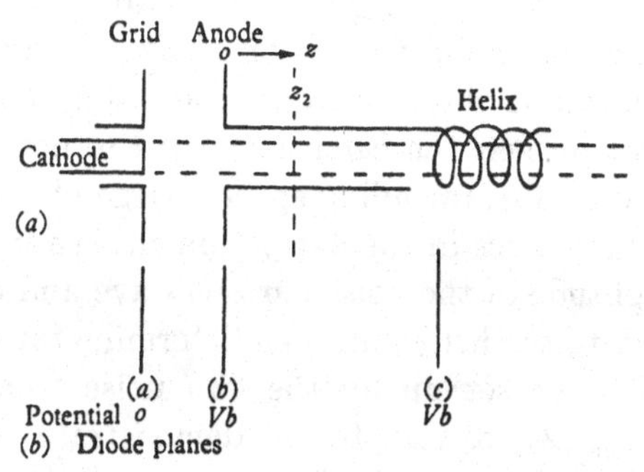

Fig. A.3.1.

diode (*a*), (*b*) if the a.c. current and velocity at plane (*a*) are known. The same quantities calculated at plane (*b*) provide the initial conditions for the diode (*b*), (*c*) and the process can be repeated indefinitely. The initial conditions for plane (*c*) are those derived by Rack and given in Chapter 8. In the simplest type of theory, it is assumed that there is complete space-charge in the first diode and zero space-charge in subsequent diodes. When this is done, it is also necessary to assume that the beam in the tunnel behaves as an infinite plane parallel beam. Both space-charge and cylindrical beam effects can be included in the analysis of the second diode by using the results of Appendix 2. This has been done by Cutler and Quate† in an experimental verification of the theory. Another refinement due to Parzen‡ is the consideration of the effect of thermal velocity spread by introducing the hydrostatic pressure term, as suggested by

† Cutler and Quate, *Phys. Rev.* **80** (1950), 875.
‡ Parzen, *J. Appl. Phys.* **22** (1951), 398.

Hahn† in his paper on kinetic theory of diodes. Effects due to the velocity distribution are also discussed by Robinson.‡ A further effect is of great practical importance, that is amplification by velocity jumps. It is easily shown that when the d.c. velocity of an electron beam is suddenly changed, the a.c. velocities before and after the jump are inversely proportional to the d.c. velocities and the a.c. current is unchanged. The effect has been made use of in an amplifier by Field, Tien and Watkins.§ It can also be used to reduce the noise velocities in a beam, and Watkins|| has given the detailed theory and has demonstrated that noise figures of 10 db. for 18 db. gain at 3000 Mc./sec. can be obtained from tubes using this device.

Let us now discuss the calculation of the noise factor in a travelling-wave tube, including space-charge effects by the use of space-charge waves in the entry tunnel. We must calculate first the amplitude of the noise current wave and noise voltage wave at entry to the helix and then determine the amplitude of the growing wave set up by the two noise components, for eqns. (23) and (24) of Chapter 14 demonstrate that non-zero velocity and conduction current components both contribute to the amplitude of the waves. We assume that the gain of the tube is sufficiently large, greater than 10 db., to ensure that the amplified wave is the only source of noise at the output. The noise factor is then defined by eqn. (58) of Chapter 14.

3.1. A.C. amplitudes at entry to the helix

In this section we proceed to calculate the noise current and velocity at entry to the helix. In the cathode to anode region (fig. 1) we assume the existence of a space-charge limited diode. The noise quantities at plane (*b*) calculated from the Llewellyn equation are used as initial values for space-charge waves in the space (*b*)–(*c*) and the values calculated at plane (*c*) are inserted as initial conditions for forced waves on the helix. This treatment differs from that of Pierce in that space-charge effects in the tunnel are included. Several authors have used this

† Hahn, *Proc. Inst. Radio Engrs., N.Y.* **36** (1948), 1115.
‡ Robinson, *Phil. Mag.* **43** (1952), 51.
§ Field, Tien and Watkins, *Proc. Inst. Radio Engrs., N.Y.* **39** (1951), 184.
|| Watkins, *Proc. Inst. Radio Engrs., N.Y.* **40** (1952), 65.

approach, particularly Smullin† and Watkins.‡ The terminology used is the same as that in Chapter 7 and Appendix 2 except that we must change the sign of the currents, since electron currents to the right are negative in Llewellyn's equations and Appendix 2 and positive in Pierce's travelling wave theory given in Chapter 14. The cross section of the beam is uniform, so we use current and not current density. We also note that, since the expression for noise figure is in terms of noise power, only the moduli of the noise current and noise velocity are important.

The grid to anode diode has been treated in § 13.8. Eqns. (46) and (47) with a change of sign for q give the following expressions for the noise conduction current and velocity at plane (b):

$$\overline{i_b^2} = \left(\frac{I\theta_1}{u_b}\right)^2 \overline{v_a^2}, \tag{A.3.1}$$

$$\overline{v_b^2} = \overline{v_a^2}, \tag{A.3.2}$$

where
$$\overline{v_a^2} = \frac{4kTe}{Im}\left(1-\frac{\pi}{4}\right)\Delta f$$

from eqn. (47) of Chapter 8.

In the tunnel space charge waves are set up according to the theory of Appendix 2. Suppose that the velocity maximum is at a plane z_2§ distant from the anode plane. Matching at the entry to the tunnel then gives

$$\overline{v_a^2} = \overline{v_{20}^2}\cos^2\theta_2, \tag{A.3.3}$$

$$\left(\frac{I\theta_1}{u_b}\right)^2 \overline{v_a^2} = \left(\frac{I}{u_b}\right)^2\left(\frac{\omega}{\omega_c}\right)^2 \overline{v_{20}^2}\sin^2\theta_2, \tag{A.3.4}$$

where
$$\theta_2 = \frac{\omega_c z_2}{u_b}.$$

Solving for $\overline{v_{20}^2}$ and θ_2 we obtain

$$\overline{v_{20}^2} = \overline{v_a^2}\left[1+\theta_1^2\left(\frac{\omega_c}{\omega}\right)^2\right], \tag{A.3.5}$$

$$\cos^2\theta_2 = \frac{1}{1+\theta_1^2(\omega_c/\omega)^2}. \tag{A.3.6}$$

† Smullin, *M.I.T. Research Lab. of Electronics Quarterly Progress Report* (July, 1950).

‡ Watkins, loc. cit.

§ Plane z must be taken inside the tunnel since the space charge wavelength is only defined therein and is indefinite in the first diode.

We then have for the wave moduli at plane (c)

$$\overline{v_c^2} = \overline{v_a^2}\left[1+\theta_1^2\left(\frac{\omega_c}{\omega}\right)^2\right]\cos^2\frac{\omega_c}{u_b}(z_c-z_2). \tag{A.3.7}$$

$$\overline{i_c^2} = \overline{v_a^2}\left[\left(\frac{\omega}{w_c}\right)^2+\theta_1^2\right]\left(\frac{I}{u_b}\right)^2\sin^2\frac{\omega_c}{u_b}(z_c-z_2). \tag{A.3.8}$$

These expressions show that the plane (c) can be chosen so that either $\overline{v_c^2}$ or $\overline{i_c^2}$ vanishes. In a klystron one might locate the input resonator at a plane where $\overline{i_c^2} = 0$, as this might eliminate amplified shot noise, which is the largest noise component in these tubes. In the travelling-wave tube both r.f. velocity and current contribute to the initial wave amplitude and the problem becomes more complicated. In the next section we therefore determine the noise figure for a travelling-wave tube in terms of the input conditions to the helix.

3.2. The noise figure in terms of the initial conditions

The noise figure of the travelling-wave tube has already been defined in eqn. (58) of Chapter 14. If the forward gain is sufficiently large only the amplified wave need be considered and we can write the following equation for the noise figure in terms of the fields at the input

$$F = \frac{|E_{iA}|^2+|E_{iN}|^2}{|E_{iA}|^2}, \tag{A.3.9}$$

where E_{iN} is now the peak input amplitude of the noise component of the amplified wave. When the impressed voltage and conduction current are not zero, eqns. (22), (23) and (24) of Chapter 14, take the following forms

$$E_1+E_2+E_3 = E_i,\dagger \tag{A.3.10}$$

$$\frac{E_1}{\zeta_1}+\frac{E_2}{\zeta_2}+\frac{E_3}{\zeta_3} = -\frac{e}{mu_0}v, \tag{A.3.11}$$

$$\frac{E_1}{\zeta_1^2}+\frac{E_2}{\zeta_2^2}+\frac{E_3}{\zeta_3^2} = -j\frac{2V_0}{\beta_e I}i. \tag{A.3.12}$$

† Watkins (loc. cit.) includes a correction term equal to $j\,.\,i\frac{8\beta_e V_0 C^3}{I_0}.Q$ in this equation. This correction takes care of the passive modes and is derived by Pierce (op. cit.), chap. 7. The inclusion of this term probably renders the theory more exact, but it complicates the discussion of the results so much that it seemed better to ignore it here. The reader should refer to Watkins' paper for detail and graphical results.

It is convenient to write $\xi_n = \beta_e C \partial_n$ from eqn. (4) of Chapter 14, where the ∂_n are, to a first approximation, the three cube roots of -1. With this notation the amplified wave E_1 is given by

$$E_1 = \frac{\dfrac{u_b \beta_e Cm}{e}\left[\dfrac{e}{u_b \beta_e Cm} E_i + (\partial_2 + \partial_3) v - j \dfrac{u_0 C_i}{I} \partial_2 \partial_3\right]}{[1-(\partial_1/\partial_2)][1-(\partial_1/\partial_3)]}. \tag{A.3.13}$$

We can now write down $|E_i a|^2$, since the Johnson noise power available at input is $kT_a \Delta f = P_n$. From the definitions of C^3 and Z, given in Chapter 14, and using the relation $V^2 = \Gamma^2 E^2 = \beta_e^2 E^2$ we obtain

$$E_i^2 = \frac{8\beta_e^2 V_b P_n}{C^3 I}, \tag{A.3.14}$$

where $\hat{E}_i$ peak value of E_i.
Since in the noise analysis we are interested in mean square values, we must replace eqn. (A.3.14) by

$$E_i^2 = \frac{4\beta_e^2 V_b P_n}{C^3 I}. \tag{A.3.15}$$

Putting eqn. (A.3.15) into eqn. (A.3.13) with $v = i = 0$ gives the required value of $|E_i|^2$, which can be used with eqn. (A.3.9) to give

$$F = 1 + \frac{Im}{2eCkT_a \Delta f}\left|(\partial_2 + \partial_3) v - j \frac{u_b Ci}{I} \partial_2 \partial_3\right|^2. \tag{A.3.16}$$

This equation can be used with the values of $\overline{v_i^2}$ and $\overline{i^2}$ given by eqns. (A.3.7) and (A.3.8) to obtain the final expression for the noise figure. When we are using space-charge wave theory the general relation can be simplified a little more using eqns. (A.2.32) and (A.2.33). The result is

$$F = 1 + \frac{Im \overline{v_{20}^2}}{2eCkT_a \Delta f}\left|(\partial_2 + \partial_3) . \cos\left(\frac{\omega_c z}{u_b}\right) - (\partial_2 \partial_3) \frac{\omega}{\omega_c} . C \sin\left(\frac{\omega_c z}{u_b}\right)\right|^2. \tag{A.3.17}$$

Eqn. (A.3.15) demonstrates that, on space-charge wave theory, it is only necessary to know the mean square noise velocity component at entry to the helix. If we now use the approximations

$\partial_2 = -\frac{\sqrt{3}}{2} - j/2$, $\partial_3 = j$, the square of the modulus becomes

$$\cos^2\left(\frac{\omega_c z}{u_b}\right) + \left(\frac{\omega}{\omega_c}\right)^2 C^2 \sin^2\left(\frac{\omega_c z}{u_b}\right). \tag{A.3.18}$$

This function oscillates with the periodicity of the space-charge waves, between 1 and $\left(\frac{\omega}{\omega_c}\right)^2 . C^2$. The latter term is of the order 0·1, so the noise figure can be improved by amounts of the order of 10 db. by the correct choice of z. It should be noted that in eqns. (A.3.7) and (A.3.18) z is measured from the plane $z = z_2$.

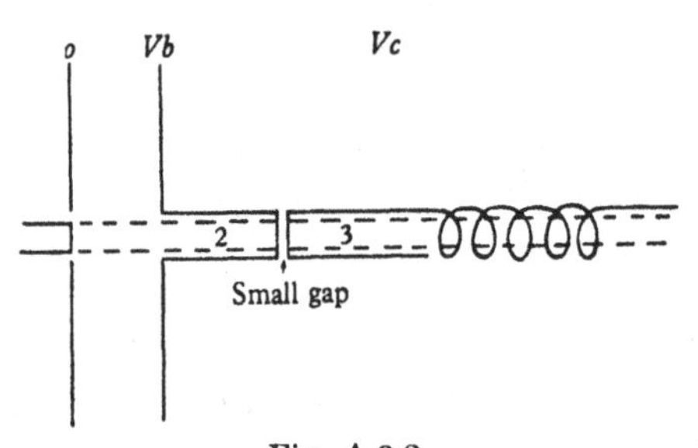

Fig. A.3.2.

Inserting the value of $\overline{v_{20}^2}$ from eqn. (A.3.5) and the expression for $\overline{v_0^2}$ the explicit expression for the noise figure is

$$F = 1 + \frac{2}{C}\frac{T}{T_a}\left(1 - \frac{\pi}{4}\right)\left[1 + \theta_1^2\left(\frac{\omega_c}{\omega}\right)^2\right]$$
$$\times \left|(\partial_2 + \partial_3)\cos\left(\frac{\omega_c z}{u_b}\right) - (\partial_2\partial_3)\frac{\omega}{\omega_c} C \sin\left(\frac{\omega_c z}{u_b}\right)\right|^2. \tag{A.3.19}$$

3.3. The use of velocity jumps to decrease noise

We now turn to the discussion of the term $[1 + \theta_1^2(\omega_c/\omega)^2]$. We have already noted that this can be reduced by including a velocity jump. At the velocity jump the r.f. conduction current is unchanged and the a.c. velocities are inversely proportional to the d.c. velocities. In fig. A.3.2 we suppose that the gap is located at a plane of max. a.c. velocity, e.g. z_2 in fig. A.3.1. Then, since $\overline{v_2^2}/\overline{v_3^2} = V_c/V_b$ we must replace $[1 + \theta_1^2(\omega_c/\omega)^2]$ by

$$[1 + \theta_1^2(\omega_c/\omega)^2]\, V_b/V_c$$

and if $V_b/V_c < 1$, the noise will be reduced. Watkins† discusses this and another scheme in which a second gap is used, the order of the voltages being high-low-high, and with the first gap at a current maximum. The second gap is at a velocity maximum.

3.4. Remarks on the adequacy of the theory

The experimental work of Cutler and Quate‡ together with work carried out at M.I.T. and Stanford Universities proves the general adequacy of the theory. The main divergences are that the noise minima are not as low as they should be. There are probably several reasons for this. First, as the major sources of noises are eliminated effects such as partition noise and ionic noise, which have been neglected above, begin to contribute to the measured noise. Secondly, we must consider effects due to the Maxwellian velocity distribution. Finally, it has been assumed that the initial conditions at the virtual cathode include no conduction current term. This is not correct, as our discussion in § 8.4 indicates that there should be a residual current fluctuation which is less than 1 per cent of the temperature limited fluctuation.

Some of these effects are discussed by Kompfner and Robinson.§ This paper was written before the Llewellyn-Rack-Peterson approach was widely accepted, but it indicates clearly the extent to which the neglected effects may contribute to the total noise.

† Watkins, loc. cit.
‡ Cutler and Quate, loc. cit.
§ Kompfner and Robinson, *Proc. Inst. Radio Engrs., N.Y.* **39** (1951), 918.

BIBLIOGRAPHY

1. APPLETON, ed. Aldous. *Thermionic Vacuum Tubes.* Methuen. 1952.
2. ARGIMBAU. *Vacuum Tube Circuits.* John Wiley. 1948.
3. BECK. *Velocity-modulated Thermionic Tubes.* Cambridge University Press. 1948.
4. CHAFFEE. *Theory of Thermionic Vacuum Tubes.* McGraw Hill. 1933.
5. COLLINS, Editor. *Microwave Magnetrons.* Vol. **6** of M.I.T. series on radar. McGraw Hill. 1948.
6. DE BOER. *Electron Emission and Adsorption Phenomena.* Cambridge University Press. 1935.
7. DOW. *Fundamentals of Engineering Electronics.* John Wiley. 1937.
8. FOWLER. *Statistical Mechanics.* Cambridge University Press. 2nd. ed., 1936.
9. HAMILTON, KUPER and KNIPP. *Klystrons and Microwave Triodes.* Vol. 7 of M.I.T. series on radar. McGraw Hill. 1948.
10. HERMANN and WAGENER. *The Oxide-coated Cathode.* Vols. I and II. Chapman and Hall. 1951.
11. JONES. *Thermionic Emission.* Methuen. 1936.
12. KÖNIG. *Laufzeittheorie der Elektronenröhren.* J. Springer, Vienna. 1948.
13. LATHAM, KING and RUSHFORTH. *The Magnetron.* Chapman and Hall. 1952.
14. LLEWELLYN. *Electron Inertia Effects.* Cambridge University Press. 1941.
15. LOVELL, Editor. *Electronics.* Pilot Press. 1947.
16. MARTON, Editor. *Advances in Electronics.* Vol. I, 1949. Vol. II, 1950. Vol. III, 1951. Academic Press Inc.
17. MOTT and GURNEY. *Electronic Processes in Ionic Crystals.* Oxford University Press. 2nd. ed., 1949.
18. MOTT and JONES. *The Theory of the Properties of Metals and Alloys.* Oxford University Press. 1936.
19. PIERCE. *Traveling-Wave Tubes.* Van Nostrand. 1950.
20. RAMO and WHINNEY. *Fields and Waves in Modern Radio.* John Wiley. 1944.

21. Reimann. *Thermionic Emission.* Chapman and Hall. 1934.
22. Slater. *Microwave Electronics.* Van Nostrand. 1950.
23. Spangenberg. *Vacuum Tubes.* McGraw Hill. 1948.
24. Stratton. *Electromagnetic Waves.* McGraw Hill. 1941.
25. Terman. *Radio Engineers Handbook.* McGraw Hill. 1943.
26. Valley and Waldman, Editors. *Vacuum Tube Amplifiers.* Vol. **18** of M.I.T. series on radar McGraw Hill. 1948.
27. Van der Bijl. *Thermionic Vacuum Tubes.* McGraw Hill. 1920.
28. Warnecke and Guénard. *Les tubes électroniques à commande par modulation de vitesse.* Gautier-Villars. 1951.
29. Weber. 'Electromagnetic Fields.' Vol. I, *Mapping of Fields.* John Wiley. 1950.

RECENT REFERENCES

CHAPTER 1

1. HERRING and NICHOLS. 'Thermionic Emission', *Rev. Mod. Phys.* **21** (1949), 187. A detailed and fully documented review of advances in the theory of thermionic emission since 1936. It is mainly concerned with the behaviour of clean metals, but a good account of patch theory is included.
2. DANFORTH. *J. Franklin Inst.* **251** (1951), 515. A review article on the thermionic properties of thoria.
3. MITCHELL and MITCHELL. 'Work functions of Cu, Ag and Al', *Proc. Roy. Soc.* **210** (1951), 70. Interesting details of modern technique.

CHAPTER 2

1. HEYDT. 'Measurement of secondary emission from dielectrics', *Rev. Sci. Inst.* **21** (1950), 639.
2. SALOW. 'Sekundärelektronenemission von Metallmischungen', *Ann. Phys., Lpz.* **5** (1950), 417.
3. ALLEN. 'Recent applications of electron multiplier tubes', *Proc. Inst. Radio Engrs.* **38** (1950), 346.
4. JACOBS. 'Field dependent secondary emission', *Phys. Rev.* **84** (1951), 877.
5. MORTON. 'Photomultipliers for scintillation counting', *R.C.A. Rev.* **10** (1949), 525.
6. HICKMAN. 'Multiplier phototubes in scintillation counters" *Electronic Engng.* **22** (1950), 474.
7. MCNARY. 'Enhanced photoelectric emission effect in BaO', *Phys. Rev.* **81** (1951), 631.
8. PAKSWER and REED. 'Photoconductivity of composite photoemissive cathodes', *J. Appl. Phys.* **22** (1951), 987.

9. Baroody. 'A theory of secondary emission from metals', *Phys. Rev.* **78** (1950), 780. Modification and extension of Kadyshevich theory.

10. Diemer and Jonker. 'The time delay of secondary emission', *Philips Res. Rep.* **5** (1950), 161. Proves experimentally that time delay $< 10^{-11}$ sec. A theoretical estimate gives 10^{-14}–10^{-15} sec.

CHAPTER 3

1. H. Leverenz. *An introduction to luminescence of solids.* John Wiley. 1950.

CHAPTER 4

Since the text was written, the use of two-dimensional resistance networks has come into vogue as a practical method for the solution of potential problems. The reader is left to decide whether or not this method is more useful than the electrolytic trough. A good account of the basic method is given by Liebmann, *Brit. J. Appl. Phys.* **1** (1950), 92.

CHAPTER 5

1. Liebmann. 'Field plotting and ray tracing in electron optics.' Review article in *Advances in Electronics*, vol. 2. 1950.

2. Burfoot. 'Numerical ray-tracing in electron lenses', *Brit. J. Appl. Phys.* **3** (1952), 22.

3. Sturrock. 'Perturbation characteristic functions applied to electron optics', *Proc. Roy. Soc.* **210** (1951), 269. Important theoretical developments.

4 Gabor. 'Electron optics at high frequencies and at relativistic velocities', *Rev. Opt.* **29** (1950), 209.

CHAPTER 6

1. H. F. Ivey. 'Cathode field in diodes under partial space charge conditions', *Phys. Rev.* **76** (1949), 554. The results enable one to calculate the cathode field in terms of the current, the saturation current and the field in the absence of charge. Thus, for example, one can correct emission data for the Schottky effect even when considerable space-charge is present.
2. Barut. *Zeits. Appl. Math. u. Phys.* **2** (1951), 35. Extends the study to the case of an electron stream with a single initial velocity.
3. Page and Adams. 'Diode space charge for any initial velocity and current', *Phys. Rev.* **76** (1949), 381.
4. Brubaker. 'Potential distribution in planar diodes at currents below the space charge limit', *Phys. Rev.* **83** (1951), 268.
5. Gamble. 'Current build-up in a planar diode', *J. Appl. Phys.* **21** (1950), 108. Deals with current build-up under pulsed voltages.
6. Fack. 'Ausgleichsvorgange in Elektronenrohren', *Frequenz*, **6** (1952), 33. A simpler treatment than item 5.

Brillouin Flow

7. Brillouin. 'A theorem of Larmor and its importance for electrons in magnetic fields', *Phys. Rev.* **67** (1945), 260.
8. Beck. *British Patent*, 674, 758.
9. Samuel. 'On the theory of axially symmetric electron beams in an axial magnetic field', *Proc. Inst. Radio Engrs.* **37** (1949), 1252.
10. Wang. 'Electron beams in axially symmetrical electric and magnetic fields', *Proc. Inst. Radio Engrs.* **38** (1950), 135.
11. Convert. 'Étude de la focalisation magnétique de faisceaux cylindriques', *J. de Physique et de Radium,* 6ème serie, **9** (1950), 551.

Note. In Brillouin type electron flow, both the space-charge repulsion force and the centrifugal force due to the rotation are exactly balanced by the magnetic force. The electrons all move with the same axial velocity, which is that corresponding to the potential on the axis and not the boundary potential. This means that large current densities may be attained without unwanted effects due to a velocity variation across the beam.

CHAPTER 7

1. TAUB and WAX. 'Theory of the planar diode', *J. Appl. Phys.* **21** (1950), 974. An extension of the Llewellyn analysis of the diode including new terms.
2. SMULLIN. 'Propagation of disturbances in one dimensional electron streams', *J. Appl. Phys.* **22** (1951), 1496. Applies small signal wave theory of Chapter 14 to the diode.
3. DIEMER. 'Microwave diode conductance in the exponential region of the characteristic', *Philips Res. Rep.* **6** (1951), 211.

CHAPTER 8

1. SCHOTTKY. 'Space charge smoothing of flicker effect', *Physica*, **4** (1937), 175.
2. KRONENBERGER. 'Experimentalle Untersuchung der Schwankungserscheinungen', *Z. Angew. Phys.* **3** (1951), 1. A careful account of measurements at very low frequencies.
3. TAUB. 'Effect of positive ions and reflected electrons on space charge reduced shot effect', *Research Lond.* **10** (1951), 391.
4. BULL. 'The space charge smoothing factor', *J.I.E.E.* **98** (1951), 149.

CHAPTER 9

The calculation of μ, etc. for very close-spaced valves has been studied by several authors.

1. HYMANN. 'Uber die strenge Berechnung des Durchgriffs ebener Systeme auf potentialtheoretischer Grundlage', *Frequenz*, **5** (1951), 57.
2. DAHLKE. 'Gittereffektivpotential und Kathodenstromdichte einer ebenen Triode unter Berucksichtigung der Inselbildung', *Telefunken Ztg.* **24** (1951), 213.

3. FORD and WALSH. 'Development of electron tubes for a new coaxial transmission system', *Bell Syst. Tech. J.* **30** (1951), 1103.
4. WALKER. 'On the electric field in a single-grid radio valve', *J.I.E.E.* **98**, pt. III (1951), 57.

CHAPTER 10

1. GUNDERT. 'Die Stromverteilungssteurung von Elektronenströmen', *Telefunken Ztg.* **24** (1951), 223.
2. WALKER. 'On the electric field in a multi-grid radio valve', *J.I.E.E.* **98**, pt. III (1951), 64.

CHAPTER 11

1. RHEAUME. 'Coaxial line power triode', *Cath. Press*, **7** (1949), 6.
2. RHEAUME. 'The ML 5682', *Cath. Press*, **8** (1951), 24.
3. ROTHE, ENGBERT and KRAFT. 'UKW-Senderöhren', *Telefunken Ztg.* **23** (1950), 175.
4. DOOLITTLE. 'Design problems in triode and tetrode tubes for U.H.F. operation', *Cath. Press*, **8** (1951), 41.
5. BEUTHERET. 'Refrigération des tubes électroniques par vaporisation d'eau', *Onde élect.* **31** (1951), 271.
6. AYER. 'Use of thoriated-W filaments in high power transmitting tubes', *Proc. Inst. Radio Engrs.* **40** (1952), 591.

CHAPTER 12

1. GUÉNARD, EPSZTEIN and CAHOUR. 'Klystron amplificateur de 5 KW a large bande passante', *Ann. Radioélectr.* **6** (1951), 109.
2. STAFF OF VARIAN ASSOCIATES. 'High-power U.H.F.-TV Klystron', *Electronics*, **24** (1951), 117.
3. BECK and CUTTING. 'Reflex klystrons for centimetre links', I.E.E. Convention on the British Contribution to Television. Paper No. 1343.
4. LAMBERT. 'Coaxial line V.M. oscillator for use in F-M radio links.' I.E.E. Convention. Paper No. 1337.
5. PEARCE and MAYO. 'The design of a reflex klystron for F–M at centimetric wavelengths.' I.E.E. Convention. Paper No. 1301.

6. VARIAN. 'Recent developments in klystrons', *Electronics*, **25** (1952), 112.

7. CHODOROW and WESTBURG. 'Space-charge effects in reflex klystrons', *Proc. Inst. Radio Engrs.* **39** (1951), 1548.

CHAPTER 13

1. MURAKAMI. 'Grounded-grid U.H.F. amplifiers', *R.C.A. Rev.* **12** (1951), 682.

2. BOWEN and MUMFORD. 'A new microwave triode: its performance as modulator and amplifier', *Bell, Syst. Tech. J.* **29** (1950), 531.

3. BELL. 'Induced grid noise and noise factor', *Proc. Inst. Radio Engrs.* **39** (1951), 1059.

CHAPTER 14

1. BRAY. 'Travelling wave valve as phase-modulator and frequency shifter', *J.I.E.E.* **99**, pt. III (1952), 15.

2. CUTLER. 'Calculation of T.W.T. gain', *Proc. Inst. Radio Engrs.* **39** (1951), 914.

3. MILLMAN. 'A spatial harmonic T.W.A. for 6 mm. wavelength', *Proc. Inst. Radio Engrs.* **39** (1951), 1035.

4. KLEEN and RUPPEL. 'Die Verzögerungsleitung als Bauelement von Elektronenröhren', *Arch. Elektrotech.* **40** (1952), 280.

5. FRIEDMAN. 'Amplification of the T.W.T.', *J. Appl. Phys.* **22** (1951), 443.

6. GOUDET et al. Series of papers entitled 'Deux tubes à onde progressive pour cables hertzien', *Ann. Télécommun.* **7** (1952), 152–204.

Some references on the more general theory of wave motion in plasmas should be included.

7. BAILEY. *Nature*, **161** (1948), 599; *J. Roy. Soc. N.S.W.* **82** (1948) 107; *Austral. J. Sci. Res.* (A), **1** (1948), 351; *Phys. Rev.* **78** (1950), 428; *Nature*, **166** (1950), 259; *Phys. Rev.* **83** (1951), 439.

8. BOHM and GROSS. *Phys. Rev.* **75** (1949), 1851 and 1864.

9. TWISS. *Proc. Phys. Soc.* (B), **64** (1951), 634.
10. SCHUMANN. *Z. Angew. Phys.* **2** (1950), 393.
11. FEINSTEIN and SEN. *Phys. Rev.* **83** (1951), 405.

CHAPTER 15

1. F. LUDI. 'Development of the turbator for radio relay equipment', *Brown Boveri Rev.* **36** (1949), 405; 'The Turbator', *Brown Boveri Rev.* **36** (1949), 315.
2. MOURIER. 'L'anticyclotron, un nouveau type de tube à propagation d'ondes a champ magnétique', *Ann. Radio-électr.*, **5** (1950), 206.
3. BUNEMANN. 'Generation and amplification of waves in dense charged beams under crossed fields', *Nature, Lond.* **165** (1950), 474.
4. WILBUR, PETERS and CHALBERG. 'Tunable miniature magnetron', *Electronics*, **25** (1952), 104.
5. REVERDIN. 'Electron optical exploration of space charge in a cut-off magnetron', *J. Appl. Phys.* **22** (1951), 257.

CHAPTER 16

1. J. RAJCHMAN. 'A selective electrostatic storage tube', *R.C.A. Rev.* **12** (1951), 53. Detailed description of a model selectron.
2. SCHAGEN. 'Mechanism of high voltage target stabilization, etc.', *Philips Res. Rep.* **6** (1951), 135.
3. VEITH. 'Le Conductron', *Vide*, **5** (1950), 887.
4. ALLARD and HILL. 'Switch and storage tubes', *Wireless Engr.* **28** (1951), 187.
5. SCHRÖTER. 'Elektronenstrahlschalter', *Telefunken Ztg.*, **24** (1951), 171.
6. KAZAN and KNOLL. 'Fundamental processes in charge controlled storage tubes', *R.C.A. Rev.* **12** (1951), 702.
7. HARRINGTON. 'Storage of small signals on a dielectric surface', *J. Appl. Phys.* **21** (1950), 1048.
8. BARTHELEMY. 'Analyseur de T.V.', *Onde élect.* **31** (1951), 415.

INDEX

Printed in the United States
By Bookmasters